Particles, Sources, and Fields

Volume I

PARTICLES, SOURCES, AND FIELDS

Volume I

JULIAN SCHWINGER
late, University of California at Los Angeles

CRC Press
Taylor & Francis Group
Boca Raton London New York

CRC Press is an imprint of the
Taylor & Francis Group, an **informa** business

ADVANCED BOOK PROGRAM

First published 1970 by Perseus Books Publishing

Published 2018 by CRC Press
Taylor & Francis Group
6000 Broken Sound Parkway NW, Suite 300
Boca Raton, FL 33487-2742

CRC Press is an imprint of the Taylor & Francis Group, an informa business

Visit the Taylor & Francis Web site at
http://www.taylorandfrancis.com

and the CRC Press Web site at
http://www.crcpress.com

Library of Congress Catalog Card Number: 98-87896

ISBN 13: 978-0-7382-0053-8 (pbk)

Cover design by Suzanne Heiser

Editor's Foreword

Perseus Books's *Frontiers in Physics* series has, since 1961, made it possible for leading physicists to communicate in coherent fashion their views of recent developments in the most exciting and active fields of physics—without having to devote the time and energy required to prepare a formal review or monograph. Indeed, throughout its nearly forty-year existence, the series has emphasized informality in both style and content, as well as pedagogical clarity. Over time, it was expected that these informal accounts would be replaced by more formal counterparts—textbooks or monographs—as the cutting-edge topics they treated gradually became integrated into the body of physics knowledge and reader interest dwindled. However, this has not proven to be the case for a number of the volumes in the series: Many works have remained in print on an on-demand basis, while others have such intrinsic value that the physics community has urged us to extend their life span.

The *Advanced Book Classics* series has been designed to meet this demand. It will keep in print those volumes in *Frontiers in Physics* or its sister series, *Lecture Notes and Supplements in Physics*, that continue to provide a unique account of a topic of lasting interest. And through a sizable printing, these classics will be made available at a comparatively modest cost to the reader.

These lecture notes by Julian Schwinger, one of the most distinguished theoretical physicists of this century, provide both beginning graduate students and experienced researchers with an invaluable introduction to the author's perspective on quantum electrodynamics and high-energy particle physics. Based on lectures delivered during the period 1966 to 1973, in which Schwinger developed a point of view (the physical source concept) and a technique that emphasized the unity of particle physics, electrodynamics, gravitational theory, and many-body theory, the notes serve as both a textbook on source theory and an informal historical record of the author's approach to many of the central problems in physics. I am most pleased that *Advanced Book Classics* will make these volumes readily accessible to a new generation of readers.

David Pines
Aspen, Colorado
July 1998

v

Vita

Julian Schwinger

University Professor, University of California, and Professor of Physics at the University of California, Los Angeles since 1972, was born in New York City on February 12, 1918. Professor Schwinger obtained his Ph.D. in physics from Columbia University in 1939. He has also received honorary doctorates in science from four institutions: Purdue University (1961), Harvard University (1962), Brandeis University (1973), and Gustavus Adolphus College (1975). In addition to teaching at the University of California, Professor Schwinger has taught at Purdue University (1941–43), and at Harvard University (1945–72). Dr. Schwinger was a Research Associate at the University of California, Berkeley, and a Staff Member of the Massachusetts Institute of Technology Radiation Laboratory. In 1965 Professor Schwinger became a co-recipient (with Richard Feynman and Sin Itiro Tomonaga) of the Nobel Prize in Physics for work in quantum electrodynamics. A National Research Foundation Fellow (1939–40) and a Guggenheim Fellow (1970), Professor Schwinger was also the recipient of the C. L. Mayer Nature of Light Award (1949); the First Einstein Prize Award (1951); a J. W. Gibbs Honorary Lecturer of the American Mathematical Society (1960); the National Medal of Science Award for Physics (1964); a Humboldt Award (1981); the Premio Citta di Castiglione de Sicilia (1986); the Monie A. Ferst Sigma Xi Award (1986); and the American Academy of Achievement Award (1987).

Special Preface

Isaac Newton used his newly invented method of fluxious (the calculus) to compare the implications of the inverse square law of gravitation with Kepler's empirical laws of planetary motion. Yet, when the time came to write the *Principia*, he resorted entirely to geometrical demonstrations. Should we conclude that calculus is superfluous?

Source theory—to which the concept of renormalization is foreign—and *renormalized operator field theory* have both been found to yield the same answers to electrodynamic problems (which disappoints some people who would prefer that source theory produce new—and wrong—answers). Should we conclude that source theory is thus superfluous?

Both questions merit the same response: the simpler, more intuitive formation, is preferable.

This edition of *Particles, Sources, and Fields* is more extensive than the original two volumes of 1970 and 1973. It now contains four additional sections that finish the chapter entitled, "Electrodynamics II." These sections were written in 1973, but remained in partially typed form for fifteen years. I am again indebted to Mr. Ronald Bohm, who managed to decipher my fading scribbles and completed the typescript. Particular attention should be directed to Section 5-9, where, in a context somewhat larger than electrodynamics, a disagreement between source theory and operator field theory finally does appear.

Readers making their first acquaintance with source theory should consult the Appendix in Volume I. This Appendix contains suggestions for threading one's way through the sometimes cluttered pages.

Los Angeles, California   J. S.
April 1988

Preface

This book is a research document, and it is a textbook. It is the record of a highly personal reaction to the crisis in high energy particle physics. The ingredients were: frustration with the mathematical ambiguities and physical remoteness of operator field theory, dissatisfaction with the overly mathematical attitude and speculative philosophy of the supposedly more physical S-matrix theory, outrage at the pretension of current algebra to be a fundamental description rather than a low energy phenomenology.

The result was a point of view and a technique that emphasizes the unity of high energy particle physics with electrodynamics, gravitational theory, and many-particle cooperative phenomena. The physical source concept, upon which it is built, had its mathematical precursor in operator field theory. But it was not until the spring of 1966, while teaching a Harvard graduate course, that I suddenly realized how the phenomenological source concept could be freed from its operator substructure and used as the basis for a completely independent development, with much closer ties to experiment.

The reconstruction of electrodynamics proceeded rapidly, at UCLA that summer, and during a repetition of the Harvard course that was, instead, devoted entirely to the new approach. Developments in pion physics that winter (1966–1967), in which the new viewpoint was most successfully applied, convinced me, if no one else, of the great advantages in mathematical simplicity and conceptual clarity that its use bestowed. The lack of appreciation of these facts by others was depressing, but understandable. Only a detailed presentation of the ideas and methods of source theory could change that situation. The writing of this book began toward the close of summer, 1968.

As a textbook, this volume is intended for use by any student, familiar with nonrelativistic quantum mechanics, who wishes to learn relativistic quantum mechanics. I think it of the utmost importance that such acquaintance with the liberating ideas of source theory occur before exposure to one of the current orthodoxies has warped him past the elastic limit. In the Preface to a volume on S-matrix theory, one author speaks of the desirability that the student have a certain innocence concerning (operator) field theory. I echo that wistful call, but widen the domain of innocence to include S-matrix theory.

In writing this book I have made no attempt to supply the traditional running historical commentary on who allegedly first did what, when. Perhaps I have been overly sensitized to the distortions inherent in the simplistic association of ideas and methods with specific individuals. But there is a more important reason. While a general critique of existing attitudes is essential in motivating this new viewpoint, it would have been too distracting if constant reference to techniques for which obsolescence is intended had accompanied the development of the new approach. The expert comes ready made with opinions about what has already been done. To the student all that matters is what is new to him and I hope that he will find much in these pages.

I am grateful to Miss Margaret Cunnane and Miss Jeri Ingerson who, at different periods, devotedly carried unaided the burden of typing the formidable manuscript. The book would never have been completed (I hold the world's record for the largest number of unfinished first chapters) without the patience and understanding of my wife. It is therefore appropriately dedicated to the C.G.S. system.

Belmont, Massachusetts   J. S.
October 1969

Contents

If you can't join 'em,
beat 'em.

Particles, Sources, and Fields

1

PARTICLES

The concept of the particle has undergone drastic changes and generalizations in the course of the historical development that led to the atom, to the nucleus, and then to subnuclear phenomena. This has also been a progression from essentially nonrelativistic behavior to an ultra-relativistic domain. It is interesting to appreciate how much of the kinematical particle attributes is implied by the assumed structure of the relativity group of transformations among equivalent coordinate systems. In preparation for this discussion we first review some properties of quantum mechanical unitary transformations.

1-1 UNITARY TRANSFORMATIONS

Quantum mechanics is a symbolic expression of the laws of microscopic measurement. States, situations of optimum information, are represented by vectors in a complex space [left vectors $\langle \ |$, right vectors $| \ \rangle$]; and physical properties by linear Hermitian operators on this space [$A| \ \rangle$, $\langle \ |A$]. The freedom in physical description corresponds to the freedom of mathematical representation associated with unitary operators. These are defined through the Hermitian adjoint operation † by

$$U^\dagger U = U U^\dagger = 1 \qquad (1\text{-}1.1)$$

or

$$U^\dagger = U^{-1}. \qquad (1\text{-}1.2)$$

Let us transform all vectors and operators according to

$$\overline{\langle \ |} = \langle \ |U, \quad \overline{| \ \rangle} = U^{-1}| \ \rangle, \qquad \overline{X} = U^{-1}XU. \qquad (1\text{-}1.3)$$

Then all numerical and adjoint relations among vectors and operators are unchanged. We verify that

$$\langle \overline{a'}|\overline{b'} \rangle = \langle a'|b' \rangle, \qquad \langle \overline{a'}|\overline{X}|\overline{b'} \rangle = \langle a'|X|b' \rangle. \qquad (1\text{-}1.4)$$

The adjoint relationship

$$\langle a'| = |a' \rangle^\dagger \qquad (1\text{-}1.5)$$

is transformed into

$$\langle \overline{a'}| = \langle a'|U = (U^{-1}|a'\rangle)^\dagger = |\overline{a'}\rangle^\dagger, \qquad (1\text{-}1.6)$$

while

$$\overline{X}^\dagger = U^{-1}X^\dagger U \qquad (1\text{-}1.7)$$

shows that the Hermitian operator A is mapped into the Hermitian operator $\overline{A}$.

1

A complete set of states $\langle a'|$ forms a basis or coordinate system in the state space. Any vector $|\ \rangle$ is represented by its components relative to this basis, $\langle a'|\ \rangle$. Another basis is produced by a unitary transformation:

$$\langle \overline{a'}| = \langle a'|U, \tag{1-1.8}$$

and the given vector has a corresponding new set of components,

$$\langle \overline{a'}|\ \rangle = \langle a'|U|\ \rangle. \tag{1-1.9}$$

These numbers are alternatively described as components, relative to the initial basis, of the new vector $U|\ \rangle$. An analogous relation for matrix elements of operators is

$$\langle \overline{a'}|X|\overline{a''}\rangle = \langle a'|UXU^{-1}|a''\rangle. \tag{1-1.10}$$

If two successive transformations are performed on the basis, the net change in the components of a vector is given by

$$\langle a'|\ \rangle \xrightarrow{U_1} \langle a'|U_1|\ \rangle \xrightarrow{U_2} \langle a'|U_2U_1|\ \rangle. \tag{1-1.11}$$

It is produced in one step by the unitary operator U_2U_1, in which the multiplication order reflects the sequence of transformations. The opposite sequence is represented by U_1U_2, and the two are compared by defining the unitary operator that is needed to convert the second sequence into the first,

$$U_{[12]}U_1U_2 = U_2U_1. \tag{1-1.12}$$

It is

$$U_{[12]} = U_2U_1U_2^{-1}U_1^{-1} = U_{[21]}^{-1}. \tag{1-1.13}$$

An infinitesimal unitary transformation is a transformation in the infinitesimal neighborhood of the identity. It is represented by

$$U = 1 + iG, \qquad U^\dagger = U^{-1} = 1 - iG, \tag{1-1.14}$$

where G is an infinitesimal Hermitian operator. When two such transformations are compared we find that

$$U_{[12]} = 1 + iG_{[12]}, \tag{1-1.15}$$

where

$$G_{[12]} = -G_{[21]} = (1/i)[G_1, G_2] \tag{1-1.16}$$

introduces the commutator of G_1 and G_2. The effect of an infinitesimal unitary transformation on an operator is given by

$$UXU^{-1} = X + \delta X, \tag{1-1.17}$$

where

$$\delta X = (1/i)[X, G]. \tag{1-1.18}$$

An equivalent form is

$$\bar{X} = U^{-1}XU = X - \delta X. \tag{1-1.19}$$

If we compare alternative evaluations of

$$U_{12}XU_{12}^{-1} = X + \delta_{[12]}X = U_2U_1U_2^{-1}U_1^{-1}XU_1U_2U_1^{-1}U_2^{-1}, \tag{1-1.20}$$

we get

$$\delta_{[12]}X = \delta_2\delta_1X - \delta_1\delta_2X. \tag{1-1.21}$$

When presented in terms of double commutators,

$$[X, [G_1, G_2]] = [[X, G_1], G_2] - [[X, G_2], G_1], \tag{1-1.22}$$

this is recognized as the Jacobi identity.

Now let us consider a group of unitary transformations with n real, continuous parameters λ_a, $a = 1, \ldots, n$, which we designate collectively as λ. If $U(\lambda_{1,2})$ are typical operators of the group, it is required that

$$U(\lambda_2)U(\lambda_1) = U(\lambda), \tag{1-1.23}$$

where

$$\lambda_a = \lambda_a(\lambda_1, \lambda_2) \tag{1-1.24}$$

are the parameters of another element of the group. For unitary operators the existence of the inverse and of the identity is assured. An infinitesimal transformation of the group with parameters $\delta\lambda_a$ is constructed from

$$G = \sum_{a=1}^{n} \delta\lambda_a G_a, \tag{1-1.25}$$

where the n finite Hermitian operators G_a are called the generators of the group. One is free to redefine the generators by real nonsingular linear transformations, with corresponding redefinitions of the parameters. On subjecting the infinitesimal transformation operator $U(\delta\lambda)$ to an arbitrary unitary transformation of the group, we must obtain another infinitesimal transformation. This is expressed by

$$U(\lambda)^{-1}G_aU(\lambda) = \sum_b u_{ab}(\lambda)G_b, \tag{1-1.26}$$

where the numbers $u_{ab}(\lambda)$ are real. We shall also use a matrix notation in the n-dimensional parameter space and write

$$U(\lambda)^{-1}GU(\lambda) = u(\lambda)G. \tag{1-1.27}$$

The unitary transformation is presented alternatively as

$$U(\lambda)G_bU(\lambda)^{-1} = \sum_a G_a\hat{u}_{ab}(\lambda) \tag{1-1.28}$$

or

$$U(\lambda)GU(\lambda)^{-1} = G\hat{u}(\lambda). \tag{1-1.29}$$

The two sets of matrices are related by

$$\hat{u} = (u^T)^{-1},\tag{1–1.30}$$

where T designates matrix transposition. Note that this is equivalent to Hermitian conjugation when applied to the u matrices, since they are real.

The correspondence established between the unitary operator $U(\lambda)$ and the matrices $u(\lambda)$, $\hat{u}(\lambda)$ is maintained under multiplication. Thus,

$$U(\lambda_1)^{-1}U(\lambda_2)^{-1}GU(\lambda_2)U(\lambda_1) = U(\lambda_1)^{-1}[u(\lambda_2)G]U(\lambda_1)$$
$$= u(\lambda_2)u(\lambda_1)G,\tag{1–1.31}$$

and

$$U(\lambda_2)U(\lambda_1)GU(\lambda_1)^{-1}U(\lambda_2)^{-1} = U(\lambda_2)[G\hat{u}(\lambda_1)]U(\lambda_2)^{-1}$$
$$= G\hat{u}(\lambda_2)\hat{u}(\lambda_1).\tag{1–1.32}$$

Since the unit operator corresponds to the unit matrix, we write

$$u(\delta\lambda) = 1 + i\sum_a \delta\lambda_a g_a, \qquad \hat{u}(\delta\lambda) = 1 + i\sum_a \delta\lambda_a \hat{g}_a,\tag{1–1.33}$$

where

$$\hat{g}_a = -g_a^T = g_a^{T*}.\tag{1–1.34}$$

This gives

$$[G, G_b] = g_b G = -G\hat{g}_b\tag{1–1.35}$$

and, if the imaginary elements of the matrix g_b are designated as

$$(g_b)_{ac} = g_{abc},\tag{1–1.36}$$

we get the explicit commutation relations of the group generators:

$$[G_a, G_b] = \sum_c g_{abc}G_c.\tag{1–1.37}$$

We see, incidentally, that

$$g_{abc} = -g_{bac}.\tag{1–1.38}$$

In view of the multiplicative correspondence between $U(\lambda)$ and $u(\lambda)[\hat{u}(\lambda)]$, the matrices $g_a[\hat{g}_a]$ also obey these commutation relations,

$$[g_a, g_b] = \sum_c g_{abc}g_c.\tag{1–1.39}$$

The latter are a set of quadratic restrictions that must be obeyed by the numbers g_{abc}, the so-called group structure constants:

$$\sum_d [g_{abd}g_{dce} + g_{bcd}g_{dae} + g_{cad}g_{dbe}] = 0.\tag{1–1.40}$$

This cyclic structure also follows immediately from the cyclic form of the

Jacobi identity:

$$[[G_a, G_b], G_c] + [[G_b, G_c], G_a] + [[G_c, G_a], G_b] = 0. \quad (1\text{-}1.41)$$

The structure constants specify the composition properties of infinitesimal parameters. Let $\delta_{[12]}\lambda_a$ be the parameters of the infinitesimal transformation that connects the two sequences in which the transformations labeled by $\delta_1\lambda_a$ and $\delta_2\lambda_a$ can be applied. According to the commutations relations of the group generators, they are given by

$$\delta_{[12]}\lambda_c = -\delta_{[21]}\lambda_c = \sum_{ab} \delta_1\lambda_a\, \delta_2\lambda_b (1/i) g_{abc}. \quad (1\text{-}1.42)$$

In the discussion to follow, group composition properties are supplied by geometrical considerations. It is important to recognize that the associated unitary group may not be an exact image of the underlying geometrical group. This is a consequence of an intrinsic arbitrariness of any quantum mechanical description whereby all states can be changed by a common phase factor, which is the unitary transformation generated by the unit operator. Consider, for example, the commutative (Abelian) group of translations in a two-dimensional space. Let the parameters of the two independent infinitesimal displacements be written δx_1, δx_2 and the corresponding Hermitian displacement operators be denoted by p_1, p_2, so that

$$G = p_1\, \delta x_1 + p_2\, \delta x_2. \quad (1\text{-}1.43)$$

The fact that successive displacements are insensitive to the order in which they are applied should imply the commutativity of p_1 and p_2. But all that is required of the commutator is that it generate a unitary transformation without physical consequences. Accordingly,

$$(1/i)[p_1, p_2] = 1, \quad (1\text{-}1.44)$$

with a suitable normalization of the displacement operators. Now the effect of the unitary transformation $U(\delta x)$ on the operators $p_{1,2}$ is given by

$$\delta p_1 = (1/i)[p_1, G] = \delta x_2, \qquad \delta p_2 = (1/i)[p_2, G] = -\delta x_1, \quad (1\text{-}1.45)$$

which shows that the displacement operators also serve as coordinate operators

$$x_1 = -p_2 \equiv q, \qquad x_2 = p_1 \equiv p. \quad (1\text{-}1.46)$$

We recognize the properties of the q,p phase space associated with a single quantum degree of freedom. Translations in this two-dimensional space are described by a three-parameter unitary group. This is made explicit in the form

$$G = p\, \delta q - q\, \delta p + \delta\varphi 1. \quad (1\text{-}1.47)$$

The correspondence between the unitary operators $U(\lambda)$ and the finite matrices $u(\lambda)$, $\hat{u}(\lambda)$ does not necessarily include the unitary character of the latter. [Note that if the $u(\lambda)$ are unitary, or real orthogonal, matrices, we have $u(\lambda) = \hat{u}(\lambda)$, and then the Hermitian, or imaginary antisymmetrical matrices $g_a = \hat{g}_a$.] Since the structure of the g matrices can be altered by changing the generator basis in the parameter space, it is useful to have a basis-independent criterion with which to judge the possibility of exhibiting Hermitian g matrices or unitary $u(\lambda)$ matrices. If the set of n G operators is replaced by the linearly independent combinations λG, the g matrices undergo the same linear transformation, together with a similarity transformation produced by the non-singular matrix λ. Because the trace of matrix products is unchanged by the latter transformation we consider the real quadratic form

$$\sum_{ab} x_a \, \text{tr} \, (g_a g_b) x_b = \text{tr} \left(\sum_a x_a g_a\right)^2, \qquad (1\text{-}1.48)$$

which must be positive-definite if the g_a are transformable into linearly independent Hermitian matrices. The failure of that property implies that such Hermitian g matrices and unitary $u(\lambda)$ matrices do not exist. This is the situation, in the example we have just discussed, of the generators q, p, 1. There the quadratic form is identically zero, since the unit operator is represented by the null matrix, while the matrices associated with q and p have only single non-diagonal entries in such a way that all matrix products vanish. The positive definiteness of the real symmetric matrix

$$\gamma_{ab} = \text{tr} \, g_a g_b \qquad (1\text{-}1.49)$$

is not only necessary but also sufficient for the g_a to be equivalent to linearly independent Hermitian matrices. Since the elements of the γ matrix are unchanged by the similarity transformation

$$u(\lambda)^{-1} g_a u(\lambda) = \sum_b u_{ab}(\lambda) g_b, \qquad (1\text{-}1.50)$$

we have

$$\gamma = u(\lambda) \gamma u(\lambda)^T. \qquad (1\text{-}1.51)$$

A real symmetrical positive-definite γ matrix can always be written as the square of another such matrix, which we designate as $\gamma^{1/2}$. Then

$$[\gamma^{-1/2} u(\lambda) \gamma^{1/2}][\gamma^{-1/2} u(\lambda) \gamma^{1/2}]^T = 1, \qquad (1\text{-}1.52)$$

which makes explicit the similarity transformation that introduces unitary (orthogonal) $u(\lambda)$ matrices and Hermitian (antisymmetrical) g matrices. In the new basis, γ is a multiple of the unit matrix.

Let us suppose that the operators G_a possess a finite-dimensional linearly independent Hermitian matrix realization. That implies the existence of the real symmetric positive-definite matrix

$$\gamma'_{ab} = \text{Tr} \, G_a G_b. \qquad (1\text{-}1.53)$$

The invariance of these numbers under unitary transformations on the operators leads again to the form

$$\gamma' = u(\lambda)\gamma'u(\lambda)^T, \tag{1-1.54}$$

with the implication that the $u(\lambda)$ can be presented as unitary matrices. The corresponding Hermitian matrices g_a, which are linearly independent if we exclude the uninteresting possibility that the group has an Abelian group as a factor, are an example of a finite-dimensional realization of the G_a, as the $u(\lambda)$ provide a finite-dimensional unitary realization of the $U(\lambda)$. Conversely, if the matrix γ_{ab} is not positive-definite, no such finite unitary [Hermitian] realization of the $U(\lambda)[G_a]$ can exist. A finite-dimensional realization of the group means that a finite number of states can be found which are transformed among themselves by all operations of the group. In general, the action of a unitary operator on a state introduces new states, and the repetition of the operation continues the process of producing additional states. This can terminate with a finite number of states only if that repetition eventually ceases to provide new operators, that is, if the group parameter space is closed. The distinction between a closed and an open group manifold is most familiar in that between rotations and translations.

If the matrices g_a are Hermitian, the structure constants g_{abc} are antisymmetrical in a and c, as well as a and b, which implies antisymmetry in b and c. This complete antisymmetry can only be realized with $n \geq 3$. For $n = 3$, a suitable normalization brings the imaginary structure constants to the unique form

$$g_{abc} = i\epsilon_{abc}, \tag{1-1.55}$$

where ϵ_{abc} is the totally antisymmetrical symbol specified by $\epsilon_{123} = +1$. The resulting group commutation relations

$$[G_1, G_2] = iG_3, \qquad [G_2, G_3] = iG_1, \qquad [G_3, G_1] = iG_2, \tag{1-1.56}$$

or

$$[G_a, G_b] = i \sum_c \epsilon_{abc}G_c, \tag{1-1.57}$$

are familiar in the theories of three-dimensional angular momentum and isotopic spin. The three-dimensional g matrices satisfy these commutation relations, and γ is a multiple of the unit matrix.

1-2 GALILEAN RELATIVITY

Space-time coordinates appear in quantum mechanics as an abstraction of the roles of the macroscopic measurement apparatus. All evidence confirms the equivalence of two coordinate systems that differ in any or all of the following ways: a translation of the spatial origin, a translation of the time origin, a rotation of the space axes, a constant relative velocity between the two systems.

These transformations constitute the relativity group, or rather the subgroup of transformations that are continuously connected with the identity. When all particles move slowly in comparison with the speed of light, the time coordinate has an absolute significance and is affected only by displacement of its origin. This is Galilean relativity. It is characterized by infinitesimal coordinate transformations $\mathbf{r}$, $t \to \bar{\mathbf{r}}$, $\bar{t}$, where

$$\bar{\mathbf{r}} = \mathbf{r} - \delta\mathbf{r}, \qquad \bar{t} = t - \delta t, \tag{1-2.1}$$

and

$$\delta\mathbf{r} = \delta\boldsymbol{\epsilon} + \delta\boldsymbol{\omega} \times \mathbf{r} + \delta\mathbf{v}t. \tag{1-2.2}$$

Note that the sign conventions are appropriate to the significance of $\delta\boldsymbol{\epsilon}$, say, as the displacement of the origin of the spatial coordinate frame to which a given point is referred. If it is the point that is moved by $\delta\mathbf{r}$, its new position relative to the fixed reference system is $\mathbf{r} + \delta\mathbf{r}$. The group composition properties of this 10-parameter group are specified by comparing the sequence of transformations

$$\begin{aligned} \bar{\mathbf{r}} &= \mathbf{r} - \delta_1\mathbf{r}, \qquad \bar{t} = t - \delta_1 t, \\ \bar{\bar{\mathbf{r}}} &= \bar{\mathbf{r}} - \delta_2\bar{\mathbf{r}}, \qquad \bar{\bar{t}} = \bar{t} - \delta_2\bar{t} \end{aligned} \tag{1-2.3}$$

with those in the opposite order. The result of performing the transformation sequence 1, 2, 1^{-1}, 2^{-1}, or, equivalently, for infinitesimal transformations, 1^{-1}, 2^{-1}, 1, 2, is

$$\begin{aligned} \delta_{[12]}\mathbf{r} &= \delta_{[12]}\boldsymbol{\epsilon} + \delta_{[12]}\boldsymbol{\omega} \times \mathbf{r} + \delta_{[12]}\mathbf{v}t, \\ \delta_{[12]}t &= 0, \end{aligned} \tag{1-2.4}$$

where

$$\begin{aligned} \delta_{[12]}\boldsymbol{\epsilon} &= \delta_1\boldsymbol{\omega} \times \delta_2\boldsymbol{\epsilon} - \delta_2\boldsymbol{\omega} \times \delta_1\boldsymbol{\epsilon} + \delta_1\mathbf{v}\,\delta_2 t - \delta_2\mathbf{v}\,\delta_1 t, \\ \delta_{[12]}\boldsymbol{\omega} &= \delta_1\boldsymbol{\omega} \times \delta_2\boldsymbol{\omega}, \\ \delta_{[12]}\mathbf{v} &= \delta_1\boldsymbol{\omega} \times \delta_2\mathbf{v} - \delta_2\boldsymbol{\omega} \times \delta_1\mathbf{v}. \end{aligned} \tag{1-2.5}$$

The infinitesimal unitary transformation, $U = 1 + iG$, that is induced by an infinitesimal coordinate transformation is given by

$$G = (1/\hbar)[\delta\boldsymbol{\epsilon} \cdot \mathbf{P} + \delta\boldsymbol{\omega} \cdot \mathbf{J} + \delta\mathbf{v} \cdot \mathbf{N} - \delta t H] + \delta\varphi 1. \tag{1-2.6}$$

The quantum unit of action, $\hbar = 1.0545 \times 10^{-27}$ erg sec, is henceforth replaced by unity on adopting suitable atomic units. The generators $\mathbf{P}$ and $\mathbf{J}$ are conventionally called the linear and angular momentum operators, while H is the energy or Hamiltonian operator. The generators of infinitesimal velocity changes have had no verbal designation. But, in this rocket conscious age, the finite velocity transformations have come to be called "boosts." Perhaps one should term $\mathbf{N}$ the booster. We need to specify group composition properties for the new scalar parameter $\delta\varphi$. A general bilinear form for $\delta_{[12]}\varphi = -\delta_{[21]}\varphi$ is

$$\begin{aligned}
\delta_{[12]}\varphi = \quad & K(\delta_1\boldsymbol{\omega}\cdot\delta_2\boldsymbol{\epsilon} - \delta_2\boldsymbol{\omega}\cdot\delta_1\boldsymbol{\epsilon}) \\
& + L(\delta_1\boldsymbol{\omega}\cdot\delta_2\mathbf{v} - \delta_2\boldsymbol{\omega}\cdot\delta_1\mathbf{v}) \\
& + M(\delta_1\boldsymbol{\epsilon}\cdot\delta_2\mathbf{v} - \delta_2\boldsymbol{\epsilon}\cdot\delta_1\mathbf{v}),
\end{aligned} \tag{1-2.7}$$

where K, L, M are constants. The Jacobi identity, applied to three sets of infinitesimal transformations, implies that

$$\begin{aligned}
& K[\delta_{[12]}\boldsymbol{\omega}\cdot\delta_3\boldsymbol{\epsilon} - \delta_3\boldsymbol{\omega}\cdot\delta_{[12]}\boldsymbol{\epsilon} + \text{cycl. perm.}] \\
& + L[\delta_{[12]}\boldsymbol{\omega}\cdot\delta_3\mathbf{v} - \delta_3\boldsymbol{\omega}\cdot\delta_{[12]}\mathbf{v} + \text{cycl. perm.}] \\
& + M[\delta_{[12]}\boldsymbol{\epsilon}\cdot\delta_3\mathbf{v} - \delta_3\boldsymbol{\epsilon}\cdot\delta_{[12]}\mathbf{v} + \text{cycl. perm.}] = 0.
\end{aligned} \tag{1-2.8}$$

One easily verifies that the coefficient of M vanishes identically, but those of K and L do not. Hence the latter coefficients must be zero, and we have, simply,

$$\delta_{[12]}\varphi = M(\delta_1\boldsymbol{\epsilon}\cdot\delta_2\mathbf{v} - \delta_2\boldsymbol{\epsilon}\cdot\delta_1\mathbf{v}). \tag{1-2.9}$$

Not as a proof but as a mnemonic, we mention that $\delta\boldsymbol{\omega}\cdot\delta\boldsymbol{\epsilon}$ and $\delta\boldsymbol{\omega}\cdot\delta\mathbf{v}$ are really pseudoscalars, while $\delta\boldsymbol{\epsilon}\cdot\delta\mathbf{v}$ is a scalar.

The full set of generator commutation relations is

$$\begin{array}{lll}
[J_k, J_l] = i\epsilon_{klm}J_m, & [P_k, J_l] = i\epsilon_{klm}P_m, & [N_k, J_l] = i\epsilon_{klm}N_m, \\
[P_k, P_l] = 0, & [N_k, N_l] = 0, & [P_k, N_l] = i\delta_{kl}M,
\end{array} \tag{1-2.10}$$

and

$$[P_k, H] = 0, \quad [J_k, H] = 0, \quad [N_k, H] = -iP_k, \tag{1-2.11}$$

where we have adopted the summation convention for repeated indices; here, the index $m = 1, 2, 3$.

The commutator of two generators can be interpreted, in two alternative ways, as the effect of an infinitesimal unitary transformation upon an operator:

$$\delta_{[12]}G = (1/i)[G_1, G_2] = \delta_2 G_1 = -\delta_1 G_2. \tag{1-2.12}$$

The commutators involving the angular momentum operator, for example, can be written as

$$\begin{aligned}
\delta_\omega\mathbf{J} &= (1/i)[\mathbf{J}, \mathbf{J}\cdot\delta\boldsymbol{\omega}] = \delta\boldsymbol{\omega}\times\mathbf{J}, \\
\delta_\omega\mathbf{P} &= (1/i)[\mathbf{P}, \mathbf{J}\cdot\delta\boldsymbol{\omega}] = \delta\boldsymbol{\omega}\times\mathbf{P}, \\
\delta_\omega\mathbf{N} &= (1/i)[\mathbf{N}, \mathbf{J}\cdot\delta\boldsymbol{\omega}] = \delta\boldsymbol{\omega}\times\mathbf{N},
\end{aligned} \tag{1-2.13}$$

which state the response of a vector to infinitesimal rotations, and

$$\delta_\omega H = (1/i)[H, \mathbf{J}\cdot\delta\boldsymbol{\omega}] = 0, \tag{1-2.14}$$

which characterizes H as a rotational scalar. Analogous linear momentum and translation response equations are

$$\delta_\epsilon \mathbf{J} = (1/i)[\mathbf{J}, \mathbf{P} \cdot \delta\epsilon] = \delta\epsilon \times \mathbf{P},$$
$$\delta_\epsilon \mathbf{P} = (1/i)[\mathbf{P}, \mathbf{P} \cdot \delta\epsilon] = 0, \qquad (1\text{--}2.15)$$
$$\delta_\epsilon \mathbf{N} = (1/i)[\mathbf{N}, \mathbf{P} \cdot \delta\epsilon] = -M\,\delta\epsilon,$$

and

$$\delta_\epsilon H = (1/i)[H, \mathbf{P} \cdot \delta\epsilon] = 0. \qquad (1\text{--}2.16)$$

The response of the angular momentum to translation is in accord with the nature of angular momentum as the moment of the linear momentum, and indicates the existence of a position vector operator $\mathbf{R}$ such that

$$(1/i)[\mathbf{R}, \mathbf{P} \cdot \delta\epsilon] = \delta\epsilon \qquad (1\text{--}2.17)$$

or

$$[R_k, P_l] = i\delta_{kl}. \qquad (1\text{--}2.18)$$

We therefore write

$$\mathbf{J} = \mathbf{R} \times \mathbf{P} + \mathbf{S}, \qquad (1\text{--}2.19)$$

where $\mathbf{S}$ is a translationally independent contribution to the angular momentum. This is internal angular momentum, or spin. The correct rotational response of $\mathbf{P}$ is assured by this construction, and that of $\mathbf{R}$ will follow if the components of $\mathbf{R}$ are mutually commutative, and if $\mathbf{S}$ commutes with $\mathbf{R}$ as well as with $\mathbf{P}$. To produce the proper rotational behavior of $\mathbf{S}$, these operators must themselves obey the angular momentum commutation relations, which we can also write as

$$\mathbf{S} \times \mathbf{S} = i\mathbf{S}. \qquad (1\text{--}2.20)$$

The translational response of $\mathbf{N}$ indicates that it can be identified with $-M\mathbf{R}$, together with a translationally invariant term. Since a boost is a translation that grows linearly with time, we infer that

$$\mathbf{N} = \mathbf{P}t - M\mathbf{R}. \qquad (1\text{--}2.21)$$

All commutators among $\mathbf{J}, \mathbf{P}, \mathbf{N}$ are reproduced by these constructions.

If $|\ \rangle$ is a dynamically possible state so is $U|\ \rangle$, where U represents a relativity transformation, since the vector $U|\ \rangle$ has the same components as does $|\ \rangle$ in the transformed description. This means that the relativity group generators are constants of the motion. Such also is the conclusion of the commutation relations involving H, if it is understood that $\mathbf{R}, \mathbf{P}, \mathbf{S}$ are not explicit function of t. Thus,

$$\frac{d\mathbf{P}}{dt} = \frac{1}{i}[\mathbf{P}, H] = 0, \qquad \frac{d\mathbf{J}}{dt} = \frac{1}{i}[\mathbf{J}, H] = 0, \qquad (1\text{--}2.22)$$

while

$$\frac{d\mathbf{N}}{dt} = \frac{\partial \mathbf{N}}{\partial t} + \frac{1}{i}[\mathbf{N}, H] = 0 \qquad \left(\frac{\partial \mathbf{N}}{\partial t} = \mathbf{P}\right). \qquad (1\text{--}2.23)$$

Of course, H is not an explicit function of t for an isolated dynamical system. The conservation of $\mathbf{N}$ also appears as

$$\frac{d\mathbf{N}}{dt} = \mathbf{P} - M\frac{d\mathbf{R}}{dt} = 0, \tag{1–2.24}$$

which clearly identifies the parameter M as the invariable mass of the system. The position vector $\mathbf{R}$ moves with constant velocity:

$$\frac{d\mathbf{R}}{dt} = \frac{\mathbf{P}}{M}. \tag{1–2.25}$$

The structure of H is determined somewhat by the various conservation laws. We note that

$$\frac{d\mathbf{P}}{dt} = \frac{1}{i}[\mathbf{P}, H] = -\frac{\partial H}{\partial \mathbf{R}} = 0, \qquad \frac{d\mathbf{R}}{dt} = \frac{1}{i}[\mathbf{R}, H] = \frac{\partial H}{\partial \mathbf{P}} = \frac{\mathbf{P}}{M}, \tag{1–2.26}$$

and

$$\frac{d\mathbf{J}}{dt} = \frac{d\mathbf{S}}{dt} = \frac{1}{i}[\mathbf{S}, H] = 0. \tag{1–2.27}$$

The consequence is

$$H = \frac{\mathbf{P}^2}{2M} + H_{\text{int}}, \tag{1–2.28}$$

which gives the decomposition into energy of motion of the whole system and internal energy. The latter will generally involve internal dynamical variables, which commute with $\mathbf{R}$ and $\mathbf{P}$, combined in such a way that H_{int} is invariant under the rotation generated by the internal angular momentum $\mathbf{S}$.

An elementary particle is a system without internal energy, or at least one for which internal energy is effectively inert under the limited physical circumstances under consideration. Let us consider n elementary particles, each described as above by variables $\mathbf{r}_a$, $\mathbf{p}_a$, $\mathbf{s}_a$ and mass m_a, $a = 1, \ldots, n$. The operators associated with different particles commute. The kinetic transformation generators of the whole system are then obtained additively as

$$\mathbf{P} = \sum_{a=1}^{n} \mathbf{p}_a,$$

$$\mathbf{J} = \sum_{a=1}^{n} (\mathbf{r}_a \times \mathbf{p}_a + \mathbf{s}_a) = \mathbf{R} \times \mathbf{P} + \mathbf{S}, \tag{1–2.29}$$

$$\mathbf{N} = \sum_{a=1}^{n} (\mathbf{p}_a t - m_a \mathbf{r}_a) = \mathbf{P}t - M\mathbf{R},$$

where

$$M = \sum_a m_a, \qquad \mathbf{R} = \sum_a \frac{m_a}{M} \mathbf{r}_a, \tag{1–2.30}$$

and

$$\mathbf{S} = \sum_a \left[(\mathbf{r}_a - \mathbf{R}) \times \left(\mathbf{p}_a - \frac{m_a}{M}\mathbf{P} \right) + \mathbf{s}_a \right]. \tag{1–2.31}$$

The operators for the total system have the required properties. Note that the internal variables introduced here are not linearly independent:

$$\sum_a m_a(\mathbf{r}_a - \mathbf{R}) = 0, \qquad \sum_a \left(\mathbf{p}_a - \frac{m_a}{M}\mathbf{P}\right) = 0, \qquad (1\text{--}2.32)$$

which is also conveyed by the commutation relation

$$\frac{1}{i}\left[(\mathbf{r}_a - \mathbf{R})_k, \left(\mathbf{p}_b - \frac{m_b}{M}\mathbf{P}\right)_l\right] = \delta_{kl}\left(\delta_{ab} - \frac{m_b}{M}\right). \qquad (1\text{--}2.33)$$

If the various particles are dynamically isolated, the energy operator is also additive. More generally, we describe interacting systems by

$$H = \sum_a \frac{\mathbf{p}_a^2}{2m_a} + V = \frac{\mathbf{P}^2}{2M} + H_{\text{int}}, \qquad (1\text{--}2.34)$$

where the internal energy of the system is

$$H_{\text{int}} = \sum_a \frac{\left(\mathbf{p}_a - \frac{m_a}{M}\mathbf{P}\right)^2}{2m_a} + V, \qquad (1\text{--}2.35)$$

and V is a scalar function of the internal coordinates $\mathbf{r}_a - \mathbf{R}$, $\mathbf{p}_a - (m_a/M)\mathbf{P}$, $\mathbf{s}_a$, and possibly others.

The number of particles being described cannot be a dynamical variable apart from rather special circumstances. Let there be several different types of particles, with masses m_α. Then

$$M = \sum_\alpha m_\alpha N_\alpha, \qquad (1\text{--}2.36)$$

where N_α is the number of particles of type α. Since there are generally no rational relations among the masses of different particles, the invariability of M implies the constancy of N_α for each type. An exception occurs for unstable particles, as in the α-instability of nuclei (α-particle kinetic energies can be sufficiently small to validate the Galilean regime). Here the mass of the unstable nucleus very closely equals the sum of α-particle and residual nuclear masses.

The general characterization of interacting systems enables one to give a simple description of the behavior of a particle that is influenced by a macroscopic, controllable environment. Since a classical theory of such interactions underlies the measurement of free particle properties, a test of self-consistency is also involved. Let the Hamiltonian operator of a system of particles be divided into two parts: H_p, comprising all terms containing the variables of a given particle; H_{-p}, being all other terms, describing the residual system after the particle of interest has been removed. We assume, for simplicity, that the interaction terms in H_p are no more than linear in the velocity and in the spin of the particle, but we do not include bilinear or spin-orbit coupling terms. Although the interaction is not necessarily electromagnetic, we use a notation designed to facilitate that identification:

$$H_p = \frac{\mathbf{p}^2}{2m} + \left[e\varphi(\mathbf{r}t) + \frac{e^2}{2mc^2} \left(\mathbf{A}(\mathbf{r}t) \right)^2 \right] - \frac{\mathbf{p}}{mc} \cdot e\mathbf{A}(\mathbf{r}t) - \mathbf{s} \cdot \mathbf{F}(\mathbf{r}t). \quad (1\text{--}2.37)$$

It is understood that the noncommuting operators $\mathbf{p}$ and $\mathbf{A}(\mathbf{r}t)$ are symmetrized in multiplication to produce a Hermitian product. The explicit time dependence appears as an effective replacement for the actual dependence on the variables of the external system, as indicated by

$$\frac{\partial}{\partial t} \mathbf{A}(\mathbf{r}t) = \frac{1}{i} [\mathbf{A}(\mathbf{r}t), H_{-p}]. \quad (1\text{--}2.38)$$

The equations of motion are

$$\mathbf{v} = \frac{d\mathbf{r}}{dt} = \frac{1}{m} \left[\mathbf{p} - \frac{e}{c} \mathbf{A}(\mathbf{r}t) \right], \qquad \frac{d\mathbf{s}}{dt} = \mathbf{s} \times \mathbf{F}(\mathbf{r}t), \quad (1\text{--}2.39)$$

and

$$m \frac{d\mathbf{v}}{dt} = e\mathbf{E} + \frac{e}{c} \mathbf{v} \times \mathbf{H} + \nabla(\mathbf{s} \cdot \mathbf{F}), \quad (1\text{--}2.40)$$

where

$$\mathbf{E} = -\nabla\varphi - \frac{1}{c} \frac{\partial}{\partial t} \mathbf{A}, \qquad \mathbf{H} = \nabla \times \mathbf{A}. \quad (1\text{--}2.41)$$

In deriving the last equation we have omitted such commutators as $[(e/c)\mathbf{A}, e\varphi]$. This is to be validated, not as a classical approximation, but through the negligibility of the dynamical reaction back on the external system. We also note the commutation relation indicated by

$$\mathbf{v} \times \mathbf{v} = i \frac{e}{m^2c} \mathbf{H}(\mathbf{r}t). \quad (1\text{--}2.42)$$

We do specialize to electromagnetism on equating $\mathbf{F}$ with a multiple of $\mathbf{H}$,

$$\mathbf{F} = g \frac{e}{2mc} \mathbf{H}, \quad (1\text{--}2.43)$$

which is the identification of the internal magnetic dipole moment,

$$\boldsymbol{\mu} = g \frac{e}{2mc} \mathbf{s}. \quad (1\text{--}2.44)$$

Analogous electric dipole moments have never been observed. One value of the gyromagnetic ratio g has a special property. In a homogeneous magnetic field the velocity and the spin vectors precess about the field axis. The two precession rates are equal if $g = 2$. The observed values of g are very slightly in excess of 2 for the electron [2(1.001160)] and the muon [2(1.001166)], but differ widely for other particles.

We shall find it interesting to consider a spinless charged particle that moves in the magnetic field of a distant stationary magnetic charge. Let the coordinate origin be placed at the position of this charge (we now use the letter g to denote its strength in Gaussian units), so that

$$\mathbf{H} = g\,\frac{\mathbf{r}}{r^3}\cdot \tag{1–2.45}$$

This system is characterized by the equation of motion

$$m\,\frac{d\mathbf{v}}{dt} = \frac{eg}{c}\,\mathbf{v}\times\frac{\mathbf{r}}{r^3}, \tag{1–2.46}$$

where a symmetrized product is understood, together with the commutation relations

$$[r_k,\,(m\mathbf{v})_l] = i\delta_{kl}, \tag{1–2.47}$$

and

$$\mathbf{v}\times\mathbf{v} = i\,\frac{eg}{m^2 c}\,\frac{\mathbf{r}}{r^3}\cdot \tag{1–2.48}$$

The last equation is inconsistent unless the particle can be controlled to remain distant from the magnetic charge, in the weak sense of $r > 0$. That follows from the Jacobi identity:

$$0 = [[\mathbf{v}_1, \mathbf{v}_2], \mathbf{v}_3] + \text{cycl. perm.} = \frac{eg}{m^3 c}\,\nabla^2\,\frac{1}{r}\cdot \tag{1–2.49}$$

The equation of moments, which uses symmetrized multiplication, is

$$\mathbf{r}\times m\,\frac{d\mathbf{v}}{dt} = \frac{eg}{c}\left(\mathbf{v}\,\frac{1}{r} - \mathbf{v}\cdot\mathbf{r}\,\frac{\mathbf{r}}{r^3}\right)\cdot \tag{1–2.50}$$

But since the Hamiltonian is no more than quadratic in the momenta, symmetrized multiplication enables one to write

$$\frac{d}{dt}\,f(\mathbf{r}) = \frac{1}{i}\,[f(\mathbf{r}), H] = \nabla f(\mathbf{r})\cdot\frac{\partial H}{\partial \mathbf{p}} = \mathbf{v}\cdot\nabla f(\mathbf{r}), \tag{1–2.51}$$

and thereby recognize the conserved angular momentum vector

$$\mathbf{J} = \mathbf{r}\times m\mathbf{v} - \frac{eg}{c}\,\frac{\mathbf{r}}{r}\cdot \tag{1–2.52}$$

One easily verifies that it is the rotation generator. There is an important consequence of that fact. Consider the coordinate wave function representing a particular state, $\langle \mathbf{r}t| \;\rangle$, and perform a coordinate system rotation about the axis provided by $\mathbf{r}$. An infinitesimal rotation is given by

$$\delta\boldsymbol{\omega} = \frac{\mathbf{r}}{r}\,\delta\varphi, \tag{1–2.53}$$

and the corresponding generator is simply

$$\delta\boldsymbol{\omega}\cdot\mathbf{J} = -\frac{eg}{c}\,\delta\varphi. \tag{1–2.54}$$

The response to a finite rotation is therefore

$$\langle \mathbf{rt}| \; \rangle \rightarrow e^{-i(eg/c)\varphi}\langle \mathbf{rt}| \; \rangle, \tag{1–2.55}$$

and the known limitation to single or double valuedness for rotation through 2π radians implies that eg/c is either an integer or an integer plus $\frac{1}{2}$.

As a discussion of magnetic charge and its implications this is quite incomplete. However, the special operator system given in Eqs. (1–2.47), (1–2.48), and (1–2.52) will soon be encountered again in a very different physical context.

1–3 EINSTEINIAN RELATIVITY

The new feature associated with the finiteness of c, the speed of light, is the abandonment of absolute simultaneity. It is replaced, for infinitesimal transformations, by

$$\delta ct = \delta\epsilon^0 + \frac{1}{c}\,\delta\mathbf{v}\cdot\mathbf{r}, \tag{1–3.1}$$

where $\delta\epsilon^0$ is the displacement of the origin for the variable ct. We now designate the space-time coordinates collectively by $x^\mu = ct$, $\mathbf{r}$, where $x^0 = -x_0 = ct$ and $x^k = x_k = r_k$. The infinitesimal coordinate transformations of the Einsteinian relativity group are

$$\bar{x}^\nu = x^\nu - \delta x^\nu, \qquad \delta x^\nu = \delta\epsilon^\nu + \delta\omega^{\mu\nu}x_\mu, \tag{1–3.2}$$

where

$$\delta\omega^{\mu\nu} = -\delta\omega^{\nu\mu}. \tag{1–3.3}$$

The six independent parameters of this four-dimensional rotation are related to $\delta\boldsymbol{\omega}$ and $\delta\mathbf{v}$ by

$$\delta\omega_{kl} = \epsilon_{klm}\delta\omega_m, \qquad \delta\omega_{0k} = \delta v_k/c. \tag{1–3.4}$$

The composition properties of the 10-parameter group are specified by

$$\begin{aligned}
\delta_{[12]}\epsilon^\nu &= \delta_1\omega^{\mu\nu}\,\delta_2\epsilon_\mu - \delta_2\omega^{\mu\nu}\,\delta_1\epsilon_\mu, \\
\delta_{[12]}\omega^{\mu\nu} &= \delta_1\omega^{\mu\lambda}\,\delta_2\omega^\nu{}_\lambda - \delta_2\omega^{\mu\lambda}\,\delta_1\omega^\nu{}_\lambda.
\end{aligned} \tag{1–3.5}$$

The generators of the unitary transformation induced by an infinitesimal coordinate transformation are comprised in

$$G = P^\mu\,\delta\epsilon_\mu + \tfrac{1}{2}J^{\mu\nu}\,\delta\omega_{\mu\nu} + \delta\varphi\mathbf{1}. \tag{1–3.6}$$

The correspondence with the Galilean generators is

$$J_{kl} = \epsilon_{klm}J_m, \qquad (1/c)\,J^{0k} = N_k, \tag{1–3.7}$$

and

$$cP^0 = H + Mc^2. \tag{1–3.8}$$

We shall recognize shortly the necessity for the shift in energy origin between the relativistic and nonrelativistic domains (to use the conventional labels for the two relativities). As to the composition law for the scalar parameter $\delta\varphi$, it is interesting that no bilinear scalar $\delta_{[12]}\varphi = -\delta_{[21]}\varphi$ can be formed in the four-dimensional Minkowski space from the vectors $\delta_{1,2}\epsilon^\mu$ and the tensors $\delta_{1,2}\omega^{\mu\nu}$. Accordingly,

$$\delta_{[12]}\varphi = 0, \tag{1-3.9}$$

and the full set of commutators for the generators is

$$[P_\mu, P_\nu] = 0,$$
$$(1/i)[P_\mu, J_{\kappa\lambda}] = g_{\mu\lambda}P_\kappa - g_{\mu\kappa}P_\lambda, \tag{1-3.10}$$
$$(1/i)[J_{\mu\nu}, J_{\kappa\lambda}] = g_{\mu\kappa}J_{\nu\lambda} - g_{\nu\kappa}J_{\mu\lambda} + g_{\nu\lambda}J_{\mu\kappa} - g_{\mu\lambda}J_{\nu\kappa},$$

where $g_{\mu\nu}$ is the metric tensor specified by

$$g_{\mu\nu}: g_{00} = -1, \qquad g_{0k} = 0, \qquad g_{kl} = \delta_{kl}. \tag{1-3.11}$$

The commutators can also be presented as

$$(1/i)[P^\nu, \tfrac{1}{2}J^{\kappa\lambda}\,\delta\omega_{\kappa\lambda}] = \delta\omega^{\mu\nu}P_\mu,$$
$$(1/i)[J^{\mu\nu}, \tfrac{1}{2}J^{\kappa\lambda}\,\delta\omega_{\kappa\lambda}] = \delta\omega^{\lambda\mu}J_\lambda{}^\nu + \delta\omega^{\lambda\nu}J^\mu{}_\lambda, \tag{1-3.12}$$

indicating the response of vectors and tensors to infinitesimal Lorentz rotations (comprising three-dimensional rotations and boosts), and

$$(1/i)[P^\nu, P^\lambda\,\delta\epsilon_\lambda] = 0,$$
$$(1/i)[J^{\mu\nu}, P^\lambda\,\delta\epsilon_\lambda] = \delta\epsilon^\mu P^\nu - \delta\epsilon^\nu P^\mu, \tag{1-3.13}$$

which gives the translational response of these operators. When written in three-dimensional notation, all these commutators reproduce the Galilean forms, with two exceptions:

$$(1/i)[P_k, N_l] = \delta_{kl}P^0/c, \qquad (1/i)[N_k, N_l] = -J_{kl}/c^2. \tag{1-3.14}$$

In Galilean relativity, then, $\mathbf{J}/c^2$ is neglected and H is neglected relative to Mc^2, giving the effective replacement of the operator P^0/c by the number M. Henceforth we adopt atomic units in which $c = 1$.

There is one obvious realization of all the commutation relations for the 10 generators. It is

$$P^\mu| \;\rangle = 0, \qquad J^{\mu\nu}| \;\rangle = 0, \tag{1-3.15}$$

which describes the total invariance of the structureless vacuum state. Any other state is an excitation of the system, characterized by $P^0 > 0$. The scalar formed from the translationally invariant P^μ,

$$M^2 = -P^\mu P_\mu, \tag{1-3.16}$$

is invariant under all operations of the Lorentz group (we are discussing only the transformations that are continuously connected with the identity, the proper orthochronous Lorentz group). According as M^2 is positive, zero, or negative, the four-vector P^μ is time-like, null, or space-like. With a time-like momentum,

$$P^0 = +(\mathbf{P}^2 + M^2)^{1/2} > 0 \tag{1-3.17}$$

is an invariant property. For $M = 0$, too,

$$P^0 = +|\mathbf{P}| > 0 \tag{1-3.18}$$

remains valid under Lorentz transformations. But the time component of a space-like vector can be given either sign by appropriately choosing the co-ordinate system. Thus $M^2 < 0$ is of no interest for physics. The nonnegative quantity $(M^2)^{1/2}$ is the mass of the system.

Another translationally invariant object is the pseudovector

$$W^\mu = {}^*J^{\mu\nu}P_\nu = P_\nu{}^*J^{\mu\nu}, \tag{1-3.19}$$

where

$$^*J^{\mu\nu} = \tfrac{1}{2}\epsilon^{\mu\nu\kappa\lambda}J_{\kappa\lambda} \tag{1-3.20}$$

forms the tensor dual to $J^{\mu\nu}$ with the aid of the totally antisymmetrical tensor specified by

$$\epsilon^{0123} = +1. \tag{1-3.21}$$

This invariance property follows from the translational response of $J_{\kappa\lambda}$, and the antisymmetry of $\epsilon^{\mu\nu\kappa\lambda}$,

$$(1/i)[W^\mu, P^\alpha \, \delta\epsilon_\alpha] = \epsilon^{\mu\nu\kappa\lambda} \, \delta\epsilon_\kappa P_\lambda P_\nu = 0. \tag{1-3.22}$$

We also note that

$$P_\mu W^\mu = 0. \tag{1-3.23}$$

The scalar

$$W^2 = W^\mu W_\mu \geq 0 \tag{1-3.24}$$

is invariant under all Lorentz transformations. As indicated, the vector W^μ, being orthogonal to P^μ, cannot be time-like. The commutation relations among the components of W^μ are

$$(1/i)[W^\mu, W^\nu] = {}^*(W^\mu P^\nu - W^\nu P^\mu). \tag{1-3.25}$$

The behavior under coordinate displacements that is presented in the equations

$$(1/i)[\mathbf{J}, \mathbf{P} \cdot \delta\epsilon] = \delta\epsilon \times \mathbf{P}, \qquad (1/i)[\mathbf{N}, \mathbf{P} \cdot \delta\epsilon] = -\dot{\delta\epsilon}P^0, \tag{1-3.26}$$

again indicates the existence of a position operator $\mathbf{R}$, obeying

$$(1/i)[\mathbf{R}, \mathbf{P} \cdot \delta\epsilon] = \delta\epsilon \tag{1-3.27}$$

or

$$(1/i)[R_k, P_l] = \delta_{kl}. \tag{1-3.28}$$

(One must stifle the impulse to introduce a time operator complementary to P^0. That would contradict the physical nature of the energy spectrum.) A particular realization of $\mathbf{J}$ and $\mathbf{N}$, in which additional displacement independent quantities do not occur, is

$$\mathbf{J} = \mathbf{R} \times \mathbf{P}, \qquad \mathbf{N} = \mathbf{P}x^0 - P^0\mathbf{R}, \tag{1-3.29}$$

where symmetrized multiplication is used for the noncommuting operators $\mathbf{R}$ and P^0,

$$(1/i)[\mathbf{R}, P^0] = \partial P^0/\partial \mathbf{P} = \mathbf{P}/P^0. \tag{1-3.30}$$

The correct three-dimensional rotational behavior of all the operators considered is obtained if

$$\mathbf{R} \times \mathbf{R} = 0. \tag{1-3.31}$$

The other characteristic commutator of Einsteinian relativity here reads

$$i\mathbf{N} \times \mathbf{N} = \mathbf{R} \times \mathbf{P}. \tag{1-3.32}$$

It is obeyed without further ado, since

$$i[P^0R_k, P^0R_l] = R_kP_l - R_lP_k. \tag{1-3.33}$$

The simplicity of this result, despite the presence of symmetrized products, depends on the fact that commutators of $\mathbf{R}$ with functions of $\mathbf{P}$ introduce no further commutators and are necessarily canceled by the Hermitian symmetrization. In this situation the information about the energy operator that can be derived from the conservation of $\mathbf{N}$,

$$d\mathbf{R}/dx^0 = \mathbf{P}/P^0, \tag{1-3.34}$$

is already contained in the relation $P^0 = (\mathbf{P}^2 + M^2)^{1/2}$.

Now let an internal angular momentum be added:

$$\mathbf{J} = \mathbf{R} \times \mathbf{P} + \mathbf{S}. \tag{1-3.35}$$

As such, $\mathbf{S}$ must commute with $\mathbf{R}$ and $\mathbf{P}$ while itself obeying the angular momentum commutation rules. It is also necessary to supplement $\mathbf{N}$ in order to generate the spin term of the commutator

$$i\mathbf{N} \times \mathbf{N} = \mathbf{R} \times \mathbf{P} + \mathbf{S}. \tag{1-3.36}$$

A suitable form is

$$\mathbf{N} = \mathbf{P}x^0 - P^0\mathbf{R} + a(P^0)\mathbf{S} \times \mathbf{P}. \tag{1-3.37}$$

The calculation of $\mathbf{N} \times \mathbf{N}$ involves

$$-iP^0\mathbf{R} \times [a(\mathbf{S} \times \mathbf{P})] - i[a(\mathbf{S} \times \mathbf{P})] \times P^0\mathbf{R} + a^2i(\mathbf{S} \times \mathbf{P}) \times (\mathbf{S} \times \mathbf{P})$$
$$= (da/dP^0)\mathbf{P} \times (\mathbf{S} \times \mathbf{P}) + 2P^0a\mathbf{S} - a^2[\mathbf{P}^2\mathbf{S} - \mathbf{P} \times (\mathbf{S} \times \mathbf{P})], \tag{1-3.38}$$

and the required result is obtained with

$$\frac{da}{dP^0} + a^2 = 0, \qquad 2P^0 a - \mathbf{P}^2 a^2 = 1 \tag{1-3.39}$$

or

$$\frac{d}{dP^0}\left(\frac{1}{a} - P^0\right) = 0, \qquad \left(\frac{1}{a} - P^0\right)^2 = M^2. \tag{1-3.40}$$

We conclude that

$$a(P^0) = (P^0 + M)^{-1}, \tag{1-3.41}$$

the alternative choice with $(P^0 - M)^{-1}$ being singular at $\mathbf{P} = 0$, $P^0 = M$. The final form is

$$\mathbf{J} = \mathbf{R} \times \mathbf{P} + \mathbf{S},$$
$$\mathbf{N} = \mathbf{P}x^0 - P^0\mathbf{R} + \frac{1}{P^0 + M}\mathbf{S} \times \mathbf{P}, \tag{1-3.42}$$

which incidentally shows that $\mathbf{S}^2$ is a Lorentz invariant. It is worth pointing out the converse, that operators with the stated properties of $\mathbf{R}$ and $\mathbf{S}$ can be constructed from the Lorentz generators ($x^0 = 0$):

$$M\mathbf{R} = -\mathbf{N} + \frac{1}{P^0(P^0 + M)}\mathbf{P}\mathbf{P} \cdot \mathbf{N} + \frac{1}{P^0 + M}\mathbf{J} \times \mathbf{P},$$
$$M\mathbf{S} = P^0\mathbf{J} - \frac{1}{P^0 + M}\mathbf{P}\mathbf{P} \cdot \mathbf{J} + \mathbf{N} \times \mathbf{P}. \tag{1-3.43}$$

The exceptional position of $M = 0$ is evident here.

The components of the pseudovector W^μ are given by

$$W^0 = \mathbf{P} \cdot \mathbf{J}, \qquad \mathbf{W} = P^0\mathbf{J} - \mathbf{P} \times \mathbf{N} \tag{1-3.44}$$

or

$$W^0 = \mathbf{P} \cdot \mathbf{S},$$
$$\mathbf{W} = P^0\mathbf{S} - \frac{1}{P^0 + M}\mathbf{P} \times (\mathbf{S} \times \mathbf{P}) = M\mathbf{S} + \frac{1}{P^0 + M}\mathbf{P}\mathbf{P} \cdot \mathbf{S}. \tag{1-3.45}$$

The last relation can also be written as

$$\mathbf{W} - \frac{\mathbf{P}}{P^0 + M}W^0 = M\mathbf{S}. \tag{1-3.46}$$

There is a connection among the several invariants:

$$W^2 = M^2\mathbf{S}^2. \tag{1-3.47}$$

This discussion refers generally to $M^2 > 0$. We next consider the limit as $M^2 \to 0$ for fixed $\mathbf{S}^2$. The resulting relation

$$\mathbf{W} = \mathbf{P}\frac{W^0}{P^0} \tag{1-3.48}$$

can be given the covariant form

$$W^\mu = \lambda P^\mu, \tag{1-3.49}$$

where

$$\lambda = \mathbf{P} \cdot \mathbf{S}/P^0 \tag{1-3.50}$$

is a Lorentz invariant. This quantity is the component of the spin along the direction of motion, or the helicity of the particle. In view of its invariance, a physical system need exhibit only one value of helicity, or, if space-reflection parity has a meaning for the interactions of that system, the pseudoscalar λ can have two values, $\pm s$. The photon, with $s = 1$, illustrates the latter situation, while for the neutrinos, with $s = \frac{1}{2}$, $\lambda = +s$ and $-s$ refer to essentially different particles. If only one helicity value is meaningful, or with $s \geq 1$, even if $\lambda = s$ and $-s$ are both realized, not all of the $2s + 1$ spin magnetic quantum number states exist. Accordingly, the operator $\mathbf{S}$ ceases to be defined (with two exceptions) in the limit $M \to 0$, and we must introduce new variables for this circumstance.

In order to delete $\mathbf{S}$ we define the new position vector

$$\hat{\mathbf{R}} = \mathbf{R} - \mathbf{S} \times \mathbf{P}/(P^0)^2, \tag{1-3.51}$$

which is such that

$$\hat{\mathbf{R}} \times \mathbf{P} = \mathbf{R} \times \mathbf{P} + \mathbf{S} - \mathbf{PP} \cdot \mathbf{S}/(P^0)^2. \tag{1-3.52}$$

Then

$$\mathbf{J} = \hat{\mathbf{R}} \times \mathbf{P} + \lambda(\mathbf{P}/P^0), \qquad \mathbf{N} = \mathbf{P}x^0 - P^0\hat{\mathbf{R}}, \tag{1-3.53}$$

and to complete the verification that only λ appears explicitly, we give the commutator

$$\hat{\mathbf{R}} \times \hat{\mathbf{R}} = -\frac{i\lambda\mathbf{P}}{(P^0)^3} = i\lambda \frac{\partial}{\partial \mathbf{P}} \frac{1}{|\mathbf{P}|}. \tag{1-3.54}$$

This is the operator system that we anticipated in discussing magnetic charge. The correspondence is

$$\hat{\mathbf{R}} \leftrightarrow m\mathbf{v}, \qquad \mathbf{P} \leftrightarrow -\mathbf{r}, \qquad \lambda \leftrightarrow eg/c, \tag{1-3.55}$$

and the restriction $r > 0$ is here validated as the Lorentz invariant energy property $P^0 > 0$. The absence of certain helicity values is now manifested by the noncommutativity of components of $\mathbf{R}$. This intrinsic nonlocality of massless particles is described by the uncertainty principle

$$\Delta \hat{R}_k \, \Delta \hat{R}_l \geq \frac{|\lambda|}{2} \left| \left\langle \frac{P_m}{(P^0)^3} \right\rangle \right|, \qquad k \neq l \neq m, \tag{1-3.56}$$

or, for a momentum state with some degree of directionality,

$$(\Delta \hat{\mathbf{R}})^2 \gtrsim |\lambda| \langle 1/(P^0)^2 \rangle, \tag{1-3.57}$$

indicating that an average wavelength roughly sets the scale of coordinate specificability. Incidentally, when the explicit constructions of $\mathbf{J}$ and $\mathbf{N}$ are inserted in the formulas for $M\mathbf{R}$ and $M\mathbf{S}$, these expressions do vanish, as does $M\hat{\mathbf{R}}$.

In the situation we have just discussed, $W^2 = 0$. There is another logical possibility. With $\lambda = \mathbf{P} \cdot \mathbf{S}/P^0$ assuming any accessible finite value, let $\mathbf{S}^2 \to \infty$ as $M^2 \to 0$ to produce the limit

$$M\mathbf{S} \to \mathbf{T}. \tag{1-3.58}$$

The characteristics of $\mathbf{S}$ give these operators the following properties:

$$\mathbf{T} \cdot \mathbf{P} = 0, \qquad \mathbf{T} \times \mathbf{T} = 0, \qquad (1/i)[\lambda, \mathbf{T}] = \mathbf{T} \times \mathbf{P}/P^0 \tag{1-3.59}$$

and

$$[\lambda, \mathbf{T}^2] = 0. \tag{1-3.60}$$

The invariant

$$\mathbf{T}^2 = W^2 \tag{1-3.61}$$

can be assigned any positive value. The components of $\mathbf{T}$ act to change λ by ± 1, and this without limit. We now have

$$W^0 = \lambda P^0, \qquad \mathbf{W} = \lambda \mathbf{P} + \mathbf{T}. \tag{1-3.62}$$

The commutation relations among the components of W_μ, which are satisfied trivially when $\mathbf{T} = 0$, here demand that

$$(1/i)[\lambda P^0, \mathbf{T}] = \mathbf{T} \times \mathbf{P}, \qquad (1/i)(\lambda \mathbf{P} + \mathbf{T}) \times (\lambda \mathbf{P} + \mathbf{T}) = P^0 \mathbf{T}, \tag{1-3.63}$$

and these are valid statements about $\mathbf{T}$. We continue to use the position vector

$$\hat{\mathbf{R}} = \mathbf{R} - \mathbf{S} \times \mathbf{P}/(P^0)^2 \tag{1-3.64}$$

and its properties:

$$[\lambda, \hat{\mathbf{R}}] = 0, \qquad \hat{\mathbf{R}} \times \hat{\mathbf{R}} = -i\lambda \mathbf{P}/(P^0)^3, \tag{1-3.65}$$

but we must be careful to note that as $M \to 0$,

$$\frac{1}{P^0 + M} \mathbf{S} \times \mathbf{P} - \frac{1}{P^0} \mathbf{S} \times \mathbf{P} \to - \frac{1}{(P^0)^2} \mathbf{T} \times \mathbf{P}. \tag{1-3.66}$$

This gives

$$\mathbf{J} = \hat{\mathbf{R}} \times \mathbf{P} + \lambda \mathbf{P}/P^0, \qquad \mathbf{N} = \mathbf{P}x^0 - P^0\hat{\mathbf{R}} - \mathbf{T} \times \mathbf{P}/(P^0)^2. \tag{1-3.67}$$

We now find, as the counterpart of $M\mathbf{S} \to \mathbf{T}$, that

$$M\mathbf{R} = \mathbf{T} \times \mathbf{P}/(P^0)^2 \tag{1-3.68}$$

and $\mathbf{R}^2 \to \infty$ with vanishing M. However, $M\hat{\mathbf{R}} = 0$. The commutation relations $i\mathbf{N} \times \mathbf{N} = \mathbf{J}$ continue to be obeyed despite the introduction of the $\mathbf{T}$

term, since

$$\hat{\mathbf{R}} \times (\mathbf{T} \times \mathbf{P}/(P^0)^2) + (\mathbf{T} \times \mathbf{P}/(P^0)^2) \times \hat{\mathbf{R}} = 0. \qquad (1\text{-}3.69)$$

This involves the commutator

$$(1/i)[\hat{R}_k, T_l] = -T_k P_l/(P^0)^2, \qquad (1\text{-}3.70)$$

which is also used to verify that $\mathbf{J}$ generates the rotations of $\mathbf{T}$. The significant observation is that λ has ceased to be a Lorentz invariant:

$$(1/i)[\lambda, \mathbf{N}] = \mathbf{T}/P^0. \qquad (1\text{-}3.71)$$

This fact, together with the unbounded nature of the λ spectrum, ranging over all integers or all integers $+\frac{1}{2}$, indicates that physically accessible states would exist for which $(\Delta \hat{\mathbf{R}})^2$ is arbitrarily large.

We suggest the following verbal principle for massless particles: A zero mass particle is not completely localizable, but a finite degree of localizability exists. The principle has the following valid consequences. There is no spinless zero mass particle, for the commutative position vector $\mathbf{R}$ would be available. The same reasoning exclude $s = \frac{1}{2}$ massless particles for which space reflection parity is meaningful. And the systems we have just discussed, with $W^2 > 0$, are condemned wholesale by the existence of states that are unlocalized without limit. There is a simple pattern for the known or strongly conjectured massless particles; their spins are given by $s = 2^\sigma$, $\sigma = -1, 0, +1$.

The concept of elementary particle in relativistic mechanics remains an operational one, that under the conditions of physical excitation available, it is consistent to assign a unique value to mass, spin, and other characteristic invariant attributes of the system. For a set of n noninteracting particles, the Lorentz generators of the whole system are given by the additive forms

$$\mathbf{P} = \sum_{a=1}^{n} \mathbf{p}_a, \qquad P^0 = \sum_{a=1}^{n} p_a^0, \qquad \mathbf{J} = \sum_a (\mathbf{r}_a \times \mathbf{p}_a + \mathbf{s}_a),$$

$$\mathbf{N} = \sum_a \left(\mathbf{p}_a x^0 - p_a^0 \mathbf{r}_a + \frac{1}{p_a^0 + m_a} \mathbf{s}_a \times \mathbf{p}_a \right). \qquad (1\text{-}3.72)$$

The operators $\mathbf{R}$ and $\mathbf{S}$ for the total system must be obtained from the construction (1–3.43); one is not likely to produce them by an *a priori* definition. Consider, for example,

$$MR = \sum_a \left(p_a^0 \mathbf{r}_a - \frac{1}{p_a^0 + m_a} \mathbf{s}_a \times \mathbf{p}_a \right)$$

$$- \frac{1}{P^0(P^0 + M)} \mathbf{PP} \cdot \sum_a \left(p_a^0 \mathbf{r}_a - \frac{1}{p_a^0 + m_a} \mathbf{s}_a \times \mathbf{p}_a \right)$$

$$+ \frac{1}{P^0 + M} [\sum_a (\mathbf{r}_a \times \mathbf{p}_a + \mathbf{s}_a)] \times \mathbf{P}. \qquad (1\text{-}3.73)$$

We approach the topic of interacting particles by giving first a relativistic generalization of the nonrelativistic treatment of a particle moving in a macroscopic environment. In order to make covariance more explicit we define a proper time derivative:

$$\frac{dF}{ds} = \frac{1}{i}\left[F, \frac{(p^0)^2}{2m}\right] + \frac{p^0}{m}\frac{\partial F}{\partial x^0},$$ (1–3.74)

with the usual symmetrization understood in the last term. Thus, for a single isolated particle, we have

$$m\frac{dx^\mu}{ds} = p^\mu, \qquad m\frac{d^2x^\mu}{ds^2} = 0,$$ (1–3.75)

and

$$-\frac{dx^\mu}{ds}\frac{dx_\mu}{ds} = 1.$$ (1–3.76)

With attention restricted to a homogeneous electromagnetic field, the covariant generalizations of Eqs. (1–2.39) and (1–2.40) are stated as

$$m\frac{d^2x^\mu}{ds^2} = eF^{\mu\nu}\frac{dx_\nu}{ds},$$
$$\frac{dw^\mu}{ds} = g\frac{e}{2m}F^{\mu\nu}w_\nu + (g-2)\frac{e}{2m}\frac{dx^\mu}{ds}\frac{dx^\lambda}{ds}F_{\lambda\nu}w^\nu.$$ (1–3.77)

The constraints

$$-\frac{dx^\mu}{ds}\frac{dx_\mu}{ds} = 1, \qquad \frac{dx^\mu}{ds}w_\mu = 0, \qquad w^\mu w_\mu = m^2s(s+1)$$ (1–3.78)

are compatible with the equations of motion, at least to terms linear in the field strengths; this involves only the commutation relations for a free particle.

Why did we not begin with a general theory of interacting particles, specified by variables r_a, p_a, s_a, $a = 1, \ldots, n$, and then proceed to follow the motion of one particle under the influence of the others, as in the nonrelativistic discussion? Quite simply, because no such general theory exists. Apart from the obviously formidable algebraic task of stating the relativistic conditions on interaction terms (small deviations from nonrelativistic behavior pose no problem), the attempt founders on the failure of the assumption that there is a fixed number of particles. The relation between relativistic and nonrelativistic energy can be exhibited as

$$P^0 = \sum_a m_a + H = \sum_\alpha m_\alpha N_\alpha + H.$$ (1–3.79)

In the nonrelativistic limit where changes in H are small compared to each m_α, the conservation of P^0 generally demands, first, the conservation of each N_α, and then, that of H. But if the kinetic and interaction energies contained in H

become comparable with individual m_α values, one can no longer conclude that the N_α remain constant. It is the characteristic feature of relativistic particle dynamics that particles can be created and annihilated in high energy encounters.

1-4 CRITIQUE OF PARTICLE THEORIES

Theory thus affirms, and experiment abundantly confirms, that the concept of the particle as an immutable object is untenable under pronounced relativistic interactions conditions. There have been two extreme reactions to this situation. They respond to the failure of a detailed space-time description in particle language by: (1) insisting on the possibility of a detailed space-time description but in terms of a concept more fundamental than particle; (2) rejecting the possibility of a detailed space-time description by denying that any concept underlies that of particle. We shall give brief descriptions of these attitudes.

1. More fundamental than particles as carriers of physical properties are the volume elements of three-dimensional space itself. If the speed of light limits every means of communication, disjoint volumes at the same time are physically independent and should contribute additively to the total energy and momentum. Using an evident limiting procedure, we write

$$P^0 = \int (d\mathbf{x}) T^{00}(x), \qquad P^k = \int (d\mathbf{x}) T^{0k}(x), \tag{1-4.1}$$

where $T^{00}(x)$, $T^{0k}(x)$ are functions of the dynamical variables at time x^0 that convey the physical situation in the infinitesimal neighborhood of the point $\mathbf{x}$. The dynamical variables, as operator functions of space and time coordinates, are operator fields, and the approach we are describing can be called operator field theory. As the above notation suggests, covariance can be made explicit by identifying the volume element $(d\mathbf{x})$ with the time component of a directed element of area on a plane space-like surface in four-dimensional space. This gives

$$P^\nu = \int_\sigma d\sigma_\mu T^{\mu\nu}(x), \tag{1-4.2}$$

which integrals are independent of the surface σ according to the conservation of P^ν. On writing the null difference of two such integrals as an equivalent volume integral,

$$0 = \left(\int_{\sigma_1} - \int_{\sigma_2} \right) d\sigma_\mu T^{\mu\nu} = \int (dx) \, \partial_\mu T^{\mu\nu}, \tag{1-4.3}$$

we recognize the sufficiency of the local condition

$$\partial_\mu T^{\mu\nu}(x) = 0. \tag{1-4.4}$$

The conservation of the six other Lorentz generators, regarded as moments of momenta,

$$J^{\mu\nu} = \int_\sigma d\sigma_\lambda (x^\mu T^{\lambda\nu} - x^\nu T^{\lambda\mu}), \tag{1-4.5}$$

is assured if

$$T^{\mu\nu}(x) = T^{\nu\mu}(x). \tag{1-4.6}$$

The three-dimensional form of these operators is

$$J_k = \int (d\mathbf{x}) \epsilon_{klm}(x_l T^0{}_m - x_m T^0{}_l),$$
$$N_k = P_k x^0 - \int (d\mathbf{x}) x_k T^{00}. \tag{1-4.7}$$

The tensor transformation response of the stress tensor $T^{\mu\nu}(x)$ to an infinitesimal Lorentz transformation is given by

$$\begin{aligned}\overline{T}^{\mu\nu}(\overline{x}) &= T^{\mu\nu}(x) + \delta\omega^{\mu\kappa} T_\kappa{}^\nu(x) + \delta\omega^{\nu\lambda} T^\mu{}_\lambda(x)\\ &= \overline{T}^{\mu\nu}(x) - \delta x^\lambda \, \partial_\lambda T^{\mu\nu}(x).\end{aligned} \tag{1-4.8}$$

The possibility of producing the new operators

$$\overline{T}^{\mu\nu}(x) = T^{\mu\nu}(x) - \delta T^{\mu\nu}(x) \tag{1-4.9}$$

by the associated unitary transformation implies the commutation relations

$$\begin{aligned}[T^{\mu\nu}(x), P_\lambda] &= (1/i)\, \partial_\lambda T^{\mu\nu}(x),\\ [T^{\mu\nu}(x), J_{\kappa\lambda}] &= (1/i)(x_\kappa \partial_\lambda - x_\lambda \partial_\kappa) T^{\mu\nu}(x)\\ &+ (1/i)(\delta_\kappa^\mu T_\lambda{}^\nu - \delta_\lambda^\mu T_\kappa{}^\nu + \delta_\kappa^\nu T^\mu{}_\lambda - \delta_\lambda^\nu T^\mu{}_\kappa)(x).\end{aligned} \tag{1-4.10}$$

Integrations over a space-like surface, employing the stated properties of $T^{\mu\nu}$, will reproduce all commutators for the 10 Lorentz generators if one uses the following integration theorem for a system that is closed in space-like directions:

$$\int_\sigma (d\sigma_\mu \partial_\nu - d\sigma_\nu \partial_\mu) f(x) = 0. \tag{1-4.11}$$

The commutators of quantum mechanics express the mutual interference of measurements on the two properties involved. The physical independence of volumes in space-like relation thus requires that

$$(x - x')^2 > 0: \qquad [T^{\mu\nu}(x), T^{\kappa\lambda}(x')] = 0. \tag{1-4.12}$$

When the coordinate system is so chosen that $x^0 = x^{0\prime}$, an everywhere-valid expression for such commutators must involve $\delta(\mathbf{x} - \mathbf{x}')$ or a finite number of derivatives of this function. For the energy and momentum densities, which are the $T^{\mu\nu}$ components used to construct the Lorentz generators, the implied form of the equal time commutation relations is

$$\frac{1}{i}[T^{00}(x),\, T^{00}(x')] = -(T^{0k}(x) + T^{0k}(x'))\partial_k\, \delta(\mathbf{x} - \mathbf{x}')$$
$$+\, \partial_m\partial_n\partial_p'\partial_q'(f^{mn,pq}(x)\, \delta(\mathbf{x} - \mathbf{x}')),$$

$$\frac{1}{i}[T^{0k}(x),\, T^{00}(x')] = -T^{00}(x)\partial^k\, \delta(\mathbf{x} - \mathbf{x}') - T^{kl}(x')\partial_l\, \delta(\mathbf{x} - \mathbf{x}')$$
$$+\, \partial_m\partial_p'\partial_q'(g^{km,pq}(x)\, \delta(\mathbf{x} - \mathbf{x}')),$$

$$\frac{1}{i}[T^{0k}(x),\, T^{0l}(x')] = -T^{0l}(x)\partial^k\, \delta(\mathbf{x} - \mathbf{x}') - T^{0k}(x')\partial^l\, \delta(\mathbf{x} - \mathbf{x}')$$
$$+\, \partial_m\partial_p'(h^{km,lp}(x)\, \delta(\mathbf{x} - \mathbf{x}')).$$

(1–4.13)

The terms involving two or more derivatives are such that they do not contribute when integrations are performed to construct one of the Lorentz generators. We have indicated only the minimum number of derivatives required; more general possibilities are introduced by appropriate generalization of f, g, and h. These three functions are symmetrical within each pair of indices, as illustrated by

$$f^{mn,pq} = f^{nm,pq} = f^{mn,qp},$$

(1–4.14)

while f and h are antisymmetrical under an exchange of the pairs, as in

$$f^{mn,pq} = -f^{pq,mn}.$$

(1–4.15)

Another relation is

$$-\partial_0 f^{mn,pq}(x) = g^{mn,pq}(x) - g^{pq,mn}(x).$$

(1–4.16)

There is a simple example of a system for which none of the additional derivative terms appear. We begin with the energy and momentum density expressions that are identified with the classical electromagnetic field:

$$T^{00} = \tfrac{1}{2}(\mathbf{E}^2 + \mathbf{H}^2), \qquad T^0_{\,k} = (\mathbf{E} \times \mathbf{H})_k.$$

(1–4.17)

The attempt to reproduce the energy density commutator

$$\frac{1}{i}[T^{00}(x),\, T^{00}(x')] = -(T^{0k}(x) + T^{0k}(x'))\partial_k\, \delta(\mathbf{x} - \mathbf{x}')$$

(1–4.18)

succeeds with the commutation relations

$$[E_k(x),\, E_l(x')] = [H_k(x),\, H_l(x')] = 0,$$
$$\frac{1}{i}[E_k(x),\, H_l(x')] = -\epsilon_{klm}\partial_m\, \delta(\mathbf{x} - \mathbf{x}'),$$

(1–4.19)

if the momentum density operator is interpreted as a symmetrized product. In arriving at the desired form we have used the formal delta function property

$$(E_k(x) - E_k(x'))(H_l(x) - H_l(x'))\partial_m\, \delta(\mathbf{x} - \mathbf{x}') = 0.$$

(1–4.20)

The commutators among the momentum density components also contain no higher derivative terms, but to reproduce the required structure it is necessary to impose the following conditions,

$$\nabla \cdot \mathbf{E}(x) = 0, \qquad \nabla \cdot \mathbf{H}(x) = 0, \tag{1-4.21}$$

which are compatible with the commutation relations. The commutators between energy and momentum densities then follow the anticipated pattern and supply familiar expressions for the stress components T_{kl}; in particular,

$$T_{kk} = T^{00}. \tag{1-4.22}$$

Although we have begun with a suggestion from classical physics, this discussion is a self-contained verification of a Lorentz invariant quantal system. Other properties are now derived from the structure of the Lorentz generators. From P^0 we deduce the equations of motion of the field operators, which are the homogeneous Maxwell equations. The Lorentz transformation behavior of the field strengths is that of the antisymmetrical tensor $F_{\mu\nu}$. As an example, consider

$$[E_k(x), J^{0l}] = (1/i)(x^0\partial^l + x^l\partial_0)E_k(x) - \left[E_k(x), \int (d\mathbf{x}')(x^{l'} - x^l)T^{00}(x')\right].$$

$$\tag{1-4.23}$$

Then, since

$$(1/i)[E_k(x), T^{00}(x')] = \epsilon_{klm}\partial_l \, \delta(\mathbf{x} - \mathbf{x}')H_m(x'), \tag{1-4.24}$$

we get the infinitesimal response

$$\delta\mathbf{E}(x) = -\delta\mathbf{v} \cdot (x^0\nabla + \mathbf{x}\partial_0)\mathbf{E}(x) - \delta\mathbf{v} \times \mathbf{H}(x). \tag{1-4.25}$$

We add brief comments about more realistic systems, in which the electromagnetic field interacts with other dynamical variables. If we are to maintain the geometrical transformation properties of $F_{\mu\nu}$, the added terms in T^{00} must not alter the computation just performed. That excludes from $[E_k(x), T^{00}(x')]$ any additional single derivative of a delta function, giving the general form,

$$\frac{1}{i} [E_k(x), T^{00}(x')] = \epsilon_{klm}\partial_l \, \delta(\mathbf{x} - \mathbf{x}')H_m(x') - j_k(x') \, \delta(\mathbf{x} - \mathbf{x}')$$
$$+ \partial_l'\partial_m'\big(\varphi_k{}^{lm}(x) \, \delta(\mathbf{x} - \mathbf{x}')\big). \tag{1-4.26}$$

It is no longer necessarily true that $\mathbf{E}$ is divergenceless, and, on writing

$$\nabla \cdot \mathbf{E} = j^0, \tag{1-4.27}$$

we have

$$\frac{1}{i} [j^0(x), T^{00}(x')] = -j^k(x')\partial_k \, \delta(\mathbf{x} - \mathbf{x}') + \partial_k\partial_l'\partial_m'\big(\varphi^{k, lm}(x) \, \delta(\mathbf{x} - \mathbf{x}')\big).$$

$$\tag{1-4.28}$$

Among the consequences of these relations that are produced by integrations over $\mathbf{x}'$ are: the inhomogeneous Maxwell equations with $j^\mu = (j^0, \mathbf{j})$ identified as the electric current vector, the local charge conservation law,

$$\partial_\mu j^\mu(x) = 0, \tag{1-4.29}$$

and a Lorentz transformation response, affirming the four-vector status of j^μ.

Examples exist of interacting systems for which the very singular terms of the energy density commutator do not appear, but there are severe restrictions in the choice of dynamical variables. Basically, only scalar, vector, and simple spinor fields are permitted. At issue here is the consistency of the operator field hypothesis, that meaning, albeit idealized, attaches to the physical properties associated with a sharply defined geometrical volume. To examine this question we consider various weighted averages of the energy density at a given time,

$$T_{1,2} = \int (d\mathbf{x}) v_{1,2}(\mathbf{x}) T^{00}(x), \tag{1-4.30}$$

and construct

$$\frac{1}{i}[T_1, T_2] = -\int (d\mathbf{x}) T^{0k}(x)(v_1(\mathbf{x})\,\partial_k v_2(\mathbf{x}) - v_2(\mathbf{x})\,\partial_k v_1(\mathbf{x}))$$

$$+ \int (d\mathbf{x}) f^{mn,pq}(x)\partial_m\,\partial_n v_1(\mathbf{x})\partial_p\,\partial_q v_2(\mathbf{x}). \tag{1-4.31}$$

This is the basis for an uncertainty principle statement about the accuracy with which values of T_1 and T_2 can be assigned in a given state. We first consider an application where $f^{mn,pq}$ does not enter, in consequence of the antisymmetry in the two sets of indices. Let T_1 and T_2 be partitions of the total energy operator, so that

$$v_1(\mathbf{x}) + v_2(\mathbf{x}) = 1. \tag{1-4.32}$$

Since derivatives of v_1 and of v_2 differ only by a minus sign, we find, simply,

$$(1/i)[T_1, T_2] = \int (d\mathbf{x}) T^{0k}(x)\,\partial_k v_1(\mathbf{x}). \tag{1-4.33}$$

Now choose $v_1(\mathbf{x})$ to be a unit step function, defining a semi-infinite region which shares a surface with the complementary volume defined by $v_2(\mathbf{x})$. With $d\mathbf{S}$ an element of area directed from the latter volume, we get

$$(1/i)[T_1, T_2] = \int dS_k T^{0k}(x). \tag{1-4.34}$$

This gives a correct account of the rate at which the energy in each partial volume changes, owing to the energy flux across the common surface. Incidentally, if the domains defined by v_1 and v_2 had been regarded as disjoint but approaching contact in a limit, the value obtained for the right-hand side would have been zero, while, if they had initially overlapped the eventual boundary and then the common volume had approached zero, the limiting value of the right-hand side would have been twice the stated one. Thus, an alternative evaluation uses the average of the two limiting definitions. Another choice of weight functions is

$$v_1(\mathbf{x}) = v(\mathbf{x}), \qquad v_2(\mathbf{x}) = x_k v(\mathbf{x}), \qquad (1\text{-}4.35)$$

which gives

$$(1/i)[T_1, T_2] = -\int (d\mathbf{x}) T^{0k}(x) (v(\mathbf{x}))^2 + 2\int (d\mathbf{x}) f^{mn,kp}(x) \partial_m \partial_n v(\mathbf{x}) \partial_p v(\mathbf{x}). \qquad (1\text{-}4.36)$$

When $v(\mathbf{x})$ is a unit step function that defines precisely a finite volume, the operators T_1, T_2 are the associated energy and its first moments. But no meaning can then be assigned to the products $\partial_m \partial_n v \partial_p v$, which calls seriously into question the consistency of any operator field theory for which $f^{mn,pq}(x) \neq 0$. This gives a privileged position to that limited class of fundamental field variables for which $f^{mn,pq}$ does vanish.

The impact of this result is only slightly weakened by the following property of physical systems that have vanishing $f^{mn,pq}$. The function

$$g^{mn,pq}(x) = g^{pq,mn}(x) \qquad (1\text{-}4.37)$$

cannot be zero, and it is correspondingly impossible to specify simultaneously, within any finite precision, the total energy and a component of total momentum that are associated with a sharply defined volume. [The term physical system occurs here as a reminder that the vacuum state, with all its attendant properties, must be compatible with the assumed characteristics of the system. In particular, the zero energy and momentum invariantly assigned to the vacuum state $| \; \rangle$ require that

$$\langle T^{00} \rangle = 0, \qquad \langle T^{0k} \rangle = 0 \qquad (1\text{-}4.38)$$

and also

$$\langle T^{kl} \rangle = 0. \qquad (1\text{-}4.39)$$

There is some freedom to adjust the definitions of the $T^{\mu\nu}$ by additive constants, but, as inspection of the $[T^{0k}(x), T^{00}(x')]$ structure will confirm, it is limited to a multiple of $g^{\mu\nu}$. A nontrivial requirement is thus given by

$$\langle (T^{00} + \tfrac{1}{3} T_{kk}) \rangle = 0. \qquad (1\text{-}4.40)$$

In the example of the electromagnetic field, with $T_{kk} = T^{00}$, it is impossible to satisfy $\tfrac{4}{3} \langle T^{00} \rangle = 0$ since $T^{00} = \tfrac{1}{2}(\mathbf{E}^2 + \mathbf{H}^2)$ is a positive-definite operator; the uncoupled electromagnetic field is not a physical system. It is at least conceivable that the vacuum properties so circumscribe the possible dynamical variables and their interactions that the real world is selected.] We consider the commutator of the energy density functional

$$T = \int (d\mathbf{x}) v(\mathbf{x}) T^{00}(x) \qquad (1\text{-}4.41)$$

with its time derivative

$$\partial_0 T = -\int (d\mathbf{x}) v(\mathbf{x}) \, \partial_k T^{0k}(x),$$

(1–4.42)

which gives the vacuum expectation value

$$\langle [i \, \partial_0 T, \, T] \rangle = \int (d\mathbf{x}) \partial_k \, \partial_m v(\mathbf{x}) \langle g^{km,pq} \rangle \partial_p \, \partial_q v(\mathbf{x})$$

$$= 2 \langle T P^0 T \rangle.$$

(1–4.43)

We recognize the necessarily positive expectation value of the energy in the nonvacuum state $T| \ \rangle$. The numbers $\langle g^{mn,pq} \rangle$ thus form a positive-definite matrix, but there is no guarantee that these numbers are bounded. It is clear from this discussion, however, that a statement about momentum density is also one concerning the time derivative of energy density, and this additional dynamical specifiability may be unnecessary to the self-consistency of the theory.

The particle, in operator field theory, is a derived dynamical concept. To construct from a few fundamental field variables a relatively large number of stable or quasi-stable excitations—particles—is the ambition of this viewpoint. A classification of particle spectra is produced as follows. Let $\chi(x)$ be an algebraic combination of the fundamental field variables, so devised that it has an elementary response to Lorentz transformations. This includes the requirements that, at $\mathbf{x} = 0$, the rotational behavior corresponds to a definite intrinsic angular momentum or spin, while translational response is parametrized by the coordinates x^μ,

$$[\chi(x), \, P_\mu] = (1/i) \, \partial_\mu \chi(x).$$

(1–4.44)

The finite unitary operator presentation of the latter is

$$\chi(x) = e^{-iPx} \chi e^{iPx},$$

(1–4.45)

where $Px = P^\mu x_\mu$ and χ is evaluated at the coordinate origin. The state $\chi(x)| \ \rangle$ is produced from the vacuum state by a localized excitation. To study the particle aspects of this excitation we examine its space-time propagation characteristics through the correlation with an analogous excitation having a different localization:

$$\langle \chi(x) \chi(x') \rangle = \langle \chi e^{iP(x-x')} \chi \rangle.$$

(1–4.46)

The unitary operator that describes the displacement from x' to x can be exhibited in terms of its eigenvalues and the associated nonnegative Hermitian projection operators,

$$e^{iP(x-x')} = \int \frac{(dp)}{(2\pi)^3} e^{ip(x-x')} F(p),$$

(1–4.47)

where

$$(dp) = dp_0 \, dp_1 \, dp_2 \, dp_3.$$

(1–4.48)

The values of p^μ that contribute to the integral, those for which $F(p) \neq 0$, must conform to the physical spectral requirements,

$$-p^2 = M^2 \geq 0, \qquad p^0 > 0. \tag{1–4.49}$$

With a given three-dimensional momentum, $(dp^0)^2 = dM^2$, so that

$$\frac{(dp)}{(2\pi)^3} = \frac{(d\mathbf{p})}{(2\pi)^3} \frac{1}{2p^0} dM^2 \equiv d\omega_p \, dM^2. \tag{1–4.50}$$

The differential $d\omega_p$ is an invariant momentum space measure on the hypersurface $-p^2 = M^2$. This gives

$$\langle \chi(x)\chi(x') \rangle = \int dM^2 \, d\omega_p e^{ip(x-x')} f(p), \tag{1–4.51}$$

where

$$f(p) = \langle \chi F(p)\chi \rangle \tag{1–4.52}$$

is a real, nonnegative function. The state $\chi| \; \rangle$ selects from $F(p)$ the subspace with the angular momentum properties implied by the rotational behavior of χ, and $f(p) \neq 0$ at $-p^2 = M^2$ asserts the existence of an excitation with those physical parameters. Merely for simplicity, we only consider a scalar field χ, which limits $f(p)$ to dependence on the scalar $-p^2$.

There are three qualitatively different possibilities that can be realized in $f(p) = f(M^2)$.

a. An isolated mass value appears in the spectrum,

$$M \sim m: \qquad f(M^2) = f \, \delta(M^2 - m^2), \qquad f > 0. \tag{1–4.53}$$

For a given spatial momentum, the time dependence of the field correlation function contains the isolated frequency $p^0 = +(\mathbf{p}^2 + m^2)^{1/2}$. This excitation is a stable particle. We note that should $\chi(x)$ obey a finite-degree differential equation,

$$\varphi(-\partial^2)\chi(x) = 0, \tag{1–4.54}$$

we have

$$\varphi(M^2)f(M^2) = 0, \tag{1–4.55}$$

and $f(M^2)$ is composed entirely from delta functions.

b. There is a pronounced increase in $f(M^2)$ above a smooth background, which is centered at $M = m$ and has a mass width measured by $\Gamma \ll m$. For a specified momentum, the time dependence that is associated with this portion of the mass spectrum is given by $(t = x^0 - x^{0'})$

$$\int dM^2 \, \exp[-ip^0(M)t] f(M^2)$$
$$\simeq \exp[-ip^0 t] \int dM^2 \, \exp[-i(t/2p^0)(M^2 - m^2)] f(M^2), \tag{1–4.56}$$

in which p^0 is the energy computed from mass m. Owing to destructive interference, the amplitude of this oscillation will drop substantially below its initial
value when

$$t \sim (p^0/m)(1/\Gamma). \tag{1-4.57}$$

This is an unstable particle decaying into several other particles with a proper
lifetime $\sim 1/\Gamma$.

c. The function $f(M^2)$ varies smoothly. In that part of the spectrum several
particles are present, with no tendency to become associated in a single unstable
particle.

Some aspects of field equal time commutation relations can be extracted
from the correlation function. Thus,

$$x^0 = x^{0'}: \qquad \langle [\chi(x), \chi(x')] \rangle = 0,$$
$$\langle [i\,\partial_0\chi(x), \chi(x')] \rangle = \delta(\mathbf{x} - \mathbf{x}') \int dM^2 f(M^2), \tag{1-4.58}$$

where

$$\delta(\mathbf{x} - \mathbf{x}') = \int d\omega_p 2p^0 \exp[i\mathbf{p} \cdot (\mathbf{x} - \mathbf{x}')]. \tag{1-4.59}$$

If the field $\chi(x)$ is a fundamental dynamical variable, its equal time commutation
relations have a kinematical basis. It is typical of a scalar field that

$$[\chi(x), \chi(x')] = 0, \qquad [i\,\partial_0\chi(x), \chi(x')] = \delta(\mathbf{x} - \mathbf{x}'), \tag{1-4.60}$$

and the latter implies the sum rule

$$1 = \int dM^2 f(M^2). \tag{1-4.61}$$

Imagine now that the field $\chi(x)$ is uncoupled from all others, and then
obeys a linear differential equation that gives $f(M^2) = \delta(M^2 - m_0^2)$. Suppose
that when the physical couplings are restored a stable particle still exists. Its
mass will be shifted by the interaction, $m_0 \to m$, and $f(M^2)$ will have multiparticle contributions in addition to the discrete mass term: $f\,\delta(M^2 - m^2)$.
The sum rule thus requires that $f < 1$. If we are not interested in the details
of the particular excitation used to generate the particle, but wish only to
describe the physical particle itself, we discard the continuous mass contribution
to $\langle \chi(x)\chi(x') \rangle$, and correspondingly adjust the scale of the correlation function
by removing the factor f. This example has supplied the designation for the
general procedure that transfers attention from the fundamental dynamical
field variables to the derived phenomenological particle level. It is called
renormalization.

More elaborate field correlation functions give information about the
interaction of particles. Consider, for example,

$$\langle \chi_a(x)\chi_b(x')\chi_c(x'')\chi_d(x''') \rangle, \tag{1-4.62}$$

where the various fields are the algebraic combinations of the fundamental dynamical variables that contain the particles, a, b, c, d, respectively, in their excitation spectra. In order to refer to the particular reaction $c + d \rightarrow a + b$ (we do not consider here properties like electric charge that introduce the distinction between particles and antiparticles), the regions in which the coordinates are placed must be appropriately chosen. The points x and x' lie far in the future of x'' and x''', while x is widely separated spatially from x' as is x'' from x'''. Under these circumstances the renormalization procedures that isolate the physical particles can be applied independently, and the resulting function of particle properties supplies the amplitude for the physical reaction. Now recall that the dynamical variables of quantum mechanics fall naturally into two sets which, at a given time, exhibit commutativity or anticommutativity, respectively, between a pair of variables referring to different degrees of freedom. It is points in space-like relation that play the latter role in operator field theory. If x and x' are sufficiently separated spatially that detailed composite structure is not involved,

$$\chi_a(x)\chi_b(x') = (-1)^{n_a n_b}\chi_b(x')\chi_a(x), \qquad (1\text{-}4.63)$$

where n_a, n_b are the respective number of anticommuting fundamental field variables used in constructing the operators $\chi_a(x)$, $\chi_b(x')$. Only the even or oddness of the integers is significant, and the sign factor is $+1$ except when both n_a and n_b are odd. Should a and b refer to the same particle type, the field operators associated with a pair of spatially separated points commute or anticommute according as the number of anticommuting fundamental variables is even or odd. The two-particle state produced from $\chi(x)\chi(x')|\ \rangle$ by renormalization is correspondingly symmetrical or antisymmetrical in the particle variables, appropriate to particles that obey Bose-Einstein or Fermi-Dirac statistics, respectively. The same statistics-labeled properties enter in relating reactions in which the roles of some initial and final particles are interchanged. Thus, the field correlation function

$$\langle \chi_a(x)\chi_c(x'')\chi_b(x')\chi_d(x''')\rangle \qquad (1\text{-}4.64)$$

can be used to derive the reaction amplitude for $b + d \rightarrow a + c$. If the two correlation functions we have mentioned are known throughout the multiple space-time domain, they are known for the regions where x' and x'' are in space-like relation. But there they are equal, or differ by a minus sign, depending upon the statistics of particles b and c. Accordingly, the two functions are space-time extrapolations or continuations of each other. The implied connections among different reaction amplitudes are usually referred to as crossing relations.

Dynamics is explicit in operator field theory. It is conveyed by the nonlinear structure of the field equations obeyed by the fundamental dynamical variables. They, in turn, imply equations connecting the various field correlation functions from which the latter can be constructed, in principle. Two radically different situations occur, in practice. In the first, the interactions are sufficiently weak that the particles of interest appear in the excitation spectra of the fundamental variables themselves. This is the assumed situation in quantum electrodynamics where the particles are the photon and the electron (or muon). The equations obeyed by the field correlation functions can be solved by perturbative or iterative methods based on the smallness of the characteristic coupling constant $\alpha = 1/137.04$. The results may be presented in two different ways, at the unrenormalized field stage, or at the renormalized particle level. The field version contains divergent integrals, the renormalized statements are finite and in exceptional agreement with experiment. The fairly rapid convergence of the renormalized expressions means that experiments do not probe to very high momenta or very short distances. The underlying hypothesis of operator field theory concerning the conceptual possibility of descriptions at arbitrarily small distances remains untested by the available evidence. This hypothesis is involved in the unrenormalized, field results, but whether the nonexistence of the expressions signifies the failure of the hypothesis, or merely the inadequacy of the perturbative calculational methods that are used, is presently unknown. It may be that operator field theory is unnecessarily dogmatic about the physical significance of arbitrarily small volume elements. Totally new concepts might enter at very large momenta, without altering the practical successes that have been obtained.

The other situation is that of strong interaction physics. Here the hypothesis that whole families of particles are the dynamical manifestation of a few fundamental field variables excludes the possibility that the excitation spectrum of the latter contains the known particles. These objects must be generated by combinations of the basic variables. Being ignorant of the underlying dynamics of the fundamental dynamical variables and lacking the computational methods that could give the consequences of that dynamics, if it were known, one must fall back on speculations concerning the composite field structure of the known particles. And such speculations become intertwined in the more immediate problems that are presented at the phenomenological level. Is it not possible to separate particle phenomenology from speculations about particle structure?

2. If the particle is the ultimate structure, no detailed description is possible in regions of intense interaction where the characteristic additivity of independent particle contributions ceases to be valid. All that can be done is to compare the state of noninteracting particles after a collision with the state of the generally different number of noninteracting particles prior to the collision. That relation is an object of calculation in operator field theory. With the present viewpoint it is the fundamental quantity which, through its postulated properties, gives the complete statement of the microscopic dynamics of particles. When we write

$$\langle f(\text{inal})| = \langle i(\text{nitial})|S, \qquad (1\text{--}4.65)$$

we introduce a transformation from one complete set of noninteracting particle states to an analogous set. Hence the operator S is unitary,

$$S^\dagger S = S S^\dagger = 1. \tag{1–4.66}$$

It is invariably referred to as the S(cattering) matrix, perhaps because one is unable to handle more than a small set of matrix elements. These matrix elements are explicit in the probability amplitude

$$\langle f | i' \rangle = \langle i | S | i' \rangle. \tag{1–4.67}$$

The absolute square $|\langle i | S | i' \rangle|^2$ is interpreted as the probability that the transition $i' \to i$ will occur during an unlimited time interval. The abhorrence of any vestige of a detailed temporal description restricts the observed particles to stable ones. In the same vein, all reference to spatial description is rejected and only the momentum specification of states, together with spin and other properties, is admitted.

If the particles happen not to interact, the appropriate S is the unit operator. Thus, more interesting than S is $S - 1$, which obeys the unitary restriction in the form

$$(S - 1) + (S - 1)^\dagger + (S - 1)^\dagger (S - 1) = 0. \tag{1–4.68}$$

Since a collision must respect the overall conservation of energy and momentum, we write

$$\langle p_1, \ldots, p_n | S - 1 | p_1', \ldots, p_{n'}' \rangle = (2\pi)^4 \, \delta \left(\sum_1^n p_a - \sum_1^{n'} p_a' \right)$$
$$\times i \langle p_1, \ldots, p_n | T | p_1', \ldots, p_{n'}' \rangle \tag{1–4.69}$$

thereby defining the transition matrix T. Only the momenta have been made explicit here, and the delta function is four-dimensional,

$$\delta(p - p') = \delta(p_0 - p_0') \, \delta(p_1 - p_1') \, \delta(p_2 - p_2') \, \delta(p_3 - p_3'). \tag{1–4.70}$$

The resulting form of the unitarity condition is nonlinear, relating matrix elements of $i(T - T^\dagger)$ to products of $T^\dagger$ and T matrix elements. The probabilities of transitions must have a Lorentz invariant significance. It is therefore asserted that the S, and the T, matrix elements must be invariant functions of their arguments. This is particularly simple when all particles are spinless since it requires that the T matrix elements be a function only of the independent scalars (we ignore the possibility of pseudoscalars) that can be formed from the $N = n + n'$ momenta, which are individually subject to the particle mass relations, $-p_a^2 = m_a^2$. The number of such scalar combinations is $3N - 10$, where the subtracted number counts the Lorentz transformation parameters. Thus, in two-particle reactions, where $N = 2 + 2$, there are only two independent scalar variables, corresponding to energy and scattering angle in the center of mass reference frame.

The constructive principle of S-matrix theory is the postulate of analyticity. It is assumed that the physical reaction amplitudes, in their dependence on the scalar variables, are boundary values of analytic functions of corresponding complex variables. Since analytic functions are specified by the nature and location of their singularities, the determination of the latter encompasses all the physics that is admitted in S-matrix theory. Here are the words of some enthusiasts: "One of the most remarkable discoveries in elementary particle physics has been that of the existence of the complex plane," ". . . the theory of functions of complex variables plays the role not of a mathematical tool, but of a fundamental description of nature inseparable from physics . . . ." From the viewpoint of analytic functions, the elements of T and $T^{\dagger}$ are distinguished as boundary values of the same analytic function that refer to opposite sides of the real axis for the relevant complex variables. The resulting discontinuity statement is the contribution of the unitarity condition toward determining the singularities of the transition matrix. But the postulate of analyticity also widens the scope of the unitarity condition to include the so-called crossed reactions. The conversion of an initial or incoming particle into a final or outgoing particle is formally expressed by the substitution $p^{\mu} \rightarrow -p^{\mu}$, as judged by the contribution to the net energy-momentum balance. This is to be achieved by analytic continuation, and the unitarity conditions for the various reactions that are in crossing relation give singularity information about one analytic function in various domains of its complex variables. As to whether this kind of information suffices: "The S-matrix is a Lorentz-invariant analytic function of all momentum variables with only those singularities required by unitarity."

There is no explicit statement of dynamics in S-matrix theory. And the possibility of regarding some particles as fundamental and deriving others as bound states is rejected as an unacceptable structuring of the particle concept, distinguishing elementary and composite particles. To prevent just such a distinction being made, it is proposed that, no matter which particles one uses to construct composites, the same total particle spectrum emerges. This view of dynamical self-consistency is usually referred to as the "bootstrap" hypothesis.

The discussion of Section 1–3 indicates that S-matrix theory is too dogmatic in dismissing all reference to microscopic space-time description. Whether or not one wishes to recognize it, the structure of the Lorentz group itself gives meaning to spatial localizability and temporal development, outside regions of intense interaction. The very nature of a collision involves a measure of space-time causal control, and the existence of even a limited microscopic space-time description indicates that causality is not likely to be restricted to macroscopic circumstances. It is widely recognized that the intuitive physical property of causality in space-time must underlie the abstract mathematical assertion of analyticity. Should one not be able to exhibit and exploit causality as a constructive principle, thereby relegating analyticity to a secondary, derived role? And as for the basic hypothesis of S-matrix theory, that the particle is the ultimate unanalyzable entity, we again ask: Is it not possible to separate particle phenomenology from speculations about particle structure?

2

SOURCES

The critical comments of the last section set the stage for the introduction of a new theory of particles. It is a phenomenological theory, designed to describe the observed particles, be they stable or unstable. No speculations about the inner structure of particles are introduced, but the road to a conceivable more fundamental theory is left open. No abstract definition of particle is devised; rather, the theory uses symbolic idealizations of the realistic procedures that give physical meaning to the particle concept. The theory is thereby firmly grounded in space-time, the arena within which the experimenter manipulates his tools, but the question of an ultimate limitation to microscopic space-time description is left open, with the decision reserved to experiment. Correspondingly, no operator fields are used. The complementary momentum-space description plays an important role, but the possibility of ultimate limitations on this space is not excluded, and there is no appeal to analyticity in momentum space. The constructive principles of the new theory are intuitive ones— causality and uniformity in space-time. What emerges is a theory intermediate in position between operator field theory and S-matrix theory, which rejects the dogmas of each, and gains thereby a calculational ease and intuitiveness that make it a worthy contender to displace the earlier formulations.

The range of the term "particle" has been systematically extended by experimental discovery. From the stable electron and proton, to the long-lived neutron, to the rapidly decaying π and Δ, to the highly unstable ρ and N^*, there has been a progression to more energetic and shorter-lived excitations. It is now the normal situation that a particle must be created in order to study it. And, in a general sense, that is also true of the very high energy stable particles produced in accelerators. One can regard all such creation acts as collisions. The essence of such a collision is that it occupies a finite, and to some degree controllable, space-time region wherein other particles combine to transmit to a particular one those properties that call it into existence and uniquely characterize it. It is part of the experimenter's creed that a new resonance not be admitted to full status as a particle until it has been observed with the same characteristics in a number of different reactions. Thus, if a particle is defined by the collisions that create it, the details of a specific reaction are not relevant and one can idealize the role of the other particles in the collision, recognizing that their function is solely to supply the needed balance of properties—they constitute the *source* for the particle of interest. What survives in the idealization is a general specification of the region in space-time where the source is effective, with some numerical measure given by a function $S(x)$, and a statement of its ability to produce various momenta, as measured by a function $S(p)$. The two source functions cannot be independent but must convey the quantum mechanical complementarity between these descriptions—the more detail that one possesses, the less is permitted to the other.

We have spoken of particle creation, but equally important is particle detection. This is invariably achieved by transmuting the particle's properties into other more easily handled forms. In a general sense, the particle is annihilated by the process of detecting it. Here too are collisions with their controllable space-time aspects which, in principle, involve the same mechanisms that create the particle. The receiving radio set unavoidably radiates, the π-meson created in nucleon collisions is captured in nuclei. Long-lived particles may decay, and thus be detected, by mechanisms too weak to be useful in creating them, but this option can be overridden at the choice of the experimenter—the neutron is not generally observed through its β-decay. The collision processes used to detect a particle can be idealized as sinks wherein the particle's properties are handed on, in a way that permits some measure of space-time and momentum description, but sink and source are clearly different aspects of the same idealization and we unite them under the general heading of "source." We now proceed to give the source concept a quantitative framework, beginning with the simple situation of stable, spinless particles.

2-1 SPIN 0 PARTICLES. WEAK SOURCE

The elementary acts to be represented as the effect of a source are the creation of a single particle where none existed previously, and the annihilation of that single particle. Since the actual presence of other particles in realistic collisions is abstractly portrayed by the source, the states appearing in corresponding quantum mechanical amplitudes, $\langle 1_p|0_-\rangle^K$ and $\langle 0_+|1_p\rangle^K$, are: $|0_-\rangle$, the vacuum state before the operation of the source K; $\langle 0_+|$, the vacuum state subsequent to the operation of the source (sink) K; and $\langle 1_p|$, $|1_p\rangle$, describing a single particle state in which the momentum is specified within a small volume element $(d\mathbf{p})$. The connection of this discrete labeling with the continuous variable specification of momentum states is

$$\langle 1_p| = (d\mathbf{p})^{1/2}\langle p|, \qquad |1_p\rangle = |p\rangle(d\mathbf{p})^{1/2}. \qquad (2\text{-}1.1)$$

The individual creation and annihilation acts are not analyzed; the source is defined as a measure of the whole process, as suggested by (we anticipate a particular variable factor)

$$\langle p|0_-\rangle^K \sim (p^0)^{-1/2}K_e(p),$$
$$\langle 0_+|p\rangle^K \sim (p^0)^{-1/2}K_a(p). \qquad (2\text{-}1.2)$$

Subscripts appear here, temporarily, to distinguish sources effective in emission or absorption. The designation "weak source" means that the definitions are appropriate when probability amplitudes referring to creation or annihilation of several particles are relatively negligible. We proceed to make these definitions more precise.

The states $\langle p|$, $|p\rangle$ refer to a particular time, or, more covariantly, a space-like surface. If the origin of the space-time coordinate frame is displaced by X^μ, corresponding states are produced by the unitary transformation:

$$\overline{\langle p|} = \langle p|e^{iP^\mu X_\mu} = e^{ipX}\langle p|, \qquad \overline{|p\rangle} = e^{-iP^\mu X_\mu}|p\rangle = |p\rangle e^{-ipX}. \quad (2\text{-}1.3)$$

Since these states play the analogous role in the new coordinate system, they are associated with a space-like surface that is displaced by X^μ in the initial coordinate system. But all that is significant in the probability amplitudes $\langle p|0_-\rangle^K$ and $\langle 0_+|p\rangle^K$ is the relation between the space-like surface and the space-time localization of the sources, for the vacuum states are invariant. Equivalent to the displacement of the surface by X^μ is the rigid displacement of the source by $-X^\mu$. This is expressed by $K \to \overline{K}$, where

$$\overline{K}(x) = K(x + X), \qquad (2\text{-}1.4)$$

or

$$\overline{K}(\overline{x}) = K(x), \qquad \overline{x}^\mu = x^\mu - X^\mu. \qquad (2\text{-}1.5)$$

Now,

$$\overline{K}_e(p) = e^{ipX}K_e(p), \qquad \overline{K}_a(p) = e^{-ipX}K_a(p), \qquad (2\text{-}1.6)$$

which shows clearly that the relation between the complementary coordinate and momentum source descriptions is given by Fourier transformation:

$$\begin{aligned} K_e(p) &= \int (dx)e^{-ipx}K_e(x), \\ K_a(p) &= \int (dx)e^{ipx}K_a(x). \end{aligned} \qquad (2\text{-}1.7)$$

The space-time coordinates in these exponential functions are referred to an origin located in the space-like surface, but we shall not usually make this explicit.

We consider next the behavior under homogeneous Lorentz transformations. The response of the single particle states to infinitesimal transformations of this nature is given by

$$\begin{aligned} \delta\langle p| &= i\langle p|(\delta\boldsymbol{\omega} \cdot \mathbf{J} + \delta\mathbf{v} \cdot \mathbf{N}), \\ \delta|p\rangle &= -i(\delta\boldsymbol{\omega} \cdot \mathbf{J} + \delta\mathbf{v} \cdot \mathbf{N})|p\rangle, \end{aligned} \qquad (2\text{-}1.8)$$

where [Eq. (1-3.29)]

$$\mathbf{J} = \mathbf{r} \times \mathbf{p}, \qquad \mathbf{N} = -(p^0)^{1/2}\mathbf{r}(p^0)^{1/2}. \qquad (2\text{-}1.9)$$

In exhibiting $\mathbf{N}$ we have set $x^0 = 0$, since this is the origin of time, and used an alternative form of the symmetrized product of $\mathbf{r}$ with p^0. The coordinate operator in the momentum description is represented by

$$\langle p|\mathbf{r} = i(\partial/\partial\mathbf{p})\langle p|, \qquad \mathbf{r}|p\rangle = -i(\partial/\partial\mathbf{p})|p\rangle, \tag{2-1.10}$$

and therefore

$$\delta((p^0)^{1/2}\langle p|) = -\left[-\delta\omega\cdot\frac{\partial}{\partial\mathbf{p}}\times\mathbf{p} + \delta\mathbf{v}\cdot p^0\frac{\partial}{\partial\mathbf{p}}\right](p^0)^{1/2}\langle p|, \tag{2-1.11}$$

with an analogous formula for $(p^0)^{1/2}|p\rangle$. Having anticipated the square root factor, we transcribe this as

$$\delta K_e(p) = \left[-\delta\omega\cdot\frac{\partial}{\partial\mathbf{p}}\times\mathbf{p} + \delta\mathbf{v}\cdot p^0\frac{\partial}{\partial\mathbf{p}}\right]K_e(p), \tag{2-1.12}$$

with a similar statement for $K_a(p)$. The implied infinitesimal change of $K_e(x)$ or $K_a(x)$ is

$$\delta K(x) = [\delta\omega\cdot\mathbf{r}\times\boldsymbol{\nabla} + \delta\mathbf{v}\cdot(\mathbf{r}\partial_0 + x^0\boldsymbol{\nabla})]K(x)$$
$$= \delta x^\nu\,\partial_\nu K(x), \tag{2-1.13}$$

where

$$\delta x^\nu = \delta\omega^{\mu\nu}x_\mu. \tag{2-1.14}$$

This result,

$$\overline{K}(x) = K(x + \delta x), \tag{2-1.15}$$

or

$$\overline{K}(\bar{x}) = K(x), \qquad \bar{x}^\mu = x^\mu - \delta x^\mu, \tag{2-1.16}$$

when combined with the displacement response, asserts that the source functions of spinless particles, $K(x)$, behave as scalar functions under the transformations of the Lorentz group.

An important corollary is that the choice of $K(x)$ as a real function would have a Lorentz-invariant meaning. That is in sharp contrast with the non-relativistic situation, where $\mathbf{N} \to -m\mathbf{r}$ and $p^0 \to \mathbf{p}^2/2m$. Then, if we consider only boosts for simplicity,

$$\delta K_e(p) = m\,\delta\mathbf{v}\cdot\partial/\partial\mathbf{p}K_e(p) \tag{2-1.17}$$

and

$$\delta K_e(\mathbf{r}, t) = \delta\mathbf{v}\cdot(-im\mathbf{r} + t\boldsymbol{\nabla})K_e(\mathbf{r}, t). \tag{2-1.18}$$

The implied form for finite transformations is

$$\overline{K}_e(\mathbf{r}, t) = \exp[\mathbf{v}\cdot(-im\mathbf{r} + t\boldsymbol{\nabla})]K_e(\mathbf{r}, t)$$
$$= \exp(-im\mathbf{v}\cdot\mathbf{r} - i\tfrac{1}{2}m\mathbf{v}^2 t)K(\mathbf{r} + \mathbf{v}t, t) \tag{2-1.19}$$

or

$$\overline{K}_e(\bar{\mathbf{r}}, t) = \exp[-i(m\mathbf{v}\cdot\mathbf{r} - \tfrac{1}{2}m\mathbf{v}^2 t)]K_e(\mathbf{r}, t),$$
$$\bar{\mathbf{r}} = \mathbf{r} - \mathbf{v}t. \tag{2-1.20}$$

Evidently, a real emission or absorption source would have no Galilean invariant meaning. Incidentally, in carrying out the evaluation of (2-1.19) we have used a simple example of the formula ($p = -i\partial/\partial q$):

$$e^{i\lambda(p+f'(q))} = e^{i[f(q+\lambda)-f(q)]}e^{i\lambda p} = e^{i\lambda p}e^{i[f(q)-f(q-\lambda)]}. \qquad (2\text{-}1.21)$$

The precise relationship between the emission and absorption abilities of a source is disclosed by combining the orthogonality of the vacuum and single particle states, prior to the intervention of sources, with the completeness of the various particle-specified states that refer to the final situation. Thus,

$$0 = \langle 0_-|1_p\rangle = \langle 0_-|0_+\rangle^K \langle 0_+|1_p\rangle^K + \sum_{p'} \langle 0_-|1_{p'}\rangle^K \langle 1_{p'}|1_p\rangle^K + \cdots, \qquad (2\text{-}1.22)$$

in which the additional terms are negligible under weak source conditions. Furthermore, it suffices to use source-free values for the factors of $\langle 0_+|1_p\rangle^K$ and $\langle 0_-|1_{p'}\rangle^K = \langle 1_{p'}|0_-\rangle^{K*}$, namely

$$K = 0: \qquad\qquad \langle 0_-|0_+\rangle = \langle 0_+|0_-\rangle^* = 1, \qquad\qquad (2\text{-}1.23)$$

expressing the invariance of the vacuum state, and

$$K = 0: \qquad\qquad\qquad \langle 1_{p'}|1_p\rangle = \delta_{pp'}, \qquad\qquad\qquad (2\text{-}1.24)$$

apart from phase factors which serve only to ensure that, in the resulting relation,

$$\langle 0_+|1_p\rangle^K = -\langle 1_p|0_-\rangle^{K*}, \qquad\qquad (2\text{-}1.25)$$

both single particle states are referred to the same space-like surface. This connection between creation and annihilation probability amplitudes can also be presented as

$$i\langle 0_+|1_p\rangle^K = [i\langle 1_p|0_-\rangle^K]^*. \qquad\qquad (2\text{-}1.26)$$

Thus, with a permissible choice of arbitrary phases, the source functions $K_e(x)$ and $K_a(x)$ are reciprocal complex conjugates. The simplest possibility, and the one with which we begin, is

$$K_e(x) = K_a(x) = K(x), \qquad\qquad (2\text{-}1.27)$$

a real function. We now unite the various details and state our explicit definitions:

$$\langle 1_p|0_-\rangle^K = iK_p, \qquad \langle 0_+|1_p\rangle^K = iK_p^*, \qquad (2\text{-}1.28)$$

where

$$K_p = (d\omega_p)^{1/2}K(p), \qquad d\omega_p = \frac{(d\mathbf{p})}{(2\pi)^3}\frac{1}{2p^0}, \qquad (2\text{-}1.29)$$

and

$$K(p) = \int (dx)e^{-ipx}K(x), \qquad K(p)^* = K(-p). \qquad (2\text{-}1.30)$$

The experimenter's basic tool is a beam of particles. A very weak beam of spinless particles has the following causal representation. We begin with the vacuum state. Then a weak source $K_2(x)$, occupying a finite space-time region, goes into action. It most often does nothing, with the associated probability amplitude $\langle 0_+|0_-\rangle^{K_2} \sim 1$, and occasionally produces a single particle, as characterized by $\langle 1_p|0_-\rangle^{K_2}$. After the emission source has ceased to operate, the resulting vacuum or single particle state persists unchanged until we enter the space-time region of an absorption source $K_1(x)$. Its effect in detecting a single particle is described by $\langle 0_+|1_p\rangle^{K_1}$, and we thus return to the vacuum state. The complete process is represented by

$$\langle 0_+|0_-\rangle^{K_1+K_2} = \langle 0_+|0_-\rangle^{K_1}\langle 0_+|0_-\rangle^{K_2} + \sum_p \langle 0_+|1_p\rangle^{K_1}\langle 1_p|0_-\rangle^{K_2} + \cdots,$$
$$(2\text{-}1.31)$$

where vacuum state subscripts designating causal sequence are always relative to the indicated sources. The individual vacuum amplitudes have the form

$$\langle 0_+|0_-\rangle^K = 1 + f(K), \qquad f(0) = 0. \qquad (2\text{-}1.32)$$

On making explicit the single particle creation and annihilation probability amplitudes, we get

$$\langle 0_+|0_-\rangle^{K_1+K_2} \cong 1 + f(K_1) + f(K_2)$$
$$+ i\int (dx)(dx')K_1(x)\left[i\int d\omega_p e^{ip(x-x')}\right]K_2(x'), \quad (2\text{-}1.33)$$

where $\cong$ refers to the weak source limitation. The functions $K_1(x)$ and $K_2(x)$ are the disjoint parts of the total source in this situation, which is given by

$$K(x) = K_1(x) + K_2(x). \qquad (2\text{-}1.34)$$

There should be nothing in the overall description to distinguish one component part of the source from another, aside from reference to the space-time region that it occupies. This is space-time uniformity. It implies that $\langle 0_+|0_-\rangle^K$ depends only upon K, and in the manner made explicit by the bilinear structure in K_1 and K_2. Accordingly, we write

$$\langle 0_+|0_-\rangle^K \cong 1 + \tfrac{1}{2}i\int (dx)(dx')K(x)\Delta_+(x-x')K(x'). \qquad (2\text{-}1.35)$$

The displacement invariant function $\Delta_+(x-x')$, as the kernel of a quadratic form, can be chosen symmetrical with no loss in generality,

$$\Delta_+(x-x') = \Delta_+(x'-x). \qquad (2\text{-}1.36)$$

The two equivalent contributions of the type K_1K_2 then supply the structure of Δ_+ for a causal arrangement:

$$x^0 > x^{0'}: \qquad \Delta_+(x-x') = i\int d\omega_p e^{ip(x-x')}. \qquad (2\text{-}1.37)$$

[We recall that p^μ is the energy-momentum vector of a particle, so that $p^0 = +(\mathbf{p}^2 + m^2)^{1/2}$.] From these characteristics of $\Delta_+(x-x')$, we deduce that

$$x^0 < x^{0'}: \qquad \Delta_+(x-x') = i\int d\omega_p e^{ip(x'-x)}. \qquad (2\text{-}1.38)$$

The explicit constructions of $\Delta_+(x - x')$ may appear to refer only to causal or time-like relations between the points x and x'. But in fact they give meaning to this function everywhere. The only possible difficulty would be that, when x and x' are in space-like relation, where causality has no invariant meaning, different values might be obtained depending upon the choice of coordinate system. This does not happen. Since $d\omega_p$ and $e^{\pm ip(x-x')}$ are invariant structures, there is no harm in choosing a coordinate system for which $x^0 = x^{0'}$, and

$$x^0 = x^{0'}: \qquad \Delta_+(x - x') = i \int d\omega_p \exp[\pm i\mathbf{p} \cdot (\mathbf{x} - \mathbf{x}')] \qquad (2\text{-}1.39)$$

is independent of the ambiguous sign, for the integral depends only upon $(\mathbf{x} - \mathbf{x}')^2 = (x - x')^2$. As a result, there is no longer any indication in (2-1.35) of the initial causal arrangement of sources, and that structure is applicable to an arbitrary disposition of sources. This space-time extrapolation must meet a severe test, however. We are now able to compute the probability that, despite the intervention of the sources, the vacuum state persists. It is

$$|\langle 0_+|0_-\rangle^K|^2 \cong 1 - \int (dx)(dx')K(x) \,\mathrm{Re}\, (1/i)\Delta_+(x - x')K(x'), \quad (2\text{-}1.40)$$

where

$$\mathrm{Re}\, (1/i)\Delta_+(x - x') = \mathrm{Re} \int d\omega_p e^{ip(x-x')} \qquad (2\text{-}1.41)$$

is valid everywhere, and the reference to the real part is unnecessary since it is implicit in the symmetry of the quadratic form. But probability considerations also demand that

$$|\langle 0_+|0_-\rangle^K|^2 \cong 1 - \sum_p |\langle 1_p|0_-\rangle^K|^2. \qquad (2\text{-}1.42)$$

The challenge is successfully met, for

$$\int (dx)(dx')K(x)\left[\int d\omega_p e^{ip(x-x')}\right]K(x') = \int d\omega_p |K(p)|^2 \cdot$$
$$= \sum_p |\langle 1_p|0_-\rangle^K|^2. \qquad (2\text{-}1.43)$$

There is one conceivable modification of $\Delta_+(x - x')$ that would appear to retain the necessary physical characteristics. It is the addition to $\Delta_+(x - x')$ of a real function, which differs from zero only when $(x - x')^\mu$ is a space-like interval. That would contribute neither to the causal exchange of particles between sources nor the computation of the vacuum persistence probability. The hypothesis of space-time uniformity, forbidding the existence of special relationships between sources, excludes that possibility. In this context, one can give the uniformity hypothesis a more precise, if rather abstract, form by considering the four-dimensional Euclidean space that is attached to the Minkowski space through the complex transformation

$$x_4 = ix^0. \qquad (2\text{-}1.44)$$

There is no analogue in Euclidean space to the Minkowski distinction between time-like and space-like intervals. Accordingly, special space-time structures would be rejected if one insisted that the invariant vacuum amplitude that describes a complete physical process continue to be meaningful and invariant on mapping the Minkowski space onto the Euclidean space. This is the Euclidean postulate.

We recognize that the Euclidean postulate is a natural one by noting that $\Delta_+(x - x')$ has the required properties; there is an associated Euclidean invariant function $\Delta_E(x - x')$ which exists almost everywhere $(x \neq x')$. It is obtained from the integral representation

$$\Delta_+(x - x') = i \int d\omega_p \exp[i\mathbf{p} \cdot (\mathbf{x} - \mathbf{x}') - ip^0|x^0 - x^{0'}|] \qquad (2\text{-}1.45)$$

by the substitution

$$i|x^0 - x^{0'}| \to |x_4 - x'_4|, \qquad (2\text{-}1.46)$$

which requires that the ordering of the real numbers x^0, $x^{0'}$ is mapped into the same ordering of the real numbers x_4, x'_4. We remove a factor of i in defining $\Delta_E(x - x')$,

$$(1/i)\Delta_+(x - x') \to \Delta_E(x - x'), \qquad (2\text{-}1.47)$$

and

$$\Delta_E(x - x') = \int \frac{(d\mathbf{p})}{(2\pi)^3} \frac{1}{2p^0} \exp[i\mathbf{p} \cdot (\mathbf{x} - \mathbf{x}') - p^0|x_4 - x'_4|]. \qquad (2\text{-}1.48)$$

An explicitly Euclidean invariant form appears on using the integral relation

$$\frac{1}{2p^0} e^{-p^0|x_4 - x'_4|} = \int_{-\infty}^{\infty} \frac{dp_4}{2\pi} \frac{e^{ip_4(x_4 - x'_4)}}{(p_4)^2 + (p^0)^2}, \qquad (2\text{-}1.49)$$

namely

$$\Delta_E(x - x') = \int \frac{(dp)_E}{(2\pi)^4} \frac{e^{ip_\mu(x - x')_\mu}}{(p_\mu)^2 + m^2}, \qquad (dp)_E = dp_1 \cdots dp_4, \qquad (2\text{-}1.50)$$

in which the notation, ignoring any distinction between contravariant and covariant components, emphasizes the Euclidean structure. With the recognition that $\Delta_E(x - x')$ is a Euclidean invariant function, dependent only upon

$$[(x - x')_\mu]^2 = R^2 \geq 0, \qquad (2\text{-}1.51)$$

we can return to **(2-1.48)** and choose the Euclidean coordinate system to get the real positive expression

$$\Delta_E(x - x') = \frac{1}{4\pi^2} \int_m^{\infty} dp^0 [(p^0)^2 - m^2]^{1/2} e^{-p^0 R}, \qquad (2\text{-}1.52)$$

which is one among a variety of single-parameter integral representations. This one immediately supplies the two limiting forms

$$mR \ll 1: \qquad \Delta_E(x - x') \simeq \frac{1}{4\pi^2 R^2},$$

$$\qquad\qquad\qquad\qquad\qquad\qquad\qquad\qquad (2\text{-}1.53)$$

$$mR \gg 1: \qquad \Delta_E(x - x') \simeq \frac{(2m)^{1/2}}{(4\pi R)^{3/2}} e^{-mR}.$$

Note also the simple inequality

$$\Delta_E(x - x') < \frac{1}{4\pi^2} \left(-\frac{d}{dR} \right) \frac{e^{-mR}}{R}. \qquad (2\text{-}1.54)$$

Even better, since it reproduces the correct limiting forms, is the inequality

$$\Delta_E(x - x') < \frac{1}{4\pi^2} \frac{e^{-mR}}{R^2} [1 + (\tfrac{1}{2}\pi m R)^{1/2}]. \qquad (2\text{-}1.55)$$

One can connect the Minkowski and Euclidean descriptions by equating the source strengths associated with corresponding volume elements

$$(dx)K(x) \rightarrow (dx)_E K_E(x),$$

$$\qquad\qquad\qquad\qquad\qquad\qquad\qquad (2\text{-}1.56)$$

$$(dx)_E = dx_1 \cdots dx_4$$

while maintaining the reality of the source function. This gives

$$\langle 0_+|0_-\rangle^K \rightarrow 1 - \tfrac{1}{2}\int (dx)(dx')K(x)\,\Delta(x - x')K(x') \Big]_E, \qquad (2\text{-}1.57)$$

and the right-hand side is a real number, which is less than unity.

The physical vacuum amplitude can also be recovered from the Euclidean version by the complex substitutions

$$x_4 \rightarrow ix^0, \qquad p_4 \rightarrow -ip_0, \qquad\qquad (2\text{-}1.58)$$

provided they are understood to mean the limit of complex rotations as the angle approaches $\pi/2$ from smaller values,

$$x_4 \rightarrow \exp\left[i\left(\frac{\pi}{2} - \epsilon\right) \right] x^0, \qquad p_4 \rightarrow \exp\left[-i\left(\frac{\pi}{2} - \epsilon\right) \right] p_0, \qquad \epsilon \rightarrow +0.$$

$$\qquad\qquad\qquad\qquad\qquad\qquad\qquad (2\text{-}1.59)$$

Such caution is necessary since the resulting Minkowski structures have singularities: $(x - x')^2 = 0$, the light cone singularity in coordinate space; $p^2 + m^2 = 0$, the particle mass shell singularity in momentum space. We find that

$$(p_\mu)^2 + m^2 \rightarrow \mathbf{p}^2 + m^2 - p_0^2(1 + 2i\epsilon) = p^\mu p_\mu + m^2 - i\epsilon,$$

$$\qquad\qquad\qquad\qquad\qquad\qquad\qquad (2\text{-}1.60)$$

$$(x_\mu - x_\mu')^2 \rightarrow (\mathbf{x} - \mathbf{x}')^2 - (x^0 - x^{0\prime})^2(1 - 2i\epsilon) = (x - x')^\mu(x - x')_\mu + i\epsilon,$$

in which, despite various scale changes, ϵ retains its meaning as a parameter that approaches zero through positive values. The resulting four-dimensional representation of $\Delta_+(x - x')$ is

$$\Delta_+(x - x') = \int \frac{(dp)}{(2\pi)^4} \frac{e^{ip(x-x')}}{p^2 + m^2 - i\epsilon}\bigg|_{\epsilon \to +0}, \qquad (dp) = dp_0 \cdots dp_3,$$

$$(2\text{-}1.61)$$

where

$$\frac{1}{x - i\epsilon}\bigg|_{\epsilon \to +0} = P\frac{1}{x} + \pi i\, \delta(x) \qquad (2\text{-}1.62)$$

introduces the Cauchy principal value for integrals. The contour integral evaluation

$$\int_{-\infty}^{\infty} \frac{dp_0}{2\pi} \frac{e^{ip_0(x^0 - x^{0'})}}{\mathbf{p}^2 + m^2 - i\epsilon - p_0^2}\bigg|_{\epsilon \to 0} = i\,\frac{e^{-i(\mathbf{p}^2 + m^2)^{1/2}|x^0 - x^{0'}|}}{2(\mathbf{p}^2 + m^2)^{1/2}} \qquad (2\text{-}1.63)$$

reproduces Eq. (2–1.45). A limiting form in coordinate space is $m^2|(x - x')^2| \ll 1$:

$$\Delta_+(x - x') \simeq \frac{i}{4\pi^2} \frac{1}{(x - x')^2 + i\epsilon} = \frac{i}{4\pi^2} P\frac{1}{(x - x')^2} + \frac{1}{4\pi} \delta[(x - x')^2].$$

$$(2\text{-}1.64)$$

Asymptotic forms appropriate to large separation will be stated for space-like intervals, $[(x - x')^2]^{1/2} = R > 0$, and for time-like intervals, $[-(x - x')^2]^{1/2} = T > 0$, although they are connected by the substitutions $R \leftrightarrow iT$:

$$mR \gg 1: \qquad \Delta_+(x - x') \simeq i\,\frac{(2m)^{1/2}}{(4\pi R)^{3/2}} e^{-mR},$$

$$(2\text{-}1.65)$$

$$mT \gg 1: \qquad \Delta_+(x - x') \simeq e^{-\pi i/4} \frac{(2m)^{1/2}}{(4\pi T)^{3/2}} e^{-imT}.$$

We turn next to the more general situation in which $K(x)$ is a complex function. Now the sources effective in emission and absorption are reciprocally complex conjugate functions. If we did no more than introduce that feature into the previous discussion, the single-particle term in the construction of the complete vacuum amplitude would become

$$i\int (dx)(dx') K_1^*(x) \left[i\int d\omega_p e^{ip(x - x')} \right] K_2(x'). \qquad (2\text{-}1.66)$$

But this is clearly incomplete, for the implied source structure, linear in

$$K(x) = K_1(x) + K_2(x) \qquad (2\text{-}1.67)$$

and linear in

$$K^*(x) = K_1^*(x) + K_2^*(x), \qquad (2\text{-}1.68)$$

also requires the contribution of the causal term

$$i \int (dx)(dx') K_1(x') \left[i \int d\omega_p e^{ip(x'-x)} \right] K_2^*(x), \qquad (2\text{--}1.69)$$

referring to the emission and subsequent absorption of another kind of particle. What is the mass of this particle? If the two masses were unequal, the structure of the new $\Delta_+(x - x')$ function in the vacuum amplitude

$$\langle 0_+|0_-\rangle^K = 1 + i \int (dx)(dx') K^*(x) \Delta_+(x - x') K(x) \qquad (2\text{--}1.70)$$

would still be given by (2–1.37) and (2–1.38), but different masses would appear in the two causal forms. Then we could no longer conclude that $\Delta_+(x - x')$ had a unique extrapolation into space-like regions. It is the principle of space-time uniformity that demands equal masses for the two kinds of particle, which are identified as particle and antiparticle. The Euclidean postulate produces the same conclusion through the absence of an invariant distinction between $x_4 - x_4' > 0$ and $x_4 - x_4' < 0$, which permits only one mass parameter to appear.

In view of these remarks the definitions that relate sources to single-particle production and annihilation probability amplitudes must be extended to

$$\langle 1_{p\pm}|0_-\rangle^K = iK_{p\pm}, \qquad \langle 0_+|1_{p\pm}\rangle^K = iK_{p\pm}^*, \qquad (2\text{--}1.71)$$

where $\pm$ distinguish particle and antiparticle, and

$$K_{p+} = (d\omega_p)^{1/2} K(p), \qquad K_{p-} = (d\omega_p)^{1/2} K^*(p). \qquad (2\text{--}1.72)$$

Note carefully the distinction between

$$K^*(p) = \int (dx) e^{-ipx} K^*(x) \qquad (2\text{--}1.73)$$

and

$$K(p)^* = \left[\int (dx) e^{-ipx} K(x) \right]^* = K^*(-p). \qquad (2\text{--}1.74)$$

Accordingly, we have

$$K_{p+}^* = (d\omega_p)^{1/2} K^*(-p), \qquad K_{p-}^* = (d\omega_p)^{1/2} K(-p). \qquad (2\text{--}1.75)$$

Thus, the explicit appearance of the p or $-p$ Fourier transform, representing the energy-momentum balance, distinguishes emission or absorption, respectively, while K and K^* identify particle and antiparticle in emission, but conversely in absorption. The function of K is to create particles and annihilate antiparticles, while K^* creates antiparticles and annihilates particles.

In analogy with the way that sources act to increase or decrease the amount of energy in the system, we can conceive of K and K^* acting to increase and decrease, respectively, the quantity of a property which must assume opposite values for particle and antiparticle. This is a familiar characteristic of electric charge, and we recognize that some charge-like property always distinguishes particle and antiparticle. The formal counterpart of these remarks stems from the invariance of the vacuum transformation function under phase transformations of the complex sources,

$$K(x) \to e^{i\varphi}K(x), \qquad K^*(x) \to e^{-i\varphi}K^*(x). \tag{2-1.76}$$

If we examine the response of the probability amplitudes $\langle 1_{p\pm}|0_-\rangle^K$ to these phase transformations, combined with a rigid displacement of the source, by X^μ, we get

$$\langle 1_{p\pm}|0_-\rangle^K \to e^{\pm i\varphi}e^{ipX}\langle 1_{p\pm}|0_-\rangle^K, \tag{2-1.77}$$

which makes explicit the mechanical and 'charge' attributes of the single-particle states.

An alternative presentation is obtained by replacing the complex source $K(x)$ with two real sources, according to

$$K(x) = 2^{-1/2}[K_{(1)}(x) - iK_{(2)}(x)], \qquad K^*(x) = 2^{-1/2}[K_{(1)}(x) + iK_{(2)}(x)]. \tag{2-1.78}$$

This gives

$$
\begin{aligned}
\langle 0_+|0_-\rangle^K \cong {}& 1 + \tfrac{1}{2}i\int (dx)(dx')K_{(1)}(x)\Delta_+(x - x')K_{(1)}(x') \\
& + \tfrac{1}{2}i\int (dx)(dx')K_{(2)}(x)\Delta_+(x - x')K_{(2)}(x') \\
= {}& \langle 0_+|0_-\rangle^{K_{(1)}}\langle 0_+|0_-\rangle^{K_{(2)}}.
\end{aligned}
\tag{2-1.79}
$$

We have now exhibited two independent sources, with their associated particles. But the fact that these particles have the same mass (and spin) implies that the decomposition can be done in an infinite variety of ways, corresponding to the phase transformations of complex sources, which now appear as two-dimensional Euclidean rotations:

$$
\begin{aligned}
K_{(1)}(x) &\to \cos\varphi K_{(1)}(x) + \sin\varphi K_{(2)}(x), \\
K_{(2)}(x) &\to -\sin\varphi K_{(1)}(x) + \cos\varphi K_{(2)}(x).
\end{aligned}
\tag{2-1.80}
$$

The latter can also be written in matrix notation as

$$K(x) \to e^{iq\varphi}K(x), \tag{2-1.81}$$

where

$$K(x) = \begin{pmatrix} K_{(1)}(x) \\ K_{(2)}(x) \end{pmatrix} \tag{2-1.82}$$

and

$$q = \begin{pmatrix} 0 & -i \\ i & 0 \end{pmatrix} \tag{2-1.83}$$

is identified as the charge matrix. Note that it is imaginary and antisymmetrical. Its eigenvalues are ± 1, and the complex sources $K(x)$, $K^*(x)$ are the corresponding eigenvectors. The real sources

$$K_{(1)}(x) = 2^{-1/2}(K(x) + K^*(x)), \qquad K_{(2)}(x) = 2^{-1/2}i(K(x) - K^*(x))$$
$$(2\text{--}1.84)$$

do not produce single-particle states of definite charge. They refer to the complementary property of charge symmetry—the states turn into themselves or their negatives, respectively, when positive and negative charges are interchanged. A matrix presentation of this transformation is

$$K(x) \rightarrow r_q K(x), \qquad (2\text{--}1.85)$$

where the real matrix

$$r_q = \begin{pmatrix} 1 & 0 \\ 0 & -1 \end{pmatrix} \qquad (2\text{--}1.86)$$

has the property

$$r_q q = -q r_q. \qquad (2\text{--}1.87)$$

The symbol C is often used for this charge reflection operation.

When two-component matrix notation is used, the vacuum amplitude has the same formal expression as with a single real source,

$$\langle 0_+|0_-\rangle^K = 1 + \tfrac{1}{2}i \int (dx)(dx') K(x) \Delta_+(x - x') K(x'). \qquad (2\text{--}1.88)$$

This remains true of its Euclidean counterpart:

$$\langle 0_+|0_-\rangle^K \rightarrow 1 - \tfrac{1}{2} \int (dx)(dx') K(x) \Delta(x - x') K(x') \Big]_E, \qquad (2\text{--}1.89)$$

which can also be written in terms of complex sources,

$$\langle 0_+|0_-\rangle^K \rightarrow 1 - \int (dx)(dx') K^*(x) \Delta(x - x') K(x') \Big]_E. \qquad (2\text{--}1.90)$$

Euclidean transformations decompose into two connected pieces, distinguished as proper and improper transformations. In contrast, the full Lorentz group contains four connected pieces, owing to the discontinuous causal distinction between $x^0 > 0$ and $x^0 < 0$. The wider invariance introduced by the Euclidean postulate thus enables one to perform some discontinuous Lorentz transformations through the intermediary of continuous Euclidean transformations. The most important example of that is

$$\bar{x}_\mu = -x_\mu, \qquad (2\text{--}1.91)$$

a proper transformation, which is a time reflection transformation in Minkowski space. The formal invariance of the vacuum amplitude under the transformation

$$K(x) \rightarrow K(-x) \qquad (2\text{--}1.92)$$

is an immediate consequence of the symmetry

$$\Delta_+(x - x') = \Delta_+(-x + x'), \qquad (2\text{--}1.93)$$

but it is the Euclidean postulate that supplies the general basis for this invariance. The reflection of the time coordinate inverts the causal order of sources and interchanges creation and annihilation. This is evident from the momentum form of the source transformation,

$$K(p) \rightarrow K(-p), \tag{2-1.94}$$

and thus

$$K_{p+} \leftrightarrow K^*_{p-}, \qquad K_{p-} \leftrightarrow K^*_{p+} \tag{2-1.95}$$

or

$$\langle 1_{p\pm}|0_-\rangle^K \leftrightarrow \langle 0_+|1_{p\mp}\rangle^K. \tag{2-1.96}$$

According to its construction as a time reflection (T) and a space reflection (P-arity), which also has the effect of interchanging particle and antiparticle—a charge reflection (C)—this process is often known as the TCP operation, but one is likely to encounter any other permutation of the three constituents. Perhaps it should be called the Shell game.

2–2 SPIN 0 PARTICLES. STRONG SOURCE

The experimenter's beam contains many particles at a given time, which exist under conditions of effective noninteraction since they are widely spaced on the scale set by microscopic interaction distances. A beam of electrically charged particles is an exception to this, in principle, but in practice the disturbance by long-range interactions can be made sufficiently small by controlling the beam density. We give a theoretical transcription of this situation by exploiting the directionality that sources possess as an aspect of the complementarity between the $K(x)$ and $K(p)$ representations. A source that is spatially diffused and suitably phased (to use antenna language) can produce a highly directional beam, sharply limiting the possible locations of a detection source if effective coupling is to be achieved. We visualize an arbitrary number of such pairs of directional weak emission and absorption sources, operating side by side with negligible cross coupling. If roughly the same causal arrangement is used for all the pairs of sources, we have produced a situation in which, at some time intermediate between the emission and absorption regions, an arbitrary number of particles can exist in circumstances of noninteraction, owing to the spatial separations among them.

We first consider real sources and designate by $K_\alpha(x)$, $\alpha = 1, 2, \ldots$, the individual weak sources that correspond to one self-contained emission and absorption process. The physical independence of these various acts, which has been achieved through our control of the sources, is expressed by simply multiplying the individual probability amplitudes to produce the vacuum amplitude for the complete arrangement:

$$\langle 0_+|0_-\rangle^K = \prod_\alpha \left[1 + \tfrac{1}{2}i \int (dx)(dx') K_\alpha(x) \Delta_+(x - x') K_\alpha(x') \right]. \tag{2-2.1}$$

The sources $K_\alpha(x)$ are disjoint parts of the total source

$$K(x) = \sum_\alpha K_\alpha(x). \qquad (2\text{-}2.2)$$

The principle of space-time uniformity requires that no specific distinctions among the components of $K(x)$ be admitted. In short, the vacuum amplitude must depend only upon $K(x)$. This is realized by incorporating the property

$$\alpha \neq \beta: \qquad \int (dx)(dx')K_\alpha(x)\Delta_+(x - x')K_\beta(x') = 0, \qquad (2\text{-}2.3)$$

which asserts the absence of coupling between different single particle exchange regions. Then, since the individual sources are weak,

$$\langle 0_+|0_-\rangle^K = \prod_\alpha \exp\left[\tfrac{1}{2}i\int (dx)(dx')K_\alpha(x)\Delta_+(x - x')K_\alpha(x')\right]$$

$$= \exp\left[\sum_\alpha \tfrac{1}{2}i\int (dx)(dx')K_\alpha(x)\Delta_+(x - x')K_\alpha(x')\right]$$

$$= \exp\left[\sum_{\alpha\beta} \tfrac{1}{2}i\int (dx)(dx')K_\alpha(x)\Delta_+(x - x')K_\beta(x')\right], \quad (2\text{-}2.4)$$

or

$$\langle 0_+|0_-\rangle^K = \exp\left[\tfrac{1}{2}i\int (dx)(dx')K(x)\Delta_+(x - x')K(x')\right]. \qquad (2\text{-}2.5)$$

The same form applies to two-component real sources, and for complex sources it becomes

$$\langle 0_+|0_-\rangle^K = \exp\left[i\int (dx)(dx')K^*(x)\Delta_+(x - x')K(x')\right]. \qquad (2\text{-}2.6)$$

We accept these exponential vacuum amplitude structures as descriptive of any arrangement of sources, with arbitrary strength, subject only to the restriction that the particles have no effective interaction. To test the consistency of this assertion we consider a simple causal arrangement, expressed by

$$K(x) = K_1(x) + K_2(x), \qquad (2\text{-}2.7)$$

in which we maintain the convention that K_1 refers to physical acts that occur after the completion of those represented by K_2. For the situation of real sources we have

$$\langle 0_+|0_-\rangle^K = \langle 0_+|0_-\rangle^{K_1} \exp\left[i\int (dx)(dx')K_1(x)\Delta_+(x - x')K_2(x')\right]\langle 0_+|0_-\rangle^{K_2}, \qquad (2\text{-}2.8)$$

where, according to the causal disposition of the sources,

$$i\int (dx)(dx')K_1(x)\Delta_+(x - x')K_2(x') = i\int (dx)(dx')K_1(x)\left[i\int d\omega_p e^{ip(x-x')}\right]K_2(x')$$

$$= \sum_p iK_{1p}^* iK_{2p}. \qquad (2\text{-}2.9)$$

The causal arrangement also enables us to analyze the complete process into an initial multiparticle emission act, represented by the probability amplitude

$\langle\{n\}|0_-\rangle^{K_2}$, and a subsequent absorption process, described by $\langle 0_+|\{n\}\rangle^{K_1}$, where $\{n\}$ indicates the collection of physical attributes that distinguish the various n-particle states. The resulting causal analysis of the vacuum amplitude is

$$\langle 0_+|0_-\rangle^K = \sum_{\{n\}} \langle 0_+|\{n\}\rangle^{K_1}\langle\{n\}|0_-\rangle^{K_2}. \qquad (2\text{-}2.10)$$

To display, in this form, the explicit structure

$$\langle 0_+|0_-\rangle^K = \langle 0_+|0_-\rangle^{K_1} \exp\left[\sum_p iK^*_{1p}iK_{2p}\right]\langle 0_+|0_-\rangle^{K_2}, \qquad (2\text{-}2.11)$$

we have only to introduce the expansion of the exponential function

$$\exp\left[\sum_p iK^*_{1p}iK_{2p}\right] = \prod_p \exp[iK^*_{1p}iK_{2p}]$$

$$= \prod_p \sum_{n_p=0} \frac{(iK^*_{1p})^{n_p}}{(n_p!)^{1/2}} \frac{(iK_{2p})^{n_p}}{(n_p!)^{1/2}}. \qquad (2\text{-}2.12)$$

This supplies the required identifications

$$\langle\{n\}|0_-\rangle^K = \langle 0_+|0_-\rangle^K \prod_p \frac{(iK_p)^{n_p}}{(n_p!)^{1/2}}, \qquad \langle 0_+|\{n\}\rangle^K = \langle 0_+|0_-\rangle^K \prod_p \frac{(iK^*_p)^{n_p}}{(n_p!)^{1/2}}, \qquad (2\text{-}2.13)$$

where the multiparticle label is realized by the collection of integers $\{n_p\}$. The evident interpretation of n_p is a particle occupation number associated with the indicated physical properties. This is confirmed by the response of the multiparticle states to the source translation $K(x) \to K(x+X)$, which gives

$$\langle\{n\}|0_-\rangle^K \to e^{iPX}\langle\{n\}|0_-\rangle^K, \qquad \langle 0_+|\{n\}\rangle^K \to \langle 0_+|\{n\}\rangle^K e^{-iPX}. \qquad (2\text{-}2.14)$$

The total energy-momentum thus obtained,

$$P^\mu = \sum_p n_p p^\mu, \qquad (2\text{-}2.15)$$

displays the additive contributions of the particles present in the state under consideration.

The probability amplitudes must meet the following total probability or completeness test:

$$\sum_{\{n\}} |\langle\{n\}|0_-\rangle^K|^2 = |\langle 0_+|0_-\rangle^K|^2 \exp\left[\sum_p |K_p|^2\right] = 1, \qquad (2\text{-}2.16)$$

and indeed

$$|\langle 0_+|0_-\rangle^K|^2 = \exp\left[-\int(dx)(dx')K(x)\,\text{Re}\,(1/i)\Delta_+(x-x')K(x')\right]$$

$$= \exp\left[-\sum_p |K_p|^2\right]. \qquad (2\text{-}2.17)$$

Note how the vacuum amplitude has been used in two distinct ways. Through the consideration of a causal arrangement, relative multiparticle amplitudes are obtained:

$$\frac{\langle\{n\}|0_-\rangle^K}{\langle 0_+|0_-\rangle^K} = \prod_p \frac{(iK_p)^{n_p}}{(n_p!)^{1/2}}, \tag{2-2.18}$$

and the assumed completeness of the multiparticle states leads to

$$1/|\langle 0_+|0_-\rangle^K|^2 = \exp\left[\sum_p |K_p|^2\right]. \tag{2-2.19}$$

Then the vacuum amplitude is applied directly as a probability amplitude, with consistent results.

The extension to a pair of real sources, or the equivalent complex source, is immediate. The summation over momenta in (2-2.12) is now supplemented by a summation over the two kinds of particles and the results are analogous, as in

$$\begin{aligned}
\langle\{n\}|0_-\rangle^K &= \langle 0_+|0_-\rangle^K \prod_p \frac{(K_{p+})^{n_{p+}}}{(n_{p+}!)^{1/2}} \frac{(K_{p-})^{n_{p-}}}{(n_{p-}!)^{1/2}} \\
&= \langle 0_+|0_-\rangle^K \prod_{pq} \frac{(K_{pq})^{n_{pq}}}{(n_{pq}!)^{1/2}},
\end{aligned} \tag{2-2.20}$$

where $q = \pm 1$ is the charge label that distinguishes particle and antiparticle. A combined source translation and phase transformation changes these states as follows:

$$\langle\{n\}|0_-\rangle^K \rightarrow e^{iQ\varphi}e^{iPX}\langle\{n\}|0_-\rangle^K, \qquad \langle 0_+|\{n\}\rangle^K \rightarrow \langle 0_+|\{n\}\rangle^K e^{-iQ\varphi}e^{-iPX} \tag{2-2.21}$$

where

$$Q = \sum_{pq} n_{pq}q, \qquad P^\mu = \sum_{pq} n_{pq}p^\mu \tag{2-2.22}$$

exhibit the total charge and energy-momentum attributes of the multiparticle state labeled $\{n\}$.

The momentum labeling of individual particle states is not the only possibility. A spherical or angular momentum specification is introduced by the transformation

$$K_p = \sum_{lm} (d\Omega)^{1/2} Y_{lm}(\mathbf{p}) K_{p^0 lm}, \tag{2-2.23}$$

where

$$\begin{aligned}
K_{p^0 lm} &= \left[\frac{|\mathbf{p}|dp^0}{\pi}\right]^{1/2} i^{-l} \int (dx) e^{ip^0 x^0} j_l(|\mathbf{p}|\,|\mathbf{x}|) Y^*_{lm}(\mathbf{x}) K(x), \\
|\mathbf{p}| &= [(p^0)^2 - m^2]^{1/2},
\end{aligned} \tag{2-2.24}$$

and j_l, Y_{lm} are standard symbols for spherical Bessel functions and spherical harmonics. The discrete angular momentum quantum numbers

$$l = 0, 1, 2, \ldots, \qquad m = l, l-1, \ldots, -l \tag{2-2.25}$$

replace the continuous orientation of the momentum vector, or rather the dense discontinuous specification within the infinitesimal solid angles $d\Omega$. Thus

$$\sum_p K_{1p}^* K_{2p} = \sum_{p^0 lm} K_{1p^0 lm}^* K_{2p^0 lm}, \tag{2-2.26}$$

and one has only to change the labels in (2–2.13) to obtain the source representation of the new multiparticle states. The generalization to complex sources is also immediate. According to the azimuthal angle dependence of $Y_{lm}(\mathbf{x})$, which is $\exp(im\varphi)$, the source rotation indicated by

$$K(\dots, \varphi) \to K(\dots, \varphi + \alpha) \tag{2-2.27}$$

induces

$$K_{p^0 lm} \to e^{im\alpha} K_{p^0 lm}, \tag{2-2.28}$$

and

$$\langle\{n\}|0_-\rangle^K \to e^{iM\alpha}\langle\{n\}|0_-\rangle^K \tag{2-2.29}$$

displays the total magnetic quantum number of the multiparticle state

$$M = \sum_{p^0 lm} n_{p^0 lm} m. \tag{2-2.30}$$

The *TCP* relation between emission and absorption now appears as

$$K_{p^0 lm} \leftrightarrow (-1)^m K_{p^0 l-m}^*, \tag{2-2.31}$$

which is stated for real sources; otherwise a reference to charge reflection is added.

There is an axial description, which is obtained from

$$K_p = \sum_{m=-\infty}^{\infty} (d\varphi/2\pi)^{1/2} e^{im\varphi} K_{p_+ p_- m}, \tag{2-2.32}$$

where

$$K_{p_+ p_- m} = \frac{1}{4\pi} (dp_+ dp_-)^{1/2} i^{-m} \int (dx) e^{i(p_+ x_- + p_- x_+)} J_m(|\mathbf{p}_\perp| \, |\mathbf{x}_\perp|) e^{-im\varphi} K(x). \tag{2-2.33}$$

The following symbols are used:

$$p_\pm = p^0 \pm p_3, \quad x_\pm = \tfrac{1}{2}(x^0 \pm x_3), \quad |\mathbf{p}_\perp| = (p_+ p_- - m^2)^{1/2}, \tag{2-2.34}$$

while $\mathbf{x}_\perp$ is the projected coordinate vector in the plane perpendicular to the third axis, and φ indicates appropriate azimuthal angles about this axis. The *TCP* operation takes the form

$$K_{p_+ p_- m} \leftrightarrow (-1)^m K_{p_+ p_- -m}^*. \tag{2-2.35}$$

The exchange of particles between sources in causal array can naturally be described in space-time language also. Direct power series expansion gives

$$\exp\left[i\int(dx)(dx')K_1(x)\Delta_+(x-x')K_2(x')\right]$$

$$= 1 + \sum_{n=1}^{\infty} i^n \int \frac{(dx_1)\cdots(dx_n)}{n!}\frac{(dx_1')\cdots(dx_n')}{n!}$$

$$\times K_1(x_1)\cdots K_1(x_n)[\text{perm}_{(n)}\Delta_+(x_i - x_j')]K_2(x_1')\cdots K_2(x_n'),\quad(2\text{-}2.36)$$

where

$$\text{perm}_{(n)}\Delta_+(x_i - x_j') = \sum_{n!\,\text{perm.}}\Delta_+(x_1 - x_{j_1}')\cdots\Delta_+(x_n - x_{j_n}')\quad(2\text{-}2.37)$$

defines the permanent, a determinant without minus signs. Clearly displayed here are the effective sources for n-particle emission and absorption, together with the function that represents the noninteracting propagation of the n particles. The latter is symmetrized among the space-time coordinates of the particles which, together with the unrestricted occupation numbers $n_p = 0, 1, 2, \ldots$, proclaims that we are describing particles that obey Bose-Einstein statistics.

The familiar characteristics of this statistics are also apparent in the answer to the following question. What are the values of the general probability amplitude $\langle\{n\}|\{n'\}\rangle^K$? This is to ask how the effectiveness of a source in emitting or absorbing particles is influenced by the prior presence of particles. We consider the following causal situation. A strong source $K_2(x)$ acts first to create particles, which are subsequently influenced by the probe source $K_0(x)$, after which the particles are annihilated by the detection source $K_1(x)$:

$$K(x) = K_1(x) + K_0(x) + K_2(x).\quad(2\text{-}2.38)$$

We use real sources and write

$$\langle 0_+|0_-\rangle^K = \langle 0_+|0_-\rangle^{K_1+K_2}\exp\left[i\int(dx)(dx')K_1(x)\Delta_+(x-x')K_0(x')\right.$$

$$\left. + i\int(dx)(dx')K_0(x)\Delta_+(x-x')K_2(x')\right]\langle 0_+|0_-\rangle^{K_0}.\quad(2\text{-}2.39)$$

The causal situation is then alternatively conveyed by

$$\langle 0_+|0_-\rangle^K = \left[\sum_{\{n\}}\langle 0_+|\{n\}\rangle^{K_1}\langle\{n\}|0_-\rangle^{K_2}\right]\exp\left[\sum_p(iK_{1p}^*iK_{0p}\right.$$

$$\left. + iK_{0p}^*iK_{2p})\right]\langle 0_+|0_-\rangle^{K_0}$$

$$= \sum_{\{n\},\{n'\}}\langle 0_+|\{n\}\rangle^{K_1}\langle\{n\}_+|\{n'\}_-\rangle^{K_0}\langle\{n'\}|0_-\rangle^{K_2},\quad(2\text{-}2.40)$$

from which the desired probability amplitudes, referring to the probe source K_0, can be obtained. We first consider a weak probe and accordingly retain only linear terms in K_0. Then, since

$$iK_p\langle\{n\}|0_-\rangle^K = (n_p + 1)^{1/2}\langle\{n + 1_p\}|0_-\rangle^K,$$

$$\langle 0_+|\{n\}\rangle^K iK_p^* = \langle 0_+|\{n + 1_p\}\rangle^K(n_p + 1)^{1/2},\qquad(2\text{-}2.41)$$

which refers very immediately to the monomial structure of these probability amplitudes, we get

$$\langle\{n+1_p\}_+|\{n\}_-\rangle^K \cong (n_p+1)^{1/2}iK_p,$$
$$\langle\{n-1_p\}_+|\{n\}_-\rangle^K \cong (n_p)^{1/2}iK_p^*, \tag{2-2.42}$$

which generalize the initial definitions (2-1.28) while retaining the weak source limitation. In particular, the probability for the emission of another particle,

$$|\langle\{n+1_p\}_+|\{n\}_-\rangle^K|^2 = (n_p+1)|K_p|^2, \tag{2-2.43}$$

shows the additional stimulated emission that is characteristic of B.E. statistics.

As a preliminary to picking out the general transition amplitudes, we construct the probability amplitude $\langle\{n\}_+|\{n\}_-\rangle^K$, which has the same initial and final configuration and, in that sense, is a generalization of the vacuum amplitude. This object is extracted by retaining only equal powers of K_{1p}^* and K_{2p} in the expansion

$$\exp\left[\sum_p (iK_{1p}^*iK_{0p} + iK_{0p}^*iK_{2p})\right] \to \prod_p [1 + iK_{1p}^*iK_{0p}iK_{0p}^*iK_{2p} + \cdots]. \tag{2-2.44}$$

We are going to introduce a useful simplification here by recognizing that for sufficiently small $d\omega_p$, higher powers in this series are negligible. It is only the dependence on the probe source that is at issue:

$$|K_{0p}|^2 \sim d\omega_p, \; (|K_{0p}|^2)^2 \sim (d\omega_p)^2, \ldots .$$

We also remark that

$$\sum_{\{n\}} \langle 0_+|\{n\}\rangle^{K_1} iK_{1p}^*K_{2p}\langle\{n\}|0_-\rangle^{K_2} = \sum_{\{n\}} \langle 0_+|\{n\}\rangle^{K_1} n_p\langle\{n\}|0_-\rangle^{K_2}, \tag{2-2.45}$$

which applies to each momentum cell independently. For the process of interest, then,

$$\exp\left[\sum_p (iK_{1p}^*iK_{0p} + iK_{0p}^*iK_{2p})\right] \to \prod_p [1 + iK_{0p}^*n_piK_{0p}]$$
$$= \exp\left[\sum_p iK_{0p}^*n_piK_{0p}\right], \tag{2-2.46}$$

and we conclude that

$$\langle\{n\}_+|\{n\}_-\rangle^K = \exp\left[\tfrac{1}{2}i\int (dx)(dx')K(x)\Delta_{\{n\}+}(x-x')K(x')\right], \tag{2-2.47}$$

where

$$\Delta_{\{n\}+}(x-x') = \Delta_+(x-x') + i\int d\omega_p n_p[e^{ip(x-x')} + e^{-ip(x-x')}]. \tag{2-2.48}$$

The last term is so written in order to maintain the symmetry in x and x'. The explicit causal forms are

$x^0 > x^{0\prime}$:
$$\Delta_{\{n\}+}(x - x') = i\int d\omega_p[(n_p + 1)e^{ip(x-x')} + n_p e^{-ip(x-x')}],$$

$x^0 < x^{0\prime}$:   (2-2.49)
$$\Delta_{\{n\}+}(x - x') = i\int d\omega_p[n_p e^{ip(x-x')} + (n_p + 1)e^{-ip(x-x')}].$$

Note that the probability amplitude $\langle\{n\}_+|\{n\}_-\rangle^K$ reduces to unity for $K = 0$, which means that the initial and final multiparticle states are referred to the same time or space-like surface, as is appropriate to a reasonably localized probe.

To find the probability amplitudes in which initial and final states are not the same, we do not return to the general construction given in (2-2.40), but use $\langle\{n\}_+|\{n\}_-\rangle^K$ directly, in the manner of the vacuum amplitude. Thus, consider the causal source arrangement

$$K(x) = K_1(x) + K_2(x),$$ (2-2.50)

which implies that

$$\langle\{n\}_+|\{n\}_-\rangle^K = \langle\{n\}_+|\{n\}_-\rangle^{K_1}$$
$$\times \exp\left[i\int(dx)(dx')K_1(x)\Delta_{\{n\}+}(x - x')K_2(x')\right]\langle\{n\}_+|\{n\}_-\rangle^{K_2}$$
$$= \sum_{\{n'\}} \langle\{n\}_+|\{n'\}_-\rangle^{K_1}\langle\{n'\}_+|\{n\}_-\rangle^{K_2}.$$ (2-2.51)

The coupling between the component sources is now given by

$$\exp\left[i\int(dx)(dx')K_1(x)\Delta_{\{n\}+}(x - x')K_2(x')\right]$$
$$= \exp\left[\sum_p \{iK_{1p}^*(n_p + 1)iK_{2p} + iK_{1p}n_p iK_{2p}^*\}\right]$$
$$= \prod_p [1 + iK_{1p}^*(n_p + 1)iK_{2p} + iK_{1p}n_p iK_{2p}^* + \cdots].$$ (2-2.52)

The explicit terms indicated for a given momentum describe the several processes in which, respectively, no change in particle number occurs, an additional particle is emitted, an incident particle is absorbed. Higher powers, containing more complicated processes involving several particles, are relatively negligible for sufficiently small $d\omega_p$; the probability for emitting two particles into the momentum range $d\omega_p$, for example, is $\sim(d\omega_p)^2$. But we must not let this apparently innocent simplification pass without comment. The infinitesimal character of $d\omega_p$ will be vitiated if there is an inordinate sensitivity to $\mathbf{p}$ produced by the appearance of very large coordinate intervals $(e^{i\mathbf{p}\cdot\mathbf{x}})$. To put it more physically, we recall that we are dealing with a beam of particles interacting with a probe source. What we have done is correct if the probe is placed well in the interior of the beam where there is no significant position dependence. It will fail if the probe is outside or near the boundaries of the beam. This is a reminder that underlying any momentum description is an appreciation of the space-time causal situation.

Having understood its limitations, we now use (2-2.52) to identify the individual probability amplitudes for processes in which, independently for any number of momentum cells, a single particle has been added or removed. The result is indicated by

$$\frac{\langle\{n + 1_e - 1_a\}_+|\{n\}_-\rangle^K}{\langle\{n\}_+|\{n\}_-\rangle^K} = \prod_e [(n_p + 1)^{1/2}iK_p] \prod_a [n_p^{1/2}iK_p^*],$$

$$\frac{\langle\{n\}_+|\{n + 1_a - 1_e\}_-\rangle^K}{\langle\{n\}_+|\{n\}_-\rangle^K} = \prod_e [n_p^{1/2}iK_p] \prod_a [(n_p + 1)^{1/2}iK_p^*],$$

(2-2.53)

where the products refer to the various particles that are (e)mitted or (a)bsorbed. The two statements are equivalent, if it is admitted that $\langle\{n + 1_p\}_+|\{n + 1_p\}_-\rangle^K$ and $\langle\{n\}_+|\{n\}_-\rangle^K$ differ negligibly, owing to the infinitesimal nature of $d\omega_p$. Now we face the consistency test associated with the alternative uses of the probability amplitude. From the completeness of the final or initial multi-particle states we deduce

$$1/|\langle\{n\}_+|\{n\}_-\rangle^K|^2 = \prod_p [1 + (n_p + 1)|K_p|^2 + n_p|K_p|^2]$$

$$= \exp\left[\sum_p (2n_p + 1)|K_p|^2\right],$$

(2-2.54)

whereas, by direct calculation,

$$|\langle\{n\}_+|\{n\}_-\rangle^K|^2 = \exp\left[-\int(dx)(dx')K(x) \operatorname{Re}(1/i)\Delta_{\{n\}+}(x - x')K(x')\right]$$

(2-2.55)

and

$$\operatorname{Re}(1/i)\Delta_{\{n\}+}(x - x') = \operatorname{Re}\int d\omega_p(2n_p + 1)e^{ip(x-x')}.$$

(2-2.56)

The test has been passed successfully.

The extension to complex sources and charged particles is straightforward, and generally the introduction of the charge label q, supplementing the momentum index p, suffices to produce the required result. We only remark on the following detail. In constructing the probability amplitude $\langle\{n\}_+|\{n\}_-\rangle^K$ we are led, as in (2-2.46), to the factor

$$\exp\left[\sum_{pq} iK_{0pq}^* n_{pq} iK_{0pq}\right]$$

(2-2.57)

and the result

$$\langle\{n\}_+|\{n\}_-\rangle^K = \exp\left[i\int(dx)(dx')K^*(x)\Delta_{\{n\}+}(x - x')K(x')\right].$$

(2-2.58)

Now, however, the propagation function $\Delta_{\{n\}+}(x - x')$ has the following meaning:

$$\Delta_{\{n\}+}(x - x') = \Delta_+(x - x') + i\int d\omega_p[n_{p+}e^{ip(x-x')} + n_{p-}e^{-ip(x-x')}],$$

(2-2.59)

which is no longer necessarily symmetrical in x and x'. There is still a TCP symmetry, in which $x^\mu \to -x^\mu$ is combined with charge reflection,

$$n_{p+} \leftrightarrow n_{p-}. \tag{2-2.60}$$

The explicit causal structure of this function is given by

$x^0 > x^{0\prime}$:
$$i \int d\omega_p [(n_{p+} + 1)e^{ip(x-x')} + n_{p-}e^{-ip(x-x')}],$$

$x^0 < x^{0\prime}$:
$$i \int d\omega_p [n_{p+}e^{ip(x-x')} + (n_{p-} + 1)e^{-ip(x-x')}]. \tag{2-2.61}$$

It should be mentioned that the propagation function is symmetrical in x and x' if the incident beam is neutral at every momentum, $n_{p+} = n_{p-}$. Then, and only then, can one introduce the real sources $K_{(1)}$, $K_{(2)}$, and associate two independent particle types with them.

We have seen causality and space-time uniformity working as creative principles. The physical requirement of completeness, or unitarity, has then been verified; it is not an independent principle.

We shall now examine this relationship in more detail. But first we return to the vacuum persistence amplitude for real sources and consider its complex conjugate:

$$\langle 0_+|0_-\rangle^{K*} = \langle 0_-|0_+\rangle^K = \exp\left[-\tfrac{1}{2}i \int (dx)(dx')K(x)\Delta_-(x - x')K(x')\right], \tag{2-2.62}$$

where

$$\Delta_-(x - x') = \Delta_+(x - x')^*. \tag{2-2.63}$$

To give a uniform presentation of the two propagation functions $\Delta_\pm$, we define, everywhere, the positive and negative frequency functions $\Delta^{(\pm)}$ as

$$\Delta^{(+)}(x - x') = \int d\omega_p e^{ip(x-x')}, \quad \Delta^{(-)}(x - x') = \int d\omega_p e^{-ip(x-x')}, \tag{2-2.64}$$

which are connected by

$$\Delta^{(-)}(x - x') = \Delta^{(+)}(x' - x) = \Delta^{(+)}(x - x')^*. \tag{2-2.65}$$

The two propagation functions then appear as

$$\Delta_+(x - x') = \begin{cases} x^0 > x^{0\prime}: & i\Delta^{(+)}(x - x'), \\ x^0 < x^{0\prime}: & i\Delta^{(-)}(x - x'); \end{cases}$$

$$\Delta_-(x - x') = \begin{cases} x^0 > x^{0\prime}: & -i\Delta^{(-)}(x - x'), \\ x^0 < x^{0\prime}: & -i\Delta^{(+)}(x - x'). \end{cases} \tag{2-2.66}$$

We note the everywhere-valid relation

$$\Delta_+(x - x') - \Delta_-(x - x') = i[\Delta^{(+)}(x - x') + \Delta^{(-)}(x - x')]. \tag{2-2.67}$$

The momentum integral derived from (2–1.45) by complex conjugation,

$$\Delta_-(x - x') = -i \int d\omega_p \exp[i\mathbf{p} \cdot (\mathbf{x} - \mathbf{x}') + ip^0|x^0 - x^{0'}|], \quad (2\text{–}2.68)$$

leads to the same Euclidean function as before,

$$i\Delta_-(x - x') \rightarrow \Delta_E(x - x'), \quad (2\text{–}2.69)$$

by means of the substitution

$$-i|x^0 - x^{0'}| \rightarrow |x_4 - x_4'|. \quad (2\text{–}2.70)$$

Then we have

$$\langle 0_-|0_+\rangle^K \rightarrow \exp\left[-\tfrac{1}{2}\int (dx)(dx')K(x)\Delta(x - x')K(x')\right]_E, \quad (2\text{–}2.71)$$

which is the same Euclidean form that is obtained from $\langle 0_+|0_-\rangle^K$, as the strong source generalization of (2–1.57). The Euclidean version of the vacuum amplitude is a real number lying in the interval between 0 and 1. The vacuum amplitude $\langle 0_-|0_+\rangle^K$ is regained through the substitution

$$x_4 \rightarrow \exp\left[-i\left(\frac{\pi}{2} - \epsilon\right)\right]x^0,$$

$$p_4 \rightarrow \exp\left[i\left(\frac{\pi}{2} - \epsilon\right)\right]p_0, \quad (2\text{–}2.72)$$

$$\epsilon \rightarrow +0$$

which, incidentally, supplies the invariant representation

$$\Delta_-(x - x') = \int \frac{(dp)}{(2\pi)^4} \frac{e^{ip(x-x')}}{p^2 + m^2 + i\epsilon}\bigg|_{\epsilon\rightarrow+0}. \quad (2\text{–}2.73)$$

Another connection between the reciprocally complex conjugate vacuum amplitudes comes from the existence of the common Euclidean transcription. Proceeding through the Euclidean form as an intermediary, we have

$$\exp\left[i\left(\frac{\pi}{2} - \epsilon\right)\right]x^0 \rightarrow x_4 \rightarrow \exp\left[-i\left(\frac{\pi}{2} - \epsilon\right)\right]x^0:$$

$$(1 - i\epsilon)x^0 \rightarrow -(1 + i\epsilon)x^0,$$

$$\exp\left[-i\left(\frac{\pi}{2} - \epsilon\right)\right]p_0 \rightarrow p_4 \rightarrow \exp\left[i\left(\frac{\pi}{2} - \epsilon\right)\right]p_0: \quad (2\text{–}2.74)$$

$$(1 + i\epsilon)p_0 \rightarrow -(1 - i\epsilon)p_0,$$

and, in response,

$$\Delta_+(x - x') \rightarrow -\Delta_-(x - x'), \qquad \langle 0_+|0_-\rangle^K \rightarrow \langle 0_-|0_+\rangle^K. \quad (2\text{–}2.75)$$

To verify this directly, we note that

$$p^2 - i\epsilon \to p^2 + i\epsilon, \qquad x^2 + i\epsilon \to x^2 - i\epsilon, \qquad \epsilon \to +0 \qquad (2\text{-}2.76)$$

with the usual adjustable scale for ϵ. One must also remember that the ordering of variables is retained throughout the transformation. Limits of integration do not change, therefore, and

$$\int (dp) \to -\int (dp), \qquad \int (dx) \to -\int (dx), \qquad (2\text{-}2.77)$$

while

$$|x^0 - x^{0\prime}| \to -|x^0 - x^{0\prime}|, \qquad (2\text{-}2.78)$$

which confirm in alternative ways the stated transformations.

We shall now use the causal structure of the theory to give a complete derivation of the unitarity property. This is done within a very limited physical context, of course, but it is clearly a general procedure. For our present purposes we replace the causal labeling K_1, K_2 by $K_{(-)}$, $K_{(+)}$. (While this may seem to be still another use of the overburdened $\pm$ signs, it will turn out to be consistent with the $\Delta_\pm$ notation.) Let the time T be located between the regions defined by the two component sources. Introduce a new time coordinate for $x^0 > T$ by reflection at T,

$$x^0 - T = T - \bar{x}^0, \qquad (2\text{-}2.79)$$

and then transform this time interval in the manner just discussed:

$$x^0 - T \to e^{-\pi i}(T - \bar{x}^0). \qquad (2\text{-}2.80)$$

The immediate effect is to replace x^0 by $\bar{x}^0$, which is earlier than T to the same extent that the original time exceeded T. The transformed $K_{(-)}$ source also occupies a reflected position, earlier than T. Before this operation is performed, the vacuum amplitude has the following composition:

$$\begin{aligned}
\langle 0_+|0_-\rangle^K = \exp\Big[&\tfrac{1}{2}i \int (dx)(dx') K_{(-)}(x) \Delta_+(x - x') K_{(-)}(x') \\
&+ \tfrac{1}{2}i \int (dx)(dx') K_{(+)}(x) \Delta_+(x - x') K_{(+)}(x') \\
&+ i \int (dx)(dx') K_{(-)}(x) i\Delta^{(+)}(x - x') K_{(+)}(x') \Big],
\end{aligned} \qquad (2\text{-}2.81)$$

where the appearance of the last term indicates the causal arrangement. When the transformation is carried out, the quadratic $K_{(+)}$ term remains unaware of what happens latter, the quadratic $K_{(-)}$ term responds in the known manner $[\Delta_+ \to -\Delta_-]$ without reference to the other terms, and the last term changes only by a minus sign arising from

$$\int (dx) \to -\int (dx), \qquad (2\text{-}2.82)$$

the influence of the $e^{-\pi i}$ factor, which is not compensated by the reflection that is also being used. In the latter the integration limits are reversed to maintain a positive measure. The result is

$$\langle 0_+|0_-\rangle^K \rightarrow \exp\left[-\tfrac{1}{2}i\int(dx)(dx')K_{(-)}(x)\Delta_-(x-x')K_{(-)}(x')\right.$$
$$+\tfrac{1}{2}i\int(dx)(dx')K_{(+)}(x)\Delta_+(x-x')K_{(+)}(x')$$
$$+\left.\int(dx)(dx')(-i)K_{(-)}(x)\Delta^{(+)}(x-x')iK_{(+)}(x')\right], \quad (2\text{-}2.83)$$

which does not depend upon T, that being any time after both sources have ceased operating. The physical meaning of this combination follows from

$$\exp\left[\int(dx)(dx')(-i)K_{(-)}(x)\Delta^{(+)}(x-x')iK_{(+)}(x')\right]$$
$$=\exp\left[\sum_p(iK_{(-)p})^*(iK_{(+)p})\right]$$
$$=\prod_p\sum_{n_p=0}\frac{[(iK_{(-)p})]^{*n_p}}{(n_p!)^{1/2}}\frac{[(iK_{(+)p})]^{n_p}}{(n_p!)^{1/2}}, \quad (2\text{-}2.84)$$

for, on using the fact that

$$\langle\{n\}|0_-\rangle^{K^*}=\langle 0_-|\{n\}\rangle^K, \quad (2\text{-}2.85)$$

we get

$$\langle 0_-|0_+\rangle^{K(-)}\exp\left[\int(dx)(dx')(-i)K_{(-)}(x)\Delta^{(+)}(x-x')iK_{(+)}(x')\right]\langle 0_+|0_-\rangle^{K(+)}$$
$$=\sum_{\{n\}}\langle 0_-|\{n\}\rangle^{K(-)}\langle\{n\}|0_-\rangle^{K(+)}$$
$$\equiv\langle 0_-|0_-\rangle^{K(-),K(+)}. \quad (2\text{-}2.86)$$

As the notation indicates, the picture has become that of a system evolving in time from the initial vacuum state under the influence of the source $K_{(+)}(x)$ and then traced back to the initial state in the presence of the source $K_{(-)}(x)$. It is not the physical system that goes back in time, of course. What is reversed is the causal order of the states that are being compared. If the two sources are identical, we must regain the initial state; that is,

$$K_{(-)}(x)=K_{(+)}(x)=K(x) \quad (2\text{-}2.87)$$

implies

$$\langle 0_-|0_-\rangle^{K,K}=\sum_{\{n\}}\langle 0_-|\{n\}\rangle^K\langle\{n\}|0_-\rangle^K=1, \quad (2\text{-}2.88)$$

a statement of completeness, or unitarity. According to the exponential structure (2-2.83), this is true if

$$i[\Delta_+(x-x')-\Delta_-(x-x')]+\Delta^{(+)}(x-x')+\Delta^{(-)}(x-x')=0, \quad (2\text{-}2.89)$$

where the last terms appear in that form to produce the necessary symmetry in x and x'. We recognize the identity (2-2.67).

The full statement of unitarity emerges on using the sources $K_{(\pm)}$ to generate arbitrary multiparticle states. We write

$$K_{(+)} = K + K_2, \qquad K_{(-)} = K + K_{2'}, \qquad (2\text{-}2.90)$$

where K_2 and $K_{2'}$ act prior to the source K, and introduce the corresponding causal analyses:

$$\langle\{n\}|0_-\rangle^{K_{(+)}} = \sum_{\{n''\}} \langle\{n\}_+|\{n''\}_-\rangle^K \langle\{n''\}|0_-\rangle^{K_2},$$

$$\langle 0_-|\{n\}\rangle^{K_{(-)}} = \sum_{\{n'\}} \langle 0_-|\{n'\}\rangle^{K_{2'}} \langle\{n'\}_-|\{n\}_+\rangle^K. \qquad (2\text{-}2.91)$$

What we must verify is that all reference to the source $K(x)$ disappears from (2–2.86), leaving only

$$\langle 0_-|0_-\rangle^{K_{2'},K_2} = \sum_{\{n'\}} \langle 0_-|\{n'\}\rangle^{K_{2'}} \langle\{n'\}|0_-\rangle^{K_2}, \qquad (2\text{-}2.92)$$

for that is the unitarity assertion about the effect of the source K:

$$\sum_{\{n\}} \langle\{n'\}_-|\{n\}_+\rangle^K \langle\{n\}_+|\{n''\}_-\rangle^K = \delta(\{n'\}, \{n''\}). \qquad (2\text{-}2.93)$$

Beyond the condition we have already used, Eq. (2–2.89), what is required is

$$\int (dx)(dx')[-iK(x)\Delta_-(x - x')K_{2'}(x') + iK(x)\Delta_+(x - x')K_2(x')$$
$$+ K(x)\Delta^{(-)}(x - x')K_{2'}(x') + K(x)\Delta^{(+)}(x - x')K_2(x')] = 0. \quad (2\text{-}2.94)$$

But, under the given causal circumstances, $\Delta_- \to -i\Delta^{(-)}$ and $\Delta_+ \to i\Delta^{(+)}$, which completes the verification.

The probability amplitude $\langle 0_-|0_-\rangle^{K_{(-)},K_{(+)}}$ is also useful for the direct computation of various expectation values. Let

$$K_{(-)}(x) = K(x), \qquad K_{(+)}(x) = K(x + X), \qquad (2\text{-}2.95)$$

for example, which replaces the unit evaluation for $X = 0$ by

$$\sum_{\{n\}} \langle 0_-|\{n\}\rangle^K e^{iPX} \langle\{n\}|0_-\rangle^K = \langle e^{iPX}\rangle_0^K. \qquad (2\text{-}2.96)$$

This is the expectation value of e^{iPX} for the states produced from the vacuum by the action of the source K. Since only the relative displacement of the two sources is significant, we have

$$\langle e^{iPX}\rangle_0^K = \exp\left[\int (dx)(dx')K(x)(\Delta^{(+)}(x - x' + X) - \Delta^{(+)}(x - x'))K(x')\right]$$
$$= \exp\left[\int d\omega_p (e^{ipX} - 1)|K(p)|^2\right] = \exp\left[\sum_p (e^{ipX} - 1)|K_p|^2\right]. \qquad (2\text{-}2.97)$$

On considering infinitesimal displacements, we learn that

$$\langle P^\mu \rangle_0^K = \int (dx)(dx') K(x)(1/i)\partial^\mu \Delta^{(+)}(x - x') K(x')$$
$$= \int d\omega_p p^\mu |K(p)|^2 = \sum_p p^\mu |K_p|^2, \quad (2\text{-}2.98)$$

or, with an obvious identification,

$$\langle n_p \rangle_0^K = |K_p|^2. \quad (2\text{-}2.99)$$

The total number of particles is indicated by

$$N = \sum_p n_p. \quad (2\text{-}2.100)$$

Thus, the average total number of particles created and the vacuum persistence probability are simply related, according to (2–2.17):

$$|\langle 0_+ | 0_- \rangle^K|^2 = \exp[-\langle N \rangle_0^K]. \quad (2\text{-}2.101)$$

The discussion of fluctuations is facilitated by writing (the various indices are omitted)

$$\langle e^{i(P - \langle P \rangle)X} \rangle = \exp\left[\int (dx)(dx') K(x)(\Delta^{(+)}(x - x' + X) - \Delta^{(+)}(x - x')\right.$$
$$\left. - X^\mu \partial_\mu \Delta^{(+)}(x - x'))K(x') \right]$$
$$= \exp\left[\sum_p (e^{ipX} - 1 - ipX)|K_p|^2 \right]. \quad (2\text{-}2.102)$$

The simplest example is

$$\langle (P - \langle P \rangle)_\mu (P - \langle P \rangle)_\nu \rangle = \langle P_\mu P_\nu \rangle - \langle P_\mu \rangle \langle P_\nu \rangle$$
$$= \int (dx)(dx') K(x)(1/i)\partial_\mu (1/i)\partial_\nu \Delta^{(+)}(x - x') K(x')$$
$$= \sum_p p_\mu p_\nu |K_p|^2, \quad (2\text{-}2.103)$$

which we can also interpret as

$$\langle n_p n_{p'} \rangle - \langle n_p \rangle \langle n_{p'} \rangle = \delta_{pp'} \langle n_p \rangle. \quad (2\text{-}2.104)$$

One consequence of the latter is

$$\langle N^2 \rangle - \langle N \rangle^2 = \langle N \rangle. \quad (2\text{-}2.105)$$

Statements about the total number of particles are derived directly by considering the sources

$$K_{(-)}(x) = K(x), \qquad K_{(+)}(x) = \lambda K(x). \quad (2.2\text{-}106)$$

According to the structure of the relative probability amplitudes, we have

$$\sum_{\{n\}} \langle 0_-|\{n\}\rangle^K \lambda^N \langle \{n\}|0_-\rangle^K = \exp\left[(\lambda - 1)\int (dx)(dx')K(x)\Delta^{(+)}(x - x')K(x')\right] \cdot$$
(2–2.107)

One differentiation with respect to λ, at $\lambda = 1$, gives

$$\langle N\rangle_0^K = \int (dx)(dx')K(x)\Delta^{(+)}(x - x')K(x') = \sum_p |K_p|^2. \quad (2–2.108)$$

The coefficient of λ^N in the summation (2–2.107) is the probability of emitting N particles, without further identification. Comparison of the power series expansion supplies it as

$$p(N, 0)^K = \frac{\langle N\rangle^N}{N!} e^{-\langle N\rangle}. \quad (2–2.109)$$

The fluctuation formula (2–2.105) is a familiar characteristic of this Poisson distribution. All such properties are derived from

$$\langle \lambda^N\rangle = \exp[(\lambda - 1)\langle N\rangle] \quad (2–2.110)$$

by differentiation with respect to λ:

$$\langle N(N - 1)\cdots(N - \nu + 1)\rangle = \langle N\rangle^\nu. \quad (2–2.111)$$

The generalization of this discussion to the amplitude $\langle\{n\}_-|\{n\}_-\rangle^{K(-),K(+)}$ only requires introducing the function $\Delta_{\{n\}+}(x - x')$ and its partners:

$$\Delta_{\{n\}-}(x - x') = \Delta_-(x - x') - i\int d\omega_p n_p(e^{ip(x-x')} + e^{-ip(x-x')}),$$

$$\Delta_{\{n\}}^{(+)}(x - x') = \int d\omega_p[(n_p + 1)e^{ip(x-x')} + n_p e^{-ip(x-x')}] \quad (2–2.112)$$

$$\Delta_{\{n\}}^{(-)}(x - x') = \int d\omega_p[n_p e^{ip(x-x')} + (n_p + 1)e^{-ip(x-x')}].$$

The causal relations among these functions are the same as in the vacuum situation although $\Delta_{\{n\}}^{(+)}(x - x')$, for example, is no longer an exclusively positive frequency function. In deriving expectation values we must note that the amplitude $\langle\{n'\}_+|\{n\}_-\rangle^K$ responds to the translation $K(x) \to K(x + X)$ with the factor

$$\exp[i(P\{n'\} - P\{n\})X], \quad (2–2.113)$$

since both initial and final states are now relevant. Some results are

$$\langle(n'_p - n_p)\rangle_n^K = |K_p|^2 \quad (2–2.114)$$

and

$$\langle n'_p n'_{p'}\rangle - \langle n'_p\rangle\langle n'_{p'}\rangle = \delta_{pp'}\langle(n'_p - n_p)\rangle(2n_p + 1). \quad (2–2.115)$$

The treatment of complex sources and charged particles is quite analogous. The vacuum amplitude describing the time cycle is

$$\langle 0_-|0_-\rangle^{K_{(-)},K_{(+)}} = \exp\Big[-i\int (dx)(dx')K^*_{(-)}(x)\Delta_-(x-x')K_{(-)}(x')$$

$$+ i\int (dx)(dx')K^*_{(+)}(x)\Delta_+(x-x')K_{(+)}(x')$$

$$+ \int (dx)(dx')K^*_{(-)}(x)\Delta^{(+)}(x-x')K_{(+)}(x')$$

$$+ \int (dx)(dx')K_{(-)}(x)\Delta^{(+)}(x-x')K^*_{(+)}(x')\Big], \quad (2\text{-}2.116)$$

which reduces to unity for $K_{(-)}(x) = K_{(+)}(x)$. By choosing the sources as

$$K_{(-)}(x) = K(x), \qquad K_{(+)}(x) = \lambda e^{i\varphi}K(x+X), \quad (2\text{-}2.117)$$

with λ real, we obtain

$$\langle \lambda^N e^{iQ\varphi} e^{iPX}\rangle_0^K$$

$$= \exp\Big[\int (dx)(dx')K^*(x)\big(\lambda e^{i\varphi}\Delta^{(+)}(x-x'+X) - \Delta^{(+)}(x-x')\big)K(x')$$

$$+ \int (dx)(dx')K(x)\big(\lambda e^{-i\varphi}\Delta^{(+)}(x-x'+X) - \Delta^{(+)}(x-x')\big)K^*(x')\Big]$$

$$= \exp\Big[\sum_{pq}(\lambda e^{iq\varphi}e^{ipX} - 1)|K_{pq}|^2\Big]. \quad (2\text{-}2.118)$$

We can also introduce the total number of positively and negatively charged particles,

$$N_+ = \tfrac{1}{2}(N+Q), \qquad N_- = \tfrac{1}{2}(N-Q), \quad (2\text{-}2.119)$$

and re-express the expectation value formulas as

$$\langle \lambda_+^{N_+}\lambda_-^{N_-}e^{iPX}\rangle_0^K = \exp\Big[\sum_p(\lambda_+ e^{ipX} - 1)|K_{p+}|^2 + \sum_p(\lambda_- e^{ipX} - 1)|K_{p-}|^2\Big]. \quad (2\text{-}2.120)$$

Accordingly,

$$\langle N_+\rangle_0^K = \sum_p |K_{p+}|^2, \qquad \langle N_-\rangle_0^K = \sum_p |K_{p-}|^2, \quad (2\text{-}2.121)$$

while individual probabilities are given by

$$p(N_+N_-,0)^K = \frac{\langle N_+\rangle^{N_+}}{N_+!}\frac{\langle N_-\rangle^{N_-}}{N_-!}e^{-\langle N\rangle}. \quad (2\text{-}2.122)$$

A simplified formula, designed to answer questions about electric charge only, is

$$\langle e^{iQ\varphi}\rangle_0^K = \exp[(e^{i\varphi} - 1)\langle N_+\rangle + (e^{-i\varphi} - 1)\langle N_-\rangle], \quad (2\text{-}2.123)$$

from which we derive the individual probabilities:

$$p(Q,0)^K = \left(\frac{\langle N_+\rangle}{\langle N_-\rangle}\right)^{(1/2)Q} I_Q[2(\langle N_+\rangle\langle N_-\rangle)^{1/2}]e^{-\langle N\rangle}, \quad (2\text{-}2.124)$$

on using a familiar Bessel function expansion. The introduction of the propagation function (2–2.59), with its attendant structures, generalizes (2–2.116) to the probability amplitude $\langle\{n\}_-|\{n\}_-\rangle^{K(-),K(+)}$.

2–3 SPIN 1 PARTICLES. THE PHOTON

Before developing the general source representation for particles of arbitrary spin, we shall give an elementary discussion of some special examples which are of great physical importance. The exponential form that has been established for the vacuum amplitude, within the context of spinless particles, embodies the physical possibility of producing any number of independent acts of single particle emission and absorption. These space-time properties are independent of the specific spin of the particle. The latter can only influence the more detailed structure of the source. It is clear that, if spin 0 particles are described by a scalar source, sources transforming as vectors and tensors of various ranks must refer to particles of unit and higher spin. A vector source, designated as $J^\mu(x)$, is the obvious candidate to describe unit spin particles. There are certain obstacles, however. This source has four components, in contrast with the three independent sources one should associate with the three spin possibilities that are accessible to a nonzero mass particle. This presumably means that $J^\mu(x)$ is a mixture of a unit spin source with a source of spinless particles, corresponding to the possibility of forming a scalar function by differentiation, $\partial_\mu J^\mu(x)$. And, independently, we observe that should we do no more than replace the real scalar source $K(x)$ by the real vector source $J^\mu(x)$:

$$\langle 0_+|0_-\rangle^J = \exp\left[\tfrac{1}{2}i\int (dx)(dx')J^\mu(x)\Delta_+(x-x')J_\mu(x')\right], \qquad (2\text{–}3.1)$$

we should encounter a serious physical difficulty, for

$$|\langle 0_+|0_-\rangle^J|^2 = \exp\left[-\int d\omega_p J^\mu(p)^* J_\mu(p)\right] \qquad (2\text{–}3.2)$$

is not guaranteed to be less than unity, since

$$J^\mu(p)^* J_\mu(p) = |\mathbf{J}(p)|^2 - |J^0(p)|^2 \qquad (2\text{–}3.3)$$

can assume either sign.

Both difficulties are overcome by the following invariant structure, which is appropriate to a particle of mass $m \neq 0$:

$$\langle 0_+|0_-\rangle^J = \exp\Big[\tfrac{1}{2}i\int (dx)(dx')\big(J^\mu(x)\Delta_+(x-x')J_\mu(x')$$
$$+ (1/m^2)\partial_\mu J^\mu(x)\Delta_+(x-x')\partial'_\nu J^\nu(x')\big)\Big]. \qquad (2\text{–}3.4)$$

In evaluating the vacuum persistence probability, we now encounter

$$J^\mu(p)^* J_\mu(p) + (1/m^2)p_\mu J^\mu(p)^* p_\nu J^\nu(p) = J^\mu(p)^* [g_{\mu\nu} + (1/m^2)p_\mu p_\nu]J^\nu(p). \qquad (2\text{–}3.5)$$

Since this is an invariant combination, it can be inspected conveniently in the rest frame of the time-like vector p^μ, where

$$p^\mu \text{ rest frame:} \qquad\qquad p^k = 0, \qquad p^0 = m. \qquad\qquad (2\text{–}3.6)$$

The components of the symmetrical tensor that appears in (2–3.5) are then given by

$$g_{\mu\nu} + (1/m^2)p_\mu p_\nu : \qquad \begin{cases} \mu = \nu = 0: 0, \\ \mu = k, \ \nu = 0: 0, \\ \mu = k, \ \nu = l: \delta_{kl}. \end{cases} \qquad\qquad (2\text{–}3.7)$$

The result is simply $|\mathbf{J}|^2$, which is positive, and which contains three independent source components, transforming among themselves under spatial rotation, as is appropriate to unit spin.

We note that $(1/m)p^\mu$ is a unit time-like vector, which can be supplemented by three orthogonal space-like vectors, $e^\mu_{p\lambda}$, obeying

$$p_\mu e^\mu_{p\lambda} = 0, \qquad e^{\mu*}_{p\lambda} e_{\mu p\lambda'} = \delta_{\lambda\lambda'}. \qquad\qquad (2\text{–}3.8)$$

They give a dyadic construction of the metric tensor,

$$g^{\mu\nu} = -(1/m^2)p^\mu p^\nu + \sum_\lambda e^\mu_{p\lambda} e^{\nu*}_{p\lambda}. \qquad\qquad (2\text{–}3.9)$$

The symmetry of $g^{\mu\nu}$ indicates that complex conjugation of the three $e^\mu_{p\lambda}$ produces a unitary transformation on the set. With the definition

$$J_{p\lambda} = (d\omega_p)^{1/2} e^{\mu*}_{p\lambda} J_\mu(p), \qquad\qquad (2\text{–}3.10)$$

the vacuum persistence probability appears as

$$|\langle 0_+|0_-\rangle^J|^2 = \exp\left[-\sum_{p\lambda} |J_{p\lambda}|^2\right]. \qquad\qquad (2\text{–}3.11)$$

We now consider a causal source arrangement,

$$J^\mu(x) = J^\mu_1(x) + J^\mu_2(x), \qquad\qquad (2\text{–}3.12)$$

which implies

$$\langle 0_+|0_-\rangle^J = \langle 0_+|0_-\rangle^{J_1} \exp\left[\int d\omega_p i J^\mu_1(p)^*(g_{\mu\nu} + m^{-2}p_\mu p_\nu) i J^\nu_2(p)\right] \langle 0_+|0_-\rangle^{J_2}$$

$$= \langle 0_+|0_-\rangle^{J_1} \exp\left[\sum_{p\lambda} i J^*_{1p\lambda} i J_{2p\lambda}\right] \langle 0_+|0_-\rangle^{J_2}. \qquad\qquad (2\text{–}3.13)$$

This standard structure identifies the multiparticle states

$$\langle \{n\}|0_-\rangle^J = \langle 0_+|0_-\rangle^J \prod_{p\lambda} \frac{(iJ_{p\lambda})^{n_{p\lambda}}}{(n_{p\lambda}!)^{1/2}},$$

$$\langle 0_+|\{n\}\rangle^J = \langle 0_+|0_-\rangle^J \prod_{p\lambda} \frac{(iJ^*_{p\lambda})^{n_{p\lambda}}}{(n_{p\lambda}!)^{1/2}}, \qquad\qquad (2\text{–}3.14)$$

where $n_{p\lambda} = 0, 1, 2, \ldots$ again indicates B.E. statistics. The consistency between the two uses of the vacuum amplitude is obvious.

One can choose the unit space-like vectors $e_{p\lambda}^{\mu}$ to be real. The orthogonality requirement

$$\mathbf{p} \cdot \mathbf{e}_{p\lambda} = p^0 e_{p\lambda}^0 \qquad (2\text{–}3.15)$$

displays the role of $\mathbf{p}$ in providing a reference direction. If $\mathbf{e}_{p\lambda}$ is perpendicular to $\mathbf{p}$, the time component $e_{p\lambda}^0$ vanishes. Let $\mathbf{e}_{p1}$ be such a real unit vector,

$$\mathbf{p} \cdot \mathbf{e}_{p1} = 0, \qquad (\mathbf{e}_{p1})^2 = 1, \qquad e_{p1}^0 = 0. \qquad (2\text{–}3.16)$$

Then

$$\mathbf{e}_{p2} = (\mathbf{p}/|\mathbf{p}|) \times \mathbf{e}_{p1}, \qquad (\mathbf{e}_{p2})^2 = 1, \qquad e_{p2}^0 = 0 \qquad (2\text{–}3.17)$$

is another one, and the set is completed by

$$\mathbf{e}_{p3} = (p^0/m)(\mathbf{p}/|\mathbf{p}|), \qquad e_{p3}^0 = |\mathbf{p}|/m. \qquad (2\text{–}3.18)$$

We note, incidentally, that

$$e_{p3}^{\mu} J_{\mu}(p) = \frac{p^0}{m|\mathbf{p}|} \left(\mathbf{p} \cdot \mathbf{J}(p) - \frac{|\mathbf{p}|^2}{p^0} J^0(p) \right)$$

$$= \frac{p^0}{|\mathbf{p}|} \left(\frac{1}{m} p^{\mu} J_{\mu}(p) + \frac{m}{p^0} J^0(p) \right). \qquad (2\text{–}3.19)$$

Alternative, complex, choices are suggested by angular momentum considerations. The response of the vector $J^{\mu}(x)$ to the homogeneous infinitesimal Lorentz transformation

$$\overline{x}^{\mu} = x^{\mu} + \delta\omega^{\mu\nu} x_{\nu} \qquad (2\text{–}3.20)$$

is

$$\overline{J}^{\mu}(\overline{x}) = J^{\mu}(x) + \delta\omega^{\mu\nu} J_{\nu}(x) \qquad (2\text{–}3.21)$$

or

$$\delta J^{\lambda}(x) = \delta\omega^{\mu\nu} x_{\mu} \partial_{\nu} J^{\lambda}(x) + \delta\omega^{\lambda\nu} J_{\nu}(x). \qquad (2\text{–}3.22)$$

For a three-dimensional rotation, this becomes

$$\begin{aligned} \delta\mathbf{J}(x) &= \delta\boldsymbol{\omega} \cdot \mathbf{x} \times \nabla \mathbf{J}(x) - \delta\boldsymbol{\omega} \times \mathbf{J}(x), \\ \delta J^0(x) &= \delta\boldsymbol{\omega} \cdot \mathbf{x} \times \nabla J^0(x) \end{aligned} \qquad (2\text{–}3.23)$$

and, equivalently,

$$\begin{aligned} \delta\mathbf{J}(p) &= \delta\boldsymbol{\omega} \cdot \mathbf{p} \times (\partial/\partial\mathbf{p})\mathbf{J}(p) - \delta\boldsymbol{\omega} \times \mathbf{J}(p) \\ \delta J^0(p) &= \delta\boldsymbol{\omega} \cdot \mathbf{p} \times (\partial/\partial\mathbf{p})J^0(p). \end{aligned} \qquad (2\text{–}3.24)$$

Now let us consider a rotation about the direction of the momentum,

$$\delta\boldsymbol{\omega} = \delta\varphi(\mathbf{p}/|\mathbf{p}|). \qquad (2\text{–}3.25)$$

We realize a single-particle state of helicity λ:

$$\delta J_{p\lambda} = i\lambda \, \delta\varphi J_{p\lambda}, \qquad (2\text{–}3.26)$$

if

$$-\mathbf{e}_{p\lambda}^{*} \times (\mathbf{p}/|\mathbf{p}|) = i\lambda \mathbf{e}_{p\lambda}^{*}. \qquad (2\text{–}3.27)$$

Zero helicity is achieved with $\mathbf{e}$ parallel to $\mathbf{p}$. Accordingly we relabel e_{p3}^{μ},

$$\mathbf{e}_{p0} = (p^0/m)(\mathbf{p}/|\mathbf{p}|), \qquad e_{p0}^0 = |\mathbf{p}|/m. \tag{2-3.28}$$

The helicity states with $\lambda = \pm 1$ correspond to the complex combinations

$$\mathbf{e}_{p+1}^* = 2^{-1/2}[-\mathbf{e}_{p1} + i\mathbf{e}_{p2}], \qquad \mathbf{e}_{p-1}^* = 2^{-1/2}[\mathbf{e}_{p1} + i\mathbf{e}_{p2}], \tag{2-3.29}$$

which are so chosen that

$$-\mathbf{e}_{p\lambda}^* \times \delta\boldsymbol{\omega} = i \sum_{\lambda'=-1}^{+1} (\delta\boldsymbol{\omega} \cdot \mathbf{S})_{\lambda\lambda'} \mathbf{e}_{p\lambda'}^* \tag{2-3.30}$$

gives the standard unit spin matrix elements.

A classification of sources and particle states in relation to total angular momentum can be introduced. As a preliminary step, we emulate the zero spin procedure and define

$$(d\omega_p)^{1/2} J^{\nu}(p) = \sum_{lm} (d\Omega)^{1/2} Y_{lm}(\mathbf{p}) J_{p^0lm}^{\nu}, \tag{2-3.31}$$

where

$$J_{p^0lm}^{\nu} = \left(\frac{|\mathbf{p}|\,dp^0}{\pi}\right)^{1/2} i^{-l} \int (dx) e^{ip^0x^0} j_l(|\mathbf{p}|\,|\mathbf{x}|) Y_{lm}^*(\mathbf{x}) J^{\nu}(x). \tag{2-3.32}$$

Nothing more need be done for the time component J^0, which is a three-dimensional scalar function. But the three components of $\mathbf{J}$ refer to the unit spin, which must be coupled appropriately with the orbital angular momentum to produce total angular momentum states. This is accomplished by the following introduction of a vector orthonormal system, replacing the scalar spherical harmonic set,

$$\sum_{lm} Y_{lm}(\mathbf{p})\mathbf{J}_{p^0lm} = \sum_{jm} \Bigg[(j(j+1))^{-1/2} \mathbf{L} Y_{jm}(\mathbf{p}) J_{p^0jm1}$$

$$+ (j(j+1))^{-1/2} \frac{\mathbf{p}}{|\mathbf{p}|} \times \mathbf{L} Y_{jm}(\mathbf{p}) J_{p^0jm2}$$

$$+ \frac{\mathbf{p}}{|\mathbf{p}|} Y_{jm}(\mathbf{p}) \left(\frac{m}{p^0} J_{p^0jm3} + \frac{|\mathbf{p}|}{p^0} J_{p^0jm}^0\right) \Bigg], \tag{2-3.33}$$

where

$$\mathbf{L} = i(\partial/\partial\mathbf{p}) \times \mathbf{p}, \tag{2-3.34}$$

and the reader is warned not to confuse the letter m, used in subscripts to denote a magnetic quantum number, with m, appearing explicitly in its role as particle mass. The above structure is such that

$$\sum_{lm} \mathbf{J}_{p^0lm}^* \cdot \mathbf{J}_{p^0lm} = \sum_{jm} \Bigg[J_{p^0jm1}^* J_{p^0jm1} + J_{p^0jm2}^* J_{p^0jm2}$$

$$+ \left(\frac{m}{p^0} J_{p^0jm3} + \frac{|\mathbf{p}|}{p^0} J_{p^0jm}^0\right)^* \left(\frac{m}{p^0} J_{p^0jm3} + \frac{|\mathbf{p}|}{p^0} J_{p^0jm}^0\right) \Bigg]. \tag{2-3.35}$$

We also note the relation

$$\sum_{lm} Y_{lm}(\mathbf{p})\left(\frac{\mathbf{p}}{m}\cdot\mathbf{J}_{p^0lm}-\frac{p^0}{m}J^0_{p^0lm}\right)=\sum_{jm} Y_{jm}(\mathbf{p})\left(\frac{|\mathbf{p}|}{p^0}J_{p^0jm3}-\frac{m}{p^0}J^0_{p^0jm}\right). \tag{2-3.36}$$

On combining the various contributions, we get the required form:

$$\int d\omega_p J^\mu(p)^*(g_{\mu\nu}+m^{-2}p_\mu p_\nu)J^\nu(p)=\sum_{p^0jm\lambda}J^*_{p^0jm\lambda}J_{p^0jm\lambda}, \tag{2-3.37}$$

where $\lambda=1,2,3$ distinguishes the three excitations with total angular momentum quantum numbers j,m.

These sources can be exhibited explicitly. The vector orthornormality property enables us to evaluate

$$J_{p^0jm1}=\left(\frac{|\mathbf{p}|\,dp^0}{\pi}\right)^{1/2}\int\frac{d\Omega}{4\pi}Y^*_{jm}(\mathbf{p})(j(j+1))^{-1/2}\mathbf{L}\cdot\mathbf{J}(\mathbf{p}), \tag{2-3.38}$$

for example, and this can be converted into

$$J_{p^0jm1}=\left(\frac{|\mathbf{p}|\,dp^0}{\pi}\right)^{1/2}i^{-j}\int(dx)e^{ip^0x^0}j_j(|\mathbf{p}||\mathbf{x}|)Y^*_{jm}(\mathbf{x})(j(j+1))^{-1/2}\mathbf{L}\cdot\mathbf{J}(x), \tag{2-3.39}$$

where now

$$\mathbf{L}=\mathbf{x}\times(1/i)\nabla. \tag{2-3.40}$$

(It is unfortunate that the combination of two well-established notational conventions produces things like j_j.) Incidentally this type of source vanishes for $j=0$. Similarly, we find that

$$J_{p^0jm2}=\left(\frac{|\mathbf{p}|\,dp^0}{\pi}\right)^{1/2}i^{-j}\int(dx)e^{ip^0x^0}j_j(|\mathbf{p}||\mathbf{x}|)$$
$$\times Y^*_{jm}(\mathbf{x})(j(j+1))^{-1/2}(1/|\mathbf{p}|)(1/i)\nabla\times\mathbf{L}\cdot\mathbf{J}(x), \tag{2-3.41}$$

which also vanishes for $j=0$, and

$$J_{p^0jm3}=\left(\frac{|\mathbf{p}|\,dp^0}{\pi}\right)^{1/2}i^{-j}$$
$$\times\int(dx)e^{ip^0x^0}j_j(|\mathbf{p}||\mathbf{x}|)Y^*_{jm}(\mathbf{x})\frac{p^0}{m|\mathbf{p}|}\left(\frac{1}{i}\nabla\cdot\mathbf{J}(x)-\frac{|\mathbf{p}|^2}{p^0}J^0(x)\right). \tag{2-3.42}$$

It is seen that the sources designated as $\lambda=1,2$ depend only upon $\mathbf{J}(x)$, and that in the form $\nabla\times\mathbf{J}(x)$, while for the third type of source we have, effectively,

$$\frac{1}{m}\left(\frac{1}{i}\nabla\cdot\mathbf{J}(x)-\frac{|\mathbf{p}|^2}{p^0}J^0(x)\right)\to\frac{1}{m}\frac{1}{i}\partial_\mu J^\mu(x)+\frac{m}{p^0}J^0(x). \tag{2-3.43}$$

There is no difficulty in implementing the same generalizations that were discussed for zero spin particles—charged particles, multiparticle initial and final states, cyclic time development—but the details are too similar to merit repetition. We turn instead to an important special situation, the limit of zero mass, as realized by the photon.

It is evident from Eq. (2–3.4) that the zero mass limit does not exist unless $\partial_\mu J^\mu(x)$ vanishes. One might write

$$\partial_\mu J^\mu(x) = mK(x), \tag{2–3.44}$$

and identify $K(x)$, in the limit $m \to 0$, as the source of a massless zero spin particle. The latter would be completely independent of the residual photon source, however, and since $m = 0$, $s = 0$ particles are unknown experimentally, in any event, we only consider photons in stating the source description:

$$\langle 0_+|0_-\rangle^J = \exp\left[\tfrac{1}{2}i \int (dx)(dx') J^\mu(x) D_+(x - x') J_\mu(x')\right],$$
$$\partial_\mu J^\mu(x) = 0. \tag{2–3.45}$$

We use the symbol D_+ to indicate the restriction to zero mass.

The source associated with a particular particle is an abstraction of the realistic processes that create or annihilate the particle. It retains what all such mechanisms have in common and ignores the specific characteristics of individual mechanisms. Any general restriction on sources that is implied by special features of the particle must be common to all mechanisms and thus states a general law of physics. In the situation of the photon we have deduced, from its zero mass, the necessity of a restriction on the vectorial source. It must be divergence-less, which is the local statement of a conservation law. We are in no doubt about the identity of this conserved physical property. It is electric charge.

The loss of one degree of excitation for massless particles is evident in the various ways of labeling particle states. Thus, as $m \to 0$ in (2–3.19), under the restriction

$$p_\mu J^\mu(p) = 0, \tag{2–3.46}$$

we arrive at

$$J_{p3} = 0, \tag{2–3.47}$$

and the two remaining sources $J_{p1,2}$ refer to the two transverse linear polarizations accessible to photons. With helicity labeling we have, equivalently,

$$J_{p0} = 0, \tag{2–3.48}$$

and $J_{p\pm1}$ represent the two circular polarizations. Turning to angular momentum states, we have analogously, from (2–3.43),

$$J_{p^0jm3} = 0. \tag{2–3.49}$$

Since $j = 0$ does not appear in the two other source types, this is the counterpart of the absence of zero helicity.

We have arrived at the restriction to two polarization or helicity states for the photon by a limiting procedure that began with massive unit spin particles.

Now let us obtain this result directly, by using the photon source description given in (2–3.45). The consideration of a causal arrangement,

$$J^\mu(x) = J^\mu_1(x) + J^\mu_2(x), \qquad (2\text{–}3.50)$$

implies

$$\langle 0_+|0_-\rangle^J = \langle 0_+|0_-\rangle^{J_1} \exp\left[\int d\omega_p iJ^\mu_1(p)^* g_{\mu\nu} iJ^\nu_2(p)\right] \langle 0_+|0_-\rangle^{J_2}. \quad (2\text{–}3.51)$$

The dyadic representation for $g_{\mu\nu}$ given in Eq. (2–3.9) is not appropriate here since p^μ is now a null vector,

$$p^2 = 0. \qquad (2\text{–}3.52)$$

Let $\bar{p}^\mu$ be obtained from p^μ by reversing the motion of the photon,

$$\bar{p}^0 = p^0, \qquad \bar{p}^k = -p^k, \qquad \bar{p}^2 = 0. \qquad (2\text{–}3.53)$$

Then $p^\mu + \bar{p}^\mu$ and $p^\mu - \bar{p}^\mu$ are, respectively, a time-like and space-like vector. They are supplemented by two orthogonal unit space-like vectors $e^\mu_{p\lambda}$,

$$e^{\mu*}_{p\lambda} e_{\mu p\lambda'} = \delta_{\lambda\lambda'}, \qquad p_\mu e^\mu_{p\lambda} = 0, \qquad \bar{p}_\mu e^\mu_{p\lambda} = 0, \qquad (2\text{–}3.54)$$

to give the dyadic construction

$$g^{\mu\nu} = \frac{(p^\mu + \bar{p}^\mu)(p^\nu + \bar{p}^\nu)}{2p\bar{p}} - \frac{(p^\mu - \bar{p}^\mu)(p^\nu - \bar{p}^\nu)}{2p\bar{p}} + \sum_\lambda e^\mu_{p\lambda} e^{\nu*}_{p\lambda}$$

$$= \frac{p^\mu\bar{p}^\nu + p^\nu\bar{p}^\mu}{p\bar{p}} + \sum_\lambda e^\mu_{p\lambda} e^{\nu*}_{p\lambda}. \qquad (2\text{–}3.55)$$

We now use the photon source restriction,

$$p_\mu J^\mu(p) = 0, \qquad p_\mu J^\mu(p)^* = 0, \qquad (2\text{–}3.56)$$

to obtain

$$\int d\omega_p J^\mu_1(p)^* g_{\mu\nu} J^\nu_2(p) = \int d\omega_p J^\mu_1(p)^* \sum_\lambda e_{\mu p\lambda} e^*_{\nu p\lambda} J^\nu_2(p)$$

$$= \sum_{p\lambda} J^*_{1p\lambda} J_{2p\lambda}, \qquad (2\text{–}3.57)$$

which is the desired particle exchange form. It is also implied by (2–3.54) that the two $e^\mu_{p\lambda}$ have zero time component and, as spatial vectors, are perpendicular to **p**. This is a self-contained description of the two transverse excitations that are permitted to photons.

It has been recognized earlier that a massless particle of definite helicity is a concept that is invariant under proper orthochronous Lorentz transformations. One should be able to make more evident this aspect of the photon helicity states. And we should like to understand why the helicity states have appeared paired although no overt reference to spatial reflection has been made. Let us begin with the remark that the conservation condition imposed on $J^\mu(x)$ is satisfied identically if

$$J^\mu(x) = \partial_\nu M^{\mu\nu}(x), \qquad (2\text{–}3.58)$$

where

$$M^{\mu\nu}(x) = -M^{\nu\mu}(x). \qquad (2\text{--}3.59)$$

We also introduce the concept of the dual to an antisymmetrical tensor:

$$*M^{\mu\nu} = \tfrac{1}{2}\epsilon^{\mu\nu\kappa\lambda}M_{\kappa\lambda}(x), \qquad (2\text{--}3.60)$$

in which $\epsilon^{\mu\nu\kappa\lambda}$ is the totally antisymmetrical tensor that is normalized by

$$\epsilon^{0123} = +1. \qquad (2\text{--}3.61)$$

The operation of forming the dual has the following repetition property,

$$**M^{\mu\nu}(x) = -M^{\mu\nu}(x). \qquad (2\text{--}3.62)$$

The dual tensor is used to write

$$J^{\mu}(x) = \partial_{\nu}M^{\mu\nu}_{+1}(x) + \partial_{\nu}M^{\mu\nu}_{-1}(x), \qquad (2\text{--}3.63)$$

where

$$M^{\mu\nu}_{\pm1}(x) = \tfrac{1}{2}(M^{\mu\nu}(x) \mp i\,{}^*M^{\mu\nu}(x)) \qquad (2\text{--}3.64)$$

are such that

$$*M^{\mu\nu}_{\pm1}(x) = \pm iM^{\mu\nu}_{\pm1}(x). \qquad (2\text{--}3.65)$$

The latter property indicates that each of these objects has only three independent components, as illustrated by

$$M^{12}_{+1} = iM^{03}_{+1}, \qquad M^{12}_{-1} = -iM^{03}_{-1}. \qquad (2\text{--}3.66)$$

These components are complex numbers, of course, and

$$M^{\mu\nu}_{-1}(x) = M^{\mu\nu}_{+1}(x)^*. \qquad (2\text{--}3.67)$$

The decomposition given in (2–3.63), and indicated by

$$J^{\mu}(x) = J^{\mu}_{+1}(x) + J^{\mu}_{-1}(x), \qquad (2\text{--}3.68)$$

is an invariant one as far as continuous changes of coordinate systems are concerned. Using three-dimensional notation, we have the explicit constructions

$$\mathbf{J}_{\pm1}(x) = \nabla \times \mathbf{M}_{\pm1}(x) \pm i\partial_0\mathbf{M}_{\pm1}(x) \qquad (2\text{--}3.69)$$

and

$$\mathbf{J}_{\pm1}(p) = i\mathbf{p} \times \mathbf{M}_{\pm1}(p) \pm p^0\mathbf{M}_{\pm1}(p). \qquad (2\text{--}3.70)$$

The effectiveness of these sources in radiating a photon of helicity λ is measured by $(p^0 = |\mathbf{p}|)$:

$$\begin{aligned}
\mathbf{e}^*_{p\lambda} \cdot \mathbf{J}_{\pm1}(p) &= p^0[i\mathbf{e}^*_{p\lambda} \times (\mathbf{p}/|\mathbf{p}|) \pm \mathbf{e}^*_{p\lambda}] \cdot \mathbf{M}_{\pm1}(p) \\
&= p^0(\lambda \pm 1)\mathbf{e}^*_{p\lambda} \cdot \mathbf{M}_{\pm1}(p), \qquad (2\text{--}3.71)
\end{aligned}$$

according to (2–3.27). The necessity of matching the value of λ with ±1 shows that the ±1 labels on the sources do indeed refer to the unique helicities of the photons that are emitted or absorbed by these component sources.

Why can one not modify this photon description by omitting $J''_{-1}(x)$, say, and thereby produce a theory with only positive helicity photons? For the same reason that one cannot have a theory in which only positively charged particles occur; it would violate the principle of space-time uniformity. To discuss this point in more detail, consider the contribution to the vacuum amplitude associated with the emission and subsequent absorption of a positive helicity photon

$$i\int (dx)(dx')J^{\mu}_{+1}(x)^{*}_{1}\left[i\int d\omega_{p}e^{ip(x-x')}\right]g_{\mu\nu}J^{\nu}_{+1}(x')_{2}, \qquad (2\text{-}3.72)$$

where the causal labels 1, 2 have been displaced for greater clarity. We are justified in writing $g_{\mu\nu}$ since the equivalent polarization vector summation reduces to the appropriate positive helicity terms, in virtue of (2-3.71). The complete source coupling should be linear in

$$J^{\nu}_{+1}(x) = J^{\nu}_{+1}(x)_{1} + J^{\nu}_{+1}(x)_{2} \qquad (2\text{-}3.73)$$

and in

$$J^{\mu}_{+1}(x)^{*} = J^{\mu}_{+1}(x)^{*}_{1} + J^{\mu}_{+1}(x)^{*}_{2}. \qquad (2\text{-}3.74)$$

One might try to resist the inference that there is another coupling involving $J^{\nu}_{+1}(x')_{1}$ and $J^{\mu}_{+1}(x)^{*}_{2}$ by introducing the space-time extrapolation of (2-3.72) with an additional factor,

$$\eta(x^{0} - x^{0\prime}) = \begin{cases} 1, & x^{0} > x^{0\prime}, \\ 0, & x^{0} < x^{0\prime}, \end{cases} \qquad (2\text{-}3.75)$$

which is designed to eliminate source arrangements where the causal roles of J^{ν}_{+1} and $J^{\mu*}_{+1}$ are reversed. This step function does have an invariant meaning when x and x' are in time-like or null relation, but it is not invariant for space-like intervals, and its introduction would violate the principle of space-time uniformity. We cannot avoid recognizing the presence of the additional causal coupling term

$$i\int (dx)(dx')J^{\nu}_{+1}(x')_{1}\left[i\int d\omega_{p}e^{ip(x'-x)}\right]g_{\mu\nu}J^{\mu}_{+1}(x')^{*}_{2}, \qquad (2\text{-}3.76)$$

and the particles described here must also be of zero mass if a unique space-time extrapolation is to be achieved. These antiparticles are the negative helicity photons,

$$J^{\mu}_{+1}(x)^{*} = J^{\mu}_{-1}(x), \qquad (2\text{-}3.77)$$

and the additional term can be rewritten as

$$i\int (dx)(dx')J^{\mu}_{-1}(x)^{*}_{1}\left[i\int d\omega_{p}e^{ip(x-x')}\right]g_{\mu\nu}J^{\nu}_{-1}(x')_{2}. \qquad (2\text{-}3.78)$$

Furthermore, the analogous structures involving $J^{\mu*}_{+1}$, J^{ν}_{-1} and $J^{\mu*}_{-1}$, J^{ν}_{+1} equal zero since one or the other factor in the polarization vector summation will vanish. The result is to reconstitute the real source

$$J^{\nu}(x) = J^{\nu}_{+1}(x) + J^{\nu}_{-1}(x), \qquad (2\text{-}3.79)$$

appearing in the coupling

$$i \int (dx)(dx') J^{\mu}(x)_1 \left[i \int d\omega_p e^{ip(x-x')} \right] g_{\mu\nu} J^{\nu}(x')_2, \tag{2-3.80}$$

which is recognized as the causal particle exchange term in

$$\tfrac{1}{2} i \int (dx)(dx') J^{\mu}(x) D_+(x - x') J_{\mu}(x'). \tag{2-3.81}$$

As a corollary of this discussion we note that the sources

$$^*J^{\mu}_{+1}(x) = i J^{\mu}_{+1}(x), \qquad ^*J^{\mu}_{-1}(x) = -i J^{\mu}_{-1}(x) \tag{2-3.82}$$

will give an equivalent description of photon emission and absorption. This new source is represented by

$$^*J^{\mu}(x) = \partial_\nu \, ^*M^{\mu\nu}(x). \tag{2-3.83}$$

The nature of the transformation is also indicated by

$$^*J_{p\lambda} = (d\omega_p)^{1/2} (\mathbf{p}/|\mathbf{p}|) \times \mathbf{e}^*_{p\lambda} \cdot \mathbf{J}(p), \tag{2-3.84}$$

which makes explicit that the polarization vectors have been rotated through the angle $\pi/2$ about the photon direction of motion. If the rotation angle is φ, the transformation becomes

$$J^{\mu}(x) \rightarrow J^{\mu}(x) \cos \varphi + \, ^*J^{\mu}(x) \sin \varphi. \tag{2-3.85}$$

The substitution of $^*J^{\mu}$ for J^{μ} also has a simple effect upon the angular momentum labeled sources, $J_{p^0 jm\lambda}$. We first remark that

$$\mathbf{J}(x) = \nabla \times \mathbf{M}(x) + \partial_0 \, ^*\mathbf{M}(x) \rightarrow \nabla \times \mathbf{M}(x) - ip^0 \, ^*\mathbf{M}(x), \tag{2-3.86}$$

where the latter substitution indicates the effective value in the integrals that compose $J_{p^0 jm\lambda}$. Similarly, we have

$$(i/p^0)\nabla \times \mathbf{J}(x) = \nabla \times \, ^*\mathbf{M}(x) + (i/p^0)(\nabla\nabla \cdot - \nabla^2)\mathbf{M}(x)$$
$$\rightarrow \nabla \times \, ^*\mathbf{M}(x) + ip^0 \mathbf{M}(x), \tag{2-3.87}$$

which uses the effective value $-\nabla^2 \rightarrow \mathbf{p}^2 = (p^0)^2$ and the property $\mathbf{L} \cdot \nabla = 0$. The substitution $J^{\mu} \rightarrow \, ^*J^{\mu}$, which is equivalent to $\mathbf{M} \rightarrow \, ^*\mathbf{M}$, $^*\mathbf{M} \rightarrow -\mathbf{M}$, interchanges these vectorial structures and transforms the two sources according to

$$J_{p^0 jm1} \rightarrow J_{p^0 jm2}, \qquad J_{p^0 jm2} \rightarrow -J_{p^0 jm1}. \tag{2-3.88}$$

The more general substitution (2-3.85) gives the rotation

$$J_{p^0 jm1} \rightarrow J_{p^0 jm1} \cos \varphi + J_{p^0 jm2} \sin \varphi,$$
$$J_{p^0 jm2} \rightarrow -J_{p^0 jm1} \sin \varphi + J_{p^0 jm2} \cos \varphi. \tag{2-3.89}$$

In the general space-time form of the vacuum amplitude, Eq. (2-3.45), sources need not emit and absorb photons and indeed may be incapable of doing so if they vary too slowly in time. It is the principle of space-time uniformity which thus asserts the physical unity between collision mechanisms that do liberate enough energy to create a particle and those otherwise analogous mechanisms that happen to have an insufficient energy supply. To illustrate the new physical information that is obtained in this manner we consider photon sources that vary very slowly in time. The way is prepared for this limit by writing

$$\int (dx)(dx') J^\mu(x) D_+(x - x') J_\mu(x')$$

$$= \int (d\mathbf{x})(d\mathbf{x}') \, dx^0 \, d\tau J^\mu(\mathbf{x}, x^0 + \tfrac{1}{2}\tau) D_+(\mathbf{x} - \mathbf{x}', \tau) J_\mu(\mathbf{x}', x^0 - \tfrac{1}{2}\tau), \quad (2\text{-}3.90)$$

where, as a consequence of (2-1.45),

$$D_+(\mathbf{x} - \mathbf{x}', \tau) = \frac{i}{4\pi^2} \frac{1}{|\mathbf{x} - \mathbf{x}'|} \int_0^\infty dp^0 \sin (p^0|\mathbf{x} - \mathbf{x}'|) e^{-ip^0|\tau|}. \quad (2\text{-}3.91)$$

This structure indicates that the scale of significant τ variation is set by $|\mathbf{x} - \mathbf{x}'|$. If the sources vary little, in the time intervals that are associated with the distances characteristic of the instantaneous distribution, one can ignore the τ dependence in $J^\mu(\mathbf{x}, x^0 \pm \tfrac{1}{2}\tau)$ and evaluate

$$\int_{-\infty}^\infty d\tau D_+(\mathbf{x} - \mathbf{x}', \tau) = \frac{1}{2\pi^2} \frac{1}{|\mathbf{x} - \mathbf{x}'|} \int_0^\infty dp^0 \frac{\sin p^0|\mathbf{x} - \mathbf{x}'|}{p^0}$$

$$= \frac{1}{4\pi} \frac{1}{|\mathbf{x} - \mathbf{x}'|}. \quad (2\text{-}3.92)$$

This gives the following form to the vacuum amplitude:

$$\langle 0_+|0_-\rangle^J = \exp\left[-i\int dx^0 E(x^0)\right], \quad (2\text{-}3.93)$$

where

$$E(x^0) = -\tfrac{1}{2}\int (d\mathbf{x})(d\mathbf{x}') J^\mu(\mathbf{x}, x^0) \frac{1}{4\pi|\mathbf{x} - \mathbf{x}'|} J_\mu(\mathbf{x}', x^0). \quad (2\text{-}3.94)$$

One recognizes here the accumulated phase change of a state that has a time variable energy, $E(x^0)$. When a steady-state regime is established, we are led to associate with it the energy value

$$E = \tfrac{1}{2}\int (d\mathbf{x})(d\mathbf{x}') \frac{J^0(\mathbf{x})J^0(\mathbf{x}') - \mathbf{J}(\mathbf{x}) \cdot \mathbf{J}(\mathbf{x}')}{4\pi|\mathbf{x} - \mathbf{x}'|}, \quad (2\text{-}3.95)$$

which is a statement of the Coulomb and Ampèrian laws of charge and current interactions. This shows how the principle of space-time uniformity provides the logical connection between the properties of photons and the characteristics of quasi-stationary charge distributions.

There is one subtlety here we should not overlook. One cannot produce a completely arbitrary static charge distribution. The local conservation condition

$\partial_\mu J^\mu = 0$ implies conservation of the total charge

$$Q = \int d\sigma_\mu J^\mu(x), \qquad (2\text{--}3.96)$$

provided the source is confined to some finite spatial region. Being zero in the initial vacuum state, the total charge remains zero. We may picture initially compensating positive and negative charge distributions being separated, moved about freely, and then recombined. But there is another way of viewing the introduction of a charge distribution into an empty region. It requires recognizing more explicitly than is usual that a physical description refers only to the finite space-time region which is under the experimenter's control. The initial and final vacuum states pertain to a bounded three-dimensional region. Life goes on, outside the walls. We thus appreciate that an arbitrary charge distribution can be produced by the transport of charge across the boundary, into the region of interest, and that this charge distribution can be dismantled ultimately by withdrawing it across the boundary.

2–4 SPIN 2 PARTICLES. THE GRAVITON

Next in complexity after scalar and vector sources is the real symmetrical tensor source

$$T^{\mu\nu}(x) = T^{\nu\mu}(x). \qquad (2\text{--}4.1)$$

It has ten components. But they include the $3 + 1$ component vector source $\partial_\mu T^{\mu\nu}$ and the scalar source

$$T(x) = g_{\mu\nu} T^{\mu\nu}(x). \qquad (2\text{--}4.2)$$

When these are removed, the residual multiplicity of five is the anticipated one for spin 2 particles of nonzero mass, m. To carry out this program we exploit our experience with unit spin particles and write directly the physically necessary structure for the vacuum persistence probability. It is

$$|\langle 0_+|0_-\rangle^T|^2 = \exp\left[-\int d\omega_p \overline{T}^{\mu\nu}(p)^* \mathfrak{g}_{\mu\kappa}(p)\mathfrak{g}_{\nu\lambda}(p)\overline{T}^{\kappa\lambda}(p)\right], \qquad (2\text{--}4.3)$$

where

$$\mathfrak{g}_{\mu\nu}(p) = g_{\mu\nu} + (1/m^2)p_\mu p_\nu, \qquad p^\mu \mathfrak{g}_{\mu\nu}(p) = 0, \qquad g^{\mu\nu}\mathfrak{g}_{\mu\nu}(p) = 3, \qquad (2\text{--}4.4)$$

and

$$\overline{T}^{\mu\nu}(p) = T^{\mu\nu}(p) - \tfrac{1}{3}g^{\mu\nu}\mathfrak{g}_{\rho\sigma}(p)T^{\rho\sigma}(p) \qquad (2\text{--}4.5)$$

obeys

$$\mathfrak{g}_{\mu\nu}(p)\overline{T}^{\mu\nu}(p) = 0. \qquad (2\text{--}4.6)$$

In the rest frame of the momentum p^μ, $\mathfrak{g}_{\mu\nu}(p)$ projects onto three-dimensional space, as detailed in (2–3.7). Accordingly, the only source components that contribute in (2–4.3) are the six $\overline{T}_{kl}$, which have a vanishing diagonal sum, in view of (2–4.6). Here is the fivefold multiplicity associated with spin 2.

An alternative writing of (2–4.3) is given by

$$\overline{T}^{\mu\nu}(p)^{*}\mathfrak{g}_{\mu\kappa}(p)\mathfrak{g}_{\nu\lambda}(p)\overline{T}^{\kappa\lambda}(p) = T^{\mu\nu}(p)^{*}\Pi_{\mu\nu,\kappa\lambda}(p)T^{\kappa\lambda}(p), \qquad (2\text{–}4.7)$$

in which

$$\Pi_{\mu\nu,\kappa\lambda}(p) = \tfrac{1}{2}(\mathfrak{g}_{\mu\kappa}(p)\mathfrak{g}_{\nu\lambda}(p) + \mathfrak{g}_{\nu\kappa}(p)\mathfrak{g}_{\mu\lambda}(p)) - \tfrac{1}{3}\mathfrak{g}_{\mu\nu}(p)\mathfrak{g}_{\kappa\lambda}(p). \qquad (2\text{–}4.8)$$

Some properties of the latter are

$$g^{\mu\nu}\Pi_{\mu\nu,\kappa\lambda}(p) = 0, \qquad g^{\mu\kappa}g^{\nu\lambda}\Pi_{\mu\nu,\kappa\lambda}(p) = 5, \qquad (2\text{–}4.9)$$

and

$$\Pi_{\mu\nu,\kappa\lambda}(p)\Pi^{\kappa\lambda}{}_{\rho\sigma}(p) = \Pi_{\mu\nu,\rho\sigma}(p). \qquad (2\text{–}4.10)$$

The projection matrix character of $\Pi_{\mu\nu,\rho\sigma}(p)$ leads to the dyadic representation

$$\Pi^{\mu\nu,\rho\sigma}(p) = \sum_{\lambda} e_{p\lambda}^{\mu\nu}e_{p\lambda}^{\rho\sigma*}, \qquad (2\text{–}4.11)$$

in which the symmetric tensors $e_{p\lambda}^{\mu\nu}$ obey

$$p_{\mu}e_{p\lambda}^{\mu\nu} = 0, \qquad g_{\mu\nu}e_{p\lambda}^{\mu\nu} = 0, \qquad e_{p\lambda}^{\mu\nu*}e_{\mu\nu p\lambda'} = \delta_{\lambda\lambda'}, \qquad (2\text{–}4.12)$$

and are five in number. The sources for specific states are then identified as

$$T_{p\lambda} = (d\omega_{p})^{1/2}e_{p\lambda}^{\mu\nu*}T_{\mu\nu}(p). \qquad (2\text{–}4.13)$$

When helicity states are used in the vector dyadic construction

$$\mathfrak{g}^{\mu\nu}(p) = \sum_{\lambda=-1}^{+1} e_{p\lambda}^{\mu}e_{p\lambda}^{\nu*}, \qquad e_{p\lambda}^{\nu*} = (-1)^{\lambda}e_{p-\lambda}^{\nu}, \qquad (2\text{–}4.14)$$

we get

$$\Pi^{\mu\nu,\rho\sigma}(p) = \sum_{\lambda\lambda'} e_{p\lambda\lambda'}^{\mu\nu}e_{p\lambda\lambda'}^{\rho\sigma*}, \qquad (2\text{–}4.15)$$

where

$$e_{p\lambda\lambda'}^{\mu\nu} = \tfrac{1}{2}(e_{p\lambda}^{\mu}e_{p\lambda'}^{\nu} + e_{p\lambda'}^{\mu}e_{p\lambda}^{\nu}) - \tfrac{1}{3}(-1)^{\lambda}\delta_{-\lambda\lambda'}\sum_{\lambda_1}(-1)^{\lambda_1}e_{p\lambda_1}^{\mu}e_{p-\lambda_1}^{\nu} \qquad (2\text{–}4.16)$$

obeys the relation

$$\sum_{\lambda}(-1)^{\lambda}e_{p-\lambda\lambda}^{\mu\nu} = -2e_{p+1-1}^{\mu\nu} + e_{p00}^{\mu\nu} = 0. \qquad (2\text{–}4.17)$$

The spin 2 helicity states are then identified as

$$
\begin{aligned}
e_{p\pm2}^{\mu\nu} &= e_{p\pm1\pm1}^{\mu\nu} = e_{p\pm1}^{\mu}e_{p\pm1}^{\nu}, \\
e_{p\pm1}^{\mu\nu} &= 2^{1/2}e_{p\pm10}^{\mu\nu} = 2^{-1/2}(e_{p\pm1}^{\mu}e_{p0}^{\nu} + e_{p0}^{\mu}e_{p\pm1}^{\nu}), \\
e_{p0}^{\mu\nu} &= (3/2)^{1/2}e_{p00}^{\mu\nu} = 6^{-1/2}(e_{p+1}^{\mu}e_{p-1}^{\nu} + e_{p-1}^{\mu}e_{p+1}^{\nu} + 2e_{p0}^{\mu}e_{p0}^{\nu}).
\end{aligned}
\qquad (2\text{–}4.18)
$$

The detailed structure of the vacuum probability amplitude that leads to the probability (2–4.3) is written out as

$$\langle 0_{+}|0_{-}\rangle = \exp[iW(T)] \qquad (2\text{–}4.19)$$

with

$$W(T) = \tfrac{1}{2} \int (dx)(dx')[T^{\mu\nu}(x)\Delta_+(x-x')T_{\mu\nu}(x')$$

$$+ (2/m^2)\partial_\nu T^{\mu\nu}(x)\Delta_+(x-x')\partial^{\lambda'}T_{\mu\lambda}(x')$$

$$+ (1/m^4)\partial_\mu\partial_\nu T^{\mu\nu}(x)\Delta_+(x-x')\partial'_\kappa\partial'_\lambda T^{\kappa\lambda}(x')$$

$$- \tfrac{1}{3}(T(x) - m^{-2}\partial_\mu\partial_\nu T^{\mu\nu}(x))\Delta_+(x-x')(T(x') - m^{-2}\partial'_\kappa\partial'_\lambda T^{\kappa\lambda}(x'))].$$

$$(2\text{-}4.20)$$

In order that this expression continue to exist in the limit as $m \to 0$, we must have

$$\partial_\nu T^{\mu\nu}(x) = m2^{-1/2}J^\mu(x), \qquad \partial_\mu J^\mu(x) = m(3^{1/2}K(x) - 2^{-1/2}T(x)), \quad (2\text{-}4.21)$$

where $J^\mu(x)$ and $K(x)$ are independent sources at $m = 0$. The particular linear combination of the two scalar sources $K(x)$, $T(x)$ is chosen to eliminate any coupling between them. This is evident in the limiting form

$$W(T) \to \tfrac{1}{2} \int (dx)(dx')[T^{\mu\nu}(x)D_+(x-x')T_{\mu\nu}(x') - \tfrac{1}{2}T(x)D_+(x-x')T(x')$$

$$+ J^\mu(x)D_+(x-x')J_\mu(x') + K(x)D_+(x-x')K(x')], \qquad (2\text{-}4.22)$$

where

$$\partial_\mu T^{\mu\nu}(x) = 0, \qquad \partial_\mu J^\mu(x) = 0. \qquad (2\text{-}4.23)$$

We see before us the invariant decomposition that the five helicity states, accessible to a massive particle of spin 2, undergo as $m \to 0$, falling into the three groups ± 2, ± 1, 0.

The massless particle of helicity ± 2 will be identified as the graviton. Its source description is given by

$$W(T) = \tfrac{1}{2} \int (dx)(dx')[T^{\mu\nu}(x)D_+(x-x')T_{\mu\nu}(x') - \tfrac{1}{2}T(x)D_+(x-x')T(x')],$$

$$(2\text{-}4.24)$$

$$\partial_\mu T^{\mu\nu}(x) = 0.$$

We insert a treatment of gravitons that begins with this characterization. The causal source arrangement

$$T^{\mu\nu}(x) = T_1^{\mu\nu}(x) + T_2^{\mu\nu}(x) \qquad (2\text{-}4.25)$$

gives the usual factorization of the vacuum amplitude:

$$\langle 0_+|0_-\rangle^T = \langle 0_+|0_-\rangle^{T_1} \exp\left[\int d\omega_p iT_1^{\mu\nu}(p)^*(g_{\mu\rho}g_{\nu\sigma} - \tfrac{1}{2}g_{\mu\nu}g_{\rho\sigma})iT_2^{\rho\sigma}(p)\right]\langle 0_+|0_-\rangle^{T_2},$$

$$(2\text{-}4.26)$$

where each component source obeys

$$p_\mu T^{\mu\nu}(p) = 0. \qquad (2\text{-}4.27)$$

This source restriction and the dyadic form (2–3.55) are combined to obtain the effective replacement

$$\tfrac{1}{2}(g^{\mu\rho}g^{\nu\sigma} + g^{\nu\rho}g^{\mu\sigma} - g^{\mu\nu}g^{\rho\sigma}) \to \sum_{\lambda\lambda'} e^{\mu\nu}_{p\lambda\lambda'} e^{\rho\sigma*}_{p\lambda\lambda'}, \qquad (2\text{–}4.28)$$

where, on using helicity states, $\lambda = \pm 1$,

$$e^{\mu\nu}_{p\lambda\lambda'} = \tfrac{1}{2}\left(e^{\mu}_{p\lambda}e^{\nu}_{p\lambda'} + e^{\mu}_{p\lambda'}e^{\nu}_{p\lambda} - \delta_{-\lambda\lambda'}\sum_{\lambda_1} e^{\mu}_{p\lambda_1}e^{\nu}_{p-\lambda_1}\right). \qquad (2\text{–}4.29)$$

The three independent tensors contained here are

$$e^{\mu\nu}_{p+1-1} = 0, \qquad (2\text{–}4.30)$$

and

$$e^{\mu\nu}_{p\pm2} = e^{\mu\nu}_{p\pm1\pm1} = e^{\mu}_{p\pm1}e^{\nu}_{p\pm1}, \qquad (2\text{–}4.31)$$

which represent the two helicity states of the graviton.

The graviton is unknown, as yet, to experimental science. Nevertheless, we shall accept it and its conjectured properties as the proper starting point for the theory of gravitational phenomena, just as the photon with its attributes initiates the theory of electromagnetic phenomena. The evidence for the existence of the graviton is indirect, but impressive. To indicate its nature we present the following parable: "The laws of quantum mechanics and relativity have been well established, but the interaction properties of electric charges are known only under quasi-static conditions. Two physicists, Max Stone and Ichirō Ido, point out that all such properties would follow from the postulated existence of a certain particle, on using the principles of source theory. They predict that the particle will one day be discovered. Others dismiss this suggestion as unwarranted speculation. The issue remains undecided."

The postulated existence of the graviton leads first and foremost to the source restriction $\partial_\mu T^{\mu\nu}(x) = 0$, which, as in the photon discussion, states the existence of a general physical law. It is a conservation law, affirming the constancy of the vector

$$P^\nu = \int d\sigma_\mu T^{\mu\nu}(x). \qquad (2\text{–}4.32)$$

The notation already indicates that there is only one conceivable identification of this vectorial property—it is energy-momentum. Unlike photon sources, which have a unique measure through the electric charge interpretation, graviton sources are confronted with an independent scale originating in the mechanical significance of $T_{\mu\nu}$. We provide an empirical conversion factor by writing

$$T^{\mu\nu}{}_{\text{grav.}} = \kappa^{1/2}T^{\mu\nu}{}_{\text{mech.}}. \qquad (2\text{–}4.33)$$

Again in contrast with electric charge, energy or mass is intrinsically positive. The establishment of a graviton source distribution in an initial vacuum situation can only be realized through the transport of energy and momentum into the region of interest, through the boundaries that delimit this domain. We consider a slowly varying distribution of graviton sources and deduce, as the analogue of (2–3.94), the energy

$$E(x^0) = -\frac{\kappa}{8\pi} \int (dx)(dx') \left[T^{\mu\nu}(\mathbf{x}, x^0) \frac{1}{|\mathbf{x} - \mathbf{x}'|} T_{\mu\nu}(\mathbf{x}', x^0) \right.$$

$$\left. - \tfrac{1}{2}T(\mathbf{x}, x^0) \frac{1}{|\mathbf{x} - \mathbf{x}'|} T(\mathbf{x}', x^0) \right], \qquad (2\text{–}4.34)$$

where the mechanical measure of graviton sources is used.

In the following astronomical applications we are concerned with the interaction between two bodies, one of which (the "Sun") has dimensions that are effectively negligible and is characterized by the single source component $T^{00}(\mathbf{x})$, such that

$$\int (dx) T^{00}(\mathbf{x}) = M. \qquad (2\text{–}4.35)$$

The interaction energy between the Sun, stationed at the origin, and the second test body with source distribution $t_{\mu\nu}(\mathbf{x}, x^0)$ is given by

$$E_{\text{int.}}(x^0) = -\frac{\kappa}{8\pi} M \int (dx) \frac{1}{|\mathbf{x}|} [2t^{00}(\mathbf{x}, x^0) + t(\mathbf{x}, x^0)], \qquad (2\text{–}4.36)$$

where

$$2t^{00} + t = t^{00} + t_{kk}. \qquad (2\text{–}4.37)$$

When a body that moves rigidly has momentum p^μ,

$$t^{\mu\nu} = \sigma p^\mu p^\nu, \qquad (2\text{–}4.38)$$

where σ is an invariant measure of the mass distribution. For a stationary body of mass m, with dimensions that are negligible compared to R, the distance from the origin, we get

$$E_{\text{int.}} = -\frac{\kappa}{8\pi} \frac{Mm}{R}. \qquad (2\text{–}4.39)$$

This is identifiable as the Newtonian potential energy of attracting masses, where

$$\kappa/8\pi = G = 6.67 \times 10^{-8} \text{ cm}^3/\text{g sec}^2$$
$$= (1.62 \times 10^{-33} \text{ cm})^2. \qquad (2\text{–}4.40)$$

The second version refers to atomic units, in which $\hbar = c = 1$.

We shall now use elementary considerations to reproduce the four observational tests of the Einsteinian modification of Newtonian theory.

1. A slowly moving atom of mass m has the total energy $m - (GM/R)m$ in the neighborhood of the body with mass M. The energy released in an internal transformation is thus reduced by the factor $1 - (GM/R)$. This is the gravitational red shift.

2. Let the test body be a light beam for which $t = \sigma p^2 = 0$. The interaction energy with the Sun thus exceeds its Newtonian value (replacing mass with total energy) by a factor of two. That is also the increase of the deflection angle of light over the Newtonian value, which is Einstein's result. For a direct calculation we compare the acquired transverse momentum with the longitudinal momentum of the beam, which passes at a distance ρ from the Sun. The deflection angle is

$$\Delta\varphi = 2GM\left(-\frac{\partial}{\partial\rho}\right)\int_{-\infty}^{\infty}\frac{dz}{(z^2+\rho)^{1/2}} = 4GM/\rho. \qquad (2\text{-}4.41)$$

3. The same interaction reduces the speed of light by the factor $1 - 2(GM/R)$, since the energy of a photon is $|\mathbf{p}|(1 - 2GM/R)$ and differentiation with respect to $\mathbf{p}$ gives the velocity. This effect has been observed by measuring time delays in radar echoes from the inner planets. We consider the superior conjunction of a planet, with the line of sight from the earth passing at distance ρ from the Sun. Then the anticipated additional time delay for the echo is

$$\Delta t = 4GM\int_0^{z_e}\frac{dz}{(z^2+\rho^2)^{1/2}} + 4GM\int_0^{z_p}\frac{dz}{(z^2+\rho^2)^{1/2}}$$
$$= 4GM\log\left(\frac{2z_e}{\rho}\frac{2z_p}{\rho}\right), \qquad (2\text{-}4.42)$$

where z_e and z_p are the distances, from the point of closest approach to the Sun, to the earth and the planet, respectively. The coefficient in the differential relation

$$-\frac{d\,\Delta t}{d\rho} = \frac{8GM}{\rho} \qquad (2\text{-}4.43)$$

has been verified with fair accuracy.

4. The most subtle and interesting test is, of course, the perihelion precession of planetary orbits. We first consider the correction to the Newtonian potential energy

$$V = -GMm/R \qquad (2\text{-}4.44)$$

that is produced by the motion of the planet. For small speeds,

$$T = \mathbf{p}^2/2m \ll m, \qquad (2\text{-}4.45)$$

we have

$$t_{kk} \approx (\mathbf{p}/m)^2 t^{00} = (2T/m)t^{00}, \qquad (2\text{-}4.46)$$

where

$$\int(d\mathbf{x})t_{\text{kin.}}^{00}(\mathbf{x}) = m + T. \qquad (2\text{-}4.47)$$

These effects correct the Newtonian potential to

$$V\left(1+\frac{2T}{m}\right)\left(1+\frac{T}{m}\right) \simeq V\left(1+\frac{3T}{m}\right). \tag{2-4.48}$$

There is a comparable relativistic modification of the kinetic energy, given by

$$(\mathbf{p}^2 + m^2)^{1/2} - m \simeq T - T^2/2m. \tag{2-4.49}$$

And, finally, there is the contribution to the energy density t^{00} that is associated with the gravitational interaction between the planet and the Sun. This is not localized on either mass, but is distributed in space in a way that can be calculated with sufficient precision from the Newtonian field strength:

$$\nabla\left[\frac{M}{|\mathbf{x}|} + \frac{m}{|\mathbf{x}-\mathbf{R}|}\right]. \tag{2-4.50}$$

The interaction energy density is proportional to the mutual term in the squared field strength. It is given by

$$t^{00}_{\text{int.}}(\mathbf{x}) = -\frac{G}{4\pi}\,\nabla\,\frac{M}{|\mathbf{x}|}\cdot\nabla\,\frac{m}{|\mathbf{x}-\mathbf{R}|}, \tag{2-4.51}$$

as one verifies by integration:

$$V = -\frac{G}{4\pi}\int(d\mathbf{x})\,\nabla\,\frac{M}{|\mathbf{x}|}\cdot\nabla\,\frac{m}{|\mathbf{x}-\mathbf{R}|} = \frac{G}{4\pi}\int(d\mathbf{x})\,\frac{M}{|\mathbf{x}|}\,\nabla^2\,\frac{m}{|\mathbf{x}-\mathbf{R}|}$$

$$= -\frac{GMm}{R}. \tag{2-4.52}$$

The energy of interaction between the mass M and this distributed mass is

$$-GM\int(d\mathbf{x})\,\frac{1}{|\mathbf{x}|}\left(-\frac{GMm}{4\pi}\right)\nabla\,\frac{1}{|\mathbf{x}|}\cdot\nabla\,\frac{1}{|\mathbf{x}-\mathbf{R}|}$$

$$= \frac{G^2M^2m}{4\pi}\int(d\mathbf{x})\,\frac{1}{2}\,\nabla\,\frac{1}{|\mathbf{x}|^2}\cdot\nabla\,\frac{1}{|\mathbf{x}-\mathbf{R}|}$$

$$= \frac{1}{2}\,\frac{G^2M^2m}{R^2} = \frac{V^2}{2m}. \tag{2-4.53}$$

All additional interaction terms are exhibited in

$$V(3T/m) - T^2/2m + V^2/2m. \tag{2-4.54}$$

This can be simplified by using the nonrelativistic energy relation, $E = T + V$, which enables T to be eliminated in favor of V. An additional constant multiple of V does not produce perihelion precession; it only changes slightly the scale of the orbit. It is the V^2 term that gives the significant deviation from the

Newtonian potential, and perihelion precession. The resulting effective potential
is

$$V_{\text{eff.}} = V - 3V^2/m. \tag{2-4.55}$$

Now, the equation of an orbit can be written

$$\frac{d^2u}{d\varphi^2} + u = -\frac{1}{L_1^2}\frac{d}{du}\frac{V_{\text{eff.}}}{m}, \tag{2-4.56}$$

where

$$u = 1/R \tag{2-4.57}$$

and L_1 is the angular momentum per unit planetary mass (it is often designated
by h). Here we have

$$\frac{d}{du}(V_{\text{eff.}}/m) = -GM - 6G^2M^2u \tag{2-4.58}$$

and

$$\frac{d^2u}{d\varphi^2} + \left(1 - \frac{6G^2M^2}{L_1^2}\right)u = \frac{GM}{L_1^2}. \tag{2-4.59}$$

We see that the essential deviation from Newtonian behavior is a reduction of
the angle scale by the factor $1 - 3G^2M^2/L_1^2$, which requires that φ increase by
more than 2π between successive perihelions. This perihelion precession angle is

$$\Delta\varphi = 6\pi G^2M^2/L_1^2 \tag{2-4.60}$$

which is exactly Einstein's result. [He gave it in terms of the semimajor
axis a, the period T, and the eccentricity e. The connection is $GM/L_1 = 2\pi(a/T)(1 - e^2)^{-1/2}$.]

2-5 PARTICLES WITH ARBITRARY INTEGER SPIN

In discussing unit spin and spin 2 particles, it was natural to replace the scalar
source of spinless particles by vector and tensor sources. The response of a vector source, for example, to a homogeneous infinitesimal Lorentz transformation
is given by (2-3.22), which we write as

$$\delta J^\lambda(x) = i\tfrac{1}{2}\,\delta\omega^{\mu\nu}[(x_\mu(1/i)\partial_\nu - x_\nu(1/i)\partial_\mu)J^\lambda(x) + (1/i)(\delta_\mu^\lambda g_{\nu\kappa} - \delta_\nu^\lambda g_{\mu\kappa})J^\kappa(x)]. \tag{2-5.1}$$

Clearly exhibited here is a four-dimensional version of orbital and spin angular
momenta. This is a particular infinitesimal illustration of the general linear
response of a multicomponent object to Lorentz transformations,

$$\bar{S}(\bar{x}) = L(l)S(x), \tag{2-5.2}$$

where

$$\bar{x}^\mu = l^\mu{}_\nu x^\nu - \epsilon^\mu, \qquad l^\mu{}_\kappa g_{\mu\nu}l^\nu{}_\lambda = g_{\kappa\lambda} \tag{2-5.3}$$

details the typical inhomogeneous Lorentz transformation. One can always choose the elements of a suitably multicomponent source to be real, and this property is maintained by a real transformation matrix $L(l)$. Corresponding to the composition property of successive Lorentz transformations,

$$\bar{\bar{x}}^\mu = l_1{}^\mu{}_\nu \bar{x}^\nu - \epsilon_1{}^\mu, \qquad \bar{x}^\nu = l_2{}^\nu{}_\lambda x^\lambda - \epsilon_2{}^\nu, \qquad (2\text{--}5.4)$$

namely

$$\bar{\bar{x}}^\mu = (l_1 l_2)^\mu{}_\lambda x^\lambda - (\epsilon_1{}^\mu + l_1{}^\mu{}_\nu \epsilon_2{}^\nu), \qquad (l_1 l_2)^\mu{}_\lambda = l_1{}^\mu{}_\nu l_2{}^\nu{}_\lambda, \qquad (2\text{--}5.5)$$

we have

$$\bar{\bar{S}}(\bar{\bar{x}}) = L(l_1)\bar{S}(\bar{x}), \qquad \bar{S}(\bar{x}) = L(l_2)S(x) \qquad (2\text{--}5.6)$$

and

$$\bar{\bar{S}}(\bar{\bar{x}}) = L(l_1)L(l_2)S(x) = L(l_1 l_2)S(x). \qquad (2\text{--}5.7)$$

It is in this sense that the finite matrices $L(l)$ give a matrix representation of the homogeneous Lorentz group.

For infinitesimal transformations

$$l^\mu{}_\nu = \delta^\mu_\nu + \delta\omega^\mu{}_\nu, \qquad \delta\omega_{\mu\nu} = -\delta\omega_{\nu\mu}, \qquad (2\text{--}5.8)$$

we write

$$L(l) = 1 + \tfrac{1}{2}i\,\delta\omega^{\mu\nu}s_{\mu\nu} \qquad (2\text{--}5.9)$$

and conclude that the matrices $s_{\mu\nu} = -s_{\nu\mu}$, which are imaginary when $L(l)$ is real, obey the commutation relations that express the composition properties of the six-parameter homogeneous group [cf. Eq. (1–3.10)]:

$$(1/i)[s_{\mu\nu}, s_{\kappa\lambda}] = g_{\mu\kappa}s_{\nu\lambda} - g_{\nu\kappa}s_{\mu\lambda} + g_{\nu\lambda}s_{\mu\kappa} - g_{\mu\lambda}s_{\nu\kappa}. \qquad (2\text{--}5.10)$$

The complete infinitesimal response of $S(x)$ is

$$\delta S(x) = i\,\delta\epsilon^\mu(1/i)\partial_\mu S(x)$$
$$+ i\tfrac{1}{2}\,\delta\omega^{\mu\nu}[x_\mu(1/i)\partial_\nu - x_\nu(1/i)\partial_\mu + s_{\mu\nu}]S(x). \qquad (2\text{--}5.11)$$

Comparison with (2–5.1) gives a (4×4)-dimensional example of the imaginary spin matrices $s_{\mu\nu}$,

$$(s_{\mu\nu})^\lambda{}_\kappa = (1/i)(\delta^\lambda_\mu g_{\nu\kappa} - \delta^\lambda_\nu g_{\mu\kappa}). \qquad (2\text{--}5.12)$$

As the following illustrations show,

$$(s_{12})^1{}_2 = -(s_{12})^2{}_1 = -i, \qquad (s_{03})^0{}_3 = +(s_{03})^3{}_0 = -i, \quad (2\text{--}5.13)$$

the matrices s_{kl} are antisymmetrical and therefore Hermitian, while the s_{0k} are symmetrical and skew-Hermitian. It was preordained that not all the $s_{\mu\nu}$ matrices could be Hermitian, for the discussion of Section 1–1 shows that the open structure of the Lorentz group precludes any finite-dimensional realization of the group. This injunction is not applicable to the attached Euclidean group, and indeed the correspondence $(x_4 = ix^0)$

$$s_{0k} = -is_{k4}, \qquad (s_{\mu\nu})^0{}_k = -i(s_{\mu\nu})_{4k}, \qquad (s_{\mu\nu})^k{}_0 = i(s_{\mu\nu})_{k4} \qquad (2\text{-}5.14)$$

does give Euclidean spin matrices,

$$(s_{\mu\nu})_{\lambda\kappa} = (1/i)(\delta_{\lambda\mu}\delta_{\nu\kappa} - \delta_{\lambda\nu}\delta_{\mu\kappa}), \qquad (2\text{-}5.15)$$

that are all imaginary, antisymmetrical, and Hermitian.

Such differential operator-matrix realizations of the Lorentz generators are in striking contrast with the operator structures given in (1–3.42) or (1–3.72), for there the finite skew-Hermitian matrices s_{0k} are replaced by Hermitian operators that are functions of momentum. How do we reconcile these very different forms? There must be a connecting transformation that preserves the commutation relations but does not maintain Hermiticity and therefore is not a unitary transformation. A suggestion of what is required comes from the following transformation, which is appropriate to a slowly moving particle $(p^0 \sim m)$:

$$\exp[\pm(\mathbf{p}\cdot\mathbf{s}/m)](-m\mathbf{r})\exp[\mp(\mathbf{p}\cdot\mathbf{s}/m)] = -m\mathbf{r} \pm [\mathbf{r}, \mathbf{p}\cdot\mathbf{s}]$$

$$- \frac{1}{2m}[[\mathbf{r}, \mathbf{p}\cdot\mathbf{s}], \mathbf{p}\cdot\mathbf{s}] + \cdots$$

$$\cong -m\mathbf{r} \pm i\mathbf{s} - \frac{1}{2m}\mathbf{s}\times\mathbf{p},$$

$$(2\text{-}5.16)$$

or

$$\exp[\pm(\mathbf{p}\cdot\mathbf{s}/m)]\left(-m\mathbf{r} + \frac{1}{2m}\mathbf{s}\times\mathbf{p}\right)\exp[\mp(\mathbf{p}\cdot\mathbf{s}/m)] \cong -m\mathbf{r} \pm i\mathbf{s}, \qquad (2\text{-}5.17)$$

illustrating how the momentum dependence is removed, at the expense of introducing skew-Hermitian operators. The following is the analogous statement for arbitrary momentum,

$$\exp[\pm\theta(\mathbf{p}\cdot\mathbf{s}/|\mathbf{p}|)]\left(-\mathbf{r}p^0 + \frac{1}{p^0+m}\mathbf{s}\times\mathbf{p}\right)\exp[\mp\theta(\mathbf{p}\cdot\mathbf{s}/|\mathbf{p}|)] = -\mathbf{r}p^0 \pm i\mathbf{s}, \qquad (2\text{-}5.18)$$

where

$$\sinh\theta = |\mathbf{p}|/m, \qquad \cosh\theta = p^0/m. \qquad (2\text{-}5.19)$$

Its verification proceeds most simply by considering components parallel and perpendicular to $\mathbf{p}$. The former reduces to the defining differential equation

$$d\theta(|\mathbf{p}|)/d|\mathbf{p}| = 1/p^0, \qquad (2\text{-}5.20)$$

and the latter to

$$\exp[\pm\theta(\mathbf{p}\cdot\mathbf{s}/|\mathbf{p}|)]\mathbf{s}\exp[\mp\theta(\mathbf{p}\cdot\mathbf{s}/|\mathbf{p}|)] = \frac{1}{m}\left[p^0\mathbf{s} - \frac{\mathbf{p}\mathbf{p}\cdot\mathbf{s}}{p^0+m} \pm i\mathbf{s}\times\mathbf{p}\right],$$

$$(2\text{-}5.21)$$

which describes the behavior of $\mathbf{s}$ under the "rotation" specified by θ. Incidentally, if a particle moves with momentum $\mathbf{p}$, along the third axis, the transformation to its rest frame is

$$\bar{x}^3 = x^3 \cosh \theta - x^0 \sinh \theta,$$
$$\bar{x}^0 = -x^3 \sinh \theta + x^0 \cosh \theta. \tag{2-5.22}$$

The appearance of $\pm i s_k$ in the role of

$$n_k = s^0{}_k \tag{2-5.23}$$

is easily understood. The commutation relations for the $s_{\mu\nu}$ are simplified by introducing the linear combinations indicated by

$$s_3^{(1)} = \tfrac{1}{2}(s_{12} + i{}^*s_{12}) = \tfrac{1}{2}(s_{12} + is_{03}),$$
$$s_3^{(2)} = \tfrac{1}{2}(s_{12} - i{}^*s_{12}) = \tfrac{1}{2}(s_{12} - is_{03}), \tag{2-5.24}$$

and their cyclic permutations. All four-dimensional commutation relations are summarized by the three-dimensional commutation properties of the two independent angular momenta,

$$\mathbf{s}^{(1)} \times \mathbf{s}^{(1)} = i\mathbf{s}^{(1)}, \qquad \mathbf{s}^{(2)} \times \mathbf{s}^{(2)} = i\mathbf{s}^{(2)}. \tag{2-5.25}$$

Conversely, we have the construction $(s_{12} = s_3, \ldots)$

$$\mathbf{s} = \mathbf{s}^{(1)} + \mathbf{s}^{(2)}, \qquad \mathbf{n} = i(\mathbf{s}^{(1)} - \mathbf{s}^{(2)}), \tag{2-5.26}$$

where the use of conventional Hermitian matrix representations for $\mathbf{s}^{(1,2)}$ gives Hermitian $\mathbf{s}$ matrices and skew-Hermitian $\mathbf{n}$ matrices. What we have encountered in (2-5.18) is the special situation in which $\mathbf{s}^{(2)} = 0$, or $\mathbf{s}^{(1)} = 0$. To deal with the more general possibility presented in (2-5.26), we must consider $\mathbf{s}$ to be the resultant of other spins. And, since it is often convenient to build the latter out of still more elementary spins, we present the following general theorem, in which the λ_α are any commuting objects that obey

$$\lambda_\alpha^2 = 1, \tag{2-5.27}$$

and

$$\mathbf{s} = \sum_{\alpha=1}^{\nu} \mathbf{s}_\alpha, \tag{2-5.28}$$

a superposition of independent spins. The theorem is

$$B^{-1}\left(-\mathbf{r}p^0 + \frac{1}{p^0 + m}\,\mathbf{s} \times \mathbf{p}\right) B = -\mathbf{r}p^0 + i\sum_\alpha \lambda_\alpha \mathbf{s}_\alpha, \tag{2-5.29}$$

where

$$B(p) = \exp\left[-\theta(\mathbf{p}/|\mathbf{p}|) \cdot \sum_\alpha \lambda_\alpha \mathbf{s}_\alpha\right]. \tag{2-5.30}$$

It is verified as before by using the individual relations

$$B^{-1}\mathbf{s}_\alpha B = \frac{1}{m}\left[p^0\mathbf{s}_\alpha - \frac{\mathbf{pp}\cdot\mathbf{s}_\alpha}{p^0+m} + i\lambda_\alpha\mathbf{s}_\alpha \times \mathbf{p}\right]. \tag{2-5.31}$$

The general matrix construction

$$\mathbf{s} = \sum_{\alpha=1}^{\nu}\mathbf{s}_\alpha, \qquad \mathbf{n} = i\sum_{\alpha=1}^{\nu}\lambda_\alpha\mathbf{s}_\alpha \tag{2-5.32}$$

clearly satisfies all the relevant commutation relations, including

$$\mathbf{n} \times \mathbf{n} = -i\mathbf{s}. \tag{2-5.33}$$

We note that the skew-Hermitian character of $\mathbf{n}$ is maintained with the use of Hermitian matrices for the λ_α, as well as with the numbers ± 1.

We have not described explicitly the symmetrization between $\mathbf{r}$ and p^0 since it would repeat the spin 0 discussion. The infinitesimal Lorentz transformation response of the state $\langle p|$ is now written as

$$\delta((p^0)^{1/2}\langle p|) = \left[i\,\delta\boldsymbol{\omega}\cdot\left(i\frac{\partial}{\partial\mathbf{p}}\times\mathbf{p}+\mathbf{s}\right)\right.$$
$$\left. + i\,\delta\mathbf{v}\cdot\left(-p^0 i\frac{\partial}{\partial\mathbf{p}} + \frac{1}{p^0+m}\mathbf{s}\times\mathbf{p}\right)\right](p^0)^{1/2}\langle p|, \tag{2-5.34}$$

where spin operators have been replaced by spin matrices, as symbolized by

$$\langle p|\mathbf{s} = \mathbf{s}\langle p|, \tag{2-5.35}$$

in which the matrices act upon unwritten indices in $\langle p|$. Next, we perform the transformation (there are constant factors to be specified)

$$(p^0)^{1/2}\langle p|0_-\rangle^S \sim B(p)S(p), \tag{2-5.36}$$

and derive

$$\delta S(p) = \left[\delta\boldsymbol{\omega}\cdot\left(\mathbf{p}\times\frac{\partial}{\partial\mathbf{p}}+i\mathbf{s}\right) + \delta\mathbf{v}\cdot\left(p^0\frac{\partial}{\partial\mathbf{p}}+i\mathbf{n}\right)\right]S(p). \tag{2-5.37}$$

On writing

$$S(p) = \int(dx)e^{-ipx}S(x), \tag{2-5.38}$$

this becomes

$$\delta S(x) = \tfrac{1}{2}\,\delta\omega^{\mu\nu}(x_\mu\partial_\nu - x_\nu\partial_\mu + is_{\mu\nu})S(x), \tag{2-5.39}$$

where the matrices $s_{\mu\nu}$ are those given in (2–5.32). The analogous construction of $\langle 0_+|p\rangle^S$ is [cf. Eq. (2–1.25)]

$$(p^0)^{1/2}\langle 0_+|p\rangle^S \sim S(p)*B(p), \tag{2-5.40}$$

which involves the Hermitian character of $B(p)$. The infinitesimal Lorentz transformation behavior of $S(x)^*$ is like that of $S(x)$, but with the matrices

$-s_{\mu\nu}^{*}$ replacing the $s_{\mu\nu}$. The use of matrix representations with imaginary $s_{\mu\nu}$ is required to be consistent with real $S(x)$, as we have mentioned before.

The compact notation used in writing (2–5.36) obscures an essential point. We are describing a particle of definite spin, but embed it in a larger system when we employ constructions like (2–5.26). Accordingly there must be present on the left of $B(p)S(p)$, say, an explicit selection of the states of interest. We shall illustrate this, and at the same time give a simple example of the connection between the present matrix approach and the earlier procedures, by choosing

$$s^{(1,2)} = \tfrac{1}{2}\sigma^{(1,2)}. \tag{2–5.41}$$

The resultant of the two spins of $\tfrac{1}{2}$ is either $s = 1$ or $s = 0$. If we wish to describe unit spin particles we must select them from the larger system. The familiar $s = 1$ triplet spin functions can be written as

$$\begin{aligned}
\chi_{\pm 1}(\sigma^{(1)}\sigma^{(2)}) &= \langle\sigma^{(1)}|\tfrac{1}{2}(1 \pm \sigma_3)|\sigma^{(2)}\rangle \\
\chi_0(\sigma^{(1)}\sigma^{(2)}) &= \langle\sigma^{(1)}|2^{-1/2}\sigma_1|\sigma^{(2)}\rangle.
\end{aligned} \tag{2–5.42}$$

An alternative version, which also involves the reality and symmetry of these functions, is

$$\chi_\lambda(\sigma^{(1)}\sigma^{(2)})^* = -i2^{-1/2}\langle\sigma^{(2)}|\sigma_2\boldsymbol{\sigma}\cdot\mathbf{e}_\lambda^*|\sigma^{(1)}\rangle, \tag{2–5.43}$$

where

$$\boldsymbol{\sigma}\cdot\mathbf{e}_{+1}^* = 2^{-1/2}(-\sigma_1 + i\sigma_2), \qquad \boldsymbol{\sigma}\cdot\mathbf{e}_{-1}^* = 2^{-1/2}(\sigma_1 + i\sigma_2),$$
$$\boldsymbol{\sigma}\cdot\mathbf{e}_0^* = \sigma_3. \tag{2–5.44}$$

The three space vectors thus defined are orthonormal in the sense

$$\mathbf{e}_\lambda^* \cdot \mathbf{e}_{\lambda'} = \delta_{\lambda\lambda'}. \tag{2–5.45}$$

We also note the singlet function:

$$\chi_0(\sigma^{(1)}\sigma^{(2)})^* = -i2^{-1/2}\langle\sigma^{(2)}|\sigma_2|\sigma^{(1)}\rangle, \tag{2–5.46}$$

which exploits the antisymmetry of the Pauli matrix σ_2, as (2–5.43) depends upon the symmetry of the three $\sigma_2\sigma_k$. The latter property is also expressed by

$$-\sigma_k{}^T = \sigma_2\sigma_k\sigma_2. \tag{2–5.47}$$

A corresponding decomposition of the four-component source into singlet and triplet functions is conveniently written as

$$S_{\sigma^{(1)}\sigma^{(2)}}(p) = -2^{-1/2}\langle\sigma^{(1)}|\sigma^\mu J_\mu(p)\sigma_2|\sigma^{(2)}\rangle, \tag{2–5.48}$$

where

$$\sigma_0 = -\sigma^0 = 1. \tag{2–5.49}$$

We now examine the unit spin particle source structure:

$$J_{p\lambda} = -i(d\omega_p)^{1/2}\chi_\lambda^* B(p)S(p), \tag{2–5.50}$$

in which appropriate factors have been supplied. A consistent use of the matrix notation gives

$$(d\omega_p)^{-1/2} J_{p\lambda} = \tfrac{1}{2} \, \mathrm{tr} \, [\boldsymbol{\sigma} \cdot \mathbf{e}_\lambda^* b(p) \sigma^\mu J_\mu(p) b(p)]$$

$$= e_{p\lambda}^{\mu*} J_\mu(p), \tag{2–5.51}$$

where

$$b(p) = \exp\left[-\tfrac{1}{2}\theta\boldsymbol{\sigma} \cdot \mathbf{p}/|\mathbf{p}|\right]. \tag{2–5.52}$$

We first note that

$$e_{p\lambda}^{\mu*} = \tfrac{1}{2} \, \mathrm{tr} \, [\boldsymbol{\sigma} \cdot \mathbf{e}_\lambda^* b(p) \sigma^\mu b(p)]$$

$$= \tfrac{1}{2} \, \mathrm{tr} \, [\sigma^\mu b(p) \boldsymbol{\sigma} \cdot \mathbf{e}_\lambda^* b(p)] \tag{2–5.53}$$

has the following property:

$$p_\mu e_{p\lambda}^{\mu*} = 0, \tag{2–5.54}$$

since

$$p^\mu \sigma_\mu = m(\cosh\theta + \sinh\theta\boldsymbol{\sigma} \cdot \mathbf{p}/|\mathbf{p}|)$$

$$= m[b(p)^{-1}]^2, \tag{2–5.55}$$

and

$$\mathrm{tr}\, \sigma_k = 0. \tag{2–5.56}$$

Next, we consider

$$e_{p\lambda}^{\mu*} e_{\mu p\lambda'} = \tfrac{1}{2} \, \mathrm{tr} \, (\sigma^\mu b\boldsymbol{\sigma} \cdot \mathbf{e}_\lambda^* b) \tfrac{1}{2} \, \mathrm{tr} \, (\sigma_\mu b\boldsymbol{\sigma} \cdot \mathbf{e}_{\lambda'} b), \tag{2–5.57}$$

in which the form of the second factor depends upon the Hermitian nature of $b(p)$ and the σ_μ. The following identity expresses the role of the four matrices $2^{-1/2}\sigma_\mu$ as an orthonormal basis for 2×2 matrices:

$$(\mathrm{tr}\, \sigma^\mu X)(\mathrm{tr}\, \sigma_\mu Y) = (\mathrm{tr}\, \sigma X) \cdot (\mathrm{tr}\, \sigma Y) - (\mathrm{tr}\, X)(\mathrm{tr}\, Y)$$

$$= 2[\mathrm{tr}\, (XY) - (\mathrm{tr}\, X)(\mathrm{tr}\, Y)]$$

$$= \det\, (X - Y) - \det\, (X + Y). \tag{2–5.58}$$

The multiplication properties of determinants, and the remark that

$$\det b(p) = 1, \tag{2–5.59}$$

which follows from (2–5.56), show that all reference to $b(p)$ disappears from (2–5.57), giving

$$e_{p\lambda}^{\mu*} e_{\mu p\lambda'} = \tfrac{1}{2} \, \mathrm{tr} \, (\boldsymbol{\sigma} \cdot \mathbf{e}_\lambda^* \boldsymbol{\sigma} \cdot \mathbf{e}_{\lambda'})$$

$$= \delta_{\lambda\lambda'}. \tag{2–5.60}$$

And, finally, let us consider

$$\sum_\lambda e_{p\lambda}^{\mu*} e_{p\lambda}^\nu = \tfrac{1}{2} \, \mathrm{tr} \, (b\sigma^\mu b\boldsymbol{\sigma}) \cdot \tfrac{1}{2} \, \mathrm{tr} \, (\boldsymbol{\sigma} b\sigma^\nu b)$$

$$= \tfrac{1}{2} \, \mathrm{tr} \, (\sigma^\mu b^2) \tfrac{1}{2} \, \mathrm{tr} \, (\sigma^\nu b^2) + \tfrac{1}{2}[\mathrm{tr} \, (\sigma^\mu \sigma^\nu) - (\mathrm{tr}\, \sigma^\mu)(\mathrm{tr}\, \sigma^\nu)]. \tag{2–5.61}$$

The individual traces here are

$$\tfrac{1}{2} \, \mathrm{tr} \, (\sigma^\mu b(p)^2) = -p^\mu/m \tag{2–5.62}$$

and

$$\tfrac{1}{2}[\text{tr } (\sigma^\mu \sigma^\nu) - (\text{tr } \sigma^\mu)(\text{tr } \sigma^\nu)] = g^{\mu\nu}, \tag{2-5.63}$$

which gives

$$\sum_\lambda e^{\mu *}_{p\lambda} e^\nu_{p\lambda} = g^{\mu\nu} + (1/m^2)p^\mu p^\nu = \bar{g}^{\mu\nu}(p). \tag{2-5.64}$$

We have now reproduced all the covariant properties of the three polarization vectors for unit spin. When the third axis in (2–5.44) is identified with the direction of the momentum vector **p**, the explicit expressions obtained from (2–5.53) are just the helicity labeled vectors (2–3.28, 29). Incidentally, on using the singlet rather than the triplet functions, we get the form

$$\tfrac{1}{2} \text{ tr } (\sigma^\mu b(p)^2) J_\mu(p) = -(1/m)p^\mu J_\mu(p), \tag{2-5.65}$$

which is the anticipated scalar combination.

As the basis for a corresponding treatment of arbitrary integer spin, we consider the spin combinations

$$\mathbf{s}^{(1)} = \sum_{\alpha=1}^n \tfrac{1}{2}\sigma_\alpha^{(1)}, \qquad \mathbf{s}^{(2)} = \sum_{\alpha=1}^n \tfrac{1}{2}\sigma_\alpha^{(2)}, \tag{2-5.66}$$

where the individual matrices act upon the appropriate index of the source

$$S_{\sigma_1^{(1)}\cdots\sigma_n^{(1)}\sigma_1^{(2)}\cdots\sigma_n^{(2)}}(p). \tag{2-5.67}$$

Since all the matrices $\sigma_\alpha^{(1,2)}$, $\alpha = 1, \ldots, n$, appear on the same footing, we impose a permissible symmetry restriction by requiring that (2–5.67) be unchanged by any permutation of the α indices, in which $\sigma_\alpha^{(1)}$ and $\sigma_\alpha^{(2)}$ are regarded as a unit. Thus, for $n = 2$, we have

$$S_{\sigma_1^{(1)}\sigma_2^{(1)}\sigma_1^{(2)}\sigma_2^{(2)}} = S_{\sigma_2^{(1)}\sigma_1^{(1)}\sigma_2^{(2)}\sigma_1^{(2)}}. \tag{2-5.68}$$

The simplest procedure is to replace each four-valued index pair $\sigma_\alpha^{(1)}\sigma_\alpha^{(2)}$ by a four-vector index μ_α, in the manner detailed for unit spin. This gives the equivalent source

$$S_{\mu_1\cdots\mu_n}(p), \tag{2-5.69}$$

which is unchanged by any permutation of the α indices.

The tensor of rank n as introduced here describes a larger system than a particle of definite spin. Part of the necessary reduction is produced by the projection factors $\bar{g}_{\mu\nu}(p)$ that appear separately for each vector index in the coupled source structure

$$S_1^{\mu_1\cdots\mu_n}(p)^* \bar{g}_{\mu_1\nu_1}(p) \cdots \bar{g}_{\mu_n\nu_n}(p) S_2^{\nu_1\cdots\nu_n}(p) \to S_{1k_1\cdots k_n}^* S_{2k_1\cdots k_n}, \tag{2-5.70}$$

where the second form refers to the rest frame of the momentum p^μ. The number of independent components possessed by the symmetrical three-dimensional tensor $S_{k_1\cdots k_n}$, namely $\tfrac{1}{2}(n+1)(n+2)$, agrees with the number of states exhibited by a symmetrical collection of n unit spins. The total spin quantum number ranges from $s = n$ through $s = n - 2, n - 4, \ldots$, terminating at 1 or 0 as n is odd or even, and $\sum (2s + 1) = \frac{1}{2}(n + 1)(n + 2)$. A combination of two unit spins into a null resultant corresponds to forming the trace of a pair of indices, as in $S_{kkk_3\cdots k_n}$. To remove this possibility and thereby select only $s = n$, we must make $S_{k_1\cdots k_n}$ traceless. The subtraction of the appropriate number of restrictions gives the independent component count

$$\tfrac{1}{2}(n + 1)(n + 2) - \tfrac{1}{2}(n - 1)n = 2n + 1, \qquad (2\text{-}5.71)$$

as expected for $s = n$.

The result of subtracting successive traces is indicated by the symmetrical form

$$\begin{aligned}
\bar{S}_{k_1\cdots k_n}x_{k_1} \cdots x_{k_n} &= S_{k_1\cdots k_n}x_{k_1} \cdots x_{k_n} \\
&\quad + c_{n1}x^2 S_{kkk_3\cdots k_n}x_{k_3} \cdots x_{k_n} \\
&\quad + c_{n2}(x^2)^2 S_{kkk'k'k_5\cdots k_n}x_{k_5} \cdots x_{k_n} + \cdots, \qquad (2\text{-}5.72)
\end{aligned}$$

where $x^2 = (x_k)^2$ and it is required that

$$\bar{S}_{kkk_3\cdots k_n} = 0. \qquad (2\text{-}5.73)$$

In view of the total symmetry of the tensor, this property guarantees the vanishing of the trace for any pair of indices. The problem thus posed is a familiar one. The polynomial of degree n given in (2-5.72) is a solution of Laplace's equation according to (2-5.73). With x^2 set equal to unity, it is a spherical harmonic of degree n. To identify the coefficients c_{nm}, it suffices to consider the single nonvanishing component $S_{33\cdots3} = 1$. With $x^2 = 1, x_3 = \mu$, we encounter the polynomial

$$\mu^n + c_{n1}\mu^{n-2} + \cdots, \qquad (2\text{-}5.74)$$

which must be proportional to Legendre's polynomial, $P_n(\mu)$. Hence,

$$m \le \tfrac{1}{2}n:$$

$$c_{nm} = \frac{(-1)^m}{m!} \frac{(2n - 2m)!}{(2n)!} \frac{n!}{(n - m)!} \frac{n!}{(n - 2m)!}. \qquad (2\text{-}5.75)$$

The reference to the rest frame is removed in

$$\begin{aligned}
\bar{S}_{\mu_1\cdots\mu_n}(p)x^{\mu_1} \cdots x^{\mu_n} &= S_{\mu_1\cdots\mu_n}(p)x^{\mu_1} \cdots x^{\mu_n} \\
&\quad + c_{n1}x^\lambda x_\lambda \bar{g}^{\mu_1\mu_2}(p)S_{\mu_1\mu_2\mu_3}\cdots(p)x^{\mu_3} \cdots x^{\mu_n} + \cdots. \\
& \qquad\qquad\qquad\qquad\qquad\qquad\qquad\qquad\qquad (2\text{-}5.76)
\end{aligned}$$

This generalizes the construction given for $n = 2$, Eq. (2-4.5), and produces symmetric tensors of rank n that obey

$$\bar{g}^{\mu_1\mu_2}(p)\bar{S}_{\mu_1\mu_2\cdots\mu_n}(p) = 0, \qquad (2\text{-}5.77)$$

provided such tensors are used in the effectively three-dimensional context of Eq. (2-5.70).

The structure of the coupling between sources can now be presented alternatively as

$$\overline{S}_1{}^{\mu_1\cdots\mu_n}(p)^* \mathfrak{g}_{\mu_1\nu_1}(p) \cdots \mathfrak{g}_{\mu_n\nu_n}(p) \overline{S}_2{}^{\nu_1\cdots\nu_n}(p)$$
$$= S_1{}^{\mu_1\cdots\mu_n}(p)^* \Pi_{\mu_1\cdots\mu_n,\nu_1\cdots\nu_n}(p) S_2{}^{\nu_1\cdots\nu_n}(p). \quad (2\text{-}5.78)$$

The form of the projection tensor Π is given by

$$x^{\mu_1} \cdots x^{\mu_n} \Pi_{\mu_1\cdots\mu_n,\nu_1\cdots\nu_n}(p) y^{\nu_1} \cdots y^{\nu_n} = (x \cdot y)^n$$
$$+ c_{n1}(x \cdot x)(x \cdot y)^{n-2}(y \cdot y) + \cdots$$
$$= [(x \cdot x)(y \cdot y)]^{(1/2)n} \frac{2^n(n!)^2}{(2n)!} P_n(\mu),$$
$$(2\text{-}5.79)$$

where, for example,

$$x \cdot y = x^\mu \mathfrak{g}_{\mu\nu}(p) x^\nu \quad (2\text{-}5.80)$$

and

$$\mu = x \cdot y / [(x \cdot x)(y \cdot y)]^{1/2}. \quad (2\text{-}5.81)$$

The addition theorem of spherical harmonics provides the factorization

$$x^{\mu_1} \cdots x^{\mu_n} \Pi_{\mu_1\cdots\mu_n,\nu_1\cdots\nu_n}(p) y^{\nu_1} \cdots y^{\nu_n} = \frac{2^n(n!)^2}{(2n)!} \frac{4\pi}{2n+1} \sum_{\lambda=-n}^{n} Y_{n\lambda}(x) Y_{n\lambda}(y)^*,$$
$$(2\text{-}5.82)$$

although we here use the symbol $Y_{n\lambda}$ to designate solid harmonics. We infer the dyadic construction

$$\Pi^{\mu_1\cdots\mu_n,\nu_1\cdots\nu_n}(p) = \sum_{\lambda=-n}^{n} e_{p\lambda}^{\mu_1\cdots\mu_n} e_{p\lambda}^{\nu_1\cdots\nu_n*}, \quad (2\text{-}5.83)$$

in which

$$e_{p\lambda}^{\mu_1\cdots\mu_n} x_{\mu_1} \cdots x_{\mu_n} = \left[\frac{2^n(n!)^2}{(2n)!} \frac{4\pi}{2n+1} \right]^{1/2} Y_{n\lambda}(x). \quad (2\text{-}5.84)$$

The solid harmonics are being used in a somewhat symbolic way. They can be removed by introducing the generating function

$$\frac{(\nu \cdot x)^n}{n!} = \left(\frac{4\pi}{2n+1} \right)^{1/2} \sum_{\lambda=-n}^{n} \frac{\xi_+^{n+\lambda} \xi_-^{n-\lambda}}{[(n+\lambda)!(n-\lambda)!]^{1/2}} Y_{n\lambda}(x), \quad (2\text{-}5.85)$$

where, in two-component matrix notation,

$$\nu_\mu = -\tfrac{1}{2} i(\xi \sigma_2 \sigma_\mu \xi). \quad (2\text{-}5.86)$$

For simplicity, we introduce the abbreviation

$$\psi_{n\lambda}(\xi) = \left[\frac{(2n)!}{(n+\lambda)!(n-\lambda)!} \right]^{1/2} \xi_+^{n+\lambda} \xi_-^{n-\lambda} \quad (2\text{-}5.87)$$

and obtain, for arbitrary n,

$$\sum_{\lambda=-n}^{n} \psi_{n\lambda}(\xi)e_{p\lambda}^{\mu_1\cdots\mu_n}x_{\mu_1}\cdots x_{\mu_n} = (2^{1/2}\nu\cdot x)^n \qquad (2\text{-}5.88)$$

and, for $n = 1$,

$$\sum_{\lambda=-1}^{1} \psi_{1\lambda}(\xi)e_{p\lambda}^{\mu}x_{\mu} = 2^{1/2}\nu\cdot x. \qquad (2\text{-}5.89)$$

Accordingly, we construct the polarization vectors for spin n, from those belonging to unit spin, by

$$\sum_{\lambda=-n}^{n} \psi_{n\lambda}(\xi)e_{p\lambda}^{\mu_1\cdots\mu_n}x_{\mu_1}\cdots x_{\mu_n} = \left[\sum_{\lambda=-1}^{1} \psi_{1\lambda}(\xi)e_{p\lambda}^{\mu}x_{\mu}\right]^n. \qquad (2\text{-}5.90)$$

The known results for $n = 2$, given in (2-4.18), are immediately reproduced. One can verify the orthonormality of the $2n + 1$ polarization tensors by multiplying one such expression with the complex conjugate of another, in which x^{μ} is replaced by $\partial/\partial x_{\mu}$. This gives

$$\sum_{\lambda\lambda'} \psi_{n\lambda}(\xi^*)e_{p\lambda}^{\mu_1\cdots\mu_n}e_{\mu_1\cdots\mu_n p\lambda'}^{*}\psi_{n\lambda'}(\eta) = (\xi^*\eta)^{2n}, \qquad (2\text{-}5.91)$$

from which we infer that

$$e_{p\lambda}^{\mu_1\cdots\mu_n}e_{\mu_1\cdots\mu_n p\lambda'}^{*} = \delta_{\lambda\lambda'}. \qquad (2\text{-}5.92)$$

It may be concluded that the source effective for emission into the specific particle state labeled $p\lambda$ is

$$S_{p\lambda} = (d\omega_p)^{1/2}e_{p\lambda}^{\mu_1\cdots\mu_n*}S_{\mu_1\cdots\mu_n}(p). \qquad (2\text{-}5.93)$$

The complete description of multiparticle emission and absorption processes for these B. E. particles is contained in the vacuum amplitude

$$\langle 0_+|0_-\rangle^S = \exp\left[iW(S)\right]. \qquad (2\text{-}5.94)$$

In order to present the structure of $W(S)$ as compactly as possible, we use the four-dimensional momentum space version of $\Delta_+(x - x')$ given in (2-1.61) and obtain

$$W(S) = \frac{1}{2}\int \frac{(dp)}{(2\pi)^4} S^{\mu_1\cdots\mu_n}(-p)\frac{\Pi_{\mu_1\cdots\mu_n,\nu_1\cdots\nu_n}(p)}{p^2 + m^2 - i\epsilon} S^{\nu_1\cdots\nu_n}(p). \qquad (2\text{-}5.95)$$

The tensor $\Pi(p)$ retains the algebraic form represented by (2-5.79) and is an even polynomial in p of degree $2n$. The corresponding coordinate space structures are illustrated, for $n = 1$ and 2, by Eqs. (2-3.4) and (2-4.20), respectively. All the generalizations discussed earlier in the context of special examples can be developed for the arbitrary integer spin situation.

No reference has been made to parity as an independent specification of particle states. That is because the particle states we have constructed are automatically endowed with a definite parity. The geometrical operation that reverses the positive sense of the three spatial axes is represented by the unitary operator R_s. Its effect upon the individual particle operators $\mathbf{r}$, $\mathbf{p}$, $\mathbf{s}$ is given by

$$R_s^{-1}\mathbf{r}R_s = -\mathbf{r}, \qquad R_s^{-1}\mathbf{p}R_s = -\mathbf{p}, \qquad R_s^{-1}\mathbf{s}R_s = \mathbf{s}. \qquad (2\text{-}5.96)$$

The transformed single particle state

$$\langle \overline{1_p}| = \langle 1_p|R_s \qquad (2\text{-}5.97)$$

refers to the spatial momentum $-\mathbf{p}$. Only for $\mathbf{p} = 0$ can one exhibit an eigen-vector of R_s, a state of definite parity. As in the discussion of continuous Lorentz transformations, what is relevant to the probability amplitude $\langle 1_p|0\rangle^S$ is the relationship between the description of the particle state and the charac-terization of the source. The transformed particle state is represented by a correspondingly transformed source which illustrates the general linear response (2–5.2):

$$\overline{S}(\overline{x}) = r_s S(x),$$
$$\overline{x}^0 = x^0, \qquad \overline{x}_k = -x_k. \qquad (2\text{-}5.98)$$

The reflection matrix r_s is required to be real if real sources are used. It acts upon the spin indices to effect the geometrical transformation

$$r_s^{-1}\mathbf{s}r_s = \mathbf{s}, \qquad r_s^{-1}\mathbf{n}r_s = -\mathbf{n}, \qquad (2\text{-}5.99)$$

or, in view of (2–5.26),

$$r_s^{-1}\mathbf{s}^{(1)}r_s = \mathbf{s}^{(2)}, \qquad r_s^{-1}\mathbf{s}^{(2)}r_s = \mathbf{s}^{(1)}. \qquad (2\text{-}5.100)$$

The corresponding action of r_s upon $S_{\sigma_1^{(1)}\ldots\sigma_n^{(1)}\sigma_1^{(2)}\ldots\sigma_n^{(2)}}$ is the interchange of the $\sigma_\alpha^{(1)}$ and $\sigma_\alpha^{(2)}$ labels, apart from the option of an additional minus sign, which is compatible with the simple geometrical property

$$r_s^2 = 1. \qquad (2\text{-}5.101)$$

The permutation of a single pair of spin indices affects oppositely the singlet and triplet combinations, corresponding to the opposite behavior, under spatial reflection, of the time and space components of a four-vector. We have ex-pressed it this way, since it leaves free the choice of overall sign in the reflec-tion response, which is the alternative between a vector, and a pseudo or axial vector. The behavior of the tensor $S_{\mu_1\ldots\mu_n}$ is that implied by the several vector indices, together with the overall $\pm$ factor. The concept of parity refers to the rest frame where the surviving source components are $S_{k_1\ldots k_n}$, which act as a unit under spatial reflection. When standard vector behavior is considered, the parity is $(-1)^n$. This gives a sequence of integer spin particles with alter-nating parities, as symbolized by 0^+, 1^-, 2^+, ... The other sequence is 0^-, 1^+, 2^-, ...

Although the only known or conjectured massless particles of integer spin have already been discussed, we shall nevertheless present a unified treatment of all integer spin massless particles. As in the special examples, it is clear that the limit $m \to 0$ in (2–5.95) cannot be performed unless

$$p_{\mu_1} S^{\mu_1 \cdots \mu_n}(p) = 0, \qquad \partial_{\mu_1} S^{\mu_1 \cdots \mu_n}(x) = 0 \qquad (2\text{–}5.102)$$

is valid at $m = 0$. Were we to carry out the limiting process in the manner already illustrated, we would be tracing the decomposition of the $2n + 1$ spin states into the helicity pairs $\lambda = \pm n, \pm(n-1), \ldots, \pm 1$, and $\lambda = 0$. This time, however, we shall directly extract $\lambda = \pm n$. The invariant form of such a source coupling is

$$S_1^{\mu_1 \cdots \mu_n}(p)^* \Pi_{\mu_1 \cdots \mu_n, \nu_1 \cdots \nu_n} S_2^{\nu_1 \cdots \nu_n}(p), \qquad (2\text{–}5.103)$$

where the projection tensor Π has a structure indicated by

$$x^{\mu_1} \cdots x^{\mu_n} \Pi_{\mu_1 \cdots \mu_n, \nu_1 \cdots \nu_n} y^{\nu_1} \cdots y^{\nu_n} = (xy)^n + d_{n1}(xx)(xy)^{n-2}(yy) + \cdots. \qquad (2\text{–}5.104)$$

The products formed from x^μ and y^μ are four-dimensional. Any use of the vector p^μ, as in (2–5.79), would give no contribution in view of the source restriction (2–5.102). We now exploit that fact to replace the tensor Π with another that is equivalent to it in the context of (2–5.103). This is accomplished by the following substitution, applied to both x^μ and y^μ,

$$x^\mu \to \left[g^{\mu\nu} - \frac{\bar{p}^\mu p^\nu}{p\bar{p}} \right] x_\nu, \qquad (2\text{–}5.105)$$

in which $\bar{p}^\mu$ is any null vector with $\bar{p}^0 > 0$, such that $p\bar{p} \neq 0$. The absence of any change when $p^\nu x_\nu = 0$ assures the equivalence of the two structures for the application of interest. The new version of Π is given by

$$x^{\mu_1} \cdots x^{\mu_n} \Pi_{\mu_1 \cdots \mu_n, \nu_1 \cdots \nu_n}(p, \bar{p}) y^{\nu_1} \cdots y^{\nu_n}$$
$$= (x \cdot y)^n + d_{n1}(x \cdot x)(x \cdot y)^{n-2}(y \cdot y) + \cdots, \qquad (2\text{–}5.106)$$

where, for example,

$$x \cdot y = x^\mu \bar{g}_{\mu\nu}(p, \bar{p}) x^\nu \qquad (2\text{–}5.107)$$

and

$$\bar{g}_{\mu\nu}(p, \bar{p}) = g_{\mu\nu} - \frac{p_\mu \bar{p}_\nu + p_\nu \bar{p}_\mu}{p\bar{p}}. \qquad (2\text{–}5.108)$$

In the discussion of the exchange of a massless particle, p^μ is also a null vector and $\bar{g}_{\mu\nu}(p, \bar{p})$ projects onto the subspace orthogonal to p^μ and $\bar{p}_\mu$:

$$p^\mu \bar{g}_{\mu\nu}(p, \bar{p}) = 0, \qquad \bar{p}^\mu \bar{g}_{\mu\nu}(p, \bar{p}) = 0. \qquad (2\text{–}5.109)$$

Considered in the rest frame of the time-like vector $p^\mu + \bar{p}^\mu$, the orthogonal vector $p^\mu - \bar{p}^\mu$ has only spatial components, doubling the particle's momentum, and we recognize that the subspace selected by $\bar{g}_{\mu\nu}$ is the two-dimensional Euclidean plane perpendicular to the momentum of the particle.

    If only helicities $\lambda = \pm n$ are to be represented in the source coupling (2–5.103), the tensor Π must be irreducible with respect to forming traces in the two-dimensional Euclidean space,

$$\bar{g}^{\mu_1\mu_2}(p, \bar{p})\Pi_{\mu_1\cdots\mu_n,\nu_1\cdots\nu_n}(p, \bar{p}) = 0. \tag{2-5.110}$$

This is equivalent to asserting that, as a function of the x variables or of the y variables in the plane, (2–5.106) is a solution of Laplace's equation, which is homogeneous of degree n. The required two-dimensional harmonic function is

$$[(x \cdot x)(y \cdot y)]^{(1/2)n}(\tfrac{1}{2})^{n-1}T_n(\mu), \tag{2-5.111}$$

where

$$\mu = x \cdot y/[(x \cdot x)(y \cdot y)]^{1/2} = \cos\phi \tag{2-5.112}$$

and $T_n(\mu)$ is the Tchebichef polynomial

$$T_n(\mu) = \cos(n\cos^{-1}\mu) = \cos n\phi. \tag{2-5.113}$$

From the coefficients of this polynomial we learn that

$$m \leq \tfrac{1}{2}n: \qquad d_{nm} = \frac{(-1)^m}{m!}\frac{n}{4^m}\frac{(n-m-1)!}{(n-2m)!}, \tag{2-5.114}$$

and, in particular,

$$n \geq 2, \qquad d_{n1} = -\tfrac{1}{4}n. \tag{2-5.115}$$

The value of $-\tfrac{1}{2}$ obtained for $n = 2$ is in agreement with (2–4.24).
    The identity

$$\cos n\phi = \tfrac{1}{2}[e^{in\varphi}e^{-in\varphi'} + e^{-in\varphi}e^{in\varphi'}], \qquad \phi = \varphi - \varphi', \tag{2-5.116}$$

provides the relevant addition theorem. It implies the dyadic construction

$$\Pi^{\mu_1\cdots\mu_n,\nu_1\cdots\nu_n}(p, \bar{p}) = \sum_{\lambda=\pm n} e_{p\lambda}^{\mu_1\cdots\mu_n} e_{p\lambda}^{\nu_1\cdots\nu_n*}, \tag{2-5.117}$$

where

$$e_{p\pm n}^{\mu_1\cdots\mu_n} x_{\mu_1}\cdots x_{\mu_n} = (\tfrac{1}{2}x \cdot x)^{(1/2)n} i^{n\pm n}e^{\pm in\varphi}. \tag{2-5.118}$$

The phases are so chosen that, for $n = 1$,

$$e_{p\pm 1}^{\mu} x_{\mu} = (\tfrac{1}{2}x \cdot x)^{1/2} i^{1\pm 1}e^{\pm i\varphi} \tag{2-5.119}$$

reproduces the conventions of (2–3.29). We now have

$$e_{p\pm n}^{\mu_1\cdots\mu_n} x_{\mu_1}\cdots x_{\mu_n} = [e_{p\pm 1}^{\mu}x_{\mu}]^n \tag{2-5.120}$$

and the explicit construction

$$e_{p\pm n}^{\mu_1\cdots\mu_n} = e_{p\pm 1}^{\mu_1}\cdots e_{p\pm 1}^{\mu_n}, \tag{2-5.121}$$

which generalizes the $n = 2$ result, Eq. (2–4.31).

The massless particle of helicity 3 is represented by the space-time source structure

$$W(S) = \tfrac{1}{2} \int (dx)(dx')[S^{\lambda\mu\nu}(x)D_+(x-x')S_{\lambda\mu\nu}(x') - \tfrac{3}{4}S^\lambda(x)D_+(x-x')S_\lambda(x')],$$

(2–5.122)

where

$$\partial_\lambda S^{\lambda\mu\nu}(x) = 0$$

(2–5.123)

and

$$S^\lambda(x) = g_{\mu\nu}S^{\lambda\mu\nu}(x).$$

(2–5.124)

Ordinary matter possesses no conserved physical properties that could be identified with the ones described by the local conservation law (2–5.123), or indeed for any $n \geq 3$. The inability to construct their sources strongly affirms the empirical absence of the particles. But perhaps one should not reject totally the possibility of eventually encountering such properties, and the associated particles, under circumstances that are presently unattainable.

2–6 SPIN ½ PARTICLES. FERMI-DIRAC STATISTICS

There are two simple alternatives for constructing a spin ½ particle description, in the sense of Eq. (2–5.26), namely

$$\begin{aligned} \mathbf{s}^{(1)} &= \tfrac{1}{2}\boldsymbol{\sigma}, & \mathbf{s}^{(2)} &= 0; \\ \mathbf{s}^{(1)} &= 0, & \mathbf{s}^{(2)} &= \tfrac{1}{2}\boldsymbol{\sigma}. \end{aligned}$$

(2–6.1)

The two possibilities are interchanged by a reflection of the spatial coordinates. This indicates the convenience of a more symmetrical treatment in which both take part. It is also advantageous to replace the complex sources upon which the 2×2 Pauli matrices act by equivalent real sources. These remarks point to the utility of a spin ½ particle description that employs four real sources. In order to retain the symbol $\boldsymbol{\sigma}$ for use in the new context, we designate the initial 2×2 matrices as τ_k, and use τ'_k for an independent set. Real, antisymmetrical $i\sigma_k$ matrices can be constructed from the $i\tau_k$ by replacing any explicit i by the algebraically equivalent real antisymmetrical matrix $i\tau'_2$. Thus,

$$i\sigma_1 = i\tau'_2\tau_1, \qquad i\sigma_2 = i\tau_2, \qquad i\sigma_3 = i\tau'_2\tau_3, \qquad (2\text{–}6.2)$$

which are indeed real, antisymmetrical 4×4 matrices. They preserve the algebraic properties of spin matrices:

$$\tfrac{1}{2}\{\sigma_k, \sigma_l\} = \delta_{kl}, \qquad i\sigma_1 i\sigma_2 i\sigma_3 = 1, \qquad (2\text{–}6.3)$$

and we identify

$$s_k = \tfrac{1}{2}\sigma_k.$$

(2–6.4)

Initially, in the role of n_k we have $i\lambda\tfrac{1}{2}\tau_k$, where λ is now a 2×2 matrix that commutes with the τ_k, and has ± 1 as eigenvalues. When the transformation

$\tau_k \to \sigma_k$ is introduced, the matrices that could be used for $i\lambda$ are just three in number:

$$ i\rho_1 = i\tau_2\tau_1', \qquad i\rho_2 = i\tau_2', \qquad i\rho_3 = i\tau_2\tau_3'. \qquad (2\text{-}6.5)$$

They are the analogues of $i\sigma_k$ with the τ and τ' matrices interchanged. The two sets of three anticommuting matrices are mutually commutative. These six antisymmetrical matrices, σ_k, ρ_k, and the ten symmetrical matrices 1, $\sigma_k\rho_l$, provide a basis for all 4×4 matrices. Since the three ρ_k are on the same footing, we arbitrarily identify λ with ρ_2 and write

$$ n_k = \tfrac{1}{2}i\alpha_k, \qquad (2\text{-}6.6)$$

where the

$$ \alpha_k = \rho_2\sigma_k \qquad (2\text{-}6.7)$$

are real, symmetrical matrices. We note their algebraic properties:

$$ \tfrac{1}{2}\{\alpha_k, \alpha_l\} = \delta_{kl}, \qquad [\alpha_k, \tfrac{1}{2}\sigma_l] = i\epsilon_{klm}\alpha_m, \qquad \boldsymbol{\alpha} \times \boldsymbol{\alpha} = 2i\boldsymbol{\sigma}, \qquad (2\text{-}6.8)$$

the last statement being the realization of Eq. (2–5.33). Since space reflection induces $\mathbf{n} \to -\mathbf{n}$ without changing $\mathbf{s}$, it is represented by a matrix that commutes with $\boldsymbol{\sigma}$ and anticommutes with ρ_2. The only matrices with those characteristics are ρ_1 and ρ_3. We choose the latter arbitrarily and multiply this antisymmetrical matrix by i to get the real space reflection matrix

$$ r_s = i\rho_3, \qquad (2\text{-}6.9)$$

which obeys

$$ r_s^2 = -1. \qquad (2\text{-}6.10)$$

The space reflection matrix appears in another role on considering the real matrix associated with an infinitesimal Lorentz transformation [cf. Eq. (2–5.9)]:

$$ \begin{aligned} L &= 1 + \tfrac{1}{2}i\, \delta\omega^{\mu\nu}s_{\mu\nu} \\ &= 1 + i\, \delta\boldsymbol{\omega} \cdot \tfrac{1}{2}\boldsymbol{\sigma} - \delta\mathbf{v} \cdot \tfrac{1}{2}\boldsymbol{\alpha}. \end{aligned} \qquad (2\text{-}6.11)$$

According to the symmetry properties of the matrices, transposition has the following effect,

$$ L^T = 1 - i\, \delta\boldsymbol{\omega} \cdot \tfrac{1}{2}\boldsymbol{\sigma} - \delta\mathbf{v} \cdot \tfrac{1}{2}\boldsymbol{\alpha}, \qquad (2\text{-}6.12)$$

whereas

$$ L^{-1} = 1 - i\, \delta\boldsymbol{\omega} \cdot \tfrac{1}{2}\boldsymbol{\sigma} + \delta\mathbf{v} \cdot \tfrac{1}{2}\boldsymbol{\alpha}. \qquad (2\text{-}6.13)$$

We express this, through the action of r_s, as

$$ L^T = r_s L^{-1} r_s^{-1} \qquad (2\text{-}6.14)$$

or

$$ L^T r_s L = r_s. \qquad (2\text{-}6.15)$$

The validity of this statement for the finite transformations of the proper orthochronous group is assured by the composition property of successive transformations,

$$(L_1 L_2)^T r_s L_1 L_2 = L_2^T L_1^T r_s L_1 L_2 = L_2^T r_s L_2 = r_s. \qquad (2\text{-}6.16)$$

The relation (2–6.15) also holds for the space-reflection transformation, since

$$r_s^T r_s = 1 \qquad (2\text{-}6.17)$$

combines the antisymmetry of r_s with the iterative property (2–6.10). The appearance of the matrix r_s in (2–6.15) exhibits it in its fundamental metric role. It is the analogue of the metric tensor in

$$l^\mu{}_\kappa g_{\mu\nu} l^\nu{}_\lambda = g_{\kappa\lambda} \qquad (2\text{-}6.18)$$

or, using matrix notation,

$$l^T g l = g, \qquad (2\text{-}6.19)$$

for $(\pm)g$, which attributes opposite signs to time and space components, is also the space-reflection matrix for vectors.

Another aspect of the infinitesimal transformation matrices (2–6.11, 12), in relation to the real symmetrical matrices α_k and

$$\alpha^0 = 1, \qquad (2\text{-}6.20)$$

is given by

$$L^T \alpha L = \alpha - \delta\omega \times \alpha - \delta\mathbf{v}\alpha^0, \qquad L^T \alpha^0 L = \alpha^0 - \delta\mathbf{v}\cdot\alpha, \quad (2\text{-}6.21)$$

which are united in

$$L^T \alpha^\mu L = (\delta^\mu_\nu + \delta\omega^\mu{}_\nu)\alpha^\nu. \qquad (2\text{-}6.22)$$

This is the response of a vector to homogeneous infinitesimal Lorentz transformations. The repetition of such transformations yields the finite transformation law

$$L^T \alpha^\mu L = l^\mu{}_\nu \alpha^\nu, \qquad (2\text{-}6.23)$$

which is also valid for the improper space-reflection transformation generated by $L = r_s$. Note that the symmetry of the α^μ and the antisymmetry of r_s, as well as their reality, is maintained by the Lorentz transformations.

We now consider the coupling between sources associated with single-particle exchange, where the individual emission and absorption acts are represented by (2–5.36) and (2–5.42), with

$$B(p) = \exp[-\tfrac{1}{2}\theta\boldsymbol{\alpha}\cdot\mathbf{p}/|\mathbf{p}|]. \qquad (2\text{-}6.24)$$

The spin $\frac{1}{2}$ particle has been placed in a larger framework, as evidenced by the existence of the three matrices ρ_k that commute with σ. Two of the four components must be rejected by interposing a spin-independent projection matrix between the two $B(p)$ factors that are associated with the individual acts. The possibilities afforded by the three ρ_k are really only two in number, depending upon whether the ρ matrix used commutes or anticommutes with $\boldsymbol{\alpha}$. In the first situation, we have

$$B(p)\tfrac{1}{2}(1+\rho_2)B(p) = \tfrac{1}{2}(1+\rho_2)B(p)^2$$
$$= \frac{1+\rho_2}{2m}(p^0 - \boldsymbol{\alpha}\cdot\mathbf{p}) = -\frac{1+\rho_2}{2m}\alpha^\mu p_\mu, \quad (2\text{-}6.25)$$

while the second one is illustrated by

$$B(p)\tfrac{1}{2}(1+\rho_3)B(p) = \tfrac{1}{2}B(p)^2 + \tfrac{1}{2}\rho_3$$
$$= \frac{1}{2m}[p^0 - \boldsymbol{\alpha}\cdot\mathbf{p} + m\rho_3] = \frac{1}{2m}[m\rho_3 - \alpha^\mu p_\mu]. \quad (2\text{-}6.26)$$

Spin $\tfrac{1}{2}$ particle sources will be designated as $\eta(x)$, $\eta(p)$ or more explicitly $\eta_\zeta(x)$, $\eta_\zeta(p)$. The space-time extrapolation of the source coupling takes two alternative forms:

$$-\tfrac{1}{2}\int (dx)(dx')\eta(x)(1+\rho_2)\alpha^\mu(1/i)\partial_\mu\Delta_+(x-x')\eta(x') \quad (2\text{-}6.27)$$

and

$$\tfrac{1}{2}\int (dx)(dx')\eta(x)[m\rho_3 - \alpha^\mu(1/i)\partial_\mu]\Delta_+(x-x')\eta(x'), \quad (2\text{-}6.28)$$

where, as we have verified on several occasions, the use of the propagation function $\Delta_+(x-x')$ is required to maintain space-time uniformity, or the Euclidean postulate. These are examples of the quadratic structure

$$\tfrac{1}{2}\int (dx)(dx')\eta_\zeta(x)K_{\zeta\zeta'}(x,x')\eta_{\zeta'}(x') = \tfrac{1}{2}\int (dx)(dx')\eta_{\zeta'}(x')K_{\zeta'\zeta}(x',x)\eta_\zeta(x).$$
$$(2\text{-}6.29)$$

As the irreducible kernel of a quadratic form, $K_{\zeta\zeta'}(x,x')$ should respond as a unit to the act of transposition, interchanging ζ and ζ', x and x'. This is not true of the first possibility, (2-6.27), since 1 and ρ_2 behave oppositely under transposition. Accordingly, the projection factor $1+\rho_2$ is spurious since only one of the terms contributes to the quadratic form. The second kernel does act as a unit under the general transposition:

$$[(m\rho_3 - \alpha^\mu(1/i)\partial'_\mu)\Delta_+(x'-x)]^T = -(m\rho_3 - \alpha^\mu(1/i)\partial_\mu)\Delta_+(x-x').$$
$$(2\text{-}6.30)$$

It is antisymmetrical!

One might try to convert this kernel to a symmetrical structure, without upsetting the spin description, by invoking particles and antiparticles. This requires an additional two-valued source index, and permits the insertion into the kernel of the antisymmetrical charge matrix q. The resulting kernel is symmetrical but indefinite, since q is converted into $-q$ by a charge reflection. That is in flat contradiction with the physical requirement on the vacuum persistence probability, which demands that the imaginary part of the quadratic form be positive,

$$|e^{iW}|^2 = e^{-2\mathrm{Im}W} \leq 1. \quad (2\text{-}6.31)$$

The conclusion is unavoidable that spin ½ presents a totally new situation. Only one course is open. Instead of trying to modify the symmetry characteristics of the kernel to suit the algebraic properties of the source, we must adapt the algebraic properties of the source to the antisymmetry of the kernel. The comparison of the two equivalent versions of (2–6.29) with the antisymmetry property

$$K_{\zeta'\zeta}(x', x) = -K_{\zeta\zeta'}(x, x') \qquad (2\text{–}6.32)$$

will cease to be a paradox and become an identity if

$$\eta_{\zeta'}(x')\eta_{\zeta}(x) = -\eta_{\zeta}(x)\eta_{\zeta'}(x'). \qquad (2\text{–}6.33)$$

We are thus forced by the characteristics of spin ½ to abandon the ordinary numerical, commutative sources of Bose-Einstein statistics and introduce a new kind of source and a new statistics. It will be verified shortly that this is Fermi-Dirac statistics.

The symmetry aspects of this discussion have been facilitated by the use of matrices with definite symmetry, the symmetrical α^{μ}, the antisymmetrical ρ_3. In later developments, however, uniformity of algebraic properties and Lorentz transformation behavior are more significant. It is algebraically awkward that the anticommuting α_k commute with α^0; the representation of a Lorentz transformation on the α^{μ} as $L^T\alpha^{\mu}L$ is not a similarity transformation, and $\alpha^{\mu}\alpha^{\nu}$ does not have tensor transformation properties. To improve the latter situation one must replace L^T with L^{-1}. That is accomplished by the relation (2–6.14) which gives the new vector transformation form

$$L^{-1}(r_s^{-1}\alpha^{\mu})L = l^{\mu}{}_{\nu}(r_s^{-1}\alpha^{\nu}). \qquad (2\text{–}6.34)$$

It is convenient to define imaginary matrices

$$\gamma^{\mu} = ir_s^{-1}\alpha^{\mu} \qquad (2\text{–}6.35)$$

that obey

$$L^{-1}\gamma^{\mu}L = l^{\mu}{}_{\nu}\gamma^{\nu}, \qquad (2\text{–}6.36)$$

together with

$$L^{-1}\gamma^{\mu}\gamma^{\nu}L = l^{\mu}{}_{\kappa}l^{\nu}{}_{\lambda}\gamma^{\kappa}\gamma^{\lambda}, \qquad (2\text{–}6.37)$$

and so forth. The algebraic property $r_s^2 = -1$, along with $\alpha^0 = 1$, shows that

$$(\gamma^0)^2 = 1, \qquad (2\text{–}6.38)$$

and

$$r_s = i\gamma^0, \qquad (2\text{–}6.39)$$

which also gives the identification

$$\gamma^0 = \rho_3. \qquad (2\text{–}6.40)$$

The γ matrices do not have a common symmetry. The definition (2–6.35) implies that

$$\gamma^{\mu T} = -i\alpha^{\mu}r_s^{-1}, \qquad (2\text{–}6.41)$$

or

$$\gamma^{\mu T} = -r_s \gamma^\mu r_s^{-1}. \tag{2-6.42}$$

This restates the antisymmetry of γ^0, which commutes with $r_s = i\gamma^0$, and shows that the γ_k are symmetrical, skew-Hermitian matrices since they anticommute with the space-reflection matrix,

$$\{\gamma^0, \gamma_k\} = 0. \tag{2-6.43}$$

Algebraic relations among the γ_k are obtained as

$$\tfrac{1}{2}\{\gamma_k, \gamma_l\} = -\tfrac{1}{2}\{\alpha_k, \alpha_l\} = -\delta_{kl}. \tag{2-6.44}$$

The various characteristics of the γ_μ contained in (2-6.38), (2-6.43), and (2-6.44) are united in

$$\tfrac{1}{2}\{\gamma_\mu \gamma_\nu\} = -g_{\mu\nu}. \tag{2-6.45}$$

This unified algebraic statement is maintained by Lorentz transformations, according to

$$L^{-1}\tfrac{1}{2}\{\gamma^\mu, \gamma^\nu\} L = -l^\mu{}_\kappa l^\nu{}_\lambda g^{\kappa\lambda} = -g^{\mu\nu}. \tag{2-6.46}$$

The γ matrices also give unified expression to

$$s_{\mu\nu} = \tfrac{1}{2}\sigma_{\mu\nu}. \tag{2-6.47}$$

We first write $\sigma = (1/2i)\alpha \times \alpha$ as

$$\sigma_{kl} = \frac{1}{2i}[\alpha_k, \alpha_l] = \tfrac{1}{2}i[\gamma_k, \gamma_l], \tag{2-6.48}$$

and then note that

$$\sigma^0{}_k = i\alpha_k = i\gamma^0\gamma_k. \tag{2-6.49}$$

These matrices are united in

$$\sigma_{\mu\nu} = \tfrac{1}{2}i[\gamma_\mu, \gamma_\nu], \tag{2-6.50}$$

which transforms as an antisymmetrical tensor,

$$L^{-1}\sigma^{\mu\nu} L = l^\mu{}_\kappa l^\nu{}_\lambda \sigma^{\kappa\lambda}. \tag{2-6.51}$$

The symmetry properties of the imaginary $\sigma_{\mu\nu}$ are given by

$$\sigma^T_{\mu\nu} = -r_s \sigma_{\mu\nu} r_s^{-1}, \tag{2-6.52}$$

which affirms that the σ_{kl} are antisymmetrical and Hermitian, while the σ_{0k} are symmetrical and skew-Hermitian.

The process of multiplying different γ matrices together terminates with

$$\gamma_5 = \gamma^0\gamma^1\gamma^2\gamma^3 \tag{2-6.53}$$

or $(\mu \neq \nu \neq \kappa \neq \lambda)$

$$\gamma^\mu\gamma^\nu\gamma^\kappa\gamma^\lambda = \epsilon^{\mu\nu\kappa\lambda}\gamma_5. \tag{2-6.54}$$

This matrix is real, and since

$$\gamma_5^T = r_s(\gamma^3\gamma^2\gamma^1\gamma^0)r_s^{-1} = r_s\gamma_5 r_s^{-1} = -\gamma_5, \tag{2-6.55}$$

it is skew-Hermitian. Each of the γ_μ matrices anticommutes with γ_5,

$$\{\gamma_\mu, \gamma_5\} = 0, \tag{2-6.56}$$

and

$$\gamma_5^2 = -1. \tag{2-6.57}$$

Alternative factorizations of γ_5 are

$$\gamma_5 = \sigma_{01}\sigma_{23} = \sigma_{02}\sigma_{31} = \sigma_{03}\sigma_{12} \tag{2-6.58}$$

or

$$\sigma_{0k} = \gamma_5\sigma_k = \sigma_k\gamma_5, \tag{2-6.59}$$

which also supplies the identification

$$i\gamma_5 = \rho_2. \tag{2-6.60}$$

The Lorentz transformation behavior of γ_5 follows from (2–6.54) as

$$L^{-1}\gamma_5 L = l^0{}_\mu l^1{}_\nu l^2{}_\kappa l^3{}_\lambda \epsilon^{\mu\nu\kappa\lambda}\gamma_5 = (\det l)\gamma_5, \tag{2-6.61}$$

which characterizes γ_5 as a pseudoscalar. It is invariant for proper transformations, $\det l = +1$, and reverses sign for improper transformations, or reflections, $\det l = -1$. The latter property follows directly from the anticommutativity of γ^0 and γ_5, as specifically noted in (2–6.55). Let us also observe the pseudo or axial vector character of $i\gamma^\mu\gamma_5$,

$$L^{-1}i\gamma^\mu\gamma_5 L = (\det l)l^\mu{}_\nu i\gamma^\nu\gamma_5. \tag{2-6.62}$$

The components of $i\gamma^\mu\gamma_5$ comprise the four ways of multiplying together three different γ matrices.

The 16 independent elements of this Clifford-Dirac algebra are organized through their Lorentz transformation behavior into the five sets

$$1, \ \gamma_\mu, \ \sigma_{\mu\nu}, \ i\gamma_\mu\gamma_5, \ \gamma_5, \tag{2-6.63}$$

for which the count is $1 + 4 + 6 + 4 + 1 = 16$. Closely related but distinct is the organization by symmetry properties. As suggested by the construction

$$\alpha^\mu = \gamma^0\gamma^\mu, \tag{2-6.64}$$

we consider $\gamma^0\Gamma$, where Γ refers to any of the sets exhibited in (2–6.63). Then,

$$(\gamma^0\Gamma)^T = -\Gamma^T\gamma^0 = -\gamma^0 r_s^{-1}\Gamma^T r_s, \tag{2-6.65}$$

and the various equivalences between transposition and space reflection show that these matrices have a definite symmetry. Indeed, the 16 matrices given by

$$\gamma^0(1, \ \gamma_\mu, \ \sigma_{\mu\nu}, \ i\gamma_\mu\gamma_5, \ \gamma_5) \tag{2-6.66}$$

comprise the $4 + 6 = 10$ symmetrical matrices $\gamma^0\gamma_\mu$, $\gamma^0\sigma_{\mu\nu}$ and the $1 + 4 + 1 = 6$ antisymmetrical matrices γ^0, $\gamma^0 i\gamma_\mu\gamma_5$, $\gamma^0\gamma_5$. All the matrices are Hermitian.

The vacuum amplitude for an arbitrary spin $\frac{1}{2}$, four component spinor source $\eta(x)$ will be stated with the matrices ρ_3 and α^μ, in (2–6.28), replaced by the appropriate γ matrices:

$$\langle 0_+|0_-\rangle^\eta = \exp[iW(\eta)],$$
$$W(\eta) = \tfrac{1}{2}\int (dx)(dx')\eta(x)\gamma^0 G_+(x - x')\eta(x'), \tag{2–6.67}$$

where

$$G_+(x - x') = (m - \gamma^\mu(1/i)\partial_\mu)\Delta_+(x - x'), \tag{2–6.68}$$

and the sources are totally anticommuting real objects,

$$\{\eta_\zeta(x), \eta_{\zeta'}(x')\} = 0, \tag{2–6.69}$$

which constitute the elements of a Grassmann or exterior algebra. Let us analyze the causal source arrangement

$$\eta(x) = \eta_1(x) + \eta_2(x). \tag{2–6.70}$$

It is important to notice that even combinations of the totally anticommuting sources are commutative objects, and that the two terms involving η_1 and η_2 are equal since the anticommutativity of the sources matches the antisymmetry of the kernel. Accordingly, we get

$$\langle 0_+|0_-\rangle^\eta = \langle 0_+|0_-\rangle^{\eta_1} \exp\left[i\int (dx)(dx')\eta_1(x)\gamma^0 G_+(x - x')\eta_2(x')\right]\langle 0_+|0_-\rangle^{\eta_2},$$

where

$$\tag{2–6.71}$$

$$x^0 > x^{0'}: \qquad G_+(x - x') = i\int d\omega_p e^{ip(x-x')}(m - \gamma p), \tag{2–6.72}$$

and therefore

$$i\int (dx)(dx')\eta_1(x)\gamma^0 G_+(x - x')\eta_2(x') = \int d\omega_p i\eta_1(p)^*\gamma^0(m - \gamma p)i\eta_2(p). \tag{2–6.73}$$

The matrix factor that occurs here is just (2–6.26) in a new notation,

$$\gamma^0(m - \gamma p) = m\gamma^0 + p^0 - \gamma^0\gamma \cdot \mathbf{p}$$
$$= \exp[-\tfrac{1}{2}\theta\gamma^0\gamma \cdot \mathbf{p}/|\mathbf{p}|]m(1 + \gamma^0)\exp[-\tfrac{1}{2}\theta\gamma^0\gamma \cdot \mathbf{p}/|\mathbf{p}|] \tag{2–6.74}$$

where, it will be recalled,

$$\cosh\theta = p^0/m, \qquad \sinh\theta = |\mathbf{p}|/m. \tag{2–6.75}$$

The projection matrix $\tfrac{1}{2}(1 + \gamma^0)$ is constructed from any two orthonormal eigenvectors v_λ,

$$\gamma^0 v_\lambda = v_\lambda, \qquad v_\lambda^* v_{\lambda'} = \delta_{\lambda\lambda'}, \tag{2–6.76}$$

in the dyadic form

$$\tfrac{1}{2}(1 + \gamma^0) = \sum_\lambda v_\lambda v_\lambda^*. \tag{2-6.77}$$

A multiplicity check is provided by the trace of this matrix equation,

$$\tfrac{1}{2}\operatorname{tr} 1 = 2 = \sum_\lambda 1, \tag{2-6.78}$$

where the relevant null trace of γ^0 expresses its antisymmetry. A more general remark follows on noting that

$$\operatorname{tr}\gamma_\mu = -\tfrac{1}{2}\operatorname{tr}\gamma_5\{\gamma_\mu, \gamma_5\} = 0. \tag{2-6.79}$$

A specific choice of the v_λ can be made as eigenvectors of a component of $\boldsymbol{\sigma}$, say σ_3,

$$\sigma_3 v_\sigma = \sigma v_\sigma, \qquad \sigma = \pm 1. \tag{2-6.80}$$

We also want these eigenvectors to be related by standard spin operations:

$$\tfrac{1}{2}(\sigma_1 + i\sigma_2)v_- = v_+, \qquad \tfrac{1}{2}(\sigma_1 - i\sigma_2)v_+ = v_-. \tag{2-6.81}$$

Other statements, expressing the use of imaginary γ^0 and $\boldsymbol{\sigma}$ matrices,

$$\gamma^0 v_\sigma^* = -v_\sigma^*, \qquad \sigma_3 v_\sigma^* = -\sigma v_\sigma^*, \qquad -\tfrac{1}{2}(\sigma_1 \pm i\sigma_2)v_\mp^* = v_\pm^*, \tag{2-6.82}$$

are satisfied by

$$v_{-\sigma}^* = i\sigma\gamma_5 v_\sigma. \tag{2-6.83}$$

Since v_σ^* is an eigenvector of γ^0 belonging to the eigenvalue -1, there are corresponding orthogonality properties,

$$v_\sigma v_{\sigma'} = 0, \qquad v_\sigma^* v_{\sigma'}^* = 0. \tag{2-6.84}$$

On inserting the eigenvector construction for $\tfrac{1}{2}(1 + \gamma^0)$ in (2-6.74), we get

$$\gamma^0(m - \gamma p) = 2m \sum_\sigma \gamma^0 u_{p\sigma} u_{p\sigma}^* \gamma^0 \tag{2-6.85}$$

where

$$\begin{aligned}
u_{p\sigma} &= \exp[\tfrac{1}{2}\theta\gamma^0\boldsymbol{\gamma}\cdot\mathbf{p}/|\mathbf{p}|]v_\sigma, \\
u_{p\sigma}^* &= v_\sigma^* \exp[\tfrac{1}{2}\theta\gamma^0\boldsymbol{\gamma}\cdot\mathbf{p}/|\mathbf{p}|],
\end{aligned} \tag{2-6.86}$$

which involves the anticommutativity of $\gamma^0\boldsymbol{\gamma}$ with γ^0 and the eigenvector significance of v_σ relative to γ^0. The same properties are used in verifying the orthonormality of these vectors in the form

$$u_{p\sigma}^* \gamma^0 u_{p\sigma'} = v_\sigma^* v_{\sigma'} = \delta_{\sigma\sigma'}. \tag{2-6.87}$$

Now, according to Eq. (2-6.59),

$$\gamma^0\boldsymbol{\gamma} = i\gamma_5\boldsymbol{\sigma}, \tag{2-6.88}$$

which, combined with the hyperbolic relations,

$$\cosh \tfrac{1}{2}\theta = \left(\frac{p^0 + m}{2m}\right)^{1/2}, \qquad \sinh \tfrac{1}{2}\theta = \left(\frac{p^0 - m}{2m}\right)^{1/2}, \qquad (2\text{-}6.89)$$

gives

$$u_{p\sigma} = \left[\left(\frac{p^0 + m}{2m}\right)^{1/2} + \left(\frac{p^0 - m}{2m}\right)^{1/2} i\gamma_5\boldsymbol{\sigma}\cdot\mathbf{p}/|\mathbf{p}|\right] v_\sigma. \qquad (2\text{-}6.90)$$

This form shows the utility of defining the v_σ with respect to $\mathbf{p}$ as a spin reference direction. Then $\boldsymbol{\sigma}\cdot\mathbf{p}/|\mathbf{p}|$ can be replaced by the eigenvalue σ, which is now a helicity label. On employing the relation (2–6.83) these vectors become, simply,

$$u_{p\sigma} = \left(\frac{p^0 + m}{2m}\right)^{1/2} v_\sigma + \left(\frac{p^0 - m}{2m}\right)^{1/2} v^*_{-\sigma}. \qquad (2\text{-}6.91)$$

They are also connected with their complex conjugates by

$$u^*_{p-\sigma} = i\sigma\gamma_5 u_{p\sigma}. \qquad (2\text{-}6.92)$$

When the following version of (2–6.85),

$$\frac{m - \gamma p}{2m} = \sum_\sigma u_{p\sigma} u^*_{p\sigma} \gamma^0 \qquad (2\text{-}6.93)$$

is combined with the orthonormality statements (2–6.87), we recognize that this non-Hermitian matrix has the algebraic projection property

$$\left(\frac{m - \gamma p}{2m}\right)^2 = \frac{m - \gamma p}{2m}. \qquad (2\text{-}6.94)$$

This is equivalent to

$$(m - \gamma p)(m + \gamma p) = 0, \qquad (2\text{-}6.95)$$

which is directly verifiable, since

$$(\gamma p)^2 = \tfrac{1}{2}\{\gamma_\mu, \gamma_\nu\} p^\mu p^\nu = -p^2 = m^2. \qquad (2\text{-}6.96)$$

We also learn that $u_{p\sigma}$ and $u^*_{p\sigma}\gamma^0$ obey

$$(\gamma p + m)u_{p\sigma} = 0, \qquad u^*_{p\sigma}\gamma^0(\gamma p + m) = 0. \qquad (2\text{-}6.97)$$

Let us return to the source coupling (2–6.73) and write

$$i\int (dx)(dx')\eta_1(x)\gamma^0 G_+(x - x')\eta_2(x') = 2m\int d\omega_p \sum_\sigma i\eta_1(p)^*\gamma^0 u_{p\sigma} u^*_{p\sigma}\gamma^0 i\eta_2(p)$$

$$= \sum_{p\sigma} i\eta^*_{1p\sigma} i\eta_{2p\sigma}, \qquad (2\text{-}6.98)$$

where

$$\eta_{p\sigma} = (2m\, d\omega_p)^{1/2} u^*_{p\sigma}\gamma^0\eta(p), \qquad \eta^*_{p\sigma} = (2m\, d\omega_p)^{1/2}\eta(p)^*\gamma^0 u_{p\sigma}; \qquad (2\text{-}6.99)$$

the consistency of the two definitions conveys the Hermitian nature of γ^0. These are the precise definitions of single particle emission and absorption sources, which have been built up from various factors. Thus $B(p)$ is contained in $\gamma^0 u_{p\sigma}$. In the rest frame of the particle, $u_{p\sigma}$ reduces to v_σ, which is an eigen-vector of γ^0 and therefore of the space-reflection matrix $r_s = i\gamma^0$. Thus the single-particle states have a definite, if imaginary, parity. Incidentally we did not prejudge this question by using the same matrix, ρ_3, in defining r_s and the projection factor $\frac{1}{2}(1 + \rho_3)$. It is now clear that the latter also performs a parity selection, and that the reflection matrix must be defined accordingly.

The particle sources $\eta_{p\sigma}$ and $\eta_{p\sigma}^*$, as linear functions of the $\eta_\zeta(x)$, are also totally anticommutative,

$$\{\eta_{p\sigma},\, \eta_{p'\sigma'}\} = \{\eta_{p\sigma},\, \eta_{p'\sigma'}^*\} = \{\eta_{p\sigma}^*,\, \eta_{p'\sigma'}^*\} = 0. \tag{2-6.100}$$

In particular,

$$(\eta_{p\sigma})^2 = 0, \qquad (\eta_{p\sigma}^*)^2 = 0. \tag{2-6.101}$$

The commutativity of even source functions is used to write

$$\exp\left[\sum_{p\sigma} i\eta_{1p\sigma}^* i\eta_{2p\sigma}\right] = \prod_{p\sigma} \exp(i\eta_{1p\sigma}^* i\eta_{2p\sigma})$$

$$= \prod_{p\sigma} \sum_{n_{p\sigma}} \frac{1}{n_{p\sigma}!} (i\eta_{1p\sigma}^* i\eta_{2p\sigma})^{n_{p\sigma}}. \tag{2-6.102}$$

All this is quite the same as with B. E. statistics. But now the power series contains just two terms: $n_{p\sigma} = 0, 1$, for, on reversing the multiplication order of two elements, we see that

$$(i\eta_{1p\sigma}^* i\eta_{2p\sigma})^2 = i\eta_{1p\sigma}^* i\eta_{2p\sigma} i\eta_{1p\sigma}^* i\eta_{2p\sigma}$$

$$= -(i\eta_{1p\sigma}^*)^2 (i\eta_{2p\sigma})^2 = 0, \tag{2-6.103}$$

and the whole series terminates at $n_{p\sigma} = 1$. In this limitation to a maximum value of unity for what are clearly particle occupation numbers we have a statement of the Exclusion Principle, which is a characteristic feature of F. D. statistics.

The causal situation is conveyed by the causal analysis of the vacuum amplitude,

$$\langle 0_+|0_-\rangle^\eta = \sum_{\{n\}} \langle 0_+|\{n\}\rangle^{\eta_1} \langle\{n\}|0_-\rangle^{\eta_2}. \tag{2-6.104}$$

It is indeed possible to factor the coupling terms in the desired way, but strict account must be kept of the minus signs that are involved. This is facilitated by the following procedure, which we illustrate with two particle states, labeled a and b,

$$(i\eta_{1a}^* i\eta_{2a})(i\eta_{1b}^* i\eta_{2b}) = (i\eta_{1b}^* i\eta_{1a}^*)(i\eta_{2a} i\eta_{2b}). \tag{2-6.105}$$

By always displacing sources through an even number of factors, one avoids the explicit appearance of minus signs. In this way we arrive at a factorization where the emission sources are multiplied in some order, read from left to right, while the absorption sources appear in the same order, but read from right to left. It is given general expression by the following identification of multi-particle states:

$$\langle\{n\}|0_-\rangle^\eta = \langle 0_+|0_-\rangle^\eta \prod_{p\sigma} (i\eta_{p\sigma})^{n_{p\sigma}},$$

$$\langle 0_+|\{n\}\rangle^\eta = \langle 0_+|0_-\rangle^\eta \prod_{p\sigma}^T (i\eta_{p\sigma}^*)^{n_{p\sigma}}, \tag{2-6.106}$$

in which $\prod^T$ symbolizes the opposite multiplication sense from $\prod$ and any standard sequence can be used for the denumerably infinite number of particle states. As in the B. E. discussion, the particle occupation number interpretation of $n_{p\sigma}$ is supported by the response to source translation, $\eta(x) \to \eta(x+X)$, which gives

$$\langle\{n\}|0_-\rangle^\eta \to e^{iPX}\langle\{n\}|0_-\rangle^\eta, \qquad \langle 0_+|\{n\}\rangle^\eta \to \langle 0_+|\{n\}\rangle^\eta e^{-iPX}, \tag{2-6.107}$$

where

$$P^\mu = \sum_{p\sigma} n_{p\sigma} p^\mu \tag{2-6.108}$$

shows the additive contributions of the various particles that are present.

The completeness requirement on the multiparticle states is stated alternatively as

$$\sum_{\{n\}} \langle 0_-|\{n\}\rangle^\eta \langle\{n\}|0_-\rangle^\eta = \langle 0_-|0_-\rangle = 1, \tag{2-6.109}$$

where

$$\langle 0_-|\{n\}\rangle^\eta = \langle\{n\}|0_-\rangle^{\eta*}, \tag{2-6.110}$$

and by

$$\sum_{\{n\}} \langle 0_+|\{n\}\rangle^\eta \langle\{n\}|0_+\rangle^\eta = \langle 0_+|0_+\rangle = 1, \tag{2-6.111}$$

with

$$\langle\{n\}|0_+\rangle^\eta = \langle 0_+|\{n\}\rangle^{\eta*}. \tag{2-6.112}$$

We have been at pains to write these more carefully than in the B. E. situation, since we are now dealing with functions of anticommuting numbers. No precautions are needed for the vacuum amplitude, which is an even function, and we present the two completeness statements as

$$1 = |\langle 0_+|0_-\rangle^\eta|^2 \sum_{\{n\}} \left[\prod_{p\sigma} (\eta_{p\sigma})^{n_{p\sigma}}\right]^* \left[\prod_{p\sigma} (\eta_{p\sigma})^{n_{p\sigma}}\right],$$

$$1 = |\langle 0_+|0_-\rangle^\eta|^2 \sum_{\{n\}} \left[\prod_{p\sigma}^T (\eta_{p\sigma}^*)^{n_{p\sigma}}\right]\left[\prod_{p\sigma}^T (\eta_{p\sigma}^*)^{n_{p\sigma}}\right]^*, \tag{2-6.113}$$

where we have omitted the compensating factors of i and $-i$. The comparison of the two forms suggests a rule of complex conjugation for F. D. sources that we shall find is a consistent one: complex conjugation also inverts the sense of multiplication, as illustrated by

$$(\eta_a\eta_b)^* = \eta_b^*\eta_a^*. \tag{2-6.114}$$

Then, the single statement of completeness is conveyed by

$$|\langle 0_+|0_-\rangle^\eta|^{-2} = \sum_{\{n\}}\left[\prod_{p\sigma}{}^T (\eta_{p\sigma}^*)^{n_{p\sigma}}\right]\left[\prod_{p\sigma}(\eta_{p\sigma})^{n_{p\sigma}}\right]$$

$$= \sum_{\{n\}}\prod_{p\sigma}(\eta_{p\sigma}^*\eta_{p\sigma})^{n_{p\sigma}} = \prod_{p\sigma}\sum_{n_{p\sigma}}(\eta_{p\sigma}^*\eta_{p\sigma})^{n_{p\sigma}}$$

$$= \prod_{p\sigma}\exp(\eta_{p\sigma}^*\eta_{p\sigma}) \tag{2-6.115}$$

or

$$|\langle 0_+|0_-\rangle^\eta|^{-2} = \exp\left[\sum_{p\sigma}\eta_{p\sigma}^*\eta_{p\sigma}\right], \tag{2-6.116}$$

which essentially reverses the factorization procedure of the causal analysis.

We must confirm this implication of completeness with a direct computation of $|\langle 0_+|0_-\rangle^\eta|^2$. It is important to recognize that the complex conjugation rule for F. D. sources implies that the product of two real sources is imaginary,

$$\left(\eta_\zeta(x)\eta_{\zeta'}(x')\right)^* = \eta_{\zeta'}(x')\eta_\zeta(x)$$
$$= -\eta_\zeta(x)\eta_{\zeta'}(x'). \tag{2-6.117}$$

Consequently, $\eta(x)\gamma^0\eta(x')$ is real since γ^0 is imaginary. This is another aspect of the matching of the statistics to the spin. Since the matrices $(1/i)\gamma^\mu$ are real, the only complex quantity in W is $\Delta_+(x - x')$, and

$$2\,\mathrm{Im}\,W = \int(dx)(dx')\eta(x)\gamma^0(m - \gamma(1/i)\partial)\,\mathrm{Re}\,(1/i)\Delta_+(x - x')\eta(x'). \tag{2-6.118}$$

The relation

$$\mathrm{Re}\,(1/i)\Delta_+(x - x') = \mathrm{Re}\int d\omega_p e^{ip(x-x')} \tag{2-6.119}$$

then gives

$$2\,\mathrm{Im}\,W = \mathrm{Re}\int d\omega_p \eta(-p)\gamma^0(m - \gamma p)\eta(p)$$

$$= \mathrm{Re}\,2m\int d\omega_p \sum_\sigma \eta(p)^*\gamma^0\, u_{p\sigma}u_{p\sigma}^*\gamma^0\eta(p)$$

$$= \mathrm{Re}\sum_{p\sigma}\eta_{p\sigma}^*\eta_{p\sigma}. \tag{2-6.120}$$

The injunction symbolized by Re is redundant, since

$$(\eta_{p\sigma}^*\eta_{p\sigma})^* = \eta_{p\sigma}^*\eta_{p\sigma}, \tag{2-6.121}$$

which makes essential use of the complex conjugation rule. This result,

$$|\langle 0_+|0_-\rangle^\eta|^2 = \exp\left[-\sum_{p\sigma}\eta_{p\sigma}^*\eta_{p\sigma}\right], \tag{2-6.122}$$

is the verification of completeness.

The Euclidean postulate was introduced as a sharpened version of the principle of space-time uniformity. It has new and interesting implications for spin ½ particles, if it is interpreted to mean that the Euclidean transcription may contain no indication of the original Minkowski space. All reference to

Minkowski space does disappear in the Euclidean description of integer spin particles, but spin $\frac{1}{2}$ introduces a new situation. In discussing unit spin, for example, we observed that the Hermitian, real, symmetrical matrices $s_{k4} = is_{0k}$ are converted to Hermitian, imaginary, antisymmetrical matrices, thus uniting them with the s_{kl}, by means of the transformation associated with $J_4 = iJ^0$. Note that it is the square root of the space-reflection matrix, or its negative, that enters this transformation. To perform an analogous operation on the real symmetrical matrices $\sigma_{k4} = \gamma^0\gamma_k$, unifying them with the imaginary antisymmetrical $\sigma_{kl} = i\gamma_k\gamma_l$, we must find a suitable unitary transformation, one that commutes with the latter set. The only possibilities available for the unitary matrix are

$$\exp[i\varphi(\gamma^0, i\gamma_5, \gamma^0\gamma_5)]. \tag{2-6.123}$$

But all these matrices are real, and cannot change the reality of the σ_{k4}. Accordingly, an inspection of the reality, or symmetry of the $\sigma_{\mu\nu}$, $\mu, \nu = 1, \ldots, 4$, leaves no doubt about which Euclidean axis is related to the Minkowski time axis. This is a violation of the Euclidean postulate.

We have already remarked that the symmetry of matrices can be reversed, without altering their space-time character, by using an independent antisymmetrical matrix

$$q = \begin{pmatrix} 0 & -i \\ i & 0 \end{pmatrix}, \tag{2-6.124}$$

which acts upon an additional two-valued source index. Its introduction enables us to form a complex unitary matrix by multiplying the real $r_s = i\gamma^0$ by the imaginary q and then taking the square root, in analogy with the unit spin procedure. The explicit transformation is

$$\sigma_{\mu\nu E} = e^{(\pi i/4)q\gamma^0}\sigma_{\mu\nu}e^{-(\pi i/4)q\gamma^0}, \tag{2-6.125}$$

and indeed

$$\sigma_{klE} = \sigma_{kl}, \qquad \sigma_{k4E} = iq\gamma_k \tag{2-6.126}$$

are all imaginary, antisymmetrical matrices. The detailed transformation from Minkowski to Euclidean source is given by

$$e^{(\pi i/4)(q\gamma^0-1)}\eta(x)(dx) \rightarrow \eta(x)(dx)\big|_E. \tag{2-6.127}$$

When this transformation is performed in the vacuum amplitude, one encounters the following matrix (note that $q\gamma^0$ is symmetrical):

$$e^{-(\pi i/4)(q\gamma^0-1)}\gamma^0\big(m - \gamma_\mu(1/i)\partial_\mu\big)e^{-(\pi i/4)(q\gamma^0-1)} = qm - \alpha_\mu\partial_\mu, \tag{2-6.128}$$

where the components of

$$\alpha_\mu = e^{-(\pi i/4)q\gamma^0}\gamma^0\gamma_\mu e^{-(\pi i/4)q\gamma^0} \tag{2-6.129}$$

are given by ($\gamma_4 = i\gamma^0$)

$$\alpha_k = \gamma^0\gamma_k, \qquad \alpha_4 = q\gamma^0. \tag{2-6.130}$$

They are all real, symmetrical matrices that obey

$$\tfrac{1}{2}\{\alpha_\mu, \alpha_\nu\} = \delta_{\mu\nu}, \tag{2-6.131}$$

and

$$\sigma_{\mu\nu E} = (1/2i)[\alpha_\mu, \alpha_\nu]. \tag{2-6.132}$$

The resulting Euclidean transcription of the spin $\frac{1}{2}$ vacuum amplitude is

$$\langle 0_+|0_-\rangle^\eta \to \exp\left[-\tfrac{1}{2}\int (dx)(dx')\eta(x)(qm - \alpha_\mu\partial_\mu)\,\Delta(x - x')\eta(x')\right]_E, \tag{2-6.133}$$

which is a real structure when real Euclidean sources are used.

The implication of the Euclidean postulate, that every spin $\frac{1}{2}$ particle possesses a charge-like attribute, is entirely compatible with the empirical situation. Although we must give special attention to the massless neutrinos, it is a general inference from the data that every fermion (F. D. particle), including electrically neutral ones, has its antiparticle counterpart, while no electrically neutral boson (B. E. particle) shows such duplexity. The charge label $q = \pm 1$ is added to the spin $\frac{1}{2}$ states by enlarging v_σ and $u_{p\sigma}$ to be eigenvectors of the charge matrix with the eigenvalue q. Since the charge matrix is imaginary, complex conjugation introduces $-q$, and some correspondingly modified statements are

$$u^*_{p\,-\sigma\,-q} = i\sigma\gamma_5 u_{p\sigma q} \tag{2-6.134}$$

and

$$u_{p\sigma q} = \left(\frac{p^0 + m}{2m}\right)^{1/2} v_{\sigma q} + \left(\frac{p^0 - m}{2m}\right)^{1/2} v^*_{-\sigma\,-q}. \tag{2-6.135}$$

The related particle source definitions are

$$\eta_{p\sigma q} = (2m\,d\omega_p)^{1/2} u^*_{p\sigma q}\gamma^0 \eta(p), \qquad \eta^*_{p\sigma q} = (2m\,d\omega_p)^{1/2}\eta(p)^*\gamma^0 u_{p\sigma q}. \tag{2-6.136}$$

This discussion of the Euclidean postulate brings the *TCP* operation to mind. Through the attached Euclidean group we produce the transformation

$$\bar{x}^\mu = -x^\mu \tag{2-6.137}$$

as

$$\bar{\eta}(\bar{x}) = r_{st}\eta(x), \tag{2-6.138}$$

where

$$r_{st} = e^{(\pi i/2)\sigma_{12}}e^{(\pi i/2)\sigma_{34}} = e^{(\pi i/2)\sigma_{23}}e^{(\pi i/2)\sigma_{14}} = e^{(\pi i/2)\sigma_{31}}e^{(\pi i/2)\sigma_{24}} = -i\gamma_5 \tag{2-6.139}$$

details the equivalent rotations through the angle π in two perpendicular planes. This matrix is antisymmetrical, imaginary, and obeys

$$r_{st}^2 = 1. \tag{2-6.140}$$

The invariance of the vacuum amplitude is verified directly on using the relation

$$i\gamma_5\gamma^0(m - \gamma^\mu(1/i)(-)\partial_\mu)(-i\gamma_5)\Delta_+(-x + x')$$
$$= \gamma^0(m - \gamma^\mu(1/i)\partial_\mu)\Delta_+(x - x'). \tag{2-6.141}$$

This is accomplished, however, at the expense of replacing the real η by an imaginary $\bar{\eta}$, since that is the nature of r_{st}. If we insist on a real $\bar{\eta}$, as in the transformation

$$\bar{\eta}(\bar{x}) = \gamma_5 \eta(x), \qquad (2\text{-}6.142)$$

W turns into $-W$. But this sign change can be compensated by reversing the multiplication order of all sources, which is in accord with the representation of causal sequence by multiplicative position.

The effect, on the individual emission and absorption sources, of the substitution

$$\eta(p) \rightarrow \gamma_5 \eta(-p) \qquad (2\text{-}6.143)$$

is given by

$$\eta_{p\sigma q} \rightarrow (2m \, d\omega_p)^{1/2} u^*_{p\sigma q} \gamma^0 \gamma_5 \eta(-p)$$
$$= (2m \, d\omega_p)^{1/2} \eta(p)^* \gamma^0 u_{p \, -\sigma \, -q}(-i\sigma) \qquad (2\text{-}6.144)$$

or

$$\eta_{p\sigma q} \rightarrow -i\sigma \eta^*_{p \, -\sigma \, -q}, \qquad (2\text{-}6.145)$$

and

$$\eta^*_{p\sigma q} \rightarrow i\sigma \eta_{p \, -\sigma \, -q}. \qquad (2\text{-}6.146)$$

The resulting correspondence between multiparticle emission and absorption processes is

$$\langle \{n\} | 0_- \rangle^\eta \rightarrow \exp\left(-\frac{\pi i}{2} \sum n_{p\sigma q} \sigma \right) \langle 0_+ | \{n'\} \rangle^\eta,$$
$$\langle 0_+ | \{n\} \rangle^\eta \rightarrow \exp\left[\frac{\pi i}{2} \sum n_{p\sigma q} \sigma \right] \langle \{n'\} | 0_- \rangle^\eta, \qquad (2\text{-}6.147)$$

where

$$n'_{p\sigma q} = n_{p \, -\sigma \, -q}, \qquad (2\text{-}6.148)$$

and the source transformation that constitutes part of the TCP operation produces the required reversal of multiplication order.

2–7 MORE ABOUT SPIN $\frac{1}{2}$ PARTICLES.   NEUTRINOS

As a preliminary to discussing the angular momentum specification of particle states we review the addition of orbital angular momentum with spin $\frac{1}{2}$. States of total angular momentum quantum number $j = l \pm \frac{1}{2}$ are selected from the subspace with orbital quantum number l by the Hermitian projection operators

$$M_{lj} = \frac{l + \frac{1}{2} \pm (\frac{1}{2} + \mathbf{L} \cdot \boldsymbol{\sigma})}{2l + 1} = \frac{j + \frac{1}{2} \pm \mathbf{L} \cdot \boldsymbol{\sigma}}{2j + 1 \mp 1}. \qquad (2\text{-}7.1)$$

They obey

$$M_{lj} M_{lj'} = \delta_{jj'} M_{lj}, \qquad \sum_j M_{lj} = 1 \qquad (2\text{-}7.2)$$

and have a trace appropriate to the multiplicity,

$$\text{tr } M_{lj} = 2j + 1. \tag{2-7.3}$$

We define orthonormal spin-angle functions

$$Z_{ljm} = M_{lj} Y_{l\,m-\frac{1}{2}} v_+ \langle l\,m - \tfrac{1}{2} + |M_{lj}|l\,m - \tfrac{1}{2} +\rangle^{-1/2}, \tag{2-7.4}$$

which are given explicitly by

$l = j - \frac{1}{2}$:

$$\left(\frac{j+m}{2j}\right)^{1/2} Y_{j-\frac{1}{2}\,m-\frac{1}{2}} v_+ + \left(\frac{j-m}{2j}\right)^{1/2} Y_{j-\frac{1}{2}\,m+\frac{1}{2}} v_-,$$

$$\tag{2-7.5}$$

$l = j + \frac{1}{2}$:

$$\left(\frac{j+1-m}{2j+2}\right)^{1/2} Y_{j+\frac{1}{2}\,m-\frac{1}{2}} v_+ - \left(\frac{j+1+m}{2j+2}\right)^{1/2} Y_{j+\frac{1}{2}\,m+\frac{1}{2}} v_-.$$

The two functions are also connected by an operator:

$$\boldsymbol{\sigma} \cdot \mathbf{n} Z_{j\mp\frac{1}{2}\,j\,m} = Z_{j\pm\frac{1}{2}\,j\,m}, \tag{2-7.6}$$

where $\mathbf{n}$ is the unit vector that supplies the angle variables of the spherical harmonics. The following properties of $\boldsymbol{\sigma} \cdot \mathbf{n}$ are involved: it commutes with the total angular momentum vector, but alters the orbital angular momentum by unity; it does not change the orthonormality of the spin-angle functions—it has unit square. All this shows that the left- and right-hand sides of (2-7.6) are the same, to within phase constants that cannot depend upon m. It then suffices to set $\mathbf{n}$ parallel to the third axis and choose $m = \frac{1}{2}$. The only surviving harmonic, $Y_{l0} = (2l + 1/4\pi)^{1/2}$, selects v_+, and (2-7.6) is confirmed.

The structure of the source coupling produced by single-particle exchange is (causal subscripts are omitted for simplicity)

$$\int d\omega_p i\eta(p)^* \exp[-\tfrac{1}{2}\theta i\gamma_5 \boldsymbol{\sigma} \cdot \mathbf{p}/|\mathbf{p}|] 2m\tfrac{1}{2}(1 + \gamma^0) \exp[-\tfrac{1}{2}\theta i\gamma_5 \boldsymbol{\sigma} \cdot \mathbf{p}/|\mathbf{p}|]\eta(p),$$

$$\tag{2-7.7}$$

restating Eqs. (2-6.73, 74). We introduce the preliminary transformation

$$(2m\,d\omega_p)^{1/2}\eta(p) = \sum_{lm} (d\Omega)^{1/2} Y_{lm}(\mathbf{p})\eta_{p^0 lm}, \tag{2-7.8}$$

where

$$\eta_{p^0 lm} = \left(\frac{2m|\mathbf{p}|\,dp^0}{\pi}\right)^{1/2} i^{-l} \int (dx) e^{ip^0 x^0} j_l(|\mathbf{p}|\,|\mathbf{x}|) Y_{lm}(\mathbf{x})^* \eta(x) \tag{2-7.9}$$

also carries an unwritten index, expressing that of the multicomponent $\eta(x)$. The projection matrix $\frac{1}{2}(1 + \gamma^0)$ makes a selection of these components, and $\frac{1}{2}(1 + \gamma^0)i\gamma_5 = i\gamma_5\frac{1}{2}(1 - \gamma^0)$ makes a complementary selection. This will be indicated by adding subscripts $+$ and $-$ to $\eta_{p^0 lm}$. The residual spin multiplicity is coupled with the $Y_{lm}(\mathbf{p})$ to produce the spin-angle functions, as in

$$\sum_{lm} (d\Omega)^{1/2} Y_{lm}(\mathbf{p}) \eta_{\pm p^0 lm} = \sum_{ljm} (d\Omega)^{1/2} Z_{ljm} \eta_{\pm p^0 ljm}, \qquad (2\text{-}7.10)$$

where we rely on context to distinguish the orbital magnetic quantum number m, which assumes integer values, from the total angular momentum magnetic quantum number m, which is an integer $+\frac{1}{2}$. The specific combination that appears in (2-7.7), for given j, m, is

$$\cos \tfrac{1}{2}\theta \sum_l (d\Omega)^{1/2} Z_{ljm} \eta_{+p^0 ljm} - \sin \tfrac{1}{2}\theta \boldsymbol{\sigma} \cdot (\mathbf{p}/|\mathbf{p}|) \sum_l (d\Omega)^{1/2} Z_{ljm} \eta_{-p^0 ljm}$$

$$= \sum_l (d\Omega)^{1/2} Z_{ljm} [\cos \tfrac{1}{2}\theta \eta_{+p^0 ljm} - \sin \tfrac{1}{2}\theta \eta_{-p^0 \bar{l}jm}], \qquad (2\text{-}7.11)$$

and its complex conjugate, where

$$\bar{l} = 2j - l, \qquad (2\text{-}7.12)$$

indicates the orbital angular momentum change that is produced by $\boldsymbol{\sigma} \cdot \mathbf{p}/|\mathbf{p}|$. The orthonormality of the Z_{ljm} in the subspace selected by $\frac{1}{2}(1 + \gamma^0)$ gives the resulting form of (2-7.7):

$$\sum_{p^0 ljm} i\eta^*_{p^0 ljm} i\eta_{p^0 ljm}, \qquad (2\text{-}7.13)$$

where

$$\eta_{p^0 ljm} = \left(\frac{p^0 + m}{2m}\right)^{1/2} \eta_{+p^0 ljm} - \left(\frac{p^0 - m}{2m}\right)^{1/2} \eta_{-p^0 \bar{l}jm} \qquad (2\text{-}7.14)$$

and the charge label is left unwritten.

On combining the various transformations, these single particle sources are exhibited as

$$\eta_{p^0 ljmq} = \int (dx) \psi_{p^0 ljmq}(x)^* \gamma^0 \eta(x), \qquad (2\text{-}7.15)$$

with

$$\psi_{p^0 ljmq}(x) = \left(\frac{|\mathbf{p}|\, dp^0}{\pi}\right)^{1/2} e^{-ip^0 x^0} [(p^0 + m)^{1/2} i^l j_l(|\mathbf{p}|\,|\mathbf{x}|) Z_{ljmq}$$

$$+ (p^0 - m)^{1/2} i^{\bar{l}} j_{\bar{l}}(|\mathbf{p}|\,|\mathbf{x}|) i\gamma_5 Z_{\bar{l}jmq}], \qquad (2\text{-}7.16)$$

wherein the Z_{ljmq} are constructed as in (2-7.5) from the eigenvectors $v_{\sigma q}$, and the spherical harmonics refer to the angles of the unit coordinate vector. The comparison of (2-7.13) and (2-7.15) with the left-hand member of (2-6.73) supplies the identification

$$x^0 > x^{0\prime}: \qquad G_+(x - x') = i \sum_{p^0 ljmq} \psi_{p^0 ljmq}(x) \psi_{p^0 ljmq}(x')^* \gamma^0 \qquad (2\text{-}7.17)$$

and the antisymmetry of $\gamma^0 G_+(x - x)$ extends this to

$$x^0 < x^{0\prime}: \qquad G_+(x - x') = -i \sum_{p^0 ljmq} \psi_{p^0 ljmq}(x)^* \psi_{p^0 ljmq}(x') \gamma^0. \qquad (2\text{-}7.18)$$

Unlike linear momentum states in general, angular momentum states permit a specification of space-reflection parity. The response of the particle sources to

$$\eta(x^0, \mathbf{x}) \rightarrow i\gamma^0 \eta(x^0, -\mathbf{x}) \qquad (2\text{-}7.19)$$

involves the transformation behavior

$$\gamma^0 \psi_{p^0 ljmq}(x^0, -\mathbf{x}) = (-1)^l \psi_{p^0 ljmq}(x^0, \mathbf{x}). \qquad (2\text{-}7.20)$$

This follows from the homogeneous nature of spherical harmonics,

$$Y_{lm}(-\mathbf{x}) = (-1)^l Y_{lm}(\mathbf{x}), \qquad (2\text{-}7.21)$$

and the significance of Z_{ljmq} as an eigenvector of γ^0 with the eigenvalue $+1$. While $i\gamma_5 Z_{\bar{l}jmq}$ assigns the eigenvalue -1 to γ^0, this sign change is compensated by $(-1)^{\bar{l}} = -(-1)^l$. The result is

$$\eta_{p^0 ljmq} \rightarrow i(-1)^l \eta_{p^0 ljmq}, \qquad (2\text{-}7.22)$$

which exhibits space parity as a product of two factors, the intrinsic parity i, and the variable orbital parity $(-1)^l$. The label l in $\psi_{p^0 ljmq}$ should be understood as $(-1)^l$, the exact parity quantum number, for both orbital angular momenta, l and $\bar{l}$, are present in this function. In the spin ½ situation the two states with common values of j, m can be distinguished by their different parity values. For spinless particles, according to (2–2.24), parity also appears as the orbital parity $(-1)^l$, multiplied into an intrinsic parity which is $+1$ for a scalar, -1 for a pseudoscalar source. Here, parity is superfluous as a label, being completely determined by the angular momentum quantum number. With unit spin particles, however, parity is insufficient to identify all three states of specified total angular momentum. In addition to an intrinsic parity factor, -1 for a vector, $+1$ for an axial vector, the state described in (2–3.39) has the orbital parity $(-1)^j$, representing $l = j$, while the two states of (2–3.41, 42) have the orbital parity $-(-1)^j$, which is common to $l = j \pm 1$. But for the massless photon there are just two types of states of a given angular momentum quantum number $j \geq 1$. The photon state with source $J_{p^0 jm1}$ has parity $-(-1)^j$, and that created by $J_{p^0 jm2}$ has parity $(-1)^j$. The two kinds of sources are conventionally called magnetic and electric multipole moments, respectively.

Before investigating the effect of the TCP operation on angular momentum states, we examine the reality properties of $\psi_{p^0 ljmq}(x)$. Let us first note that

$$Z^*_{ljmq} = (-1)^{l+j+m} i\gamma_5 Z_{l\ j\ -m\ -q}, \qquad (2\text{-}7.23)$$

which uses the spherical harmonic property

$$Y^*_{lm} = (-1)^m Y_{l\ -m} \qquad (2\text{-}7.24)$$

and the complex conjugation behavior of $v_{\sigma q}$, being Eq. (2–6.134) with $p^0 = m$. On forming the complex conjugate of $\psi_{p^0 ljmq}$, the additional minus signs that are produced by the explicit appearance of i may be compensated through the space-time reflection $x^\mu \to -x^\mu$,

$$\psi_{p^0 ljmq}(-x)^* = (-1)^{l+j+m} i\gamma_5 \psi_{p^0 l\, j\, -m\, -q}(x). \qquad (2\text{-}7.25)$$

The connection which this relation establishes between the two causal forms of $G_+(x - x')$ is conveyed by the invariance property

$$\gamma^0 G_+(x - x') = -i\gamma_5 \gamma^0 G_+(-x + x') i\gamma_5, \qquad (2\text{-}7.26)$$

in agreement with (2-6.141). The effect, in Eq. (2-7.15), of the substitution

$$\eta(x) \to \gamma_5 \eta(-x) \qquad (2\text{-}7.27)$$

is given by

$$\eta_{p^0 ljmq} \to \int (dx)\psi_{p^0 ljmq}(-x)^* \gamma^0 \gamma_5 \eta(x) = i(-1)^{l+j+m} \eta^*_{p^0 l\, j\, -m\, -q}, \qquad (2\text{-}7.28)$$

and then

$$\eta^*_{p^0 ljmq} \to -i(-1)^{l+j+m} \eta_{p^0 l\, j\, -m\, -q}. \qquad (2\text{-}7.29)$$

This gives the detailed correspondence between single-particle emission and absorption acts. The multiparticle correspondence is analogous, with the reversal of multiplication order appearing as an aspect of the TCP transformation.

A space-time description of the multiparticle exchange between sources is produced by the power series expansion:

$$\exp\left[i \int (dx)(dx')\eta_1(x)\gamma^0 G_+(x - x')\eta_2(x')\right]$$
$$= 1 + \sum_{n=1} i^n \int \frac{(dx_1)\cdots(dx_n)}{n!} \frac{(dx_1')\cdots(dx_n')}{n!} \eta_1(x_n)\cdots\eta_1(x_1)$$
$$\times [\det_{(n)} \gamma^0 G_+(x_i - x_j')]\eta_2(x_1')\cdots\eta_2(x_n'), \qquad (2\text{-}7.30)$$

where the discrete indices on sources and propagation functions are regarded as combined with the explicit space-time coordinates. In contrast with the permanent of (2-2.37), the symmetry of which conveys the commutativity of B. E. sources, the antisymmetrical determinant

$$\det_{(n)} \gamma^0 G_+(x_i - x_j') = \sum_{n!\,\text{perm.}} \epsilon_{j_1\cdots j_n} \gamma^0 G_+(x_1 - x_{j_1}') \cdots \gamma^0 G_+(x_n - x_{j_n}') \qquad (2\text{-}7.31)$$

expresses the anticommutativity of F. D. sources. We see here the simple and necessary connection between the symmetry properties that characterize the two statistics and the elementary algebraic properties that distinguish the two kinds of sources. Appropriately symmetrized products of individual propagation functions give the space-time representation of the noninteracting multiparticle situation.

Let us discuss now those generalizations in which the terminal vacuum states are replaced by multiparticle states. A causal situation is considered, containing emission source η_2, probe source η_0, detection source η_1:

$$\eta(x) = \eta_1(x) + \eta_0(x) + \eta_2(x). \qquad (2\text{-}7.32)$$

The vacuum amplitude is given by

$$\langle 0_+|0_-\rangle^{\eta} = \langle 0_+|0_-\rangle^{\eta_1+\eta_2} \exp\left[i\int (dx)(dx')\eta_1(x)\gamma^0 G_+(x-x')\eta_0(x')\right.$$
$$\left. + i\int (dx)(dx')\eta_0(x)\gamma^0 G_+(x-x')\eta_2(x')\right]\langle 0_+|0_-\rangle^{\eta_0}$$
$$= \left[\sum_{\{n\}} \langle 0_+|\{n\}\rangle^{\eta_1}\langle\{n\}|0_-\rangle^{\eta_2}\right] \exp\left[\sum_r (i\eta_{1r}^*i\eta_{0r} + i\eta_{0r}^*i\eta_{2r})\right]\langle 0_+|0_-\rangle^{\eta_0},$$
$$(2\text{--}7.33)$$

where the index r represents any set of single-particle labels, say $p\sigma q$. The causal analysis of this vacuum amplitude is

$$\langle 0_+|0_-\rangle^{\eta} = \sum_{\{n\},\{n'\}} \langle 0_+|\{n\}\rangle^{\eta_1}\langle\{n\}_+|\{n'\}_-\rangle^{\eta_0}\langle\{n'\}|0_-\rangle^{\eta_2}, \quad (2\text{--}7.34)$$

from which the detailed effect of the probe source can be inferred. To describe a weak probe one must interpret the product $i\eta_r\langle\{n\}|0_-\rangle^{\eta}$. If the single-particle state or mode r is initially occupied, $n_r = 1$, the result is zero, $(\eta_r)^2 = 0$. This is the Exclusion Principle, forbidding the introduction of an additional particle into an already occupied mode. Otherwise,

$$n_r = 0: \qquad i\eta_r\langle\{n\}|0_-\rangle^{\eta} = (-1)^{n_{<r}} \langle\{n+1_r\}|0_-\rangle^{\eta}, \quad (2\text{--}7.35)$$

where $n_{<r}$ counts the number of occupied modes that precede r in the standard sequence, which is the number of source factors in $\langle\{n\}|0_-\rangle^{\eta}$ that $i\eta_r$ must be moved through in order to place it in proper position. Similarly,

$$n_r = 0: \qquad \langle 0_+|\{n\}\rangle^{\eta}i\eta_r^* = \langle 0_+|\{n+1_r\}\rangle^{\eta}(-1)^{n_{<r}}, \quad (2\text{--}7.36)$$

and we get the weak source results

$$n_r = 0: \qquad \langle\{n+1_r\}_+|\{n\}_-\rangle^{\eta} \cong (-1)^{n_{<r}}i\eta_r, $$
$$n_r = 1: \qquad \langle\{n-1_r\}_+|\{n\}_-\rangle^{\eta} \cong (-1)^{n_{<r}}i\eta_r^*. \qquad (2\text{--}7.37)$$

To construct the probability amplitude $\langle\{n\}_+|\{n\}_-\rangle^{\eta}$, one must retain only equal powers of η_{1r}^* and η_{2r} in the expansion of (2–7.33),

$$\exp\left[\sum_r (i\eta_{1r}^*i\eta_{0r} + i\eta_{0r}^*i\eta_{2r})\right] \to \prod_r [1 + i\eta_{1r}^*i\eta_{0r}i\eta_{0r}^*i\eta_{2r}]. \quad (2\text{--}7.38)$$

In contrast with the B. E. situation, the series terminates with the indicated product. On referring to (2–7.35, 36) we see that

$$\sum_{\{n\}} \langle 0_+|\{n\}\rangle^{\eta_1}i\eta_{1r}^*i\eta_{2r}\langle\{n\}|0_-\rangle^{\eta_2} = \sum_{\{n\}} \langle 0_+|\{n\}\rangle^{\eta_1}n_r\langle\{n\}|0_-\rangle^{\eta_2}, \quad (2\text{--}7.39)$$

where the factor n_r indicates the absence of the term $n_r = 0$. The effective substitution is, then,

$$\exp\left[\sum_r (i\eta_{1r}^*i\eta_{0r} + i\eta_{0r}^*i\eta_{2r})\right] \to \prod_r [1 + i\eta_{0r}n_r i\eta_{0r}^*]$$
$$= \exp\left[\sum_r i\eta_{0r}n_r i\eta_{0r}^*\right]. \quad (2\text{--}7.40)$$

The linear relation between $\eta(x)$ and emission and absorption sources for any type of mode specification can be written as

$$\eta_r = \int (dx)\psi_r(x)^* \gamma^0 \eta(x), \qquad \eta_r^* = \int (dx)\eta(x)\gamma^0 \psi_r(x). \qquad (2\text{-}7.41)$$

Thus,

$$\psi_{p\sigma q}(x) = (2m\, d\omega_p)^{1/2} u_{p\sigma q} e^{ipx}, \qquad (2\text{-}7.42)$$

and Eq. (2–7.16) supplies another example. The related construction of the propagation function is that illustrated in (2–7.17, 18):

$$x^0 > x^{0\prime}: \qquad G_+(x - x') = i \sum_r \psi_r(x)\psi_r(x')^* \gamma^0$$

$$x^0 < x^{0\prime}: \qquad G_+(x - x') = -i \sum_r \psi_r(x)^* \psi_r(x')\gamma^0. \qquad (2\text{-}7.43)$$

We now get

$$\langle \{n\}_+ | \{n\}_- \rangle^\eta = \exp\left[\tfrac{1}{2}i \int (dx)(dx')\eta(x)\gamma^0 G_{\{n\}+}(x - x')\eta(x') \right] \qquad (2\text{-}7.44)$$

with

$$G_{\{n\}+}(x - x') = G_+(x - x') - i \sum_r n_r [\psi_r(x)\psi_r(x')^* - \psi_r(x)^*\psi_r(x')]\gamma^0. \qquad (2\text{-}7.45)$$

The form of the second term assures the antisymmetry of $\gamma^0 G_{\{n\}+}(x - x')$. Explicit causal structures are

$$x^0 > x^{0\prime}: \qquad i \sum_r [(1 - n_r)\psi_r(x)\psi_r(x')^* + n_r\psi_r(x)^*\psi_r(x')]\gamma^0,$$

$$x^0 < x^{0\prime}: \qquad -i \sum_r [n_r\psi_r(x)\psi_r(x')^* + (1 - n_r)\psi_r(x)^*\psi_r(x')]\gamma^0. \qquad (2\text{-}7.46)$$

The comparison with (2–2.49) emphasizes the essential role of the statistics in stimulating (B. E.) or suppressing (F. D.) additional particle emission.

We return to the vacuum amplitude expression (2–7.33) and observe that, in general,

$$\exp\left[\sum_r (i\eta_{1r}^* i\eta_{0r} + i\eta_{0r}^* i\eta_{2r}) \right] = \prod_r [1 + i\eta_{1r}^* i\eta_{0r} + i\eta_{0r}^* i\eta_{2r}]$$

$$\times \prod_r [1 + i\eta_{1r}^* i\eta_{0r} i\eta_{0r}^* i\eta_{2r}], \qquad (2\text{-}7.47)$$

which converts (2–7.33) into

$$\langle 0_+ | 0_- \rangle^\eta = \sum_{\{n\}} \langle 0_+ | \{n\} \rangle^{\eta_1} \langle \{n\}_+ | \{n\}_- \rangle^{\eta_0} \prod_r [1 + i\eta_{1r}^* i\eta_{0r} + i\eta_{0r}^* i\eta_{2r}]\langle \{n\} | 0_- \rangle^{\eta_2}. \qquad (2\text{-}7.48)$$

A typical term of the product $\prod_r$ in (2–7.48) appears as

$$\prod_e^T (i\eta_{1r}^*) \prod_e (i\eta_{0r}) \prod_a^T (i\eta_{0r}^*) \prod_a (i\eta_{2r}), \qquad (2\text{-}7.49)$$

where the two sets of modes labeled a and e are disjoint, since the individual mode factors are linear in η_{2r} and η_{1r}^*. If a nonvanishing term is to result in

(2-7.48), it is necessary that $n_a = n_e = 0$. Then the a(bsorbed) modes are those, occupied in the initial state,

$$\prod_a (i\eta_r)\langle\{n\}|0_-\rangle^\eta = \prod_a (-1)^{n<r}\langle\{n+1_a\}|0_-\rangle^\eta, \qquad (2\text{-}7.50)$$

that are not occupied finally, while the e(mitted) modes are those occupied finally,

$$\langle 0_+|\{n\}\rangle^\eta \prod_e^T (i\eta_r^*) = \langle 0_+|\{n+1_e\}\rangle^\eta \prod_e (-1)^{n<r}, \qquad (2\text{-}7.51)$$

which were initially unoccupied. The outcome is

$$n_a = n_e = 0:$$
$$\langle\{n+1_e\}_+|\{n+1_a\}_-\rangle^\eta = \langle\{n\}_+|\{n\}_-\rangle^\eta \prod_e [(-1)^{n<r}i\eta_r] \prod_a^T [(-1)^{n<r}i\eta_r^*],$$
$$(2\text{-}7.52)$$

which is the generalization of (2-7.37).

In order to test the completeness of the multiparticle states in this general context, we multiply (2-7.52) on the left by its complex conjugate and present the result in a form that reinstates $\{n\}$ as an arbitrary initial state:

$$n_a = 1, n_e = 0:$$
$$\langle\{n\}_-|\{n+1_e-1_a\}_+\rangle^\eta\langle\{n+1_e-1_a\}_+|\{n\}_-\rangle^\eta$$
$$= |\langle\{n\}_+|\{n\}_-\rangle^\eta|^2 \prod_e (\eta_r^*\eta_r) \prod_a (\eta_r\eta_r^*). \quad (2\text{-}7.53)$$

The summation over all final states is represented by

$$\sum_{\{n'\}} \langle\{n\}_-|\{n'\}_+\rangle^\eta\langle\{n'\}_+|\{n\}_-\rangle^\eta$$
$$= |\langle\{n\}_+|\{n\}_-\rangle|^2 \prod_r [(1+(1-n_r)\eta_r^*\eta_r)(1+n_r\eta_r\eta_r^*)]$$
$$= |\langle\{n\}_+|\{n\}_-\rangle|^2 \exp\left[\sum_r (1-2n_r)\eta_r^*\eta_r\right], \qquad (2\text{-}7.54)$$

keeping in mind the anticommutativity of sources. But, according to (2-7.44, 45),

$$\langle\{n\}_+|\{n\}_-\rangle^\eta = \langle 0_+|0_-\rangle^\eta \exp\left[\sum_r n_r\eta_r^*\eta_r\right], \qquad (2\text{-}7.55)$$

and on utilizing the analogue of (2-6.122) for a general mode specification:

$$|\langle 0_+|0_-\rangle^\eta|^2 = \exp\left[-\sum_r \eta_r^*\eta_r\right], \qquad (2\text{-}7.56)$$

we get

$$|\langle\{n\}_+|\{n\}_-\rangle^\eta|^2 = \exp\left[-\sum_r (1-2n_r)\eta_r^*\eta_r\right]. \qquad (2\text{-}7.57)$$

This confirms that the left-hand side of (2-7.54) equals unity. To derive (2-7.56) directly from the propagation function construction (2-7.43) we note that, as a statement about individual elements,

$$\text{Re } \gamma^0 G_+(x - x') = \text{Re } i \sum_r \gamma^0 \psi_r(x)\psi_r(x')^* \gamma^0 \qquad (2\text{-}7.58)$$

is valid for all $x - x'$. But $i\eta(x)\eta(x')$ is real, and therefore

$$|\langle 0_+|0_-\rangle^\eta|^2 = \exp\left[-\text{Re }\sum_r \int (dx)(dx')\eta(x)\gamma^0\psi_r(x)\psi_r(x')^*\gamma^0\eta(x')\right]$$
$$= \exp\left[-\sum_r \eta_r^*\eta_r\right], \qquad (2\text{-}7.59)$$

as anticipated.

The reduction of unitarity to causality for spin $\frac{1}{2}$ particles imitates the pattern already established with spinless particles. We follow the development of the system from the initial vacuum state, under the influence of the source $\eta_{(+)}(x)$, and then trace it back to the initial state, using the source $\eta_{(-)}(x)$. The structure of the propagation function $\Delta_+(x - x')$ governs this evolution and we get, as the analogue of (2–2.83),

$$\langle 0_-|0_-\rangle^{\eta_{(-)},\eta_{(+)}} = \exp\left[-\tfrac{1}{2}i\int (dx)(dx')\eta_{(-)}(x)\gamma^0 G_-(x - x')\eta_{(-)}(x')\right.$$
$$+ \tfrac{1}{2}i\int (dx)(dx')\eta_{(+)}(x)\gamma^0 G_+(x - x')\eta_{(+)}(x')$$
$$\left.+ \int (dx)(dx')\eta_{(-)}(x)\gamma^0 G^{(+)}(x - x')\eta_{(+)}(x')\right], \qquad (2\text{-}7.60)$$

with

$$G_-(x - x') = (m - \gamma^\mu(1/i)\partial_\mu)\Delta_-(x - x'),$$
$$G^{(\pm)}(x - x') = (m - \gamma^\mu(1/i)\partial_\mu)\Delta^{(\pm)}(x - x'). \qquad (2\text{-}7.61)$$

Some relations among these matrix functions are:

$$G_-(x - x') = G_+(x - x')^*, \qquad G^{(-)}(x - x') = G^{(+)}(x - x')^*$$
$$-[\gamma^0 G^{(+)}(x' - x)]^T = \gamma^0 G^{(-)}(x - x'), \qquad (2\text{-}7.62)$$

and

$$G_+(x - x') = \begin{cases} x^0 > x^{0\prime}: & iG^{(+)}(x - x'), \\ x^0 < x^{0\prime}: & iG^{(-)}(x - x'); \end{cases}$$

$$\qquad (2\text{-}7.63)$$

$$G_-(x - x') = \begin{cases} x^0 > x^{0\prime}: & -iG^{(-)}(x - x'), \\ x^0 < x^{0\prime}: & -iG^{(+)}(x - x'). \end{cases}$$

When arbitrary mode functions are used,

$$G^{(+)}(x - x') = \sum_r \psi_r(x)\psi_r(x')^*\gamma^0, \qquad G^{(-)}(x - x') = -\sum_r \psi_r(x)^*\psi_r(x')\gamma^0.$$
$$\qquad (2\text{-}7.64)$$

The various functions are also connected by the identity

$$i[G_+(x - x') - G_-(x - x')] + G^{(+)}(x - x') + G^{(-)}(x - x') = 0. \quad (2\text{-}7.65)$$

According to the causal analysis

$$\langle 0_-|0_-\rangle^{\eta(-),\eta(+)} = \sum_{\{n\}} \langle 0_-|\{n\}\rangle^{\eta(-)}\langle\{n\}|0_-\rangle^{\eta(+)}, \qquad (2\text{-}7.66)$$

a check of completeness or unitarity is performed by verifying that (2-7.60) reduces to unity on identifying $\eta_{(-)}(x)$ and $\eta_{(+)}(x)$. This is just the content of the identity (2-7.65), combined with the third statement of (2-7.62). The generalization to the amplitude $\langle\{n\}_-|\{n\}_-\rangle^{\eta(-),\eta(+)}$ involves the replacement of $G_+(x - x')$ by $G_{\{n\}+}(x - x')$. It is expressed by retaining the same set of relations but based on the new definitions,

$$G^{(+)}_{\{n\}}(x - x') = \sum_r [(1 - n_r)\psi_r(x)\psi_r(x')^* + n_r\psi_r(x)^*\psi_r(x')]\gamma^0,$$

$$G^{(-)}_{\{n\}}(x - x') = -\sum_r [n_r\psi_r(x)\psi_r(x')^* + (1 - n_r)\psi_r(x)^*\psi_r(x')]\gamma^0. \qquad (2\text{-}7.67)$$

As in the spin 0 discussion, the general unitarity proof uses the sources $\eta_{(\pm)}$ to generate arbitrary terminal states. The complete removal of reference to a subsequently acting source $\eta(x)$ requires the additional relations

$$\int (dx)(dx')\eta(x)\gamma^0[iG_+(x - x') + G^{(+)}(x - x')]\eta_{(+)}(x') = 0,$$

$$\int (dx)(dx')\eta(x)\gamma^0[-iG_-(x - x') + G^{(-)}(x - x')]\eta_{(-)}(x') = 0, \qquad (2\text{-}7.68)$$

which are correct statements under the assigned causal circumstances.

The charge property that is required by the Euclidean postulate played no role in the initial treatment of spin ½ particles. This raises a question concerning the possible existence of other kinds of spin ½ particles for which the charge matrix does make an explicit appearance. We shall take for granted in this discussion that the charge attribute remains unaltered during the travel of the particle between emission and absorption sources. That permits a dependence of the propagation function upon the charge matrix q, but not upon the charge-reflection matrix r_q. The general form of such a propagation function is

$$\gamma^0 G_+(x - x') = \gamma^0[m_1 + m_2\gamma_5 - a\gamma^\mu(1/i)\partial_\mu - a\lambda\gamma^\mu iq\gamma_5(1/i)\partial_\mu]\Delta_+(x - x'), \qquad (2\text{-}7.69)$$

which lacks only the matrices $\sigma_{\mu\nu} = -\sigma_{\nu\mu}$. Since γ^0 and $\gamma^0\gamma_5$ are antisymmetrical and $\gamma^0\gamma^\mu$ symmetrical, the antisymmetrical matrix q cannot multiply them. But q must be used to reverse the antisymmetry of $\gamma^0\gamma^\mu i\gamma_5$. This structure should fit into the mode function pattern detailed in (2-7.63, 64) since the latter refers only to the combination of individual emission and absorption acts. An essential aspect is the positiveness property

$$\int (dx)(dx')\varphi(x)^*\gamma^0 G^{(+)}(x - x')\varphi(x') = \sum_r \left|\int dx\varphi(x)^*\gamma^0\psi_r(x)\right|^2 \geq 0, \qquad (2\text{-}7.70)$$

where $\varphi(x)$ is an arbitrary complex-valued numerical function. Its implication

for (2–7.69) is the following Hermitian matrix positiveness requirement associated with any particle momentum p^μ,

$$\gamma^0[m_1 + m_2\gamma_5 - a\gamma p - a\lambda iq\gamma_5\gamma p] \geq 0, \qquad (2\text{–}7.71)$$

which incidentally asserts the reality of m_1, m_2, a, and λ. When a particle of
mass $m > 0$ is viewed in its rest frame, this condition reads

$$ma + ma\lambda iq\gamma_5 + m_1\gamma^0 + m_2\gamma^0\gamma_5 \geq 0. \qquad (2\text{–}7.72)$$

The three matrices $iq\gamma_5$, γ^0, $\gamma^0\gamma_5$ anticommute and are of unit square, from
which we infer the numerical requirements

$$ma \pm [(ma\lambda)^2 + m_1^2 + m_2^2]^{1/2} \geq 0. \qquad (2\text{–}7.73)$$

In addition to the conclusion that a is positive, we note that the zero value must
be attained if a projection matrix is to be produced, and accordingly

$$m^2a^2(1 - \lambda^2) = m_1^2 + m_2^2. \qquad (2\text{–}7.74)$$

Throughout the open interval $\lambda^2 < 1$, it is permissible to normalize a by

$$a^2(1 - \lambda^2) = 1, \qquad (2\text{–}7.75)$$

with the consequence

$$m_1^2 + m_2^2 = m^2. \qquad (2\text{–}7.76)$$

This is represented as

$$a = \cosh\theta, \qquad a\lambda = \sinh\theta; \qquad m_1 = m\cos\varphi, \qquad m_2 = m\sin\varphi. \qquad (2\text{–}7.77)$$

It is then easily seen that

$$\gamma^0 G_+(x - x') = e^{-(1/2)\varphi\gamma_5}e^{(1/2)\theta iq\gamma_5}$$
$$\times [\gamma^0(m - \gamma^\mu(1/i)\partial_\mu)\Delta_+(x - x')]e^{(1/2)\theta iq\gamma_5}e^{(1/2)\varphi\gamma_5}. \qquad (2\text{–}7.78)$$

In view of the antisymmetry of γ_5 and the symmetry of $q\gamma_5$, these matrix factors can be consistently transferred to the two source functions in W. Thus,
despite the initial appearance of q and of γ_5, they disappear after suitable redefinition of the source and we restore the structure of (2–6.67, 68).

The remaining possibility is

$$\lambda^2 = 1, \qquad m_1 = m_2 = 0. \qquad (2\text{–}7.79)$$

If we now choose

$$a = \tfrac{1}{2}, \qquad (2\text{–}7.80)$$

there emerges

$$\gamma^0 G_+(x - x') = -\frac{1 + i\lambda q\gamma_5}{2}\gamma^0\gamma^\mu(1/i)\partial_\mu\Delta_+(x - x')$$

$$= \frac{1 + i\lambda q\gamma_5}{2}\gamma^0(m - \gamma^\mu(1/i)\partial_\mu)\Delta_+(x - x')\frac{1 + i\lambda q\gamma_5}{2}.$$

$$(2\text{–}7.81)$$

The first version indicates that we have regained (2–6.27), with the objection to the antisymmetry of $\rho_2 = i\gamma_5$ removed by the presence of the additional antisymmetrical matrix λq. The second version is related to the standard form of $\gamma^0 G_+$ by symmetrical matrix factors, which could be transferred to the sources. These are singular projection matrices, however, and the new sources will be subject to the restrictive condition

$$(1 - i\lambda q\gamma_5)\eta(x) = 0. \tag{2–7.82}$$

A source constraint states a universal characteristic of all realistic mechanisms that contribute to the creation or annihilation of the given particle. Whether the interaction mechanisms of spin ½ particles might be compatible with such a restriction cannot be examined at this point, save for one exceptional class of spin ½ particles—the neutrinos.

Only one kind of neutrino interaction has been observed, processes in which they are created or annihilated in company with a charged lepton (electron or muon). The neutrino associated with an electron is a different particle from that associated with a muon. The masses of the neutrinos are small on the scale set by their leptonic partner, but they are not known to be zero in the same sense as are the photon and graviton masses, where inverse masses must exceed very large macroscopic distances. Both neutrinos are observed with a unique helicity, which is determined only by the electric charge of its partner and reverses sign with the latter. The spin ½ possibility just discussed provides a natural framework for the representation of these properties. (Indeed, it was in essence proposed long before the experimental disclosure of the two neutrinos, and constituted a prediction of that fact.) To avoid confusion with electric charge, let the charge property carried by the neutrinos be designated by l, the leptonic charge. The alternatives contained in $\lambda^2 = 1$ give two kinds of particles which are distinguished by the corresponding projection factor:

$$\tfrac{1}{2}(1 \pm il\gamma_5). \tag{2–7.83}$$

On considering a neutrino of momentum p^μ, the other matrix factor in (2–7.81) becomes

$$-\gamma^0\gamma^\mu p_\mu = p^0 - i\gamma_5\boldsymbol{\sigma}\cdot\mathbf{p}$$
$$= p^0 \mp l|\mathbf{p}|(\boldsymbol{\sigma}\cdot\mathbf{p}/|\mathbf{p}|), \tag{2–7.84}$$

which uses the effective equivalence between $i\gamma_5$ and $\pm l$ that is enforced by (2–7.83). When the neutrino has an energy that is large in comparison with its mass, which need not be zero, nor the same for the two neutrinos, a unique helicity is selected:

$$\boldsymbol{\sigma}\cdot\mathbf{p}/|\mathbf{p}| \to \mp l. \tag{2–7.85}$$

The conservation of leptonic charge requires that the electrically charged lepton accompanying a given neutrino carry a leptonic charge opposite to that of the neutrino:

$$l + l_{\text{ch. lept.}} = 0. \tag{2-7.86}$$

We now put forward the natural hypothesis that one role of leptonic charge is to distinguish, and label the two leptons with a common electric charge q:

$$l_{\text{ch. lept.}} = \mp q. \tag{2-7.87}$$

Its consequence is the empirical equivalence between neutrino helicity and the accompanying electric charge,

$$\boldsymbol{\sigma} \cdot \mathbf{p}/|\mathbf{p}| \to -q. \tag{2-7.88}$$

In the interest of completeness we shall exhibit the sources for specific neutrino states, under the simplifying assumption of zero neutrino mass. One uses the dyadic construction

$$\tfrac{1}{2}(1 + il\gamma_5)\tfrac{1}{2}(1 - l\boldsymbol{\sigma} \cdot \mathbf{p}/|\mathbf{p}|) = \sum_{l'=-1}^{+1} u_{pl'}u_{pl'}^*, \tag{2-7.89}$$

where

$$u_{pl'}^* u_{pl''} = \delta_{l'l''} \tag{2-7.90}$$

and

$$u_{pl}^* = u_{p-l}. \tag{2-7.91}$$

These eigenvectors also obey

$$(i\gamma_5 - l)u_{pl} = 0, \qquad u_{pl}^*(i\gamma_5 - l) = 0$$
$$(\boldsymbol{\sigma} \cdot \mathbf{p}/|\mathbf{p}| + l)u_{pl} = 0, \qquad u_{pl}^*(\boldsymbol{\sigma} \cdot \mathbf{p}/|\mathbf{p}| + l) = 0. \tag{2-7.92}$$

The sources for this kind of neutrino are

$$\dot{\eta}_{pl} = (2p^0\,d\omega_p)^{1/2}u_{pl}^*\eta(p), \qquad \eta_{pl}^* = (2p^0\,d\omega_p)^{1/2}\eta(p)^*u_{pl}, \tag{2-7.93}$$

and those for the second type are obtained by the reflection

$$u_{pl} \to r_l u_{pl}, \qquad u_{pl}^* \to u_{pl}^* r_l, \tag{2-7.94}$$

where r_l is the real symmetrical leptonic charge-reflection matrix. The TCP substitution

$$\eta(p) \to \gamma_5\eta(-p) \tag{2-7.95}$$

interchanges the neutrino emission and antineutrino absorption sources (2-7.93) in the following manner:

$$\eta_{pl} \to (-il)\eta_{p}^*{}_{-l}, \qquad \eta_{pl}^* \to (il)\eta_{p}{}_{-l}. \tag{2-7.96}$$

Additional minus signs appear for the other neutrino type.

Finally, here is a brief comment on the Euclidean transcription of the vacuum amplitude associated with (2-7.81) ($\pm l$ replace λq). It is

$$\langle 0_+|0_-\rangle^{\eta} \to \exp\left[-\tfrac{1}{2}\int (dx)(dx')\eta(x)\tfrac{1}{2}(1 \pm l\alpha_5)(-\alpha_\mu\partial_\mu)\Delta(x - x')\eta(x')\right]_E, \tag{2-7.97}$$

where

$$\alpha_5 = \alpha_1\alpha_2\alpha_3\alpha_4 \qquad (2\text{-}7.98)$$

extends the set of real symmetrical, anticommuting matrices of unit square. Unlike the Minkowski form, $l\alpha_5$ is antisymmetrical, and correspondingly anticommutes with the α_μ. The explicit appearance of the imaginary matrix l means that the individual Euclidean forms are not real. But if the masses of the two neutrinos are the same (zero?), complex conjugation interchanges the two Euclidean structures. Then one can regard the two neutrino sources as obtained by projection from one general source and the complete Euclidean vacuum amplitude is real, being (2-7.97) without the factor $\frac{1}{2}(1 \pm l\alpha_5)$.

2-8 PARTICLES OF INTEGER $+\frac{1}{2}$ SPIN

One can describe spin $\frac{3}{2}$ particles by combining the four-vector treatment of unit spin with the four-component spinor aspect of spin $\frac{1}{2}$. The resulting vector-spinor source $\eta_\zeta^\mu(x)$ has 16 components, apart from additional charge multiplicity. The reduction of this larger system to the one of interest is partly produced by the projection matrices appropriate to the constituents; $\bar{g}_{\mu\nu}(p)$ for spin 1, $m - \gamma p$ for spin $\frac{1}{2}$. On considering the source coupling associated with single-particle exchange, in the rest frame of the particle, this procedure supplies the effective source $\frac{1}{2}(1 + \gamma^0)\eta_k$, which has six components. The final reduction to the four components characteristic of spin $\frac{3}{2}$ is accomplished by the projection matrix (2-7.1), with $l = 1$, $j = \frac{3}{2}$. In the context of three-component vectors and two-component spinors it is represented by

$$(M_{3/2})_{kl} = \delta_{kl} - \tfrac{1}{3}\sigma_k\sigma_l. \qquad (2\text{-}8.1)$$

The projector character of this matrix is equivalent to the property

$$\sigma_k(M_{3/2})_{kl} = (M_{3/2})_{kl}\sigma_l = 0, \qquad (2\text{-}8.2)$$

and its specific identification is confirmed by evaluating the trace over the six-dimensional space. The resulting form of the source coupling in the rest frame is

$$\eta_k^*\tfrac{1}{2}(1 + \gamma^0)(\delta_{kl} - \tfrac{1}{3}\sigma_k\sigma_l)\eta_l = \bar{\eta}_k^*\tfrac{1}{2}(1 + \gamma^0)\bar{\eta}_k, \qquad (2\text{-}8.3)$$

where

$$\bar{\eta}_k = \eta_k - \tfrac{1}{3}\sigma_k\sigma_l\eta_l \qquad (2\text{-}8.4)$$

obeys

$$\sigma_k\bar{\eta}_k = 0, \qquad (2\text{-}8.5)$$

which makes explicit the rejection of the spin $\frac{1}{2}$ composite system.
The removal of the rest frame specification is facilitated by writing

$$\sigma_k\sigma_l = \gamma_k i\gamma_5\gamma_l i\gamma_5, \qquad (2\text{-}8.6)$$

and the straightforward generalization of (2-8.3) (apart from a factor of $2m$) is

$$\eta^\mu(p)^*\gamma^0(m - \gamma p)[\bar{g}_{\mu\nu}(p) - \tfrac{1}{3}\bar{g}_{\mu\kappa}(p)\gamma^\kappa i\gamma_5\bar{g}_{\nu\lambda}(p)\gamma^\lambda i\gamma_5]\eta^\nu(p). \qquad (2\text{-}8.7)$$

The second term is given a somewhat simpler form by noting, successively,

$$(m - \gamma p)\tilde{g}_{\mu\kappa}(p)\gamma^\kappa i\gamma_5 \tilde{g}_{\nu\lambda}(p)\gamma^\lambda i\gamma_5 = \tilde{g}_{\mu\kappa}(p)\gamma^\kappa i\gamma_5 (m - \gamma p)\tilde{g}_{\nu\lambda}(p)\gamma^\lambda i\gamma_5$$

$$= -\tilde{g}_{\mu\kappa}(p)\gamma^\kappa (m + \gamma p)\tilde{g}_{\nu\lambda}(p)\gamma^\lambda$$

$$= -\left(\gamma_\mu + \frac{1}{m}p_\mu\right)(m + \gamma p)\left(\gamma_\nu + \frac{1}{m}p_\nu\right).$$

$$(2\text{-}8.8)$$

The first rearrangement restates the commutativity of $1 + \gamma^0$ with σ_k; the second invokes the properties of $i\gamma_5$; and the last uses the presence of the factor $m + \gamma p$ to substitute m for γp. An alternative form replaces $\gamma_\mu + (1/m)p_\mu$ with $-(i/m)\sigma_{\mu\lambda}p^\lambda$. The implied expression for the vacuum amplitude

$$\langle 0_+ | 0_- \rangle^\eta = \exp[iW(\eta)], \qquad (2\text{-}8.9)$$

written in four-dimensional momentum space for conciseness, is given by

$$W(\eta) = \frac{1}{2}\int \frac{(dp)}{(2\pi)^4}\, \eta^\mu(-p)\gamma^0$$

$$\times \left[(m - \gamma p)\left(g_{\mu\nu} + \frac{1}{m^2}p_\mu p_\nu\right) + \frac{1}{3}\left(\gamma_\mu + \frac{1}{m}p_\mu\right)(m + \gamma p)\right.$$

$$\left.\times\left(\gamma_\nu + \frac{1}{m}p_\nu\right)\right]\frac{1}{p^2 + m^2 - i\epsilon}\, \eta^\nu(p). \quad (2\text{-}8.10)$$

The kernel of this quadratic form is antisymmetrical, under transposition of the matrix and vector indices combined with $p^\mu \to -p^\mu$, which demands a corresponding anticommutativity of the sources or F. D. statistics.

To identify particle emission and absorption sources, specifically those referring to the four helicity states associated with a given momentum, we consider a causal arrangement and examine the coupling term in iW:

$$\int d\omega_p i\eta_1^\mu(p)^* \gamma^0 \left[(m - \gamma p)\sum_\lambda e_{p\lambda\mu}e_{p\lambda\nu}^* - \tfrac{1}{3}\sum_\lambda e_{p\lambda\mu}e_{p\lambda}^{\alpha*}\gamma_\alpha i\gamma_5(m - \gamma p)\right.$$

$$\left.\times \sum_{\lambda'}\gamma_\beta i\gamma_5 e_{p\lambda'}^\beta e_{p\lambda'\nu}^*\right]i\eta_2^\nu(p), \quad (2\text{-}8.11)$$

in which we have returned to the version given in (2–8.7), and used the dyadic construction for $\tilde{g}_{\mu\nu}(p)$. The introduction of the dyadic spinor realization for $(m - \gamma p)/2m$ gives the form

$$\sum_{\lambda\sigma} e_{p\lambda}^\mu u_{p\sigma}u_{p\sigma}^{*}\gamma^0 e_{p\lambda}^{\nu*} - \tfrac{1}{3}\sum_{\lambda\sigma\lambda'} e_{p\lambda}^\mu e_{p\lambda}^{\alpha*}\gamma_\alpha i\gamma_5 u_{p\sigma}u_{p\sigma}^{*}\gamma^0\gamma_\beta i\gamma_5 e_{p\lambda'}^\beta e_{p\lambda'}^{\nu*}$$

$$= \sum_{\lambda=-3/2}^{3/2} u_{p\lambda}^\mu u_{p\lambda}^{\nu*}\gamma^0, \quad (2\text{-}8.12)$$

which serves to identify the four eigenvectors $u_{p\lambda}^\mu$, $\lambda = \frac{3}{2}, \ldots, -\frac{3}{2}$. In order to exhibit them explicitly we use the following property,

$$e_{p\lambda}^{\alpha*}\gamma_\alpha i\gamma_5 u_{p\sigma} = (\sigma - 2\lambda)(1 - \lambda^2 + 2^{1/2}\lambda^2)u_{p\sigma-2\lambda}, \quad (2\text{-}8.13)$$

which is to be understood in the limited context, $\sigma = \pm 1$. This formula is easily checked in the rest frame, where the left-hand side reduces to $e_\lambda^* \cdot \sigma v_\sigma$. Its general validity is assured, according to (2–6.86), if

$$e_{p\lambda}^{\mu*}\gamma^0\gamma_\mu = \exp[-\tfrac{1}{2}\theta i\gamma_5\sigma \cdot \mathbf{p}/|\mathbf{p}|]i\gamma_5 e_\lambda^* \cdot \sigma \exp[-\tfrac{1}{2}\theta i\gamma_5\sigma \cdot \mathbf{p}/|\mathbf{p}|]. \quad (2\text{–}8.14)$$

It will be recognized that this is equivalent to the construction (2–5.53), with the σ matrices replaced by the algebraically indistinguishable set $i\gamma_5\sigma$.

The explicit forms of the helicity labeled eigenvectors are

$$\begin{aligned}
u_{p\frac{3}{2}}^\nu &= e_p^\nu{}_{+1}u_{p+}, \\
u_{p\frac{1}{2}}^\nu &= 3^{-1/2}[e_p^\nu{}_{+1}u_{p-} + 2^{1/2}e_{p0}^\nu u_{p+}], \\
u_{p\,-\frac{1}{2}}^\nu &= 3^{-1/2}[e_p^\nu{}_{-1}u_{p+} + 2^{1/2}e_{p0}^\nu u_{p-}], \\
u_{p\,-\frac{3}{2}}^\nu &= e_p^\nu{}_{-1}u_{p-},
\end{aligned} \qquad (2\text{–}8.15)$$

which are standard combinations of states for unit and $\frac{1}{2}$ angular momentum. Their orthonormality properties are given by

$$u_{p\lambda}^{\nu*}u_{p\lambda'\nu} = \delta_{\lambda\lambda'}. \qquad (2\text{–}8.16)$$

The resulting source identification is

$$\eta_{p\lambda} = (2m\,d\omega_p)^{1/2}u_{p\lambda}^{\nu*}\gamma^0\eta_\nu(p), \qquad \eta_{p\lambda}^* = (2m\,d\omega_p)^{1/2}\eta_\nu(p)^*\gamma^0 u_{p\lambda}^\nu, \quad (2\text{–}8.17)$$

to which a charge label can be added. The complex conjugation properties of the constituents supply the relation

$$u_{p-\lambda}^{\nu*} = (-1)^{(3/2)+\lambda}i\gamma_5 u_{p\lambda}^\nu. \qquad (2\text{–}8.18)$$

The *TCP* substitution

$$\eta^\mu(p) \to \gamma_5\eta^\mu(-p) \qquad (2\text{–}8.19)$$

therefore induces

$$\eta_{p\lambda} \to i(-1)^{(3/2)-\lambda}\eta_{p-\lambda}^*, \qquad \eta_{p\lambda}^* \to -i(-1)^{(3/2)-\lambda}\eta_{p-\lambda}, \quad (2\text{–}8.20)$$

which can be supplemented by a charge index, in the usual way.

The treatment of massless spin $\frac{3}{2}$ particles generally follows the unit spin pattern. On writing

$$\partial_\mu\eta^\mu(x) = m\eta(x), \qquad (2\text{–}8.21)$$

one recognizes that, as $m \to 0$, helicity $\pm\frac{3}{2}$ decouples from helicity $\pm\frac{1}{2}$, which is represented by the spin $\frac{1}{2}$ source, $\eta(x) - \tfrac{1}{2}i\gamma_\nu\eta^\nu(x)$. Massless particles of helicity $\pm\frac{3}{2}$ are described by

$$\begin{aligned}
W = \tfrac{1}{2}\int (dx)(dx')\eta^\mu(x)\gamma^0[g_{\mu\nu}(-\gamma^\lambda(1/i)\partial_\lambda)D_+(x-x') \\
- \tfrac{1}{2}\gamma_\mu(-\gamma^\lambda(1/i)\partial_\lambda)D_+(x-x')\gamma_\nu]\eta^\nu(x'),
\end{aligned} \qquad (2\text{–}8.22)$$

where

$$\partial_\mu\eta^\mu(x) = 0. \qquad (2\text{–}8.23)$$

The coupling term in iW for a causal arrangement is

$$\int d\omega_p i\eta_1^\mu(p)^* \gamma^0 [g_{\mu\nu}(-\gamma p) - \tfrac{1}{2} g_{\mu\alpha}\gamma^\alpha(-\gamma p)\gamma^\beta g_{\beta\nu}] i\eta_2^\nu(p). \qquad (2\text{-}8.24)$$

One can replace $g_{\mu\nu}$ everywhere by just the two terms of the dyadic that refer to helicity ± 1:

$$g^{\mu\nu} \to \sum_{\lambda=\pm 1} e_{p\lambda}^\mu e_{p\lambda}^{\nu*}, \qquad (2\text{-}8.25)$$

since

$$p_\mu \eta^\mu(p) = 0, \qquad (\gamma p)^2 = 0. \qquad (2\text{-}8.26)$$

We also use

$$\gamma^0(-\gamma p) = 2p^0 \tfrac{1}{2}(1 - i\gamma_5 \boldsymbol{\sigma} \cdot \mathbf{p}/|\mathbf{p}|)$$
$$= 2p^0 \sum_{\sigma'=\pm} u_{p\sigma'} u_{p\sigma'}^*, \qquad (2\text{-}8.27)$$

where

$$(\boldsymbol{\sigma} \cdot \mathbf{p}/|\mathbf{p}| - \sigma')u_{p\sigma'} = (i\gamma_5 + \sigma')u_{p\sigma'} = 0, \qquad (2\text{-}8.28)$$

and the algebraic properties

$$\{e_{p\lambda}^{\alpha*}\gamma_\alpha, \gamma p\} = 0,$$
$$e_{p\lambda}^{\alpha*}\gamma_\alpha \gamma_\beta e_{p\lambda'}^\beta = -\delta_{\lambda\lambda'}(1 - \lambda\boldsymbol{\sigma} \cdot \mathbf{p}/|\mathbf{p}|). \qquad (2\text{-}8.29)$$

The outcome is the replacement of (2-8.24) with

$$\int d\omega_p i\eta_1^\mu(p)^* 2p^0 \left[\sum_{\lambda\sigma} e_{p\lambda\mu} u_{p\sigma} \tfrac{1}{2}(1 + \lambda\sigma) e_{p\lambda\nu}^* u_{p\sigma}^* \right] i\eta_2^\nu(p), \qquad (2\text{-}8.30)$$

in which $\tfrac{1}{2}(1 + \lambda\sigma)$ selects only the states of helicity $\pm\tfrac{3}{2}$. The two mode functions are

$$u_{p\pm\frac{3}{2}}^\nu = e_{p\pm 1}^\nu u_{p\pm}, \qquad (2\text{-}8.31)$$

and the corresponding sources are given by

$$\lambda = \pm\tfrac{3}{2}: \qquad \eta_{p\lambda} = (2p^0 \, d\omega_p)^{1/2} u_{p\lambda}^{\nu*} \eta_\nu(p),$$
$$\eta_{p\lambda}^* = (2p^0 \, d\omega_p)^{1/2} \eta_\nu(p)^* u_{p\lambda}^\nu. \qquad (2\text{-}8.32)$$

As in the spin $\tfrac{1}{2}$ neutrino discussion, one can introduce an additional decomposition in which helicity is coupled to charge in a unique way.

Preparatory to generalizing this approach to all particles of spin $s = n + \tfrac{1}{2}$, $n = 1, 2, \ldots$, we return to the rest frame spin projection

$$\bar{\eta}_k = \eta_k - \tfrac{1}{3}\sigma_k \sigma_l \eta_l \qquad (2\text{-}8.33)$$

and remark that

$$\bar{\eta}_k = \tfrac{2}{3}\sigma_p \Pi_{kp,lq}\sigma_q \eta_l, \qquad (2\text{-}8.34)$$

where

$$\Pi_{kp,lq} = \tfrac{1}{2}(\delta_{kl}\delta_{pq} + \delta_{kq}\delta_{lp}) - \tfrac{1}{3}\delta_{kp}\delta_{lq} \qquad (2\text{-}8.35)$$

is the rest frame version of the $n = 2$ projection tensor that is defined generally by (2–5.79). The properties of this tensor assure that

$$\sigma_k \bar\eta_k = \tfrac{2}{5}\sigma_k\sigma_p\Pi_{kp,lq}\sigma_q\eta_l = \tfrac{2}{5}\Pi_{kk,lq}\sigma_q\eta_l = 0. \tag{2–8.36}$$

Here is the generalization of this rest-frame treatment to symmetric tensor-spinor sources:

$$\bar\eta_{k_1\cdots k_n} = \frac{n+1}{2n+3}\,\sigma_p\Pi_{k_1\cdots k_n p,\,l_1\cdots l_n q}\sigma_q\eta_{l_1\cdots l_n}. \tag{2–8.37}$$

Although it is evident that the $\bar\eta_{k_1\cdots k_n}$ are traceless, this property can be regarded as a consequence of

$$\sigma_{k_1}\bar\eta_{k_1\cdots k_n} = 0, \tag{2–8.38}$$

according to

$$0 = \sigma_{k_2}\sigma_{k_1}\bar\eta_{k_1 k_2 k_3\cdots k_n} = \bar\eta_{k k k_3\cdots k_n}. \tag{2–8.39}$$

Keeping in mind the restriction to two-component spinors, we see that the count of independent components is

$$2[\tfrac{1}{2}(n+1)(n+2) - \tfrac{1}{2}n(n+1)] = 2(n+\tfrac{1}{2})+1, \tag{2–8.40}$$

consistent with the description of spin $s = n + \tfrac{1}{2}$. The numerical factor in (2–8.37) can be derived by noting that the latter should be an identity if $\eta_{l_1\cdots l_n}$ is replaced by $\bar\eta_{l_1\cdots l_n}$. In that circumstance the projection tensor reduces to the symmetrized unit matrix appropriate to $n+1$ indices. There are two classes of terms; those that select $p = q$, which are $n!$ in number, and those with $p = l_j$, $q = k_i$, being the remaining $n(n!)$ terms. The first set is multiplied by $(\sigma_p)^2 = 3$, and the second set by 2, since

$$\sigma_p\sigma_q = 2\delta_{qp} - \sigma_q\sigma_p \tag{2–8.41}$$

and $\sigma_{l_j}\bar\eta_{l_1\cdots l_n} = 0$. Accordingly,

$$\sigma_p\Pi_{k_1\cdots k_n p,\,l_1\cdots l_n q}\sigma_q\bar\eta_{l_1\cdots l_n} = \frac{1}{(n+1)!}\,[3(n!) + 2n(n!)]\bar\eta_{k_1\cdots k_n}$$
$$= \frac{2n+3}{n+1}\,\bar\eta_{k_1\cdots k_n}, \tag{2–8.42}$$

as stated in (2–8.37). Alternatively, one can verify that the trace of the projection matrix that is defined on the space of n three-component vectors and two-component spinors has the required value of $2(n+\tfrac{1}{2})+1$:

$$\operatorname{tr}_{(2)}\left[\frac{n+1}{2n+3}\,\sigma_p\Pi_{k_1\cdots k_n p,\,k_1\cdots k_n q}\sigma_q\right] = \frac{2n+2}{2n+3}\,\Pi_{k_1\cdots k_{n+1},\,k_1\cdots k_{n+1}}$$
$$= 2n+2, \tag{2–8.43}$$

since the trace of the projection matrix that refers to $n+1$ three-vector indices equals $2(n+1)+1$.

A particle of spin $s = n + \frac{1}{2}$ can be described by the symmetrical tensor-spinor source $\eta_\zeta^{\mu_1\cdots\mu_n}(x)$. The four-dimensional momentum space version of W is

$$W = \frac{1}{2}\int\frac{(dp)}{(2\pi)^4}\,\eta^{\mu_1\cdots\mu_n}(-p)\gamma^0\,\frac{n+1}{2n+3}\,i\gamma^\alpha\gamma_5(m-\gamma p)\Pi_{\mu_1\cdots\mu_n\alpha,\nu_1\cdots\nu_n\beta}(p)$$

$$\times\, i\gamma^\beta\gamma_5\,\frac{1}{p^2+m^2-i\epsilon}\,\eta^{\nu_1\cdots\nu_n}(p). \tag{2-8.44}$$

This is not to be taken literally, however, for an algebraic simplification should be performed before the space-time extrapolation embodied in the four-dimensional version is carried out. In the initial causal situation, where $(m+\gamma p)/2m$ is a projection matrix selecting $\gamma p = m$, two powers of the momentum, appearing in the form $\gamma^\kappa p_\kappa(m+\gamma p)p_\lambda\gamma^\lambda$, can be removed. Thus, the matrix polynomial in p that occurs in (2–8.44) is of degree $2n+1 = 2s$. That is illustrated for $n = 1$, $s = \frac{3}{2}$, by (2–8.10).

In the direct application of (2–8.44) to a causal arrangement, the introduction of the dyadic constructions for the spinor and tensor projection matrices must supply the tensor-spinor dyadic

$$\frac{n+1}{2n+3}\sum_{\lambda\sigma}e_{p\lambda}^{\mu_1\cdots\mu_n\alpha}\gamma_\alpha i\gamma_5 u_{p\sigma}u_{p\sigma}^*\gamma^0\gamma_\beta i\gamma_5 e_{p\lambda}^{\nu_1\cdots\nu_n\beta*} = \sum_{\lambda=-s}^{s}u_{p\lambda}^{\mu_1\cdots\mu_n}u_{p\lambda}^{\nu_1\cdots\nu_n*}\gamma^0,$$

$$\tag{2-8.45}$$

where

$$u_{p\lambda}^{\nu_1\cdots\nu_n*}\gamma^0 u_{\nu_1\cdots\nu_n p\lambda'} = \delta_{\lambda\lambda'} \tag{2-8.46}$$

is consistent with the properties of this structure. It is not difficult to pick out the term of highest helicity, $\lambda = s$,

$$u_{ps}^{\nu_1\cdots\nu_n} = e_{pn}^{\nu_1\cdots\nu_n}u_{p+}, \tag{2-8.47}$$

which appears on the left-hand side of (2–8.45) with the coefficient

$$\frac{n+1}{2n+3}\left(2+\frac{1}{n+1}\right) = 1. \tag{2-8.48}$$

The other helicity functions are produced most simply from this one, by rotation, as effectively realized in the algebraic construction

$$\sum_{\lambda=-s}^{s}\psi_{s\lambda}(\xi)u_{p\lambda}^{\nu_1\cdots\nu_n} = \left(\sum_{\lambda=-n}^{n}\psi_{n\lambda}(\xi)e_{p\lambda}^{\nu_1\cdots\nu_n}\right)\left(\sum_\sigma\xi_\sigma u_{p\sigma}\right), \tag{2-8.49}$$

where $\psi_{s\lambda}(\xi)$ is defined as in (2–5.87) but with n replaced by s. The results for $s = \frac{3}{2}$ that are given in (2–8.15) are immediately reproduced in this way. The sources for the helicity labeled states of these F. D. particles are identified as

$$\eta_{p\lambda} = (2m \, d\omega_p)^{1/2} u_{p\lambda}^{\nu_1 \cdots \nu_n *} \gamma^0 \eta_{\nu_1 \cdots \nu_n}(p),$$

$$\eta_{p\lambda}^* = (2m \, d\omega_p)^{1/2} \eta_{\nu_1 \cdots \nu_n}(p)^* \gamma^0 u_{p\lambda}^{\nu_1 \cdots \nu_n}. \tag{2-8.50}$$

To close this section, we consider the massless particles of helicity $\pm (n + \frac{1}{2})$, $n = 1, 2 \ldots$ (although no example comes to mind). It seems evident that the necessary tensor-spinor restriction

$$\partial_{\mu_1} \eta^{\mu_1 \cdots \mu_n}(x) = 0 \tag{2-8.51}$$

must be accompanied by the corresponding projection tensor, which is described by (2–5.104), and indeed the general form is

$$W = \frac{1}{2} \int (dx)(dx') \eta^{\mu_1 \cdots \mu_n}(x) \gamma^0 \gamma^\alpha (-\gamma^\lambda (1/i) \partial_\lambda) D_+(x - x')$$

$$\times \frac{1}{2} \Pi_{\mu_1 \cdots \mu_n \alpha, \nu_1 \cdots \nu_n \beta} \gamma^\beta \eta^{\nu_1 \cdots \nu_n}(x'). \tag{2-8.52}$$

For $n = 1$, where

$$\Pi_{\mu\alpha,\nu\beta} = \frac{1}{2}(g_{\mu\nu}g_{\alpha\beta} + g_{\mu\beta}g_{\nu\alpha}) - \frac{1}{3}g_{\mu\alpha}g_{\nu\beta}, \tag{2-8.53}$$

one verifies that (2–8.22) is reproduced, while becoming aware of the equivalent form

$$W = \frac{1}{2} \int (dx)(dx') \eta^\mu(x) \gamma^0 \frac{1}{2} \gamma_\nu (-\gamma^\lambda (1/i) \partial_\lambda) D_+(x - x') \gamma_\mu \eta^\nu(x'). \tag{2-8.54}$$

The coupling term in iW for a causal arrangment is obtained from (2–8.52) as

$$\int d\omega_p i \eta_1^{\mu_1 \cdots \mu_n}(p)^* \gamma^0 \gamma^\alpha (-\gamma p) \frac{1}{2} \Pi_{\mu_1 \cdots \mu_n \alpha, \nu_1 \cdots \nu_n \beta}(p, \bar{p}) \gamma^\beta i \eta_2^{\nu_1 \cdots \nu_n}(p), \tag{2-8.55}$$

where the introduction of the new projection tensor, defined in (2–5.106), is justified by the properties

$$p_{\nu_1} \eta^{\nu_1 \cdots \nu_n}(p) = 0, \qquad (\gamma p)^2 = 0. \tag{2-8.56}$$

The dyadic construction (2–5.117), combined with (2–5.121), converts the tensor-matrix of (2–8.55) into (the tensor indices are raised, for clarity)

$$\frac{1}{2} \sum_{\pm} e_{p \pm n}^{\mu_1 \cdots \mu_n} \gamma_\alpha e_{p \pm 1}^\alpha \gamma^0 \gamma p \gamma_\beta e_{p \pm 1}^{\beta *} e_{p \pm n}^{\nu_1 \cdots \nu_n *}, \tag{2-8.57}$$

where, utilizing (2–8.29) and (2–8.27),

$$\frac{1}{2} \gamma_\alpha e_{p \pm 1}^\alpha \gamma^0 \gamma p \gamma_\beta e_{p \pm 1}^{\beta *} = -\frac{1}{2} \gamma_\alpha e_{p \pm 1}^\alpha \gamma_\beta e_{p \pm 1}^{\beta *} \gamma^0 (-\gamma p)$$

$$= 2p^0 \sum_{\sigma'} \frac{1}{2}(1 \pm \sigma') u_{p\sigma'} u_{p\sigma'}^*. \tag{2-8.58}$$

The factor $\frac{1}{2}(1 \pm \sigma')$ locks the spin $\frac{1}{2}$ helicity to the others, and we recognize the generalization of (2-8.31) for $s = n + \frac{1}{2}$:

$$u_{p\pm s}^{\nu} = e_{p\pm n}^{\nu_1 \cdots \nu_n} u_{p\pm}, \qquad (2\text{-}8.59)$$

with the associated source definitions

$$\eta_{p\lambda} = (2p^0 \, d\omega_p)^{1/2} u_{p\lambda}^{\nu_1 \cdots \nu_n *} \eta_{\nu_1 \cdots \nu_n}(p),$$

$$\lambda = \pm s: \qquad\qquad\qquad\qquad\qquad\qquad\qquad (2\text{-}8.60)$$

$$\overset{*}{\eta}_{p\lambda} = (2p^0 \, d\omega_p)^{1/2} \eta_{\nu_1 \cdots \nu_n}(p) \overset{*}{u}_{p\lambda}^{\nu_1 \cdots \nu_n}.$$

2-9 UNIFICATION OF ALL SPINS AND STATISTICS

The procedures we have followed for describing the various spin possibilities exploit elementary angular momentum properties. The spin $s = n = 2, 3, \ldots$ can be compounded from n unit spins. And it suffices to add a single spin of $\frac{1}{2}$ to produce the sequence $s = n + \frac{1}{2}$, $n = 1, 2, \ldots$. But all spin possibilities can be constructed by combining the fundamental spin $\frac{1}{2}$ system a sufficient number of times—even for integer spin, odd for integer $+ \frac{1}{2}$ spin. Accordingly, we replace the tensor or multivector, expressing the composition of unit spins, and the related tensor-spinor, by the universal multispinor that is appropriate to the composition of a number of spin $\frac{1}{2}$ constituents. A multispinor source will be explicitly written as $S_{\zeta_1 \cdots \zeta_n}(x)$, but the indices will often be suppressed. All component spins are on the same footing and additional symmetry requirements can be imposed on the multispinor. The most important of these is the requirement of total symmetry:

$$S_{\zeta_1 \cdots \zeta_n}(x) = S_{\zeta_{\alpha_1} \cdots \zeta_{\alpha_n}}(x), \qquad (2\text{-}9.1)$$

where $\alpha_1 \ldots \alpha_n$ is any permutation of $1 \ldots n$.

The multispinor refers to a larger system than desired and projection matrices are required, even as in the simplest situation $n = 1$, $s = \frac{1}{2}$. Indeed, it suffices to use such spin $\frac{1}{2}$ projection matrices on each spinor index to obtain the required reduction to the physical system of spin s,

$$s = \tfrac{1}{2}n, \qquad (2\text{-}9.2)$$

for a symmetrical multispinor. In the rest frame of a massive particle, that projection matrix is

$$\prod_{\alpha=1}^{n} \tfrac{1}{2}(1 + \gamma^0)_\alpha, \qquad (2\text{-}9.3)$$

where α designates the spinor index on which the corresponding matrix acts. Its effect is to reduce the range of each spinor index to two values. A symmetrical function of n two-valued indices has a number of independent components equal to

$$n + 1 = 2s + 1, \qquad (2\text{-}9.4)$$

as anticipated. The spin values obtained in this way are

$$n = 1, 2, 3, \ldots, \qquad s = \tfrac{1}{2}, 1, \tfrac{3}{2}, \ldots . \qquad (2\text{-}9.5)$$

Only $s = 0$ is missing. For that it suffices to consider $n = 2$ and choose the antisymmetrical spinor

$$S_{\zeta_1\zeta_2}(x) = -S_{\zeta_2\zeta_1}(x). \qquad (2\text{-}9.6)$$

An antisymmetrical function of a pair of effectively two-valued indices has only one independent component.

The general expression of these remarks is given by the following vacuum amplitude,

$$\langle 0_+|0_-\rangle^S = \exp[iW(S)], \qquad (2\text{-}9.7)$$

where, using four-dimensional momentum space, we have

$$W(S) = \frac{1}{2} \int \frac{(dp)}{(2\pi)^4} \, S(-p) \prod_{\alpha=1}^{n} [\gamma^0(m - \gamma p)]_\alpha \, \frac{1}{p^2 + m^2 - i\epsilon} \, S(p). \qquad (2\text{-}9.8)$$

The kernel of this quadratic form has a definite symmetry under matrix transposition combined with the substitution $p^\mu \to -p^\mu$,

$$\prod_{\alpha=1}^{n} [\gamma^0(m + \gamma p)]_\alpha^T = (-1)^n \prod_{\alpha=1}^{n} [\gamma^0(m - \gamma p)]_\alpha. \qquad (2\text{-}9.9)$$

Accordingly, if the algebraic properties of the source are to match the symmetry properties of the kernel, we must have

$$\begin{aligned} n \text{ even, } s = \text{ integer:} \quad & [S(x), S(x')] = 0, \quad \text{B. E. statistics,} \\ n \text{ odd, } s = \text{ integer} + \tfrac{1}{2}: \quad & \{S(x), S(x')\} = 0, \quad \text{F. D. statistics,} \end{aligned} \qquad (2\text{-}9.10)$$

which is the general statement of the connection between spin and statistics. This proof will be complete, however, only when we have shown that any attempt to reverse these natural connections does violence to the completeness of the multiparticle states.

Let us consider the causal arrangement

$$S(x) = S_1(x) + S_2(x), \qquad (2\text{-}9.11)$$

which implies

$$\langle 0_+|0_-\rangle^S = \langle 0_+|0_-\rangle^{S_1} \exp\left[\int d\omega_p i S_1(p)^* \prod_\alpha [\gamma^0(m - \gamma p)]_\alpha i S_2(p) \right] \langle 0_+|0_-\rangle^{S_2}. \qquad (2\text{-}9.12)$$

On using (2–6.93) for each spinor index, we have

$$\prod_{\alpha=1}^{n} [\gamma^0(m - \gamma p)]_\alpha = (2m)^n \prod_{\alpha=1}^{n} \left[\sum_\sigma \gamma^0 u_{p\sigma} u_{p\sigma}^* \gamma^0 \right]_\alpha, \qquad (2\text{-}9.13)$$

which, in general, must be projected onto the space of symmetrical spinors. Employing helicity labeled spin functions, for definiteness, we recognize that the highest helicity contained in (2–9.13), $\lambda = \tfrac{1}{2}n$, is represented by the function

$$u_{ps} = \prod_{\alpha=1}^{n} (u_{p+})_\alpha, \tag{2-9.14}$$

and the whole set is generated by

$$\sum_{\lambda=-s}^{s} \psi_{s\lambda}(\xi) u_{p\lambda} = \prod_{\alpha=1}^{n} \left(\sum_\sigma \xi_\sigma u_{p\sigma} \right)_\alpha. \tag{2-9.15}$$

In the special situation of the antisymmetrical spinor with $n = 2$, the single eigenvector is

$$u_p = 2^{-1/2}[(u_{p+})_1(u_{p-})_2 - (u_{p-})_1(u_{p+})_2]. \tag{2-9.16}$$

The orthonormality of the helicity functions, in the form

$$u_{p\lambda}^* \left[\prod_\alpha \gamma_\alpha^0 \right] u_{p\lambda'} = \delta_{\lambda\lambda'}, \tag{2-9.17}$$

is derived from (2–9.15) as

$$\sum_{\lambda\lambda'} \psi_{s\lambda}(\eta^*) u_{p\lambda}^* \left[\prod_\alpha \gamma_\alpha^0 \right] u_{p\lambda'} \psi_{s\lambda'}(\xi) = (\eta^*\xi)^{2s}$$

$$= \sum_{\lambda=-s}^{s} \psi_{s\lambda}(\eta^*) \psi_{s\lambda}(\xi). \tag{2-9.18}$$

With the definitions

$$S_{p\lambda} = ((2m)^n \, d\omega_p)^{1/2} u_{p\lambda}^* \left[\prod_\alpha \gamma_\alpha^0 \right] S(p),$$
$$S_{p\lambda}^* = ((2m)^n \, d\omega_p)^{1/2} S(p)^* \left[\prod_\alpha \gamma_\alpha^0 \right] u_{p\lambda}, \tag{2-9.19}$$

Eq. (2–9.12) becomes

$$\langle 0_+|0_-\rangle^S = \langle 0_+|0_-\rangle^{S_1} \exp\left[\sum_{p\lambda} iS_{1p\lambda}^* iS_{2p\lambda} \right] \langle 0_+|0_-\rangle^{S_2}$$

$$= \langle 0_+|0_-\rangle^{S_1} \prod_{p\lambda} \sum_{n_{p\lambda}} \frac{(iS_{1p\lambda}^* iS_{2p\lambda})^{n_{p\lambda}}}{n_{p\lambda}!} \langle 0_+|0_-\rangle^{S_2}, \tag{2-9.20}$$

which uses the fact that even functions of the sources are commutative for either statistics. The causal analysis

$$\langle 0_+|0_-\rangle^S = \sum_{\{n\}} \langle 0_+|\{n\}\rangle^{S_1} \langle\{n\}|0_-\rangle^{S_2} \tag{2-9.21}$$

leads to the identifications

$$\langle\{n\}|0_-\rangle^S = \langle 0_+|0_-\rangle^S \prod_{p\lambda} \frac{(iS_{p\lambda})^{n_{p\lambda}}}{(n_{p\lambda}!)^{1/2}},$$

$$\langle 0_+|\{n\}\rangle^S = \langle 0_+|0_-\rangle^S \prod_{p\lambda}^{T} \frac{(iS_{p\lambda}^*)^{n_{p\lambda}}}{(n_{p\lambda}!)^{1/2}}, \tag{2-9.22}$$

where opposite multiplication order is used in the two products. It is only through the implicit algebraic properties of the sources,

$$
\begin{array}{ll}
\text{B. E.:} & [S_{p\lambda}, S_{p'\lambda'}] = 0, \\
\text{F. D.:} & \{S_{p\lambda}, S_{p'\lambda'}\} = 0,
\end{array}
\tag{2-9.23}
$$

that the two statistics are distinguished. In particular, the algebraic property

$$
\text{F. D.:} \qquad S_{p\lambda}^2 = 0 \tag{2-9.24}
$$

leads to the characteristic F. D. limitation, $n_{p\lambda} = 0, 1$.

The two expressions of completeness,

$$
\sum_{\{n\}} \langle 0_-|\{n\}\rangle^S \langle\{n\}|0_-\rangle^S = 1, \qquad \sum_{\{n\}} \langle 0_+|\{n\}\rangle^S \langle\{n\}|0_+\rangle^S = 1, \tag{2-9.25}
$$

become, respectively,

$$
|\langle 0_+|0_-\rangle^S|^2 \sum_{\{n\}} \left[\prod_{p\lambda} \frac{(iS_{p\lambda})^{n_{p\lambda}}}{(n_{p\lambda}!)^{1/2}}\right]^* \left[\prod_{p\lambda} \frac{(iS_{p\lambda})^{n_{p\lambda}}}{(n_{p\lambda}!)^{1/2}}\right] = 1, \tag{2-9.26}
$$

and

$$
|\langle 0_+|0_-\rangle^S|^2 \sum_{\{n\}} \left[\prod_{p\lambda}^T \frac{(iS_{p\lambda}^*)^{n_{p\lambda}}}{(n_{p\lambda}!)^{1/2}}\right] \left[\prod_{p\lambda}^T \frac{(iS_{p\lambda}^*)^{n_{p\lambda}}}{(n_{p\lambda}!)^{1/2}}\right]^* = 1. \tag{2-9.27}
$$

These are identical if the real sources $S(x)$ obey

$$
(S(x)S(x'))^* = S(x')S(x). \tag{2-9.28}
$$

Then the single statement of completeness is given by

$$
\begin{aligned}
|\langle 0_+|0_-\rangle^S|^{-2} &= \sum_{\{n\}} \left[\prod_{p\lambda}^T \frac{(S_{p\lambda}^*)^{n_{p\lambda}}}{(n_{p\lambda}!)^{1/2}}\right] \left[\prod_{p\lambda} \frac{(S_{p\lambda})^{n_{p\lambda}}}{(n_{p\lambda}!)^{1/2}}\right] \\
&= \sum_{\{n\}} \prod_{p\lambda} \frac{(S_{p\lambda}^* S_{p\lambda})^{n_{p\lambda}}}{n_{p\lambda}!} = \exp\left[\sum_{p\lambda} S_{p\lambda}^* S_{p\lambda}\right]. \tag{2-9.29}
\end{aligned}
$$

For a direct computation of $|\langle 0_+|0_-\rangle^S|^2$ we return to (2-9.8) and note that complex conjugation interchanges $S(p)$ and $S(-p)$ while reversing their multiplication order. Therefore,

$$
\left[S(-p) \prod_{\alpha=1}^n [\gamma^0(m - \gamma p)]_\alpha S(p)\right]^* = S(-p) \prod_{\alpha=1}^n [\gamma^0(m - \gamma p)]_\alpha S(p), \tag{2-9.30}
$$

according to the Hermitian nature of each $\gamma^0(m - \gamma p)$ matrix. This reality property gives

$$
|\langle 0_+|0_-\rangle^S|^2 = \exp\left[-\int \frac{(dp)}{(2\pi)^4} S(-p) \right.
$$
$$
\left. \times \prod_\alpha [\gamma^0(m - \gamma p)]_\alpha S(p) \operatorname{Im} \frac{1}{p^2 + m^2 - i\epsilon}\right], \tag{2-9.31}
$$

where, as a statement about integrals,

$$\int \frac{(dp)}{(2\pi)^4} \, \mathrm{Im} \, \frac{1}{p^2 + m^2 - i\epsilon} = \int \frac{(dp)}{(2\pi)^4} \, \pi\delta(p^2 + m^2)$$

$$= \tfrac{1}{2}\int d\omega_p \bigg|_{p^2+m^2=0, p^0>0} + \tfrac{1}{2}\int d\omega_p \bigg|_{p^2+m^2=0, p^0<0}$$

$$(2\text{-}9.32)$$

indicates the restriction to $p^0 = \pm(\mathbf{p}^2 + m^2)^{1/2}$. The two terms are interchanged by the substitution $p^\mu \to -p^\mu$, under which the integrand of (2-9.31) remains unaltered. Accordingly, with p^μ designating a physical momentum, $p^0 > 0$, we get (all this is the four-dimensional momentum space equivalent of an often repeated space-time computation)

$$|\langle 0_+|0_-\rangle^S|^2 = \exp\left[-\int d\omega_p S(p)^* \prod_\alpha [\gamma^0(m - \gamma p)]_\alpha S(p)\right]$$

$$= \exp\left[-\sum_{p\lambda} S_{p\lambda}^* S_{p\lambda}\right], \qquad (2\text{-}9.33)$$

in conformity with the requirement of completeness.

Now let us examine how this consistency would be affected if we intervened in (2-9.8) to reverse the natural connection between spin and statistics, by injecting an antisymmetrical matrix

$$q = \begin{pmatrix} 0 & -i \\ i & 0 \end{pmatrix} \qquad (2\text{-}9.34)$$

which acts on an independent index and thus preserves the spin classification that has been achieved. The identification of multiparticle states from a causal arrangement proceeds analogously, with the helicity vectors $u_{p\lambda}$ extended to $u_{p\lambda q}$, $q = \pm 1$. The result is given by the following replacement in (2-9.22):

$$S_{p\lambda} \to e^{(\pi i/4)(1-q)}S_{p\lambda q}, \qquad S_{p\lambda}^* \to e^{(\pi i/4)(1-q)}S_{p\lambda q}^*, \qquad (2\text{-}9.35)$$

where the product of the additional phase constants reproduces q. In the direct consideration of completeness, however, these phase constants disappear along with the factors of i, and the outcome is just (2-9.29) with the q index added,

$$|\langle 0_+|0_-\rangle^S|^{-2} = \exp\left[\sum_{p\lambda q} S_{p\lambda q}^* S_{p\lambda q}\right]. \qquad (2\text{-}9.36)$$

Turning to the vacuum amplitude itself, we observe that the reality property (2-9.30) persists with the Hermitian matrix q inserted, and that matrix survives in (2-9.33) to give

$$|\langle 0_+|0_-\rangle^S|^2 = \exp\left[-\sum_{p\lambda q} S_{p\lambda q}^* q S_{p\lambda q}\right]. \qquad (2\text{-}9.37)$$

The clear contradiction with (2-9.36) completes the unified proof of the connection between spin and statistics.

The TCP operation is defined for every spin by the substitution

$$S(p) \rightarrow \left[\prod_{\alpha=1}^{n} \gamma_{5\alpha}\right] S(-p), \tag{2-9.38}$$

combined with reversing the multiplication order of all sources. The effect of the substitution on W comes down to the minus sign induced on each γ^0 by the γ_5 transformation, and thus W is multiplied by $(-1)^n$. The reversal in the sense of multiplication introduces a plus or minus sign, in accordance with the statistics. Through the connection between spin and statistics, W, (and the vacuum amplitude), is left invariant under the complete TCP operation.

To study the effect of TCP on individual emission and absorption sources, we first notice the generalization of the spin $\frac{1}{2}$ complex conjugation property (2-6.92), which depends upon the multiplicative composition of the $u_{p\lambda}$,

$$u_{p\,-\lambda}^* = e^{\pi i \lambda} \left[\prod_{\alpha} \gamma_{5\alpha}\right] u_{p\lambda}. \tag{2-9.39}$$

Then we find that

$$S_{p\lambda} \rightarrow e^{-\pi i \lambda} S_{p\,-\lambda}^*, \qquad S_{p\lambda}^* \rightarrow e^{\pi i \lambda} S_{p\,-\lambda}, \tag{2-9.40}$$

to which a charge index can be added in the known way. The corresponding multiparticle transformation is

$$\langle\{n\}|0_-\rangle^S \rightarrow e^{-\pi i \Sigma \lambda n_{p\lambda}}\langle 0_+|\{n'\}\rangle^S, \qquad \langle 0_+|\{n\}\rangle^S \rightarrow e^{\pi i \Sigma \lambda n_{p\lambda}}\langle\{n'\}|0_-\rangle^S, \tag{2-9.41}$$

where

$$n_{p\lambda}' = n_{p\,-\lambda}. \tag{2-9.42}$$

We have been discussing particle aspects in which unification is achieved, the specific nature of the system being implicit in the particular value of n, the number of multispinor indices. But when we turn to the Euclidean postulate in the context of multispinor sources, the fundamental difference between the statistics, or between integer and integer $+ \frac{1}{2}$ spin, becomes explicit. The Euclidean transformation

$$\rho^{-1}S(p) \rightarrow S(p)_E, \tag{2-9.43}$$

where ρ is a matrix to be specified, replaces the kernel of (2-9.8) with

$$\rho^T\left[\prod_{\alpha=1}^{n} \gamma_{\alpha}^0\right]\rho \prod_{\alpha=1}^{n}(m - \rho^{-1}\gamma_\mu\rho p_\mu)_\alpha. \tag{2-9.44}$$

The γ^0 matrices, which mirror the indefinite Minkowski metric, must be removed in the transformation to the Euclidean description. This is accomplished, for n even, by the symmetrical matrix

$$\rho = \exp\left[\frac{\pi i}{4}\left(1 - \prod_{\alpha=1}^{n} \gamma_{\alpha}^0\right)\right], \tag{2-9.45}$$

which is such that

$$\rho^T\left[\prod_{\alpha} \gamma_{\alpha}^0\right]\rho = 1, \tag{2-9.46}$$

while the matrices

$$\rho^{-1}\gamma_{k\alpha}\rho = \left[\prod_{\beta=1}^{n}\gamma_\beta^0\right]i\gamma_{k\alpha}, \qquad \rho^{-1}\gamma_{4\alpha}\rho = i\gamma_\alpha^0 \qquad (2\text{-}9.47)$$

are all real, antisymmetrical matrices that obey

$$\tfrac{1}{2}\{\gamma_\mu, \gamma_\nu\} = -\delta_{\mu\nu}, \qquad (2\text{-}9.48)$$

in which we continue to designate the transformed matrices as γ_μ. When this is combined with the transformation of momentum integrals,

$$-i\int\frac{(dp)}{(2\pi)^4} \rightarrow \int\frac{(dp)}{(2\pi)^4}\bigg|_E, \qquad (2\text{-}9.49)$$

we get the correspondence

s integer:

$$\langle 0_+|0_-\rangle^S \rightarrow \exp\left[-\frac{1}{2}\int\frac{(dp)}{(2\pi)^4}\, S(-p)\prod_{\alpha=1}^{n}(m-\gamma_\mu p_\mu)_\alpha\,\frac{1}{p^2+m^2}\,S(p)\right]_E. \qquad (2\text{-}9.50)$$

The possibility of producing the transformation (2–9.46) is contingent on the left-hand side being a symmetrical matrix. For n odd, it is an antisymmetrical matrix. If the latter is to represent the Euclidean metric, it must be unaltered by Euclidean transformations and is therefore in the nature of a charge matrix q. The Euclidean postulate requires that every integer $+\tfrac{1}{2}$ spin particle carry a charge-like attribute. Were (2–9.45) applied unaltered with n odd, we would get $\rho^T\rho = i$, which does not eliminate the γ^0 matrices. The appropriate definition of ρ for odd n is

$$\rho = \exp\left[\frac{\pi i}{4}\left(1 - q\prod_{\alpha=1}^{n}\gamma_\alpha^0\right)\right], \qquad (2\text{-}9.51)$$

and now

$$\rho^T\left[\prod_\alpha\gamma_\alpha^0\right]\rho = q. \qquad (2\text{-}9.52)$$

The transformed γ_μ matrices are

$$\rho^{-1}\gamma_{k\alpha}\rho = q\left[\prod_{\beta=1}^{n}\gamma_\beta^0\right]i\gamma_{k\alpha}, \qquad \rho^{-1}\gamma_{4\alpha}\rho = i\gamma_\alpha^0, \qquad (2\text{-}9.53)$$

which continue to be real, antisymmetrical, and governed algebraically by (2–9.48). Thus the Euclidean correspondence for odd n is

s integer $+\tfrac{1}{2}$:

$$\langle 0_+|0_-\rangle^S \rightarrow \exp\left[-\frac{1}{2}\int\frac{(dp)}{(2\pi)^4}\, S(-p)q\prod_{\alpha=1}^{n}(m-\gamma_\mu p_\mu)_\alpha\,\frac{1}{p^2+m^2}\,S(p)\right]_E. \qquad (2\text{-}9.54)$$

Incidentally, for $n = 1$ the connection with the real, symmetrical α_μ matrices of (2-6.129) is

$$\alpha_\mu = -iq\gamma_\mu, \tag{2-9.55}$$

where these γ_μ matrices are the transformed ones of (2-9.53).

The space-reflection transformation is defined generally by

$$\bar{S}(\bar{x}) = r_s S(x), \qquad \bar{x}^0 = x^0, \qquad \bar{x}_k = -x_k, \tag{2-9.56}$$

with

$$r_s = (\pm) \prod_{\alpha=1}^{n} (i\gamma_\alpha^0). \tag{2-9.57}$$

Some properties of this real matrix are given by

$$r_s^T = (-1)^n r_s, \qquad r_s^2 = (-1)^n, \tag{2-9.58}$$

which distinguish integer from integer $+ \frac{1}{2}$ spin, and the generally valid

$$r_s^T r_s = 1. \tag{2-9.59}$$

The uniform selection of $\gamma^{0\prime} = +1$ in the rest frame gives the definite parity $(\pm)i^n$, which is real for integer spin. With $n = 2$, the alternatives of anti-symmetrical and symmetrical spinors give the spin-parity properties 0^-, 1^- and 0^+, 1^+, corresponding to the sign option in (2-9.57). Otherwise, with the general use of symmetrical spinors, integer spin particles fall into the two sequences of parity $(\pm)(-1)^s$.

No rest frame is available for massless particles. In this circumstance, the kernel of (2-9.12), referring to causal conditions, becomes

$$\prod_{\alpha=1}^{n} (-\gamma^0\gamma p)_\alpha = (2p^0)^n \prod_{\alpha=1}^{n} [\tfrac{1}{2}(1 - i\gamma_5\boldsymbol{\sigma} \cdot \mathbf{p}/|\mathbf{p}|)]_\alpha. \tag{2-9.60}$$

Now it is the values of the individual helicity matrices $\boldsymbol{\sigma} \cdot \mathbf{p}/|\mathbf{p}|$ and the associated γ_5 matrices that specify a particle state. For a systematic classification of almost all helicities, using symmetrical spinor sources, it suffices to identify the value of every $i\gamma_5$ matrix and thereby of the individual helicity matrices. This is accomplished by inserting the following symmetrical real projection matrix:

$$\Pi_{\gamma_5} = \tfrac{1}{2}(1 + i\gamma_{51}i\gamma_{52}) \ldots \tfrac{1}{2}(1 + i\gamma_{5n-1}i\gamma_{5n}). \tag{2-9.61}$$

Then we have

$$\prod_{\alpha=1}^{n} [\tfrac{1}{2}(1 - i\gamma_5\boldsymbol{\sigma} \cdot \mathbf{p}/|\mathbf{p}|)]_\alpha \Pi_{\gamma_5} = \sum_{\lambda = \pm(1/2)n} u_{p\lambda} u_{p\lambda}^*, \tag{2-9.62}$$

with

$$u_{p\lambda}^* u_{p\lambda'} = \delta_{\lambda\lambda'}. \tag{2-9.63}$$

The limitation to a pair of helicity states is confirmed by evaluating the trace of the left-hand side in (2–9.62), for which one can use the full 4^n-dimensional multispinor space: $4^n(1/2^n)(1/2^{n-1}) = 2$. The list of all helicities obtained in this way, $\lambda = \pm\frac{1}{2}, \pm 1, \pm\frac{3}{2}, \ldots$, only lacks $\lambda = 0$. For that, one can choose $n = 2$, replace the γ_5 projection factor by $\frac{1}{2}(1 - i\gamma_{51}i\gamma_{52})$, and use an antisymmetrical spinor. The emission and absorption sources are identified as

$$\lambda = \pm\tfrac{1}{2}n: \qquad S_{p\lambda} = [(2p^0)^n \, d\omega_p]^{1/2} u^*_{p\lambda} S(p)$$
$$S^*_{p\lambda} = [(2p^0)^n \, d\omega_p]^{1/2} S(p)^* u_{p\lambda}. \qquad (2\text{–}9.64)$$

Although this discussion applies to n both even and odd, the necessary existence of a charge property in the latter situation, of λ an integer $+ \frac{1}{2}$, invites a further classification in which the helicity is tied to the charge value. This is produced by replacing (2–9.61) with the symmetrical real projection matrix (the common $\pm$ sign gives two alternatives)

$$\Pi\gamma_5 = \prod_{\alpha=1}^{n} [\tfrac{1}{2}(1 \pm qi\gamma_5)]_\alpha. \qquad (2\text{–}9.65)$$

For a given value of q the trace of the complete projection matrix now equals $4^n(1/2^n)(1/2^n) = 1$. There are only two states, labeled by $q = \pm 1$, and the helicity is

$$\lambda = (\mp)q\tfrac{1}{2}n, \qquad (2\text{–}9.66)$$

where the sign option refers to the alternatives of (2–9.65). In each situation

$$\prod_{\alpha=1}^{n} [\tfrac{1}{2}(1 - i\gamma_5\boldsymbol{\sigma} \cdot \mathbf{p}/|\mathbf{p}|)\tfrac{1}{2}(1 \pm qi\gamma_5)]_\alpha = \sum_{q'=\pm 1} u_{pq'} u^*_{pq'}, \qquad (2\text{–}9.67)$$

and

$$S_{pq} = [(2p^0)^n \, d\omega_p]^{1/2} u^*_{pq} S(p),$$
$$S^*_{pq} = [(2p^0)^n \, d\omega_p]^{1/2} S(p)^* u_{pq}. \qquad (2\text{–}9.68)$$

This treatment is less general than the earlier neutrino discussion for $n = 1$, since that did not require the assumption of zero mass.

We shall close this section by examining the connection between the multispinor description and the tensor treatment of integer spin particles, in the simplest situation of a second rank spinor $S_{\zeta_1\zeta_2}$. It is convenient to regard the latter as a matrix, and to correspondingly rewrite the structure of W as

$$W = \frac{1}{2} \int \frac{(dp)}{(2\pi)^4} S_{\zeta_1\zeta_2}(-p)[\gamma^0(m - \gamma p)]_{\zeta_1\zeta_1'}$$
$$\times [\gamma^0(m - \gamma p)]_{\zeta_2\zeta_2'} S_{\zeta_1'\zeta_2'}(p) \, \frac{1}{p^2 + m^2 - i\epsilon} \qquad (Cont.)$$

$$= \frac{1}{2} \int \frac{(dp)}{(2\pi)^4} S_{\bar{s}_1\bar{s}_2}(-p)[\gamma^0(m - \gamma p)]_{\bar{s}_1\bar{s}_1'} S_{\bar{s}_1'\bar{s}_2'}(p)$$

$$\times [\gamma^0(-m - \gamma p)]_{\bar{s}_2'\bar{s}_2} \frac{1}{p^2 + m^2 - i\epsilon}$$

$$= \frac{1}{2} \int \frac{(dp)}{(2\pi)^4} \operatorname{tr}\left[(-m - \gamma p)S(-p)^T \gamma^0(m - \gamma p)S(p)\gamma^0\right] \frac{1}{p^2 + m^2 - i\epsilon}.$$

$$(2\text{-}9.69)$$

The general antisymmetrical and symmetrical matrix can be presented as, respectively,

$$2S_a(p) = i\gamma^0 S_1(p) + i\gamma_5\gamma^0 S_2(p) + \gamma_5\gamma^\mu\gamma^0 S_\mu(p),$$
$$2S_s(p) = \gamma^\mu\gamma^0 S_\mu(p) + \tfrac{1}{2}\sigma^{\mu\nu}\gamma^0 S_{\mu\nu}(p),$$

$$(2\text{-}9.70)$$

in which the individual matrices are real. As a useful algebraic rearrangement, we note that

$$\operatorname{tr}\left[(-m - \gamma p)S(-p)^T\gamma^0(m - \gamma p)S(p)\gamma^0\right] = -(p^2 + m^2)\operatorname{tr} S(-p)^T\gamma^0 S(p)\gamma^0$$
$$+ m\operatorname{tr}\left(S(-p)^T\gamma^0[\gamma p, S(p)\gamma^0]\right)$$
$$+ \tfrac{1}{2}\operatorname{tr}\left([\gamma p, S(-p)^T\gamma^0][\gamma p, S(p)\gamma^0]\right)$$

$$(2\text{-}9.71)$$

where, for the two symmetries,

$$[\gamma p, S_a(p)\gamma^0] = \gamma^\mu i\gamma_5 p_\mu S_2(p) + \gamma_5 p_\mu S^\mu(p),$$
$$[\gamma p, S_s(p)\gamma^0] = -\tfrac{1}{2}\sigma^{\mu\nu}i(p_\mu S_\nu(p) - p_\nu S_\mu(p)) - \gamma^\nu i p^\mu S_{\mu\nu}(p).$$

$$(2\text{-}9.72)$$

The evaluation is reduced to computing the traces of matrices formed by multiplying linear combinations of the Dirac matrices. These 16 matrices are orthogonal in the sense of the product defined by the trace. Their normalizations vary in sign with the Hermitian or skew-Hermitian nature of the matrix, as dictated by the space-time metric. Thus, the algebraic properties of the γ_μ imply that

$$\tfrac{1}{4}\operatorname{tr}\gamma_\mu\gamma_\nu = \tfrac{1}{4}\operatorname{tr}\gamma_\mu\gamma_5\gamma_\nu\gamma_5 = -g_{\mu\nu},$$

$$(2\text{-}9.73)$$

while

$$\tfrac{1}{4}\operatorname{tr}\sigma_{\mu\nu}\sigma_{\kappa\lambda} = g_{\mu\kappa}g_{\nu\lambda} - g_{\mu\lambda}g_{\nu\kappa}.$$

$$(2\text{-}9.74)$$

The results are

$$\operatorname{tr}\left[(m + \gamma p)S_a(-p)\gamma^0(m - \gamma p)S_a(p)\gamma^0\right]$$
$$= -(p^2 + m^2)[S_1(-p)S_1(p) + S_2(-p)S_2(p) + S^\mu(-p)S_\mu(p)] + K(-p)K(p),$$

$$(2\text{-}9.75)$$

with

$$K(x) = 2^{1/2}(mS_2(x) + \partial_\mu S^\mu(x)),$$

$$(2\text{-}9.76)$$

and

$$\text{tr}\,[(-m - \gamma p)S_s(-p)\gamma^0(m - \gamma p)S(p)]$$
$$= -(p^2 + m^2)[S^\mu(-p)S_\mu(p) + \tfrac{1}{2}S^{\mu\nu}(-p)S_{\mu\nu}(p)]$$
$$+ J^\mu(-p)J_\mu(p) + \frac{1}{m^2}\, p_\mu J^\mu(-p)p_\nu J^\nu(p), \qquad (2\text{-}9.77)$$

in which

$$J^\nu(x) = 2^{1/2}\big(mS^\nu(x) + \partial_\mu S^{\mu\nu}(x)\big). \qquad (2\text{-}9.78)$$

The K and J structures are the anticipated ones for spin 0 and spin 1. There are additional terms, however, which modify the vacuum amplitude by the typical factor (S stands for S_1, S_2, S_μ, $S_{\mu\nu}$)

$$\exp\left[-i\int (dx)S(x)S(x)\right]. \qquad (2\text{-}9.79)$$

This is an equivalent description. The additional phase factor does not change the vacuum persistence probability nor does it contribute to the coupling between sources in a causal arrangement. And it has no implication for the observable aspects of the energy associated with a quasi-static source distribution, for they refer to the effect of a relative displacement of two disjoint parts. Physical considerations that are sensitive to such source overlap terms can appear only in the further development and specialization of the general source formalism.

For $m = 0$, unit helicity particles should be selected by inserting the projection matrix Π_{γ_5}. Its action upon the second-rank spinor is given by the matrix transcription

$$\Pi_{\gamma_5}S(p) \rightarrow \tfrac{1}{2}[S(p) - i\gamma_5 S(p)i\gamma_5]. \qquad (2\text{-}9.80)$$

The two terms in the symmetrical spinor of (2–9.70) commute and anticommute, respectively, with γ_5. Only the latter is retained by the projection matrix, which effectively sets $S_\mu(p)$ equal to zero. As we recognize from (2–9.78), the divergence of the vector source $J^\nu(x)$ then vanishes identically and the photon description is regained. It would not have sufficed to merely let $m \rightarrow 0$ in (2–9.78), since it is also necessary that $(1/m)\partial_\nu J^\nu \rightarrow 0$. We have remarked that the antisymmetrical spinor should be supplied analogously with a γ_5 projection factor that differs from (2–9.80) in the relative sign of the two terms. This selects terms in S_a that commute with γ_5, which is uniquely the axial vector contribution of (2–9.70). Now, however, it is sufficient to set $m = 0$ in the effective source (2–9.76). It seems to be a specific property of the second-rank spinor representation that the source of massless spin 0 particles acquires the special form of the divergence of a vector.

3

FIELDS

3–1 THE FIELD CONCEPT. SPIN 0 PARTICLES

Sources are introduced to give an idealized description of the creation and the detection of particles. But the purpose of this activity is to study the properties of the particles, and this takes place in some region intermediate between the locations of the terminal acts of creation and detection. Thus one needs a convenient measure of the strength of the excitation that is produced in a region that may be far from its sources. The natural way to obtain such a measure is by investigating the effect on a probe or test source that is introduced into the region of interest. Accordingly, considering spin 0 particles and their real scalar sources, as represented by

$$W(K) = \tfrac{1}{2} \int (dx)(dx') K(x) \Delta_+(x - x') K(x'), \qquad (3\text{–}1.1)$$

we examine the effect of adding an additional weak source $\delta K(x)$. It is given by

$$\delta W(K) = \int (dx)\, \delta K(x) \phi(x), \qquad (3\text{–}1.2)$$

where

$$\phi(x) = \int (dx') \Delta_+(x - x') K(x'). \qquad (3\text{–}1.3)$$

This combination of source and propagation function, measuring the effect of pre-existing sources on a weak test source, is the *field* of the sources. It is defined in an analogous way for any type of particle, as indicated by

$$\delta W(S) = \int (dx)\, \delta S(x) \chi(x). \qquad (3\text{–}1.4)$$

The propagation function $\Delta_+(x - x')$ is characterized by the statements

$$x^0 > x^{0\prime}: \qquad \Delta_+(x - x') = i \int d\omega_p e^{ip(x-x')},$$

$$\qquad\qquad\qquad\qquad\qquad\qquad\qquad\qquad\qquad (3\text{–}1.5)$$

$$x^0 < x^{0\prime}: \qquad \Delta_+(x - x') = i \int d\omega_p e^{ip(x'-x)},$$

or, equivalently, by the four-dimensional momentum integral

$$\Delta_+(x - x') = \int \frac{(dp)}{(2\pi)^4} \frac{e^{ip(x-x')}}{p^2 + m^2 - i\epsilon}\bigg|_{\epsilon \to +0}. \qquad (3\text{–}1.6)$$

We now recognize that it obeys a simple inhomogeneous differential equation. That is most evident from the second expression since the application of the differential operator that produces $p^2 + m^2$, when acting on $\exp[ip(x - x')]$, cancels the denominator and leaves the four-dimensional delta function

$$\delta(x - x') = \int \frac{(dp)}{(2\pi)^4} e^{ip(x-x')},\tag{3-1.7}$$

or

$$(-\partial^2 + m^2)\Delta_+(x - x') = \delta(x - x').\tag{3-1.8}$$

Alternatively, one uses Eq. (3-1.5) and notes that

$$x \neq x': \qquad (-\partial^2 + m^2)\Delta_+(x - x') = 0,\tag{3-1.9}$$

since $p^2 + m^2 = 0$ in these integrals, while the discontinuity of the time derivative across $x^0 = x^{0\prime}$,

$$\partial_0 \Delta_+(x - x')\Big]_{x^0 = x^{0\prime} - 0}^{x^0 = x^{0\prime} + 0} = \int d\omega_p 2p^0 \exp[i\mathbf{p} \cdot (\mathbf{x} - \mathbf{x}')]$$

$$= \delta(\mathbf{x} - \mathbf{x}'),\tag{3-1.10}$$

is equivalent to the presence of the four-dimensional delta function in Eq. (3-1.8).

The differential equation that it obeys identifies $\Delta_+(x - x')$ as a Green's function of the differential operator $-\partial^2 + m^2$. It is the particular solution that has only positive frequencies for $x^0 > x^{0\prime}$ ($e^{-ip^0x^0}$, $p^0 > 0$) and only negative frequencies for $x^0 < x^{0\prime}$ ($e^{ip^0x^0}$). This boundary condition is more simply stated by considering the associated Euclidean Green's function. The latter obeys the differential equation

$$[-(\partial_\mu)^2 + m^2]\Delta_E(x - x') = \delta_E(x - x'),\tag{3-1.11}$$

where

$$(dx)\,\delta(x - x') \leftrightarrow (dx)\,\delta(x - x')]_E\tag{3-1.12}$$

or ($x_4 = ix^0$)

$$(1/i)\,\delta(x - x') \leftrightarrow \delta_E(x - x')\tag{3-1.13}$$

restates the correspondence

$$(1/i)\Delta_+(x - x') \leftrightarrow \Delta_E(x - x').\tag{3-1.14}$$

Unlike the Minkowski situation, the two fundamental solutions of the Euclidean differential equation are sharply distinguished by their asymptotic behavior: $\sim e^{\pm mR}$. Thus the requirement of boundedness, for $x \neq x'$, uniquely selects one solution, the one that is produced automatically by the Fourier integral solution of (3-1.11),

$$\Delta_E(x - x') = \int \frac{(dp)_E}{(2\pi)^4} \frac{e^{ip_\mu(x-x')_\mu}}{(p_\mu)^2 + m^2},\tag{3-1.15}$$

and $\Delta_+(x - x')$ is recovered by the previously explored procedures. The alternative methods of imposing boundary conditions can also be applied directly to the differential equation that describes the field of an arbitrary source,

$$(-\partial^2 + m^2)\phi(x) = K(x). \tag{3-1.16}$$

Other kinds of fields and Green's functions are introduced on considering the time cycle description that is associated with an initial vacuum state: $\langle 0_- | 0_- \rangle^{K(-),K(+)}$. It is characterized by

$$
\begin{aligned}
W(K_{(-)}, K_{(+)}) = \tfrac{1}{2} &\int (dx)(dx') K_{(+)}(x) \Delta_+(x - x') K_{(+)}(x') \\
- \tfrac{1}{2} &\int (dx)(dx') K_{(-)}(x) \Delta_-(x - x') K_{(-)}(x') \\
+ &\int (dx)(dx') K_{(-)}(x)(-i) \Delta^{(+)}(x - x') K_{(+)}(x').
\end{aligned} \tag{3-1.17}
$$

Now the test source response is written,

$$\delta W = \int (dx)\, \delta K_{(+)}(x) \phi_{(+)}(x) - \int (dx)\, \delta K_{(-)}(x) \phi_{(-)}(x), \tag{3-1.18}$$

where the minus sign of the second term recalls the opposite sense of time development that is involved. The two fields encountered here are

$$
\begin{aligned}
\phi_{(+)}(x) &= \int (dx') \Delta_+(x - x') K_{(+)}(x') - i \int (dx') \Delta^{(-)}(x - x') K_{(-)}(x'), \\
\phi_{(-)}(x) &= \int (dx') \Delta_-(x - x') K_{(-)}(x') + i \int (dx') \Delta^{(+)}(x - x') K_{(+)}(x').
\end{aligned} \tag{3-1.19}
$$

Let us examine these fields for the particular situation in which

$$K_{(-)}(x) = K_{(+)}(x) = K(x). \tag{3-1.20}$$

Then

$$
\begin{aligned}
\phi_{(+)}(x) &= \int (dx')[\Delta_+(x - x') - i\Delta^{(-)}(x - x')] K(x'), \\
\phi_{(-)}(x) &= \int (dx')[\Delta_-(x - x') + i\Delta^{(+)}(x - x')] K(x')
\end{aligned} \tag{3-1.21}
$$

and, on using the relations

$$\Delta_+(x - x') = \begin{cases} x^0 > x^{0\prime}: & i\Delta^{(+)}(x - x'), \\ x^0 < x^{0\prime}: & i\Delta^{(-)}(x - x'), \end{cases} \tag{3-1.22}$$

$$\Delta_-(x - x') = \begin{cases} x^0 > x^{0\prime}: & -i\Delta^{(-)}(x - x'), \\ x^0 < x^{0\prime}: & -i\Delta^{(+)}(x - x'), \end{cases} \tag{3-1.23}$$

in which

$$\Delta^{(+)}(x - x') = \Delta^{(-)}(x' - x) = \int d\omega_p e^{ip(x-x')}, \tag{3-1.24}$$

we see that

$$\phi_{(-)}(x) = \phi_{(+)}(x) = \phi_{\text{ret.}}(x) \tag{3-1.25}$$

where

$$\phi_{\text{ret.}}(x) = \int (dx') \Delta_{\text{ret.}}(x - x') K(x') \tag{3-1.26}$$

and

$$\Delta_{\text{ret.}}(x - x') = \Delta_+(x - x') - i\Delta^{(-)}(x - x') = \Delta_-(x - x') + i\Delta^{(+)}(x - x')$$
$$= \begin{cases} x^0 > x^{0\prime}: & i\Delta^{(+)}(x - x') - i\Delta^{(-)}(x - x'), \\ x^0 < x^{0\prime}: & 0. \end{cases} \tag{3-1.27}$$

As the last property proclaims, this is a retarded Green's function. It is real, since complex conjugation interchanges the functions $\Delta^{(+)}$ and $\Delta^{(-)}$. The three Green's functions Δ_+, Δ_-, $\Delta_{\text{ret.}}$ refer to the same inhomogeneous differential equation since $\Delta^{(\pm)}$ are solutions of the homogeneous equation

$$(-\partial^2 + m^2)\Delta^{(\pm)}(x - x') = 0, \tag{3-1.28}$$

and therefore

$$(-\partial^2 + m^2)\phi_{\text{ret.}}(x) = K(x). \tag{3-1.29}$$

The retarded field of an arbitrary source can be found by solving this equation with the boundary condition that the field be zero prior to the intervention of the sources. It is remarkable that this classic boundary condition requires the device of the closed time path for its appearance. Incidentally, the form assumed by δW for the circumstances stated in (3–1.20), namely

$$\delta W = \int (dx)(\delta K_{(+)}(x) - \delta K_{(-)}(x))\phi_{\text{ret.}}(x), \tag{3-1.30}$$

is a reminder that $W = 0$ for $K_{(-)}(x) = K_{(+)}(x)$, and this property will persist if the equality of the sources $K_{(\pm)}(x)$ is maintained by the test source.

To return to the general situation given in Eq. (3–1.19), it is seen that these fields obey the differential equations

$$(-\partial^2 + m^2)\phi_{(+)}(x) = K_{(+)}(x), \qquad (-\partial^2 + m^2)\phi_{(-)}(x) = K_{(-)}(x). \tag{3-1.31}$$

The solution is characterized by the following boundary conditions. Before any sources are operative, $\phi_{(+)}(x)$ contains only negative frequencies and $\phi_{(-)}(x)$ only positive frequencies; after all sources have ceased to function, $\phi_{(+)}(x)$ and $\phi_{(-)}(x)$ are equal. The initial boundary conditions are made explicit by writing

$$\phi_{(+)}(x) = \int (dx')\Delta_+(x - x')[K_{(+)}(x') - K_{(-)}(x')]$$
$$+ \int (dx')\Delta_{\text{ret.}}(x - x')K_{(-)}(x'),$$
$$\phi_{(-)}(x) = \int (dx')\Delta_-(x - x')[K_{(-)}(x') - K_{(+)}(x')]$$
$$+ \int (dx')\Delta_{\text{ret.}}(x - x')K_{(+)}(x'). \tag{3-1.32}$$

And, if it is observed that

$$\Delta_+(x - x') + \Delta_-(x - x') = \Delta_{\text{ret.}}(x - x') + \Delta_{\text{adv.}}(x - x'), \tag{3-1.33}$$

where

$$\Delta_{\text{adv.}}(x - x') = \Delta_+(x - x') - i\Delta^{(+)}(x - x') = \Delta_-(x - x') + i\Delta^{(-)}(x - x')$$

$$= \begin{cases} x^0 > x^{0\prime}: & 0, \\ x^0 < x^{0\prime}: & i\Delta^{(-)}(x - x') - i\Delta^{(+)}(x - x'), \end{cases} \tag{3-1.34}$$

one gets

$$\phi_{(+)}(x) - \phi_{(-)}(x) = \int (dx') \Delta_{\text{adv.}}(x - x')[K_{(+)}(x') - K_{(-)}(x')], \tag{3-1.35}$$

which makes explicit the final boundary condition.

Still other kinds of fields and Green's functions appear on replacing the vacuum state with a general multiparticle state. Rather than use any specific one, we consider a parametrized mixture, as in

$$\sum_{\{n\}} \langle \{n\}_+ | \{n\}_- \rangle^K p_\beta(\{n\}) = \exp[iW_\beta(K)], \tag{3-1.36}$$

where

$$p_\beta(\{n\}) = C \exp\left[-\sum_p \beta_\mu p^\mu n_p\right], \qquad \sum_{\{n\}} p_\beta(\{n\}) = 1, \tag{3-1.37}$$

and β_μ is an arbitrary time-like vector with $\beta_0 > 0$. On reviewing the discussion of $\langle \{n\}_+ | \{n\}_- \rangle^K$, particularly Eq. (2–2.46), we recognize that this probability amplitude is linear in each occupation number n_p, which is merely replaced by an average value in (3–1.36), namely:

$$\langle n_p \rangle_\beta = \sum_{\{n\}} n_p p_\beta(\{n\}) = \frac{\partial}{\partial(-\beta p)} \log\left[\sum_{\{n\}} \exp\left(-\sum_p \beta p n_p\right)\right] = \frac{1}{e^{\beta p} - 1}. \tag{3-1.38}$$

Accordingly, we have

$$W_\beta(K) = \tfrac{1}{2} \int (dx)(dx') K(x) \Delta_\beta(x - x') K(x'), \tag{3-1.39}$$

with

$$\Delta_\beta(x - x') = \Delta_+(x - x') + i \int d\omega_p \langle n_p \rangle_\beta [e^{ip(x-x')} + e^{-ip(x-x')}]. \tag{3-1.40}$$

By expanding $\exp[iW_\beta(K)]$ into the form (3–1.36) the individual $\langle \{n\}_+ | \{n\}_- \rangle^K$ can be recovered. The field

$$\phi_\beta(x) = \int (dx') \Delta_\beta(x - x') K(x'), \tag{3-1.41}$$

which is defined by

$$\delta W_\beta(K) = \int (dx)\, \delta K(x) \phi_\beta(x), \tag{3-1.42}$$

obeys the same differential equation,

$$(-\partial^2 + m^2)\phi_\beta(x) = K(x), \tag{3-1.43}$$

since $\Delta_\beta(x - x')$ is another Green's function of the differential operator $-\partial^2 + m^2$.

To examine the boundary conditions that characterize this Green's function, we write

$$\Delta_\beta(x - x') = \begin{cases} x^0 > x^{0'}: & i\Delta_\beta^{(+)}(x - x'), \\ x^0 < x^{0'}: & i\Delta_\beta^{(-)}(x - x'), \end{cases} \tag{3-1.44}$$

where

$$\Delta_\beta^{(+)}(x - x') = \int d\omega_p [(\langle n_p \rangle_\beta + 1)e^{ip(x-x')} + \langle n_p \rangle_\beta e^{-ip(x-x')}],$$

$$\Delta_\beta^{(-)}(x - x') = \int d\omega_p [\langle n_p \rangle_\beta e^{ip(x-x')} + (\langle n_p \rangle_\beta + 1)e^{-ip(x-x')}] \tag{3-1.45}$$

are related by

$$\Delta_\beta^{(-)}(x - x') = \Delta_\beta^{(+)}(x' - x). \tag{3-1.46}$$

On noting that

$$\langle n_p \rangle_\beta + 1 = e^{\beta p} \langle n_p \rangle_\beta, \tag{3-1.47}$$

we can assert the formal connections

$$\Delta_\beta^{(+)}(x - x') = \Delta_\beta^{(-)}(x - x' - i\beta), \qquad \Delta_\beta^{(-)}(x - x') = \Delta_\beta^{(+)}(x - x' + i\beta), \tag{3-1.48}$$

and these are all combined in

$$\begin{aligned} \Delta_\beta^{(+)}(x - x' + \tfrac{1}{2}i\beta) &= \Delta_\beta^{(+)}(x' - x + \tfrac{1}{2}i\beta) \\ &= \Delta_\beta^{(-)}(x - x' - \tfrac{1}{2}i\beta) \\ &= \Delta_\beta^{(-)}(x' - x - \tfrac{1}{2}i\beta), \end{aligned} \tag{3-1.49}$$

which is a real function. In analogy with the transformation from the Euclidean to the Minkowski description, it is useful to introduce an extrapolation to a real time-like displacement,

$$i\beta_\mu \rightarrow X_\mu, \tag{3-1.50}$$

where

$$X^0 > 0. \tag{3-1.51}$$

For the restricted domain specified by

$$0 < x^0 - x^{0'} < X^0 \tag{3-1.52}$$

or

$$-X^0 < x^0 - x^{0'} < 0, \tag{3-1.53}$$

which are united in

$$|x^0 - x^{0'}| < X^0, \tag{3-1.54}$$

the relations (3–1.48) become statements about the propagation function

$$\Delta_\beta(x - x') = \Delta_\beta(x' - x), \tag{3-1.55}$$

namely

$$\Delta_\beta(x - x') = \Delta_\beta(x - x' \pm X). \tag{3-1.56}$$

This is an assertion of periodicity, which is the boundary condition for these functions.

If one wishes to verify that the periodicity condition does produce the desired solution of the Green's function differential equation, it is convenient to adopt the rest frame of the time-like vector X^μ, with $X^0 = T$, and satisfy the periodicity requirement in x^0 by using Fourier series while retaining the Fourier integral treatment of the spatial coordinates. That gives the Green's function representation

$$\Delta_T(x - x') = \int \frac{(d\mathbf{p})}{(2\pi)^3} e^{i\mathbf{p}\cdot(\mathbf{x}-\mathbf{x}')} \frac{1}{T} \sum_{n=-\infty}^{\infty} \frac{e^{-(2\pi i n/T)(x^0 - x^{0'})}}{\mathbf{p}^2 + m^2 - (2\pi n/T)^2}. \quad (3\text{-}1.57)$$

The Fourier series that occurs here is not unfamiliar:

$$\frac{1}{T} \sum_{n=-\infty}^{\infty} \frac{e^{-(2\pi i n/T)(x^0 - x^{0'})}}{(p^0)^2 - (2\pi n/T)^2} = \frac{i}{2p^0} \frac{e^{(1/2)ip^0 T} e^{-ip^0|x^0 - x^{0'}|} + e^{-(1/2)ip^0 T} e^{ip^0|x^0 - x^{0'}|}}{e^{(1/2)ip^0 T} - e^{-(1/2)ip^0 T}}$$

$$(3\text{-}1.58)$$

and the substitution inverse to (3–1.50) in the rest frame,

$$T \to -i\beta_0, \quad (3\text{-}1.59)$$

followed by removal of the reference to the rest frame, does indeed produce $\Delta_\beta(x - x')$. The same results are obtained directly from the differential equation for $\phi_\beta(x)$ by imposing the periodicity boundary condition

$$\phi_\beta(x) = \phi_\beta(x \pm X). \quad (3\text{-}1.60)$$

To extend this discussion to the time cycle function,

$$\sum_{\{n\}} \langle \{n\}_-|\{n\}_-\rangle^{K_{(-)}, K_{(+)}} p_\beta(\{n\}) = \exp[iW_\beta(K_{(-)}, K_{(+)})], \quad (3\text{-}1.61)$$

we must also consider the function that replaces $\Delta_{\{n\}-}(x - x')$:

$$\Delta_{-\beta}(x - x') = \begin{cases} x^0 > x^{0'}: & -i\Delta_\beta^{(-)}(x - x'), \\ x^0 < x^{0'}: & -i\Delta_\beta^{(+)}(x - x'). \end{cases} \quad (3\text{-}1.62)$$

The designation that we have given it exploits the following formal property of the averaged occupation numbers,

$$\langle n_p \rangle_{-\beta} = -(\langle n \rangle_\beta + 1), \quad (3\text{-}1.63)$$

and thus

$$\Delta_{-\beta}^{(+)}(x - x') = -\Delta_\beta^{(-)}(x - x'), \qquad \Delta_{-\beta}^{(-)}(x - x') = -\Delta_\beta^{(+)}(x - x'). \quad (3\text{-}1.64)$$

The required generalization of (3–1.17) is

$$W_\beta(K_{(-)}, K_{(+)}) = \tfrac{1}{2}\int (dx)(dx') K_{(+)}(x)\Delta_\beta(x - x')K_{(+)}(x')$$

$$- \tfrac{1}{2}\int (dx)(dx') K_{(-)}(x)\Delta_{-\beta}(x - x')K_{(-)}(x')$$

$$+ \int (dx)(dx') K_{(-)}(x)(-i)\Delta_\beta^{(+)}(x - x')K_{(+)}(x'). \quad (3\text{-}1.65)$$

The fields defined by

$$\delta W_\beta(K_{(-)}, K_{(+)}) = \int (dx)\ \delta K_{(+)}(x)\phi_{\beta(+)}(x) - \int (dx)\ \delta K_{(-)}(x)\phi_{\beta(-)}(x)$$

$$(3\text{-}1.66)$$

are

$$\phi_{\beta(+)}(x) = \int (dx')\Delta_\beta(x - x')K_{(+)}(x') - i\int (dx')\Delta_\beta^{(-)}(x - x')K_{(-)}(x'),$$

$$\phi_{\beta(-)}(x) = \int (dx')\Delta_{-\beta}(x - x')K_{(-)}(x') + i\int (dx')\Delta_\beta^{(+)}(x - x')K_{(+)}(x').$$

$$(3\text{-}1.67)$$

In the particular situation

$$K_{(-)}(x) = K_{(+)}(x) = K(x), \qquad (3\text{-}1.68)$$

these fields become

$$\phi_{\beta(-)}(x) = \phi_{\beta(+)}(x) = \phi_{\text{ret.}}(x), \qquad (3\text{-}1.69)$$

since all the causal relations among the various Δ functions are independent of β, as is

$$\Delta_\beta^{(+)}(x - x') - \Delta_\beta^{(-)}(x - x') = \Delta^{(+)}(x - x') - \Delta^{(-)}(x - x'), \quad (3\text{-}1.70)$$

according to (3–1.45). Removing the restriction (3–1.68), the same reasons operate to produce

$$\phi_{\beta(+)}(x) - \phi_{\beta(-)}(x) = \int (dx')\Delta_{\text{adv.}}(x - x')(K_{(+)}(x') - K_{(-)}(x')). \quad (3\text{-}1.71)$$

One can also write

$$\phi_{\beta(+)}(x) - \phi_{(+)}(x)$$
$$= \phi_{\beta(-)}(x) - \phi_{(-)}(x)$$
$$= i\int (dx')(\Delta_\beta^{(+)}(x - x') - \Delta^{(+)}(x - x'))(K_{(+)}(x') - K_{(-)}(x')), \quad (3\text{-}1.72)$$

or an equivalent expression using $\Delta^{(-)}$ functions.

In obtaining these results directly from the differential equations

$$(-\partial^2 + m^2)\phi_{\beta(+)}(x) = K_{(+)}(x), \qquad (-\partial^2 + m^2)\phi_{\beta(-)}(x) = K_{(-)}(x), \quad (3\text{-}1.73)$$

continuity between the two functions is required after the sources have ceased operation. Prior to the action of any source, the two fields are connected by

$$\phi_{\beta(-)}(x + X) = \phi_{\beta(+)}(x), \qquad (3\text{-}1.74)$$

which is the appropriate form of the periodicity condition.

Some of this discussion retains its formal appearance when charged particles are considered and described by a pair of real sources, since the field is correspondingly generalized. But the extension to terminal multiparticle states is most conveniently performed when complex sources are used. Accordingly, we present that treatment in some detail. Beginning with the vacuum amplitude,

$$\langle 0_+|0_-\rangle^K = \exp[iW(K)],$$

$$W(K) = \int (dx)(dx')K^*(x)\Delta_+(x - x')K(x'), \tag{3-1.75}$$

the introduction of probe sources defines two fields, according to

$$\delta W(K) = \int (dx)[\delta K^*(x)\phi(x) + \delta K(x)\phi^*(x)]. \tag{3-1.76}$$

They are

$$\phi(x) = \int (dx')\Delta_+(x - x')K(x'),$$

$$\phi^*(x) = \int (dx')\Delta_+(x - x')K^*(x'). \tag{3-1.77}$$

One must not be misled by the notation and conclude that these fields are in complex conjugate relation. That is a correct assertion about the differential equations they obey,

$$(-\partial^2 + m^2)\phi(x) = K(x), \qquad (-\partial^2 + m^2)\phi^*(x) = K^*(x), \tag{3-1.78}$$

but these equations are to be solved with the same boundary conditions—that of outgoing waves in time, by which is meant positive frequencies in the future and negative frequencies in the past of the source.

Let us examine the structure of these fields in the two asymptotic time regions. If the fields are evaluated at a time after the sources have ceased operation, that causal circumstance is expressed by replacing $\Delta_+(x - x')$ with $i\Delta^{(+)}(x - x')$. Thus ($x > K$ suggests the causal arrangement):

$$x > K: \qquad \phi(x) = i\int (dx')\left[\int d\omega_p e^{ip(x-x')}\right]K(x')$$

$$= \sum_p (d\omega_p)^{1/2}e^{ipx}iK_{p+} \tag{3-1.79}$$

and

$$x > K: \qquad \phi^*(x) = i\int (dx')\left[\int d\omega_p e^{ip(x-x')}\right]K^*(x)$$

$$= \sum_p (d\omega_p)^{1/2}e^{ipx}iK_{p-}, \tag{3-1.80}$$

according to the definitions (2–1.72). In the other situation, the evaluation of fields prior to the functioning of the source, $\Delta_+(x - x')$ is replaced by $i\Delta^{(-)}(x - x')$:

$$x < K: \qquad \phi(x) = i\int (dx')\left[\int d\omega_p e^{-ip(x-x')}\right]K(x')$$

$$= \sum_p (d\omega_p)^{1/2}e^{-ipx}iK^*_{p-},$$

$$x < K: \qquad \phi^*(x) = i\int (dx')\left[\int d\omega_p e^{-ip(x-x')}\right]K^*(x')$$

$$= \sum_p (d\omega_p)^{1/2}e^{-ipx}iK^*_{p+}. \tag{3-1.81}$$

The two causal evaluations of the fields are associated with particle emission and absorption processes, respectively. They assign the field $(d\omega_p)^{1/2}e^{ipx}$ to an individual emission act and the field $(d\omega_p)^{1/2}e^{-ipx}$ to an individual absorption act. As in the interpretation of complex sources these fields produce definite charge changes. Depending upon the causal situation, $\phi(x)$ describes emitted positively charged particles or absorbed negatively charged particles, while $\phi^*(x)$ represents emitted negatively charged particles or absorbed positively charged particles.

The time cycle vacuum amplitude is

$$\langle 0_-|0_-\rangle^{K_{(-)},K_{(+)}} = \exp[iW(K_{(-)}, K_{(+)})],$$

$$W(K_{(-)}, K_{(+)}) = \int (dx)(dx')K^*_{(+)}(x)\Delta_+(x - x')K_{(+)}(x')$$
$$- \int (dx)(dx')K^*_{(-)}(x)\Delta_-(x - x')K_{(-)}(x')$$
$$- i\int (dx)(dx')K^*_{(-)}(x)\Delta^{(+)}(x - x')K_{(+)}(x')$$
$$- i\int (dx)(dx')K_{(-)}(x)\Delta^{(+)}(x - x')K^*_{(+)}(x'). \quad (3\text{-}1.82)$$

Fields are defined by

$$\delta W(K_{(-)}, K_{(+)}) = \int (dx)[\delta K^*_{(+)}(x)\phi_{(+)}(x) + \delta K_{(+)}(x)\phi^*_{(+)}(x)$$
$$- \delta K^*_{(-)}(x)\phi_{(-)}(x) - \delta K_{(-)}(x)\phi^*_{(-)}(x)]. \quad (3\text{-}1.83)$$

The explicit forms are

$$\phi_{(+)}(x) = \int (dx')\Delta_+(x - x')K_{(+)}(x') - i\int (dx')\Delta^{(-)}(x - x')K_{(-)}(x'),$$

$$\phi^*_{(+)}(x) = \int (dx')\Delta_+(x - x')K^*_{(+)}(x') - i\int (dx')\Delta^{(-)}(x - x')K^*_{(-)}(x')$$

$$(3\text{-}1.84)$$

and

$$\phi_{(-)}(x) = \int (dx')\Delta_-(x - x')K_{(-)}(x') + i\int (dx')\Delta^{(+)}(x - x')K_{(+)}(x'),$$

$$\phi^*_{(-)}(x) = \int (dx')\Delta_-(x - x')K^*_{(-)}(x') + i\int (dx')\Delta^{(+)}(x - x')K^*_{(+)}(x').$$

$$(3\text{-}1.85)$$

It is seen that the field structure already given in Eq. (3-1.19) is duplicated here, and the earlier discussion can be applied, enlarged by the substitutions $K \to K^*$, $\phi \to \phi^*$. In particular, when

$$K_{(-)}(x) = K_{(+)}(x) = K(x), \quad (3\text{-}1.86)$$

which implies the analogous complex conjugate equations, we have

$$\phi_{(-)}(x) = \phi_{(+)}(x) = \phi_{\text{ret.}}(x) \quad (3\text{-}1.87)$$

and

$$\phi^*_{(-)}(x) = \phi^*_{(+)}(x) = \phi^*_{\text{ret.}}(x), \quad (3\text{-}1.88)$$

where the two retarded fields *are* complex conjugates since $\Delta_{\text{ret.}}(x - x')$ is a real function. One implication of this property is that any small deviation of

$W(K_{(-)}, K_{(+)})$ from zero is real,

$$\delta W(K_{(-)}, K_{(+)}) = \int (dx)[(\delta K_{(+)}^*(x) - \delta K_{(-)}^*(x))\phi_{\text{ret.}}(x)$$
$$+ (\delta K_{(+)}(x) - \delta K_{(-)}(x))\phi_{\text{ret.}}^*(x)]$$
$$= \delta W(K_{(-)}, K_{(+)})^*. \qquad (3\text{-}1.89)$$

That is a special example of the following statement, which comes immediately from the interpretation of the time cycle vacuum amplitude that is given explicitly in Eq. (2-2.86):

$$[iW(K_{(-)}, K_{(+)})]^* = iW(K_{(+)}, K_{(-)}). \qquad (3\text{-}1.90)$$

The replacement of the vacuum state with a general multiparticle state can be parametrized, as in (3-1.36), with the weight function

$$p_{\alpha\beta}(\{n\}) = C \exp\left[-\sum_{pq}(\alpha q + \beta_\mu p^\mu)n_{pq}\right],$$
$$\sum_{\{n\}} p_{\alpha\beta}(\{n\}) = 1. \qquad (3\text{-}1.91)$$

The averaged occupation numbers are, analogously,

$$\langle n_{pq} \rangle_{\alpha\beta} = \sum_{\{n\}} n_{pq} p_{\alpha\beta}(\{n\}) = \frac{1}{e^{\alpha q + \beta p} - 1}, \qquad (3\text{-}1.92)$$

and $\Delta_+(x - x')$ becomes

$$\Delta_{\alpha\beta}(x - x') = \Delta_+(x - x') + i\int d\omega_p[\langle n_{p+} \rangle_{\alpha\beta} e^{ip(x-x')} + \langle n_{p-} \rangle_{\alpha\beta} e^{-ip(x-x')}], \qquad (3\text{-}1.93)$$

according to (2-2.59). We note again that this function is not symmetrical in x and x', but there is a symmetry in which positive and negative charges are also interchanged. The latter is accomplished by the parameter transformation $\alpha \rightarrow -\alpha$, and

$$\Delta_{\alpha\beta}(x - x') = \Delta_{-\alpha\beta}(x' - x). \qquad (3\text{-}1.94)$$

The functions $\Delta_{\alpha\beta}^{(\pm)}(x - x')$, defined as always by

$$\Delta_{\alpha\beta}(x - x') = \begin{cases} x^0 > x^{0'}: & i\Delta_{\alpha\beta}^{(+)}(x - x'), \\ x^0 < x^{0'}: & i\Delta_{\alpha\beta}^{(-)}(x - x') \end{cases} \qquad (3\text{-}1.95)$$

are

$$\Delta_{\alpha\beta}^{(+)}(x - x') = \int d\omega_p[(\langle n_{p+} \rangle_{\alpha\beta} + 1)e^{ip(x-x')} + \langle n_{p-} \rangle_{\alpha\beta} e^{-ip(x-x')}],$$
$$\Delta_{\alpha\beta}^{(-)}(x - x') = \int d\omega_p[\langle n_{p+} \rangle_{\alpha\beta} e^{ip(x-x')} + (\langle n_{p-} \rangle_{\alpha\beta} + 1)e^{-ip(x-x')}], \qquad (3\text{-}1.96)$$

and

$$\Delta_{\alpha\beta}^{(-)}(x - x') = \Delta_{-\alpha\beta}^{(+)}(x' - x). \qquad (3\text{-}1.97)$$

Other relations that follow from

$$\langle n_{p+}\rangle_{\alpha\beta} + 1 = e^{\alpha+\beta p}\langle n_{p+}\rangle_{\alpha\beta},$$
$$\langle n_{p-}\rangle_{\alpha\beta} + 1 = e^{-\alpha+\beta p}\langle n_{p-}\rangle_{\alpha\beta}, \tag{3-1.98}$$

are

$$\Delta_{\alpha\beta}^{(+)}(x - x') = e^{\alpha}\Delta_{\alpha\beta}^{(-)}(x - x' - i\beta),$$
$$\Delta_{\alpha\beta}^{(-)}(x - x') = e^{-\alpha}\Delta_{\alpha\beta}^{(+)}(x - x' + i\beta). \tag{3-1.99}$$

The periodicity property is correspondingly modified to

$$\Delta_{\alpha\beta}(x - x') = e^{\alpha}\Delta_{\alpha\beta}(x - x' - X)$$
$$= e^{-\alpha}\Delta_{\alpha\beta}(x - x' + X). \tag{3-1.100}$$

Although we shall not give the details, this boundary condition on the propagation function $\Delta_{\alpha\beta}(x - x')$, in conjunction with its differential equation, does reproduce the original function.

The counterpart of the multiparticle replacement of $\Delta_+(x - x')$ with $\Delta_{\alpha\beta}(x - x')$ is

$$\Delta_-(x - x') \rightarrow \Delta_{\alpha\beta}(x' - x)^* = \Delta_{-\alpha\,-\beta}(x - x'). \tag{3-1.101}$$

The time cycle function

$$\sum_{\{n\}} \langle\{n\}_-|\{n\}_-\rangle^{K_{(-)},K_{(+)}} p_{\alpha\beta}(\{n\}) = \exp[iW_{\alpha\beta}(K_{(-)}, K_{(+)}] \tag{3-1.102}$$

is given by

$$W_{\alpha\beta}(K_{(-)}, K_{(+)}) = \int (dx)(dx')K_{(+)}^*(x)\Delta_{\alpha\beta}(x - x')K_{(+)}(x')$$
$$- \int (dx)(dx')K_{(-)}^*(x)\Delta_{-\alpha\,-\beta}(x - x')K_{(-)}(x')$$
$$- i\int (dx)(dx')K_{(-)}^*(x)\Delta_{\alpha\beta}^{(+)}(x - x')K_{(+)}(x')$$
$$- i\int (dx)(dx')K_{(+)}^*(x)\Delta_{\alpha\beta}^{(-)}(x - x')K_{(-)}(x'). \tag{3-1.103}$$

Note that if we wished to write the last term in the alternative way that uses the $\Delta^{(+)}$ function, it is necessary to change α into $-\alpha$, according to Eq. (3-1.97). The fields defined in the manner of (3-1.83) are now given by

$$\phi_{\alpha\beta(+)}(x) = \int (dx')\Delta_{\alpha\beta}(x - x')K_{(+)}(x') - i\int (dx')\Delta_{\alpha\beta}^{(-)}(x - x')K_{(-)}(x'),$$
$$\phi_{\alpha\beta(+)}^*(x) = \int (dx')\Delta_{-\alpha\beta}(x - x')K_{(+)}^*(x') - i\int (dx')\Delta_{-\alpha\beta}^{(-)}(x - x')K_{(-)}^*(x'),$$
$$\tag{3-1.104}$$

and

$$\phi_{\alpha\beta(-)}(x) = \int (dx')\Delta_{-\alpha\ -\beta}(x - x')K_{(-)}(x') + i\int (dx')\Delta^{(+)}_{\alpha\beta}(x - x')K_{(+)}(x'),$$

$$\phi^*_{\alpha\beta(-)}(x) = \int (dx')\Delta_{\alpha\ -\beta}(x - x')K^*_{(-)}(x') + i\int (dx')\Delta^{(+)}_{-\alpha\beta}(x - x')K^*_{(+)}(x').$$

$$(3\text{-}1.105)$$

This is the structure already presented in (3–1.67) with all Δ functions carrying the additional index α or $-\alpha$, which are interchanged as $K \to K^*$, $\phi \to \phi^*$. By adding the index α to the fields, the statements (3–1.69) and (3–1.71) become applicable, together with the results of the substitution $K \to K^*$, $\phi \to \phi^*$. The index α can be introduced everywhere in (3–1.72), but it must be replaced by $-\alpha$ in $\Delta^{(+)}_{\alpha\beta}(x - x')$, when sources and fields are assigned an asterisk. Finally, we note that the field boundary conditions to be imposed on the differential equations are

$$e^{-\alpha}\phi_{\alpha\beta(-)}(x + X) = \phi_{\alpha\beta(+)}(x), \qquad e^{\alpha}\phi^*_{\alpha\beta(-)}(x + X) = \phi^*_{\alpha\beta(+)}(x). \quad (3\text{-}1.106)$$

3–2 THE FIELD CONCEPT.   SPIN ½ PARTICLES

The vacuum amplitude for this system is

$$\langle 0_+|0_-\rangle^\eta = \exp[iW(\eta)],$$

$$W(\eta) = \tfrac{1}{2}\int (dx)(dx')\eta(x)\gamma^0 G_+(x - x')\eta(x'),$$

$$(3\text{-}2.1)$$

where

$$G_+(x - x') = (m - \gamma^\mu(1/i)\partial_\mu)\Delta_+(x - x'). \quad (3\text{-}2.2)$$

Let us observe immediately that the algebraic properties of the γ^μ matrices imply:

$$(\gamma^\mu(1/i)\partial_\mu + m)G_+(x - x') = (-\partial^2 + m^2)\Delta_+(x - x') = \delta(x - x'). \quad (3\text{-}2.3)$$

This identifies $G_+(x - x')$ as a Green's function of the Dirac matrix differential operator. According to the structure of $\Delta_+(x - x')$, it is the one that obeys outgoing wave time boundary conditions. The field definition to be used here is

$$\delta W(\eta) = \int (dx)\ \delta\eta(x)\gamma^0\psi(x) = \int (dx)\psi(x)\gamma^0\ \delta\eta(x). \quad (3\text{-}2.4)$$

The presence of the antisymmetrical matrix γ^0 compensates the anticommutativity of the probe source $\delta\eta(x)$ with $\psi(x)$, which is formed from and is of the same nature as the totally anticommuting sources $\eta(x)$,

$$\psi(x) = \int (dx')G_+(x - x')\eta(x'). \quad (3\text{-}2.5)$$

It is also useful to note that

$$\psi(x)\gamma^0 = \int (dx')\eta(x')\gamma^0 G_+(x' - x), \quad (3\text{-}2.6)$$

as follows directly from the definition, or by using the antisymmetry property

$$[\gamma^0 G_+(x' - x)]^T = -\gamma^0 G_+(x - x'). \tag{3-2.7}$$

The solution of the Green's function differential equation (3-2.3) is given by the equivalent four-dimensional momentum integrals

$$
\begin{aligned}
G_+(x - x') &= \int \frac{(dp)}{(2\pi)^4} \frac{e^{ip(x-x')}}{\gamma p + m - i\epsilon}\bigg|_{\epsilon \to +0} \\
&= \int \frac{(dp)}{(2\pi)^4} \frac{m - \gamma p}{p^2 + m^2 - i\epsilon} e^{ip(x-x')}\bigg|_{\epsilon \to +0},
\end{aligned} \tag{3-2.8}
$$

where ϵ has a different scale in the two expressions. Alternatively, the Green's function is constructed from solutions of the homogeneous Dirac equation,

$$
G_+(x - x') = \begin{cases} x^0 > x^{0'}: & iG^{(+)}(x - x'), \\ x^0 < x^{0'}: & iG^{(-)}(x - x'), \end{cases} \tag{3-2.9}
$$

where [Eq. (2-7.43)]

$$
\begin{aligned}
G^{(+)}(x - x') &= \int d\omega_p (m - \gamma p) e^{ip(x-x')} \\
&= \sum_{p\sigma q} \psi_{p\sigma q}(x) \psi_{p\sigma q}(x')^* \gamma^0, \\
G^{(-)}(x - x') &= \int d\omega_p (m + \gamma p) e^{-ip(x-x')} \\
&= -\sum_{p\sigma q} \psi_{p\sigma q}(x)^* \psi_{p\sigma q}(x') \gamma^0,
\end{aligned} \tag{3-2.10}
$$

and [Eq. (2-7.42)]

$$\psi_{p\sigma q}(x) = (2m \, d\omega_p)^{1/2} u_{p\sigma q} e^{ipx}. \tag{3-2.11}$$

A charge label appears since this is a general attribute of spin $\frac{1}{2}$ particles. The inhomogeneous term of the differential equation (3-2.3) is equivalent to the time discontinuity

$$
\gamma^0 \left(\frac{1}{i}\right) G_+(x - x') \bigg]_{x^0 - x^{0'} = -0}^{x^0 - x^{0'} = +0} = \gamma^0 [G^{(+)}(x - x') - G^{(-)}(x - x')]_{x^0 = x^{0'}}
$$

$$
= \delta(\mathbf{x} - \mathbf{x}'). \tag{3-2.12}
$$

This requirement is obeyed by the explicit momentum integrals of (3-2.10), and is also expressed by

$$
x^0 = x^{0'}:
$$

$$
\sum_{p\sigma q} [\psi_{p\sigma q}(x)\psi_{p\sigma q}(x')^* + \psi_{p\sigma q}(x)^*\psi_{p\sigma q}(x')] = \delta(\mathbf{x} - \mathbf{x}'), \tag{3-2.13}
$$

which is a statement of completeness for the wavefunctions of positive and negative frequency, $\psi_{p\sigma q}(x)$ and $\psi_{p\sigma q}(x)^*$.

The evaluation of the fields in causal situations—after the source has ceased functioning, or prior to its introduction—is given by

$$x > \eta: \qquad \psi(x) = \int (dx') \left[i \sum_{p\sigma q} \psi_{p\sigma q}(x) \psi_{p\sigma q}(x')^* \gamma^0 \right] \eta(x')$$

$$= \sum_{p\sigma q} \psi_{p\sigma q}(x) i \eta_{p\sigma q},$$

(3-2.14)

$$x < \eta: \qquad \psi(x) = \int (dx') \left[-i \sum_{p\sigma q} \psi_{p\sigma q}(x)^* \psi_{p\sigma q}(x') \gamma^0 \right] \eta(x')$$

$$= \sum_{p\sigma q} \psi_{p\sigma q}(x)^* i \eta_{p\sigma q}^*,$$

where [Eq. (2-7.41)]

$$\eta_{p\sigma q}^* = \int (dx) \eta(x) \gamma^0 \psi_{p\sigma q}(x) = -\int (dx) \psi_{p\sigma q}(x) \gamma^0 \eta(x). \qquad (3\text{-}2.15)$$

The $x < \eta$ structure also emerges directly from (3-2.6) as

$$x < \eta: \qquad \psi(x)\gamma^0 = \int (dx') \eta(x') \gamma^0 \left[i \sum_{p\sigma q} \psi_{p\sigma q}(x') \psi_{p\sigma q}(x)^* \gamma^0 \right]$$

$$= \sum_{p\sigma q} i \eta_{p\sigma q}^* \psi_{p\sigma q}(x)^* \gamma^0. \qquad (3\text{-}2.16)$$

The field that follows the action of a source describes the previously emitted particles, and associates the wave function $\psi_{p\sigma q}(x)$ with an individual emission act; the field that precedes the action of a source describes subsequently absorbed particles and associates the wave function $\psi_{p\sigma q}(x)^*$ with an individual absorption act.

It will be noticed that positive and negatively charged particles have been given a uniform treatment. That is because we used real sources, and assigned the task of selecting a specific charge to the multicomponent $u_{p\sigma q}$ or $\psi_{p\sigma q}(x)$. This is natural since, unlike the spin 0 situation, spin ½ already demands the presence of the four-component $u_{p\sigma}$. One can, however, also follow the procedure of preselecting the charge by using complex sources. From the pair of four-component real sources $\eta_{(1)}(x)$, $\eta_{(2)}(x)$, we construct

$$\eta(x) = 2^{-1/2} (\eta_{(1)}(x) - i\eta_{(2)}(x)), \qquad \eta(x)^* = 2^{-1/2} (\eta_{(1)}(x) + i\eta_{(2)}(x)).$$

(3-2.17)

Then the vacuum amplitude is represented by

$$W(\eta) = \int (dx)(dx') \bar{\eta}(x) G_+(x - x') \eta(x'), \qquad (3\text{-}2.18)$$

which uses the definition

$$\bar{\eta}(x) = \eta(x)^* \gamma^0, \qquad (3\text{-}2.19)$$

and fields are defined through

$$\delta W(\eta) = \int (dx) [\delta \bar{\eta}(x) \psi(x) + \bar{\psi}(x) \delta \eta(x)]. \qquad (3\text{-}2.20)$$

These fields are

$$\psi(x) = \int (dx') G_+(x - x')\eta(x'),$$

$$\bar{\psi}(x) = \int (dx') \bar{\eta}(x') G_+(x' - x). \tag{3-2.21}$$

They obey the differential equations

$$(\gamma^\mu(1/i)\partial_\mu + m)\psi(x) = \eta(x),$$

$$\bar{\psi}(x)(\gamma^\mu(1/i)\partial_\mu^T + m) = \bar{\eta}(x), \tag{3-2.22}$$

the second of which uses the notation

$$\bar{\psi}(x)\partial_\mu^T = -\partial_\mu\bar{\psi}(x), \tag{3-2.23}$$

and the observation that

$$G_+(x' - x) = (m + \gamma(1/i)\partial)\Delta_+(x - x') \tag{3-2.24}$$

obeys

$$G_+(x' - x)(\gamma(1/i)\partial^T + m) = (m + \gamma(1/i)\partial)(m - \gamma(1/i)\partial)\Delta_+(x - x')$$

$$= \delta(x - x'). \tag{3-2.25}$$

To conform with the new notation, we now write

$$G^{(+)}(x - x') = \sum_{p\sigma} \psi_{p\sigma}(x)\bar{\psi}_{p\sigma}(x'), \tag{3-2.26}$$

with

$$\psi_{p\sigma}(x) = (2m\,d\omega_p)^{1/2} u_{p\sigma} e^{ipx}, \qquad \bar{\psi}_{p\sigma}(x) = (2m\,d\omega_p)^{1/2} e^{-ipx}\bar{u}_{p\sigma}, \tag{3-2.27}$$

and

$$\bar{u}_{p\sigma} = u^*_{p\sigma}\gamma^0. \tag{3-2.28}$$

Then we must also write

$$G^{(-)}(x - x') = \gamma^0 \left[\sum_{p\sigma} \bar{\psi}_{p\sigma}(x)\psi_{p\sigma}(x') \right] \gamma^0, \tag{3-2.29}$$

which restates the property (3-2.7). The fields in the two causal situations are obtained as

$$x > \eta: \qquad \psi(x) = \int (dx') \left[i \sum_{p\sigma} \psi_{p\sigma}(x)\bar{\psi}_{p\sigma}(x') \right] \eta(x')$$

$$= \sum_{p\sigma} \psi_{p\sigma}(x) i\eta_{p\sigma+},$$

$$\bar{\psi}(x) = \int (dx') \bar{\eta}(x')\gamma^0 \left[i \sum_{p\sigma} \bar{\psi}_{p\sigma}(x')\psi_{p\sigma}(x) \right] \gamma^0$$

$$= \sum_{p\sigma} \bar{\psi}_{p\sigma}(x)\gamma^0 i\eta_{p\sigma-}, \tag{3-2.30}$$

with

$$\eta_{p\sigma+} = \int (dx) \bar{\psi}_{p\sigma}(x)\eta(x), \qquad \eta_{p\sigma-} = \int (dx) \bar{\eta}(x)\gamma^0 \bar{\psi}_{p\sigma}(x), \tag{3-2.31}$$

and

$$x < \eta: \qquad \psi(x) = \int (dx')\gamma^0 \Big[i \sum_{p\sigma} \psi_{p\sigma}(x)\psi_{p\sigma}(x')\Big]\gamma^0 \eta(x')$$

$$= \sum_{p\sigma} \psi_{p\sigma}(x)\gamma^0 i\eta^*_{p\sigma-},$$

$$\bar{\psi}(x) = \int (dx')\bar{\eta}(x')\Big[i \sum_{p\sigma} \psi_{p\sigma}(x')\bar{\psi}_{p\sigma}(x)\Big]$$

$$= \sum_{p\sigma} \bar{\psi}_{p\sigma}(x) i\eta^*_{p\sigma+}, \tag{3-2.32}$$

where

$$\eta^*_{p\sigma+} = \int (dx)\bar{\eta}(x)\psi_{p\sigma}(x), \qquad \eta^*_{p\sigma-} = \int (dx)\eta(x)\gamma^0\psi_{p\sigma}(x). \tag{3-2.33}$$

The specifications of the particle sources follow from the earlier discussion by identifying $\psi(x)$ and $\bar{\psi}(x)\gamma^0$ with the projections of the eight-component field onto the positive and negative charge space, respectively. To use the fields $\psi(x)$ and $\bar{\psi}(x)$ is surely the most familiar and the most popular way of applying the Dirac equation. Nevertheless, we regard the asymmetry of the forms (3-2.30, 32), in contrast with (3-2.14), as justifying, in general, the employment of the real sources and the multicomponent fields that are defined in charge and spin space, rather than the pairs of complex sources and their associated fields.

The time cycle vacuum amplitude

$$\langle 0_+|0_-\rangle^{\eta_{(-)},\eta_{(+)}} = \exp[iW(\eta_{(-)}, \eta_{(+)})] \tag{3-2.34}$$

is characterized by [Eq. (2-7.60)]

$$W(\eta_{(-)}, \eta_{(+)}) = \tfrac{1}{2}\int (dx)(dx')[\eta_{(+)}(x)\gamma^0 G_+(x - x')\eta_{(+)}(x')$$
$$- \eta_{(-)}(x)\gamma^0 G_-(x - x')\eta_{(-)}(x')$$
$$- i\eta_{(-)}(x)\gamma^0 G^{(+)}(x - x')\eta_{(+)}(x')$$
$$- i\eta_{(+)}(x)\gamma^0 G^{(-)}(x - x')\eta_{(-)}(x')]. \tag{3-2.35}$$

The fields defined by the differential expression

$$\delta W(\eta_{(-)}, \eta_{(+)}) = \int (dx)[\delta\eta_{(+)}(x)\gamma^0\psi_{(+)}(x) - \delta\eta_{(-)}(x)\gamma^0\psi_{(-)}(x)] \tag{3-2.36}$$

are given as

$$\psi_{(+)}(x) = \int (dx')G_+(x - x')\eta_{(+)}(x') - i\int (dx')G^{(-)}(x - x')\eta_{(-)}(x'),$$
$$\psi_{(-)}(x) = \int (dx')G_-(x - x')\eta_{(-)}(x') + i\int (dx')G^{(+)}(x - x')\eta_{(+)}(x'). \tag{3-2.37}$$

Since $G_-(x - x')$ obeys the same inhomogeneous differential equation as $G_+(x - x')$, according to [Eq. (2-7.65)]

$$i[G_+(x - x') - G_-(x - x')] + G^{(+)}(x - x') + G^{(-)}(x - x') = 0, \tag{3-2.38}$$

these fields satisfy the differential equations

$$[\gamma(1/i)\partial + m]\psi_{(+)}(x) = \eta_{(+)}(x), \qquad [\gamma(1/i)\partial + m]\psi_{(-)}(x) = \eta_{(-)}(x). \quad (3\text{--}2.39)$$

An alternative presentation of the solution is

$$\psi_{(+)}(x) = \int (dx') G_{\text{ret.}}(x - x')\eta_{(+)}(x')$$
$$+ i\int (dx') G^{(-)}(x - x')(\eta_{(+)}(x') - \eta_{(-)}(x')),$$
$$\psi_{(-)}(x) = \int (dx') G_{\text{ret.}}(x - x')\eta_{(-)}(x') \qquad (3\text{--}2.40)$$
$$+ i\int (dx') G^{(+)}(x - x')(\eta_{(+)}(x') - \eta_{(-)}(x'))$$

while

$$\psi_{(+)}(x) - \psi_{(-)}(x) = \int (dx') G_{\text{adv.}}(x - x')(\eta_{(+)}(x') - \eta_{(-)}(x')). \quad (3\text{--}2.41)$$

Introduced here are the real Green's functions

$$G_{\text{ret.}}(x - x') = G_+(x - x') - iG^{(-)}(x - x') = G_-(x - x') + iG^{(+)}(x - x')$$
$$= \begin{cases} x^0 > x^{0'}: & iG^{(+)}(x - x') - iG^{(-)}(x - x'), \\ x^0 < x^{0'}: & 0, \end{cases} \quad (3\text{--}2.42)$$

and

$$G_{\text{adv.}}(x - x') = G_+(x - x') - iG^{(+)}(x - x') = G_-(x - x') + iG^{(-)}(x - x')$$
$$= \begin{cases} x^0 > x^{0'}: & 0, \\ x^0 < x^{0'}: & iG^{(-)}(x - x') - iG^{(+)}(x - x'). \end{cases} \quad (3\text{--}2.43)$$

It is evident from these forms that, prior to the introduction of the sources, $\psi_{(+)}(x)$ contains only negative frequencies and $\psi_{(-)}(x)$ only positive frequencies, and that $\psi_{(+)}(x) = \psi_{(-)}(x)$ after the sources have ceased operation. These are the boundary conditions that accompany the differential equations (3–2.39). Incidentally, as a glance at Eqs. (3–1.27) and (3–1.34) will confirm, the retarded and advanced Green's functions are also constructed as

$$G_{\text{ret.,adv.}}(x - x') = (m - \gamma(1/i)\partial)\Delta_{\text{ret.,adv.}}(x - x'). \quad (3\text{--}2.44)$$

In the special situation

$$\eta_{(-)}(x) = \eta_{(+)}(x) = \eta(x) \quad (3\text{--}2.45)$$

one evidently has

$$\psi_{(-)}(x) = \psi_{(+)}(x) = \psi_{\text{ret.}}(x), \quad (3\text{--}2.46)$$

with

$$\psi_{\text{ret.}}(x) = \int (dx') G_{\text{ret.}}(x - x')\eta(x') \quad (3\text{--}2.47)$$

appearing as that real solution of the differential equation

$$(\gamma(1/i)\partial + m)\psi_{\text{ret.}}(x) = \eta(x) \quad (3\text{--}2.48)$$

which vanishes before the source comes into action. The form of W for small deviations from the situation (3–2.45) is

$$\delta W = \int (dx)(\delta \eta_{(+)}(x) - \delta \eta_{(-)}(x))\gamma^0 \psi_{\text{ret.}}(x). \qquad (3\text{–}2.49)$$

When the multiparticle mixture given in (3–1.91) is applied to a F. D. system, where $n_{pq} = 0, 1$, the averaged occupation numbers are

$$\langle n_{pq}\rangle_{\alpha\beta} = \frac{e^{-(\alpha q + \beta p)}}{1 + e^{-(\alpha q + \beta p)}} = \frac{1}{e^{\alpha q + \beta p} + 1}. \qquad (3\text{–}2.50)$$

On referring to Eq. (2–7.45), we see that $G_+(x - x')$ is replaced by

$$G_{\alpha\beta}(x - x') = G_+(x - x')$$
$$- i \sum_{p\sigma q} \langle n_{pq}\rangle_{\alpha\beta} \, [\psi_{p\sigma q}(x)\psi_{p\sigma q}(x')^* - \psi_{p\sigma q}(x)^*\psi_{p\sigma q}(x')]\gamma^0, \qquad (3\text{–}2.51)$$

and this Green's function appears in

$$W_{\alpha\beta}(\eta) = \tfrac{1}{2}\int (dx)(dx')\eta(x)\gamma^0 G_{\alpha\beta}(x - x')\eta(x') \qquad (3\text{–}2.52)$$

to determine

$$\sum \langle\{n\}_+|\{n\}_-\rangle^\eta p_{\alpha\beta}(\{n\}) = \exp[iW_{\alpha\beta}(\eta)]. \qquad (3\text{–}2.53)$$

For simplicity, no parameter has been introduced to distinguish the various spin states. The functions defined by

$$G_{\alpha\beta}(x - x') = \begin{cases} x^0 > x^{0\prime}: & iG_{\alpha\beta}^{(+)}(x - x'), \\ x^0 < x^{0\prime}: & iG_{\alpha\beta}^{(-)}(x - x') \end{cases} \qquad (3\text{–}2.54)$$

are, explicitly,

$$G_{\alpha\beta}^{(+)}(x - x') = \sum_{p\sigma q} [(1 - \langle n_{pq}\rangle_{\alpha\beta})\psi_{p\sigma q}(x)\psi_{p\sigma q}(x')^*$$
$$+ \langle n_{pq}\rangle_{\alpha\beta}\psi_{p\sigma q}(x)^*\psi_{p\sigma q}(x')]\gamma^0,$$
$$G_{\alpha\beta}^{(-)}(x - x') = -\sum_{p\sigma q} [\langle n_{pq}\rangle_{\alpha\beta}\psi_{p\sigma q}(x)\psi_{p\sigma q}(x')^* \qquad (3\text{–}2.55)$$
$$+ (1 - \langle n_{pq}\rangle_{\alpha\beta})\psi_{p\sigma q}(x)^*\psi_{p\sigma q}(x')]\gamma^0,$$

and

$$[\gamma^0 G_{\alpha\beta}^{(+)}(x' - x)]^T = -\gamma^0 G_{\alpha\beta}^{(-)}(x - x'), \qquad (3\text{–}2.56)$$

which expresses the necessary antisymmetry of $\gamma^0 G_{\alpha\beta}(x - x')$ under the complete transposition of space-time coordinates and discrete indices. The relation

$$1 - \langle n_{pq}\rangle_{\alpha\beta} = e^{\alpha q + \beta p}\langle n_{pq}\rangle_{\alpha\beta} \qquad (3\text{–}2.57)$$

implies

$$G_{\alpha\beta}^{(+)}(x - x') = -e^{\alpha q} G_{\alpha\beta}^{(-)}(x - x' - i\beta),$$
$$G_{\alpha\beta}^{(-)}(x - x') = -e^{-\alpha q} G_{\alpha\beta}^{(+)}(x - x' + i\beta), \qquad (3\text{–}2.58)$$

where the symbol q now indicates the antisymmetrical charge matrix. The corresponding Green's function property is

$$G_{\alpha\beta}(x - x') = -e^{\alpha q}G_{\alpha\beta}(x - x' - X)$$
$$= -e^{-\alpha q}G_{\alpha\beta}(x - x' + X). \tag{3–2.59}$$

If α is set equal to zero, or, more generally, is accommodated in $G_{\alpha\beta}$ by a coordinate dependent redefinition, the boundary condition on $G_{\alpha\beta}(x - x')$ appears as a sign change in response to a coordinate displacement by X^μ; this property might be called antiperiodicity. Concerning the time cycle generalization of these results, we shall only remark that the multiparticle replacement for $G_-(x - x')$ is $G_{-\alpha\ -\beta}(x - x')$. This statement is equivalent to the relations

$$G^{(+)}_{-\alpha\ -\beta}(x - x') = -G^{(-)}_{\alpha\beta}(x - x'), \qquad G^{(-)}_{-\alpha\ -\beta}(x - x') = -G^{(+)}_{\alpha\beta}(x - x'),$$
$$\tag{3–2.60}$$

and they follow from the average occupation number property

$$\langle n_{pq}\rangle_{-\alpha\ -\beta} = 1 - \langle n_{pq}\rangle_{\alpha\beta}. \tag{3–2.61}$$

3–3 SOME OTHER SPIN VALUES

Spin 1. According to Eq. (2–3.4), unit spin particles of mass $m \neq 0$ are described by

$$W(J) = \tfrac{1}{2}\int (dx)(dx')[J^\mu(x)\Delta_+(x - x')J_\mu(x')$$
$$+ (1/m^2)\partial_\mu J^\mu(x)\Delta_+(x - x')\partial'_\nu J^\nu(x')]. \tag{3–3.1}$$

The consideration of a test source defines a vector field:

$$\delta W(J) = \int (dx)\ \delta J^\mu(x)\phi_\mu(x), \tag{3–3.2}$$

which is presented as

$$\phi_\mu(x) = \int (dx')\Delta_+(x - x')J_\mu(x') - (1/m^2)\partial_\mu\int (dx')\Delta_+(x - x')\partial'_\nu J^\nu(x').$$
$$\tag{3–3.3}$$

The divergence of the vector field is

$$\partial_\mu\phi^\mu(x) = \int (dx')\Delta_+(x - x')\partial'_\mu J^\mu(x') - (1/m^2)\partial^2\int (dx')\Delta_+(x - x')\partial'_\nu J^\nu(x')$$
$$= (1/m^2)\partial_\mu J^\mu(x), \tag{3–3.4}$$

and this derived scalar field vanishes outside the region occupied by the source. The differential equation that is inferred from (3–3.3),

$$(-\partial^2 + m^2)\phi_\mu(x) = J_\mu(x) - (1/m^2)\partial_\mu\partial_\nu J^\nu(x), \tag{3–3.5}$$

can also be presented as

$$(-\partial^2 + m^2)\phi_\mu(x) + \partial_\mu\partial_\nu\phi^\nu(x) = J_\mu(x), \tag{3–3.6}$$

on using the relation (3–3.4). Another version of this differential field equation is

$$\partial_\nu G^{\mu\nu}(x) + m^2\phi^\mu(x) = J^\mu(x), \tag{3-3.7}$$

where

$$G_{\mu\nu}(x) = -G_{\nu\mu}(x) = \partial_\mu\phi_\nu(x) - \partial_\nu\phi_\mu(x). \tag{3-3.8}$$

The differential equations that relate the vector field to its source conversely determine the field when appropriate boundary conditions are added. The divergence of the vector equation regains the relation (3–3.4), and thereby the form (3–3.5). The solution of the latter with the outgoing wave boundary condition is just our starting point of Eq. (3–3.3). In this and other examples of B. E. systems, different boundary conditions can also be used, in straightforward generalization of the spin 0 discussion.

Spin 2. Massive particles of spin 2 are described by [Eq. (2–4.20)]

$$W(T) = \tfrac{1}{2}\int(dx)(dx')[T^{\mu\nu}(x)\Delta_+(x-x')T_{\mu\nu}(x')$$
$$+ (2/m^2)\partial_\nu T^{\mu\nu}(x)\Delta_+(x-x')\partial^{\lambda'}T_{\mu\lambda}(x')$$
$$+ (1/m^4)\partial_\mu\partial_\nu T^{\mu\nu}(x)\Delta_+(x-x')\partial'_\kappa\partial'_\lambda T^{\kappa\lambda}(x')$$
$$- \tfrac{1}{3}(T(x) - (1/m^2)\partial_\mu\partial_\nu T^{\mu\nu}(x))$$
$$\times \Delta_+(x-x')(T(x') - (1/m^2)\partial'_\kappa\partial'_\lambda T^{\kappa\lambda}(x'))], \tag{3-3.9}$$

in which

$$T(x) = g_{\mu\nu}T^{\mu\nu}(x). \tag{3-3.10}$$

The symmetrical tensor field that is introduced through

$$\delta W(T) = \int(dx)\,\delta T^{\mu\nu}(x)\phi_{\mu\nu}(x) \tag{3-3.11}$$

is obtained as

$$\phi_{\mu\nu}(x) = \int(dx')\Delta_+(x-x')T_{\mu\nu}(x')$$
$$- (1/m^2)\partial_\mu\int(dx')\Delta_+(x-x')\partial^{\lambda'}T_{\lambda\nu}(x')$$
$$- (1/m^2)\partial_\nu\int(dx')\Delta_+(x-x')\partial^{\lambda'}T_{\mu\lambda}(x')$$
$$+ (1/m^4)\partial_\mu\partial_\nu\int(dx')\Delta_+(x-x')\partial'_\kappa\partial'_\lambda T^{\kappa\lambda}(x')$$
$$- \tfrac{1}{3}(g_{\mu\nu} - (1/m^2)\partial_\mu\partial_\nu)\int(dx')\Delta_+(x-x')(T(x') - (1/m^2)\partial'_\kappa\partial'_\lambda T^{\kappa\lambda}(x')). \tag{3-3.12}$$

The divergence of this tensor field is the vector

$$\partial^\mu\phi_{\mu\nu}(x) = (1/m^2)\partial^\mu T_{\mu\nu}(x) - (1/3m^2)\partial_\nu(T(x) + (2/m^2)\partial_\kappa\partial_\lambda T^{\kappa\lambda}(x)), \tag{3-3.13}$$

which vanishes in source-free regions. That is also true of the scalar field

$$\phi(x) = g_{\mu\nu}\phi^{\mu\nu}(x) = -(1/3m^2)(T(x) + (2/m^2)\partial_\kappa\partial_\lambda T^{\kappa\lambda}(x)), \tag{3-3.14}$$

and of the combination

$$\partial^\mu \phi_{\mu\nu}(x) - \partial_\nu \phi(x) = (1/m^2)\partial^\mu T_{\mu\nu}(x). \tag{3–3.15}$$

The differential equation derived from Eq. (3–3.12) is

$$(-\partial^2 + m^2)\phi_{\mu\nu}(x) = T_{\mu\nu}(x) - (1/m^2)(\partial_\mu\partial^\lambda T_{\lambda\nu}(x) + \partial_\nu\partial^\lambda T_{\mu\lambda}(x))$$
$$+ g_{\mu\nu}(1/m^2)\partial_\kappa\partial_\lambda T^{\kappa\lambda}(x)$$
$$- \tfrac{1}{3}(g_{\mu\nu} - (1/m^2)\partial_\mu\partial_\nu)(T(x) + (2/m^2)\partial_\kappa\partial_\lambda T^{\kappa\lambda}(x)). \tag{3–3.16}$$

On replacing the vector and scalar source combinations by the equivalent field structures, this differential equation becomes

$$(-\partial^2 + m^2)\phi_{\mu\nu}(x) + \partial_\mu\partial^\lambda\phi_{\lambda\nu}(x) + \partial_\nu\partial^\lambda\phi_{\mu\lambda}(x) - \partial_\mu\partial_\nu\phi(x)$$
$$- g_{\mu\nu}[(-\partial^2 + m^2)\phi(x) + \partial_\kappa\partial_\lambda\phi^{\kappa\lambda}(x)] = T_{\mu\nu}(x). \tag{3–3.17}$$

Or again, on using the information supplied by its trace,

$$(-\partial^2 + m^2)\phi(x) + \partial_\kappa\partial_\lambda\phi^{\kappa\lambda}(x) = -\tfrac{1}{2}(T(x) + m^2\phi(x)), \tag{3–3.18}$$

the latter can be presented as

$$(-\partial^2 + m^2)\phi_{\mu\nu}(x) + \partial_\mu\partial^\lambda\phi_{\lambda\nu}(x) + \partial_\nu\partial^\lambda\phi_{\mu\lambda}(x) - \partial_\mu\partial_\nu\phi(x)$$
$$+ g_{\mu\nu}\tfrac{1}{2}m^2\phi(x) = T_{\mu\nu}(x) - \tfrac{1}{2}g_{\mu\nu}T(x). \tag{3–3.19}$$

Other versions of this differential field equation are

$$\partial^\lambda G_{\mu\nu\lambda}(x) - \partial_\nu G_{\mu\lambda}{}^\lambda(x) + m^2(\phi_{\mu\nu}(x) + \tfrac{1}{2}g_{\mu\nu}\phi(x)) = T_{\mu\nu}(x) - \tfrac{1}{2}g_{\mu\nu}T(x), \tag{3–3.20}$$

with

$$G_{\mu\lambda\nu}(x) = -G_{\nu\lambda\mu}(x) = \partial_\mu\phi_{\lambda\nu}(x) - \partial_\nu\phi_{\lambda\mu}(x), \tag{3–3.21}$$

and

$$\partial^\lambda H_{\mu\nu\lambda}(x) - \partial_\nu H_{\mu\lambda}{}^\lambda(x) + m^2(\phi_{\mu\nu}(x) + \tfrac{1}{2}g_{\mu\nu}\phi(x)) = T_{\mu\nu}(x) - \tfrac{1}{2}g_{\mu\nu}T(x), \tag{3–3.22}$$

where

$$H_{\mu\nu\lambda}(x) = H_{\nu\mu\lambda}(x) = \partial_\mu\phi_{\nu\lambda}(x) + \partial_\nu\phi_{\mu\lambda}(x) - \partial_\lambda\phi_{\mu\nu}(x),$$
$$H_{\mu\lambda}{}^\lambda(x) = \partial_\mu\phi(x). \tag{3–3.23}$$

Conversely, the differential field equations, supplemented by the outgoing wave boundary condition, have the unique solution given in (3–3.12).

The explicit space-time relation between field and source is still quite manageable for spin 2, but becomes increasingly unwieldy for higher spin values. This is eased to some extent by using four-dimensional momentum notation, as we first illustrate for spin 2. The appropriate specialization of the general expression (2–5.95) is

$$W(T) = \frac{1}{2}\int \frac{(dp)}{(2\pi)^4} T^{\mu\nu}(-p) \frac{\bar{g}_{\mu\kappa}(p)\bar{g}_{\nu\lambda}(p) - \tfrac{1}{3}\bar{g}_{\mu\nu}(p)\bar{g}_{\kappa\lambda}(p)}{p^2 + m^2 - i\epsilon} T^{\kappa\lambda}(p), \tag{3–3.24}$$

where

$$\bar{g}_{\mu\nu}(p) = g_{\mu\nu} + (1/m^2)p_\mu p_\nu. \qquad (3\text{-}3.25)$$

Some properties of this tensor are:

$$p^\mu \bar{g}_{\mu\nu}(p) = \frac{p^2 + m^2}{m^2} p_\nu, \qquad \bar{g}^\nu{}_\nu(p) = 3 + \frac{p^2 + m^2}{m^2},$$

$$\bar{g}_{\mu\lambda}(p)\bar{g}^\lambda{}_\nu(p) = \bar{g}_{\mu\nu}(p) + \frac{p^2 + m^2}{m^2} \frac{1}{m^2} p_\mu p_\nu. \qquad (3\text{-}3.26)$$

The field defined by

$$\delta W(T) = \int \frac{(dp)}{(2\pi)^4} \, \delta T^{\mu\nu}(-p)\phi_{\mu\nu}(p) \qquad (3\text{-}3.27)$$

appears as

$$\phi_{\mu\nu}(p) = \frac{1}{p^2 + m^2 - i\epsilon} \left[\bar{g}_{\mu\kappa}(p)\bar{g}_{\nu\lambda}(p) - \tfrac{1}{3}\bar{g}_{\mu\nu}(p)\bar{g}_{\kappa\lambda}(p)\right]T^{\kappa\lambda}(p). \quad (3\text{-}3.28)$$

The derived vector and scalar fields are indicated by

$$p^\mu \phi_{\mu\nu}(p) = \frac{1}{m^2} \left[p_\kappa \bar{g}_{\nu\lambda}(p) - \tfrac{1}{3}p_\nu \bar{g}_{\kappa\lambda}(p)\right]T^{\kappa\lambda}(p) \qquad (3\text{-}3.29)$$

and

$$\phi(p) = \frac{1}{m^2} \left[\frac{1}{m^2} p_\kappa p_\lambda - \tfrac{1}{3}\bar{g}_{\kappa\lambda}(p)\right]T^{\kappa\lambda}(p). \qquad (3\text{-}3.30)$$

The algebraic combination

$$(p^2 + m^2)\phi_{\mu\nu}(p) - p_\mu p^\lambda \phi_{\lambda\nu}(p) = \left[g_{\mu\kappa}\bar{g}_{\nu\lambda}(p) - \tfrac{1}{3}g_{\mu\nu}\bar{g}_{\kappa\lambda}(p)\right]T^{\kappa\lambda}(p), \quad (3\text{-}3.31)$$

its trace,

$$(p^2 + m^2)\phi(p) - p^\kappa p^\lambda \phi_{\kappa\lambda}(p) = -\tfrac{1}{3}\bar{g}_{\kappa\lambda}(p)T^{\kappa\lambda}(p), \qquad (3\text{-}3.32)$$

and the additional combination

$$p_\nu p^\lambda \phi_{\mu\lambda}(p) - p_\mu p_\nu \phi(p) = g_{\mu\kappa} \frac{1}{m^2} p_\nu p_\lambda T^{\kappa\lambda}(p) \qquad (3\text{-}3.33)$$

then lead directly to the equation

$$(p^2 + m^2)\phi_{\mu\nu}(p) - p_\mu p^\lambda \phi_{\lambda\nu}(p) - p_\nu p^\lambda \phi_{\mu\lambda}(p) + p_\mu p_\nu \phi(p)$$
$$- g_{\mu\nu}[(p^2 + m^2)\phi(p) - p^\kappa p^\lambda \phi_{\kappa\lambda}(p)] = T_{\mu\nu}(p), \quad (3\text{-}3.34)$$

which is the momentum space transcription of (3-3.17).

Spin 3. In three examples, of spins 0, 1, 2, the use of the structure indicated generally by Eq. (2-5.95) has produced fields that obey second-order differential equations, with an inhomogeneous term that is just the source function. This ceases to be true for higher spin values, in the sense that derivatives of the source function cannot be completely eliminated. There is, however, the possibility of restoring the simpler situation by using the freedom to modify W by adding real terms in which sources are multiplied together at the same space-time point.

As we have noted in Section 2–9, such contact terms contribute neither to the vacuum persistence probability nor to the coupling of causally disposed sources. It is only through the additional consideration involving the structure of field equations that a reason for their presence and specific appearance can be adduced. The maintenance of the inhomogeneous field equation form that we now regard as standard also requires the introduction of certain auxiliary fields, which vanish outside the regions occupied by sources.

The discussion of spin 3, which is intended to illustrate these remarks, will be facilitated by first examining the simpler situation of massless particles. We have not done this for spins 1 and 2, since these physically important examples will receive individual and extended treatments. Let it be remarked, however, that the appropriate field equations are obtained merely by setting $m = 0$ in, say, Eqs. (3–3.6) and (3–3.19). Also, the physically necessary source restrictions, of vanishing divergences, are algebraic consequences of the field equations, as we can recognize from the $m = 0$ limit of Eqs. (3–3.4) and (3–3.15). But more of this later. The massless spin 3 situation is represented by [Eq. (2–5.122)]

$$W(S) = \frac{1}{2} \int \frac{(dp)}{(2\pi)^4} \frac{1}{p^2 - i\epsilon} [S^{\lambda\mu\nu}(-p)S_{\lambda\mu\nu}(p) - \tfrac{3}{4}S^\lambda(-p)S_\lambda(p)], \quad (3\text{–}3.35)$$

with

$$S^\lambda(p) = S^{\lambda\nu}{}_\nu(p) \qquad (3\text{–}3.36)$$

and the following restriction on the symmetrical third-rank tensor:

$$p_\lambda S^{\lambda\mu\nu}(p) = 0. \qquad (3\text{–}3.37)$$

The latter implies a lack of uniqueness for the field defined by

$$\delta W(S) = \int \frac{(dp)}{(2\pi)^4} \, \delta S^{\lambda\mu\nu}(-p)\phi_{\lambda\mu\nu}(p), \qquad (3\text{–}3.38)$$

since any additional term containing p_λ, p_μ, or p_ν as factors will not contribute in (3–3.38), owing to the source restriction (3–3.37). Hence, the general form of the field is

$$\phi_{\lambda\mu\nu}(p) = \frac{1}{p^2 - i\epsilon} [S_{\lambda\mu\nu}(p) - \tfrac{1}{4}(g_{\mu\nu}S_\lambda(p) + g_{\nu\lambda}S_\mu(p) + g_{\lambda\mu}S_\nu(p))]$$
$$+ p_\lambda \phi_{\mu\nu}(p) + p_\mu \phi_{\nu\lambda}(p) + p_\nu \phi_{\lambda\mu}(p), \qquad (3\text{–}3.39)$$

in which the cyclically related sets of terms are required by the total symmetry of the tensor $\phi_{\lambda\mu\nu}$. The new symmetrical tensor $\phi_{\mu\nu}(p)$ is determined by the source restriction. In order to use the latter, we first note that

$$\phi_\lambda(p) = \phi_{\lambda\nu}{}^\nu(p) = \frac{1}{p^2 - i\epsilon} (-\tfrac{1}{2})S_\lambda(p) + p_\lambda\phi(p) + 2p^\nu\phi_{\lambda\nu}(p), \quad (3\text{–}3.40)$$

where

$$\phi(p) = \phi^\nu{}_\nu(p). \qquad (3\text{–}3.41)$$

Multiplication of Eq. (3-3.39) with p^λ then introduces just the combination that is evaluated as $\frac{1}{2}(\phi_\lambda - p_\lambda \phi)$, and we get

$$p^2 \phi_{\mu\nu} = p^\lambda \phi_{\lambda\mu\nu} - \tfrac{1}{2}(p_\mu \phi_\nu + p_\nu \phi_\mu) + p_\mu p_\nu \phi. \qquad (3\text{-}3.42)$$

An equation for $\phi(p)$ is produced by combining

$$p^\lambda p^\mu p^\nu \phi_{\lambda\mu\nu} = 3p^2 p^\mu p^\nu \phi_{\mu\nu} \qquad (3\text{-}3.43)$$

with

$$p^2 p^\mu p^\nu \phi_{\mu\nu} = p^\lambda p^\mu p^\nu \phi_{\lambda\mu\nu} - p^2 p^\lambda \phi_\lambda + (p^2)^2 \phi, \qquad (3\text{-}3.44)$$

namely

$$(p^2)^2 \phi = p^2 p^\lambda \phi_\lambda - \tfrac{2}{3} p^\lambda p^\mu p^\nu \phi_{\lambda\mu\nu}. \qquad (3\text{-}3.45)$$

The momentum space version of the field equation obeyed by $\phi_{\lambda\mu\nu}(p)$ is now obtained as

$$p^2 \phi_{\lambda\mu\nu} - p_\lambda p^{\lambda'} \phi_{\lambda'\mu\nu} - p_\mu p^{\mu'} \phi_{\lambda\mu'\nu} - p_\nu p^{\nu'} \phi_{\lambda\mu\nu'} + p_\mu p_\nu \phi_\lambda + p_\nu p_\lambda \phi_\mu$$
$$+ p_\lambda p_\mu \phi_\nu - 3 p_\lambda p_\mu p_\nu \phi = S_{\lambda\mu\nu} - \tfrac{1}{4}(g_{\mu\nu} S_\lambda + g_{\nu\lambda} S_\mu + g_{\lambda\mu} S_\nu). \quad (3\text{-}3.46)$$

The construction of $\phi(p)$ given in (3-3.45) is derived directly from this equation by multiplication with $p^\lambda p^\mu p^\nu$. It might seem that we have failed to meet the objective of providing second-order differential field equations. Three derivatives act upon ϕ and the scalar field obeys a fourth-order differential equation. But notice that the field equation involves only this combination of fields:

$$\phi_{\lambda\mu\nu}(p) - 3 \frac{p_\lambda p_\mu p_\nu}{p^2 - i\epsilon} \, \phi(p), \qquad (3\text{-}3.47)$$

since

$$p_\lambda p^{\lambda'}(p_{\lambda'} p_\mu p_\nu) - p_\mu p_\nu (p_\lambda p^2) = 0. \qquad (3\text{-}3.48)$$

The combination given in (3-3.47) is an acceptable redefinition of $\phi_{\lambda\mu\nu}(p)$, which means that $\phi(p)$ can be transformed away. Thus, our final set of field equations is (3-3.46), with $\phi = 0$, or equivalently,

$$p^2 \phi_{\lambda\mu\nu} - p_\lambda p^{\lambda'} \phi_{\lambda'\mu\nu} - \cdots + p_\mu p_\nu \phi_\lambda + \cdots$$
$$- g_{\mu\nu}(p^2 \phi_\lambda + \tfrac{1}{2} p_\lambda p^{\lambda'} \phi_{\lambda'} - p^{\mu'} p^{\nu'} \phi_{\lambda\mu'\nu'}) - \cdots = S_{\lambda\mu\nu}, \quad (3\text{-}3.49)$$

where dots represent the terms that are generated by cyclic permutation from the given ones. The following is an algebraic consequence of this equation,

$$p^\lambda S_{\lambda\mu\nu}(p) = g_{\mu\nu} \tfrac{1}{4} p^\lambda S_\lambda(p); \qquad (3\text{-}3.50)$$

it is consistent with the vanishing divergence of the source, but does not imply it.

Now let us consider a massive particle of spin 3, first using unaltered the description given by Eq. (2-5.95):

$$W(S) = \frac{1}{2} \int \frac{(dp)}{(2\pi)^4} \, S^{\lambda\mu\nu}(-p) \, \frac{\Pi_{\lambda\mu\nu,\lambda'\mu'\nu'}(p)}{p^2 + m^2 - i\epsilon} \, S^{\lambda'\mu'\nu'}(p), \qquad (3\text{-}3.51)$$

where, according to (2–5.79) with $c_{31} = -\frac{3}{5}$,

$$\Pi_{\lambda\mu\nu,\lambda'\mu'\nu'} = \bar{g}_{\lambda\lambda'}\bar{g}_{\mu\mu'}\bar{g}_{\nu\nu'} - \tfrac{2}{5}[\bar{g}_{\lambda\mu}\bar{g}_{\nu\nu'}\bar{g}_{\lambda'\mu'} + \bar{g}_{\mu\nu}\bar{g}_{\lambda\lambda'}\bar{g}_{\mu'\nu'} + \bar{g}_{\nu\lambda}\bar{g}_{\mu\mu'}\bar{g}_{\nu'\lambda'}], \quad (3\text{–}3.52)$$

in which the necessary symmetrization in λ', μ', ν' has not been made explicit. Some properties of this tensor are

$$p^\lambda\Pi_{\lambda\mu\nu,\lambda'\mu'\nu'} = \frac{p^2 + m^2}{m^2}[p_{\lambda'}\bar{g}_{\mu\mu'}\bar{g}_{\nu\nu'} - \tfrac{2}{5}(p_\mu\bar{g}_{\nu\nu'}\bar{g}_{\lambda'\mu'} + p_{\lambda'}\bar{g}_{\mu\nu}\bar{g}_{\mu'\nu'} + p_\nu\bar{g}_{\mu\mu'}\bar{g}_{\nu'\lambda'})],$$

$$(3\text{–}3.53)$$

and

$$g^{\mu\nu}\Pi_{\lambda\mu\nu,\lambda'\mu'\nu'} = \frac{p^2 + m^2}{m^2}\left[\frac{p_{\mu'}p_{\nu'}}{m^2}\,\bar{g}_{\lambda\lambda'} - \frac{1}{5}\left(\frac{p_\lambda p_{\nu'}}{m^2}\,\bar{g}_{\lambda'\mu'}\right.\right.$$

$$\left.\left. + \bar{g}_{\lambda\lambda'}\bar{g}_{\mu'\nu'} + \frac{p_\lambda p_{\mu'}}{m^2}\,\bar{g}_{\nu'\lambda'}\right)\right]. \quad (3\text{–}3.54)$$

The field defined by

$$\delta W(S) = \int \frac{(dp)}{(2\pi)^4}\,\delta S^{\lambda\mu\nu}(-p)\phi_{\lambda\mu\nu}(p) \quad (3\text{–}3.55)$$

is obtained as

$$\phi_{\lambda\mu\nu}(p) = \frac{1}{p^2 + m^2 - i\epsilon}\,\Pi_{\lambda\mu\nu,\lambda'\mu'\nu'}(p)S^{\lambda'\mu'\nu'}(p). \quad (3\text{–}3.56)$$

The presence of the symmetrical source function enforces the symmetrization in λ', μ', ν'. To avoid doing this explicitly for λ, μ, ν, we introduce an auxiliary symmetrical tensor $s^{\lambda\mu\nu}$, and display the field as

$$s^{\lambda\mu\nu}(p^2 + m^2)\phi_{\lambda\mu\nu} = s^{\lambda\mu\nu}[\bar{g}_{\lambda\lambda'}\bar{g}_{\mu\mu'}\bar{g}_{\nu\nu'} - \tfrac{3}{5}\bar{g}_{\lambda\mu}\bar{g}_{\nu\nu'}\bar{g}_{\lambda'\mu'}]S^{\lambda'\mu'\nu'}. \quad (3\text{–}3.57)$$

Similarly, we have

$$s^{\lambda\mu\nu}p_\lambda p^{\lambda'}\phi_{\lambda'\mu\nu} = s^{\lambda\mu\nu}\left[\frac{p_\lambda p^{\lambda'}}{m^2}\,\bar{g}_{\mu\mu'}\bar{g}_{\nu\nu'} - \frac{2}{5}\frac{p_\lambda p_\mu}{m^2}\,\bar{g}_{\nu\nu'}\bar{g}_{\lambda'\mu'} - \frac{1}{5}\frac{p_\lambda p_{\lambda'}}{m^2}\,\bar{g}_{\mu\nu}\bar{g}_{\mu'\nu'}\right]S^{\lambda'\mu'\nu'}$$

$$(3\text{–}3.58)$$

and

$$s^{\lambda\mu\nu}p_\mu p_\nu\phi_\lambda = s^{\lambda\mu\nu}\left[\frac{p_\mu p_{\mu'}}{m^2}\frac{p_\nu p_{\nu'}}{m^2}\,\bar{g}_{\lambda\lambda'} - \frac{2}{5}\frac{p_\lambda p_\mu}{m^2}\frac{p_\nu p_{\nu'}}{m^2}\,\bar{g}_{\lambda'\mu'}\right.$$

$$\left. - \frac{1}{5}\frac{p_\mu p_\nu}{m^2}\,\bar{g}_{\lambda\lambda'}\bar{g}_{\mu'\nu'}\right]S^{\lambda'\mu'\nu'}. \quad (3\text{–}3.59)$$

The combination suggested by the left-hand side of (3–3.46) (with $\phi = 0$) is

$$s^{\lambda\mu\nu}[(p^2 + m^2)\phi_{\lambda\mu\nu} - 3p_\lambda p^{\lambda'}\phi_{\lambda'\mu\nu} + 3p_\mu p_\nu\phi_\lambda]$$

$$= s^{\lambda\mu\nu}\left[\bar{g}_{\lambda\lambda'}\bar{g}_{\mu\mu'}\bar{g}_{\nu\nu'} - \tfrac{3}{5}g_{\mu\nu}\bar{g}_{\lambda\lambda'}\bar{g}_{\mu'\nu'} - \tfrac{3}{5}g_{\mu\nu}\bar{g}_{\lambda\lambda'}\,\frac{p_{\mu'}p_{\nu'}}{m^2}\right.$$

$$\left. - \frac{3}{5}\frac{p_\mu p_\nu}{m^2}\frac{p_\lambda p_{\lambda'}}{m^2}\,g_{\mu'\nu'} + \frac{2}{5}\frac{p_\lambda p_\mu p_\nu}{m^3}\frac{p_{\lambda'}p_{\mu'}p_{\nu'}}{m^3}\right]S^{\lambda'\mu'\nu'}. \quad (3\text{–}3.60)$$

Quite apart from the explicit appearance of numerous derivatives acting on the source, the coefficient of S_λ does not have the value given in Eq. (3–3.46). That is rectified, however, by considering

$$s^{\lambda\mu\nu}m^2 g_{\mu\nu}\phi_\lambda = s^{\lambda\mu\nu}\left[-\tfrac{1}{5}g_{\mu\nu}g_{\lambda\lambda'}g_{\mu'\nu'} + \tfrac{4}{5}g_{\mu\nu}g_{\lambda\lambda'}\frac{p_{\mu'}p_{\nu'}}{m^2}\right.$$
$$\left. -\tfrac{3}{5}g_{\mu\nu}\frac{p_\lambda p_{\lambda'}}{m^2}g_{\mu'\nu'} + \tfrac{2}{5}g_{\mu\nu}\frac{p_\lambda p_{\lambda'}}{m^2}\frac{p_{\mu'}p_{\nu'}}{m^2}\right]S^{\lambda'\mu'\nu'}, \quad (3\text{–}3.61)$$

for then

$$s^{\lambda\mu\nu}[(p^2 + m^2)\phi_{\lambda\mu\nu} - 3p_\lambda p^{\lambda'}\phi_{\lambda'\mu\nu} + 3p_\mu p_\nu\phi_\lambda + \tfrac{3}{4}m^2 g_{\mu\nu}\phi_\lambda]$$
$$= s^{\lambda\mu\nu}[g_{\lambda\lambda'}g_{\mu\mu'}g_{\nu\nu'} - \tfrac{3}{4}g_{\mu\nu}g_{\lambda\lambda'}g_{\mu'\nu'}]S^{\lambda'\mu'\nu'} + s^{\lambda\mu\nu}\left(\tfrac{3}{4}g_{\mu\nu}p_\lambda + \frac{p_\lambda p_\mu p_\nu}{m^2}\right)\Sigma, \quad (3\text{–}3.62)$$

with

$$m^2\Sigma(p) = \frac{2}{5}\left(\frac{p_{\lambda'}p_{\mu'}p_{\nu'}}{m^2} - \tfrac{3}{2}p_{\lambda'}g_{\mu'\nu'}\right)S^{\lambda'\mu'\nu'}(p). \quad (3\text{–}3.63)$$

And, from the equation for ϕ_λ we get

$$(p^2 + 2m^2)\Sigma(p) - m^2 p^\lambda\phi_\lambda(p) = -p^\lambda S_\lambda(p). \quad (3\text{–}3.64)$$

As a differential equation, (3–3.62) contains third derivatives of the auxiliary scalar field Σ. If we reduced this to second derivatives by regarding the gradient of the scalar field as the appropriate auxiliary field, the second-order differential equation that the vector field obeys contains second derivatives of the source function.

This is the background for the addition to W of a specific contact coupling term:

$$W_{\text{cont.}}(S) = \frac{1}{2}\int\frac{(dp)}{(2\pi)^4}\left[\frac{1}{4m^2}\Sigma(-p)p_\lambda S^\lambda(p) - \frac{1}{4m^2}p_\lambda S^\lambda(-p)\Sigma(p)\right.$$
$$\left. -\frac{1}{40m^6}(5m^2 - p^2)p_\kappa S^\kappa(-p)p_\lambda S^\lambda(p)\right]. \quad (3\text{–}3.65)$$

The associated supplement to the field is given by

$$s^{\lambda\mu\nu}\phi_{\lambda\mu\nu}\Big|_{\text{cont.}} = s^{\lambda\mu\nu}\left[\frac{1}{10m^6}\left(-p_\lambda p_\mu p_\nu + \frac{p^2 + m^2}{4}g_{\mu\nu}p_\lambda\right)p_\kappa S^\kappa - \frac{1}{4m^2}g_{\mu\nu}p_\lambda\Sigma\right], \quad (3\text{–}3.66)$$

and its contribution to the field equation is

$$s^{\lambda\mu\nu}[(p^2 + m^2)\phi_{\lambda\mu\nu} - 3p_\lambda p^{\lambda'}\phi_{\lambda'\mu\nu} + 3p_\mu p_\nu\phi_\lambda + \tfrac{3}{4}m^2 g_{\mu\nu}\phi_\lambda]_{\text{cont.}}$$
$$= s^{\lambda\mu\nu}\left[-\left(\tfrac{3}{4}g_{\mu\nu}p_\lambda + \frac{1}{m^2}p_\lambda p_\mu p_\nu\right)\Sigma - g_{\mu\nu}p_\lambda\tfrac{1}{2}im\Phi\right], \quad (3\text{–}3.67)$$

where

$$im\Phi = -\tfrac{1}{4}\Sigma - \frac{1}{40m^4}(5m^2 - p^2)p_\lambda S^\lambda$$

$$= \frac{1}{10m^4}[\tfrac{1}{4}(p^2 + m^2)p_\lambda S^\lambda - p_\lambda p_\mu p_\nu S^{\lambda\mu\nu}]. \tag{3-3.68}$$

The addition of these terms to the right-hand side of Eq. (3–3.62) removes the third derivatives of Σ, and replaces them by first derivatives of Φ. And there is an additional contribution in (3–3.64), which can be added to the left-hand side as

$$m^2 p_\lambda \phi^\lambda \Big|_{\text{cont.}} = -\tfrac{1}{2}p^2\Sigma + \frac{1}{20m^4}p^2(m^2 - p^2)p_\lambda S^\lambda. \tag{3-3.69}$$

When Σ is replaced by Φ in that equation, all source terms disappear. This successful realization of our program is displayed in the pair of differential equations:

$$(-\partial^2 + m^2)\phi_{\lambda\mu\nu} + \partial_\lambda\partial^{\lambda'}\phi_{\lambda'\mu\nu} + \cdots - \partial_\mu\partial_\nu\phi_\lambda - \cdots + \tfrac{1}{4}m^2 g_{\mu\nu}\phi_\lambda + \cdots$$
$$+ \tfrac{1}{6}m(g_{\mu\nu}\partial_\lambda + \cdots)\Phi = S_{\lambda\mu\nu} - \tfrac{1}{4}(g_{\mu\nu}S_\lambda + \cdots),$$
$$(-\partial^2 + 4m^2)\Phi(x) = \tfrac{1}{2}m\partial_\lambda\phi^\lambda(x). \tag{3-3.70}$$

Note that the scalar field effectively vanishes as $m \to 0$, and (3–3.46) is recovered. When supplemented by outgoing wave boundary conditions, the unique solution of the set (3–3.70) is the field $\phi_{\lambda\mu\nu}$ given in Eq. (3–3.56).

Spin $\tfrac{3}{2}$. According to Eq. (2–8.10), this system is described by

$$W(\eta) = \frac{1}{2}\int\frac{(dp)}{(2\pi)^4}\,\eta^\mu(-p)\gamma^0\left[(m - \gamma p)\bar{g}_{\mu\nu}(p)\right.$$

$$\left.+ \frac{1}{3}\left(\gamma_\mu + \frac{1}{m}p_\mu\right)(m + \gamma p)\left(\gamma_\nu + \frac{1}{m}p_\nu\right)\right]\frac{1}{p^2 + m^2 - i\epsilon}\,\eta^\nu(p). \tag{3-3.71}$$

The field defined by

$$\delta W(\eta) = \int\frac{(dp)}{(2\pi)^4}\,\delta\eta^\mu(-p)\gamma^0\psi_\mu(p) \tag{3-3.72}$$

is, therefore,

$$\psi_\mu(p) = \frac{1}{p^2 + m^2 - i\epsilon}\left[(m - \gamma p)\bar{g}_{\mu\nu}(p)\right.$$

$$\left.+ \frac{1}{3}\left(\gamma_\mu + \frac{1}{m}p_\mu\right)(m + \gamma p)\left(\gamma_\nu + \frac{1}{m}p_\nu\right)\right]\eta^\nu(p). \tag{3-3.73}$$

Two derived fields are

$$p^\mu\psi_\mu = \frac{1}{m^2}[(m - \gamma p)p_\nu + \tfrac{1}{3}\gamma p(m\gamma_\nu + p_\nu)]\eta^\nu \tag{3-3.74}$$

and

$$\gamma^\mu \psi_\mu = \frac{1}{m^2} [p_\nu - \tfrac{1}{3}(m\gamma_\nu + p_\nu)]\eta^\nu, \tag{3-3.75}$$

which supply the linear combinations

$$p^\mu \psi_\mu + \gamma p \gamma^\mu \psi_\mu = (1/m)p_\nu \eta^\nu \tag{3-3.76}$$

and

$$-p^\mu \psi_\mu + (m - \gamma p)\gamma^\mu \psi_\mu = -\frac{1}{3m} (m\gamma_\nu + p_\nu)\eta^\nu. \tag{3-3.77}$$

The use of either of the identities

$$\begin{aligned}
(m + \gamma p)(m\gamma_\mu + p_\mu) &= (m\gamma_\mu - p_\mu)(m - \gamma p), \\
(m\gamma_\mu + p_\mu)(m + \gamma p) &= (m - \gamma p)(m\gamma_\mu - p_\mu),
\end{aligned} \tag{3-3.78}$$

produces the equation

$$(\gamma p + m)\psi_\mu = \eta_\mu + \tfrac{1}{3}\gamma_\mu \left(\gamma_\nu + \frac{1}{m} p_\nu\right)\eta^\nu + \frac{1}{m^2} p_\mu[p_\nu - \tfrac{1}{3}(m\gamma_\nu + p_\nu)]\eta^\nu, \tag{3-3.79}$$

from which we immediately obtain

$$(\gamma p + m)\psi_\mu - p_\mu \gamma^\nu \psi_\nu - \gamma_\mu p^\nu \psi_\nu + \gamma_\mu (m - \gamma p)\gamma^\nu \psi_\nu = \eta_\mu. \tag{3-3.80}$$

Like the field equation for spin $\tfrac{1}{2}$, this is (the momentum space equivalent of) a first-order differential equation with the source appearing as the inhomogeneous term. The solution of the equation under outgoing wave boundary conditions is the field given by (3-3.73). As is indicated by the $m \to 0$ limit of (3-3.76), the necessary source restriction for massless particles,

$$p_\mu \eta^\mu(p) = 0, \tag{3-3.81}$$

is an algebraic consequence of the $m = 0$ field equation. There is a remarkably compact way of presenting Eq. (3-3.80), which is reached by commuting γ_μ with $(m - \gamma p)$ and noting that

$$\psi_\mu + \gamma_\mu \gamma^\nu \psi_\nu = -i\sigma_{\mu\nu}\psi^\nu, \qquad p_\mu \gamma_\nu - p_\nu \gamma_\mu = -\frac{i}{2}[\sigma_{\mu\nu}, \gamma p]. \tag{3-3.82}$$

It is

$$-\frac{i}{2}\{\sigma_{\mu\nu}, \gamma p + m\}\psi^\nu = \eta_\mu. \tag{3-3.83}$$

With the examples of spins 2, 3, and $\tfrac{3}{2}$ before us we can recognize the possibility of simple algebraic redefinitions of the sources that preserve the general structure of the field equations, but introduce or modify contact terms in the expression for W. Thus, let

$$T_{\mu\nu}(x) = T'_{\mu\nu}(x) - ag_{\mu\nu}T'(x), \tag{3-3.84}$$

which has the inverse

$$T'_{\mu\nu}(x) = T_{\mu\nu}(x) + \frac{a}{1 - 4a}\, g_{\mu\nu} T(x). \tag{3-3.85}$$

On introducing this redefinition into the expression for $W(T)$, say Eq. (3-3.24), the additional $g_{\mu\nu}$ terms supply only contact contributions. The explicit statement is

$$W(T) = W(T') - \frac{a}{3m^4} \int \frac{(dp)}{(2\pi)^4} \left[T'(-p) p^\mu p^\nu T'_{\mu\nu}(p) + p^\mu p^\nu T'_{\mu\nu}(-p) T'(p) \right]$$

$$+ \frac{1}{3m^2} \int \frac{(dp)}{(2\pi)^4} T'(-p) \left[\frac{p^2}{m^2}\, a^2 - a(2a - 1) \right] T'(p), \tag{3-3.86}$$

where $W(T')$ has the same functional form as $W(T)$. The implied field transformation inferred from

$$\delta W(T) = \int \frac{(dp)}{(2\pi)^4}\, \delta T^{\mu\nu}(-p) \phi_{\mu\nu}(p) = \int \frac{(dp)}{(2\pi)^4}\, \delta T'^{\mu\nu}(-p) \phi'_{\mu\nu}(p) \tag{3-3.87}$$

appears as

$$\phi'_{\mu\nu} = \phi_{\mu\nu} - a g_{\mu\nu}\phi \tag{3-3.88}$$

or

$$\phi_{\mu\nu} = \phi'_{\mu\nu} + \frac{a}{1 - 4a}\, g_{\mu\nu}\phi'. \tag{3-3.89}$$

Since field and source are transformed linearly, and locally in space-time, the differential field equation maintains its general form, but with changed coefficients. This is illustrated for $a = \frac{1}{2}$, where

$$\phi'_{\mu\nu} = \phi_{\mu\nu} - \tfrac{1}{2} g_{\mu\nu}\phi, \qquad \phi_{\mu\nu} = \phi'_{\mu\nu} - \tfrac{1}{2} g_{\mu\nu}\phi', \tag{3-3.90}$$

and similarly for the sources, by the differential equation

$$(m^2 - \partial^2)\phi'_{\mu\nu} + \partial_\mu \partial^\lambda \phi'_{\lambda\nu} + \partial_\nu \partial^\lambda \phi'_{\mu\lambda} - \tfrac{1}{2} g_{\mu\nu}(2m^2 - \partial^2)\phi' = T'_{\mu\nu}. \tag{3-3.91}$$

If this is applied to massless particles, the differential equation demands the vanishing divergence, not of the source $T'_{\mu\nu}$, but of the combination

$$T'_{\mu\nu} - \tfrac{1}{2} g_{\mu\nu} T' = T_{\mu\nu}. \tag{3-3.92}$$

Concerning spin 3 we remark only that the redefinition indicated by

$$\begin{aligned} S_{\lambda\mu\nu} &= S'_{\lambda\mu\nu} - a(g_{\mu\nu}S'_\lambda + g_{\nu\lambda}S'_\mu + g_{\lambda\mu}S'_\nu), \\ \phi'_{\lambda\mu\nu} &= \phi_{\lambda\mu\nu} - a(g_{\mu\nu}\phi_\lambda + g_{\nu\lambda}\phi_\mu + g_{\lambda\mu}\phi_\nu), \end{aligned} \tag{3-3.93}$$

which maintains the general structure of the field equations but changes the contact terms, cannot be used to remove the latter. As we observed, second-order differential equations lacking in source derivatives are not obtained if contact terms are omitted.

Returning to spin $\frac{3}{2}$, we note that the linear source transformation

$$\eta_\mu(p) = \eta_\mu'(p) + a\gamma_\mu\gamma^\nu\eta_\nu'(p) \tag{3-3.94}$$

produces contact terms:

$$W(\eta) = W(\eta') + \frac{a}{3m^2}\int\frac{(dp)}{(2\pi)^4}\,[\eta'^\mu(-p)\gamma^0\gamma_\mu p_\nu\eta'^\nu(p) + p_\mu\eta'^\mu(-p)\gamma^0\gamma_\nu\eta'^\nu(p)]$$

$$+\frac{1}{3m^2}\int\frac{(dp)}{(2\pi)^4}\,\eta'^\mu(-p)\gamma^0\gamma_\mu[\gamma p a^2 + ma(2a-1)]\gamma_\nu\eta'^\nu(p). \tag{3-3.95}$$

The field transformation implied by

$$\delta W(\eta) = \int\frac{(dp)}{(2\pi)^4}\,\delta\eta^\mu(-p)\psi_\mu(p) = \int\frac{(dp)}{(2\pi)^4}\,\delta\eta'^\mu(-p)\psi_\mu'(p) \tag{3-3.96}$$

can be written as

$$\psi_\mu' = \psi_\mu + a\gamma_\mu\gamma^\nu\psi_\nu \tag{3-3.97}$$

or

$$\psi_\mu = \psi_\mu' + \tfrac{1}{2}(1-\alpha)\gamma_\mu\gamma^\nu\psi_\nu', \tag{3-3.98}$$

where

$$\alpha = (1-2a)/(1-4a). \tag{3-3.99}$$

The transformed version of the field equation (3-3.80) is

$$(\gamma p + m)\psi_\mu' - \alpha(p_\mu\gamma^\nu\psi_\nu' + \gamma_\mu p^\nu\psi_\nu') + \gamma_\mu(\alpha'm - \alpha''\gamma p)\gamma^\nu\psi_\nu' = \eta_\mu' \tag{3-3.100}$$

in which

$$\alpha' = 3\alpha(\alpha-1)+1, \qquad \alpha'' = \tfrac{1}{2}(3\alpha^2 - 2\alpha + 1). \tag{3-3.101}$$

Notice that there are just two situations in which

$$\alpha = \alpha' = \alpha'', \tag{3-3.102}$$

namely

$$\alpha = 1, \quad a = 0; \qquad \alpha = \tfrac{1}{3}, \quad a = 1. \tag{3-3.103}$$

The first gives the original field equation, and the second produces

$$(\gamma p + m)\psi_\mu' - \tfrac{1}{3}(p_\mu\gamma^\nu\psi_\nu' + \gamma_\mu p^\nu\psi_\nu') + \tfrac{1}{3}\gamma_\mu(m - \gamma p)\gamma^\nu\psi_\nu' = \eta_\mu'. \tag{3-3.104}$$

Another simple choice is

$$\alpha = 0, \quad a = \tfrac{1}{2}, \tag{3-3.105}$$

which gives

$$(\gamma p + m)\psi_\mu' + \tfrac{1}{2}\gamma_\mu(2m - \gamma p)\gamma^\nu\psi_\nu' = \eta_\mu'. \tag{3-3.106}$$

For $m = 0$, the field equations imply the vanishing divergence of a quantity that is the transformed statement of the original source function.

Spin $\frac{5}{2}$. Again, we turn first to the massless situation in order to get the simplest indication of the field equation structure. The appropriate specialization of the general form given in (2-8.52) is

$$W(\eta) = \frac{1}{2} \int \frac{(dp)}{(2\pi)^4} \frac{1}{p^2 - i\epsilon} [\eta^{\mu\nu}(-p)\gamma^0(-\gamma p)\eta_{\mu\nu}(p)$$
$$- \tfrac{1}{2}\eta^{\mu\nu}(-p)\gamma^0\gamma_\mu(-\gamma p)\gamma^\lambda\eta_{\lambda\nu}(p) - \tfrac{1}{4}\eta(-p)\gamma^0(-\gamma p)\eta(p)], \quad (3\text{-}3.107)$$

where

$$\eta(p) = g_{\mu\nu}\eta^{\mu\nu}(p) \qquad (3\text{-}3.108)$$

and

$$p_\mu\eta^{\mu\nu}(p) = 0. \qquad (3\text{-}3.109)$$

The field defined by

$$\delta W(\eta) = \int \frac{(dp)}{(2\pi)^4} \delta\eta^{\mu\nu}(-p)\psi_{\mu\nu}(p) \qquad (3\text{-}3.110)$$

is not unique, since the source is restricted by (3-3.109). Its general form is

$$\psi_{\mu\nu} = \frac{1}{p^2 - i\epsilon}[-\gamma p\eta_{\mu\nu} - \tfrac{1}{4}\gamma_\mu(-\gamma p)\gamma^\lambda\eta_{\lambda\nu} - \tfrac{1}{4}\gamma_\nu(-\gamma p)\gamma^\lambda\eta_{\mu\lambda}$$
$$- \tfrac{1}{4}g_{\mu\nu}(-\gamma p)\eta(p)] + p_\mu\psi_\nu + p_\nu\psi_\mu, \qquad (3\text{-}3.111)$$

where the vector-spinor ψ_μ is to be determined through the source restriction.
We shall not detail the extraction of the field equation, except to remark that,
as in the spin 3 discussion, higher derivatives seem to appear but can be removed
through a permissible redefinition of the field, namely

$$\psi_{\mu\nu} + 2p_\mu p_\nu(\gamma p/p^2)\gamma^\lambda\psi_\lambda \rightarrow \psi_{\mu\nu}. \qquad (3\text{-}3.112)$$

The final form of the field equation is

$$\gamma p\psi_{\mu\nu} - (p_\mu\gamma^{\mu'}\psi_{\mu'\nu} + \gamma_\mu p^{\mu'}\psi_{\mu'\nu} + p_\nu\gamma^{\nu'}\psi_{\mu\nu'} + \gamma_\nu p^{\nu'}\psi_{\mu\nu'})$$
$$- (\gamma_\mu\gamma p\gamma^{\mu'}\psi_{\mu'\nu} + \gamma_\nu\gamma p\gamma^{\nu'}\psi_{\mu\nu'}) + g_{\mu\nu}\gamma^{\mu'}p^{\nu'}\psi_{\mu'\nu'}$$
$$+ \tfrac{1}{2}(\gamma_\mu p_\nu + \gamma_\nu p_\mu - g_{\mu\nu}\gamma p)\psi = \eta_{\mu\nu}. \qquad (3\text{-}3.113)$$

The algebraic consequence of this equation,

$$p^\mu\eta_{\mu\nu} + \tfrac{1}{4}\gamma_\nu\gamma^{\mu'}p^{\nu'}\eta_{\mu'\nu'} = 0, \qquad (3\text{-}3.114)$$

is consistent with the source restriction, but does not imply it.

By following the instructions given in Section 2-8, we arrive at the following
source description for a massive particle of spin $\tfrac{5}{2}$:

$$W(\eta) = \frac{1}{2} \int \frac{(dp)}{(2\pi)^4} \eta^{\mu\nu}(-p)\gamma^0 \frac{1}{p^2 + m^2 - i\epsilon}$$

$$\times \left[(m - \gamma p)(\bar{g}_{\mu\mu'}\bar{g}_{\nu\nu'} - \tfrac{1}{5}\bar{g}_{\mu\nu}\bar{g}_{\mu'\nu'}) \right.$$

$$\left. + \frac{2}{5}\left(\gamma_\mu + \frac{1}{m}p_\mu\right)(m + \gamma p)\left(\gamma_{\mu'} + \frac{1}{m}p_{\mu'}\right)\bar{g}_{\nu\nu'} \right]\eta^{\mu'\nu'}(p). \quad (3\text{-}3.115)$$

Let us also state here the additional contact terms that are required to bring the field equations into first-order form without source derivatives. They are derived from

$$W_{\text{cont.}}(\eta) = \frac{1}{10m^4} \int \frac{(dp)}{(2\pi)^4} \, \eta^{\mu\nu}(-p)\gamma^0 p_\mu p_\nu \left[\gamma_{\mu'} p_{\nu'} \eta^{\mu'\nu'}(p) + \frac{m}{4} \, \eta(p) \right]$$

$$+ \frac{1}{20m^4} \int \frac{(dp)}{(2\pi)^4} \, \eta^{\mu\nu}(-p)\gamma^0 \gamma_\mu p_\nu \tfrac{1}{4}(\gamma p - 5m)\gamma_{\mu'} p_{\nu'} \eta^{\mu'\nu'}(p)$$

$$+ \frac{1}{10m^3} \int \frac{(dp)}{(2\pi)^4} \, \eta(-p)\tfrac{1}{16}(\gamma p - 9m)\gamma_\mu p_\nu \eta^{\mu\nu}(p)$$

$$+ \frac{1}{20m^2} \int \frac{(dp)}{(2\pi)^4} \, \eta(-p)\gamma^0 \tfrac{1}{64}(\gamma p - 13m)\eta(p). \qquad (3\text{-}3.116)$$

The field equations are

$$(\gamma p + m)\psi_{\mu\nu} - (p_\mu \gamma^{\mu'} \psi_{\mu'\nu} + \gamma_\mu p^{\mu'} \psi_{\mu'\nu} + p_\nu \gamma^{\nu'} \psi_{\mu\nu'} + \gamma_\nu p^{\nu'} \psi_{\mu\nu'})$$
$$+ \gamma_\mu(m - \gamma p)\gamma^{\mu'} \psi_{\mu'\nu} + \gamma_\nu(m - \gamma p)\gamma^{\nu'} \psi_{\mu\nu'} + g_{\mu\nu}\gamma^{\mu'} p^{\nu'} \psi_{\mu'\nu'}$$
$$+ \tfrac{1}{2}[\gamma_\mu p_\nu + \gamma_\nu p_\mu - g_{\mu\nu}(\gamma p + m)]\psi - g_{\mu\nu}m\Psi = \eta_{\mu\nu},$$
$$(\gamma p - 3m)\Psi + \tfrac{5}{12}m\psi = 0, \quad (3\text{-}3.117)$$

which reduce to (3–3.113) as $m \to 0$. The diligent reader who is desirous of confirming the statement of Eq. (3–3.117) could solve these field equations in conjunction with outgoing wave boundary conditions and reproduce the field $\psi_{\mu\nu}$ that is derived from (3–3.115) and (3–3.116), together with the following expression for the auxiliary field,

$$\Psi(p) = \frac{1}{8m^3}\left[p_\mu p_\nu \eta^{\mu\nu}(p) + \tfrac{1}{4}(\gamma p - m)\gamma_\mu p_\nu \eta^{\mu\nu}(p) + \frac{m}{16}(\gamma p - 5m)\eta(p) \right].$$
$$(3\text{-}3.118)$$

3–4 MULTISPINOR FIELDS

The multispinor description provides a unified approach to all spin values, with the first-order differential equation of spin $\tfrac{1}{2}$ setting the pattern of field equations. In order to realize this standard, however, contact terms must be introduced in all situations save that of spin $\tfrac{1}{2}$. Consider, for example, the description of unit spin that is provided by the symmetric spinor of the second rank, as contained in

$$W(\eta) = \frac{1}{2} \int \frac{(dp)}{(2\pi)^4} \, \eta(-p)\gamma_1^0 \gamma_2^0 \frac{1}{2m}\left[\frac{(m - \gamma_1^\mu p_\mu)(m - \gamma_2^\mu p_\mu)}{p^2 + m^2 - i\epsilon} + 1 \right]\eta(p).$$
$$(3\text{-}4.1)$$

The appropriate contact term has already been introduced, and the source differs in normalization from that used in (2–9.8), as indicated generally by

$$\eta(p) = (2m)^{1/2(n-1)}S(p). \qquad (3\text{-}4.2)$$

The field definition is

$$\delta W(\eta) = \int \frac{(dp)}{(2\pi)^4}\, \delta\eta(-p)\gamma_1^0\gamma_2^0\psi(p),\qquad (3\text{-}4.3)$$

and

$$\psi(p) = \frac{1}{2m}\left[\frac{(m - \gamma_1 p)(m - \gamma_2 p)}{p^2 + m^2 - i\epsilon} + 1\right]\eta(p).\qquad (3\text{-}4.4)$$

Now we observe that

$$(\gamma_1 p + m)\psi(p) = \frac{1}{2m}\,[m - \gamma_2 p + m + \gamma_1 p]\eta(p),$$

$$(\gamma_2 p + m)\psi(p) = \frac{1}{2m}\,[m - \gamma_1 p + m + \gamma_2 p]\eta(p),\qquad (3\text{-}4.5)$$

and addition gives

$$[\tfrac{1}{2}(\gamma_1^\mu + \gamma_2^\mu)p_\mu + m]\psi(p) = \eta(p).\qquad (3\text{-}4.6)$$

Written in coordinate space, this is the first-order differential field equation

$$[\tfrac{1}{2}(\gamma_1^\mu + \gamma_2^\mu)(1/i)\partial_\mu + m]\psi(x) = \eta(x).\qquad (3\text{-}4.7)$$

Another consequence of the equation pair (3–4.5),

$$(\gamma_1^\mu - \gamma_2^\mu)p_\mu\psi(p) = \frac{1}{m}\,(\gamma_1^\mu - \gamma_2^\mu)p_\mu\eta(p),\qquad (3\text{-}4.8)$$

is also contained in (3–4.6), as revealed by multiplication with $\gamma_1 p - \gamma_2 p$ and the use of

$$(\gamma_1 p)^2 = (\gamma_2 p)^2.\qquad (3\text{-}4.9)$$

The simple algebraic property just recorded provides an elementary and generally useful control over our procedures. The eigenvalue relationships, $\gamma_1 p = \pm\gamma_2 p$, are invariant statements of the rest-frame possibilities, $\gamma_1^0 = \pm\gamma_2^0$, where the plus sign selects the appropriate subspace for the description of the particle. Correspondingly, setting $\gamma_2 p = \gamma_1 p$ in (3–4.6) reduces the latter to the form of the spin $\frac{1}{2}$ Dirac equation, while the choice $\gamma_2 p = -\gamma_1 p$ effectively removes the coordinate derivatives and supplies a field that vanishes outside the source. The structure given in (3–4.1) becomes more obvious, for, with $\gamma_2 p = \gamma_1 p$,

$$\frac{(m - \gamma_1 p)^2}{p^2 + m^2 - i\epsilon} = \frac{m - \gamma_1 p}{\gamma_1 p + m - i\epsilon} = \frac{2m}{\gamma_1 p + m - i\epsilon} - 1,\qquad (3\text{-}4.10)$$

which puts into evidence the contact term and normalization that are needed to attain (3–4.6).

The effectiveness of such considerations becomes clearer on turning to the multispinor of rank 3, where the choice $\gamma_1 p = \gamma_2 p = \gamma_3 p$ leads to the examination of

$$\frac{(m - \gamma_1 p)^3}{p^2 + m^2 - i\epsilon} = \frac{(m - \gamma_1 p)^2}{\gamma_1 p + m - i\epsilon} = \frac{(2m)^2}{\gamma_1 p + m - i\epsilon} - 3m + \gamma_1 p.\qquad (3\text{-}4.11)$$

This indicates the proper starting point:

$$W(\eta) = \frac{1}{2} \int \frac{(dp)}{(2\pi)^4} \, \eta(-p)\gamma_1^0\gamma_2^0\gamma_3^0 \frac{1}{(2m)^2}$$
$$\times \left[\frac{(m - \gamma_1 p)(m - \gamma_2 p)(m - \gamma_3 p)}{p^2 + m^2 - i\epsilon} + 3m \right.$$
$$\left. - \tfrac{1}{3}(\gamma_1 p + \gamma_2 p + \gamma_3 p) \right] \eta(p), \quad (3\text{-}4.12)$$

and displays the field

$$\psi(p) = \frac{1}{(2m)^2} \left[\frac{\prod(m - \gamma p)_\alpha}{p^2 + m^2 - i\epsilon} + 3m - \tfrac{1}{3}\sum \gamma_\alpha p \right] \eta(p) \quad (3\text{-}4.13)$$

that is defined by

$$\delta W(\eta) = \int \frac{(dp)}{(2\pi)^4} \, \delta\eta(-p) \prod (\gamma_\alpha^0)\psi(p). \quad (3\text{-}4.14)$$

There are only two general alternatives open to the $\gamma_\alpha p$. Either they are all equal, or one of them has a sign opposite to the other two. These situations are characterized by

$\gamma_1 p = \gamma_2 p = \gamma_3 p$:

$$\psi = \frac{1}{\gamma_1 p + m - i\epsilon} \, \eta, \qquad [\tfrac{1}{3}\sum \gamma_\alpha p + m]\psi = \eta, \quad (3\text{-}4.15)$$

and, for example,

$\gamma_2 p = -\gamma_3 p$:

$$\psi = \frac{1}{m^2}(m - \tfrac{1}{3}\gamma_1 p)\eta, \qquad [\tfrac{1}{3}\sum \gamma_\alpha p + m]\psi = \left(1 + \frac{1}{(3m)^2} p^2\right)\eta. \quad (3\text{-}4.16)$$

An expression that covers all contingencies is

$$[\tfrac{1}{3}\sum \gamma_\alpha p + m]\psi = \left[1 - \frac{1}{(3m)^2} \frac{1}{2}\sum_{\alpha<\beta}(\tfrac{1}{2}(\gamma_\alpha p - \gamma_\beta p))^2 \right]\eta; \quad (3\text{-}4.17)$$

it can be verified directly. Note that the field statement of Eq. (3-4.16) can also be presented as the unrestricted equation

$$\tfrac{1}{2}(\gamma_2 p - \gamma_3 p)\psi = \frac{1}{m^2}(m - \tfrac{1}{3}\gamma_1 p)\tfrac{1}{2}(\gamma_2 p - \gamma_3 p)\eta. \quad (3\text{-}4.18)$$

Then, with the definitions illustrated by

$$\psi_{[23]} = \frac{1}{(6m)^2}(\gamma_2 p - \gamma_3 p)\eta, \quad (3\text{-}4.19)$$

which is antisymmetrical in the indicated pair of indices, we arrive at the system of first-order differential equations,

$$[\tfrac{1}{3}\sum \gamma_\alpha p + m]\psi + \sum_{\alpha<\beta}\tfrac{1}{2}(\gamma_\alpha p - \gamma_\beta p)\psi_{[\alpha\beta]} = \eta,$$
$$(\gamma_1 p - 3m)\psi_{[23]} + (\tfrac{1}{6})\tfrac{1}{2}(\gamma_2 p - \gamma_3 p)\psi = 0, \quad (3\text{-}4.20)$$

where the latter illustrates the set of three equations that are related by cyclic permutation.

It is also interesting to eliminate the three auxiliary fields and present the field equation in the form

$$\left\{\frac{1}{3}\sum \gamma_\alpha p + m - \frac{1}{6}\left[\frac{(\frac{1}{2}(\gamma_2 p - \gamma_3 p))^2}{\gamma_1 p - 3m} + \cdots\right]\right\}\psi = \eta, \qquad (3\text{-}4.21)$$

where the dots indicate the two analogous expressions produced by cyclic permutation. To verify that this single equation permits the reconstruction of the original field, it suffices to examine it in two situations:

$$\gamma_1 p = \gamma_2 p = \gamma_3 p: \qquad (\gamma_1 p + m)\psi = \eta,$$

$$(3\text{-}4.22)$$

$$\gamma_2 p = -\gamma_3 p: \quad \left[\tfrac{1}{3}\gamma_1 p + m - \frac{1}{3}\frac{(\gamma_1 p)^2}{\gamma_1 p - 3m}\right]\psi = \eta,$$

which do indeed contain the results of (3–4.15) and (3–4.16). There is another way to convey (3–4.21), which follows from the observation that

$$[m - \tfrac{1}{3}\sum \gamma_\alpha p]\tfrac{1}{2}(\gamma_2 p - \gamma_3 p) = (m - \tfrac{1}{3}\gamma_1 p)\tfrac{1}{2}(\gamma_2 p - \gamma_3 p). \quad (3\text{-}4.23)$$

It is the second-order differential equation

$$\left[(3m)^2 - \left(\sum \gamma_\alpha p\right)^2 + \tfrac{1}{2}\sum_{\alpha<\beta}\left(\tfrac{1}{2}(\gamma_\alpha p - \gamma_\beta p)\right)^2\right]\psi = 3(3m - \sum \gamma_\alpha p)\eta,$$

$$(3\text{-}4.24)$$

or, alternatively,

$$[p^2 + m^2 - \tfrac{9}{8}\{p^2 + (\tfrac{1}{3}\sum \gamma_\alpha p)^2\}]\psi = (m - \tfrac{1}{3}\sum \gamma_\alpha p)\eta. \quad (3\text{-}4.25)$$

The latter may be compared with the second-order differential equation for unit spin:

$$[p^2 + m^2 - \{p^2 + (\tfrac{1}{2}\sum \gamma_\alpha p)^2\}]\psi = (m - \tfrac{1}{2}\sum \gamma_\alpha p)\eta, \quad (3\text{-}4.26)$$

and for spin $\tfrac{1}{2}$:

$$(p^2 + m^2)\psi = (m - \gamma p)\eta. \qquad (3\text{-}4.27)$$

The discussion of the fourth-rank multispinor begins with the algebraic statement

$$\frac{(m - \gamma_1 p)^4}{p^2 + m^2 - i\epsilon} = \frac{(2m)^3}{\gamma_1 p + m - i\epsilon} - 7m^2 + 4m\gamma_1 p - (\gamma_1 p)^2. \quad (3\text{-}4.28)$$

But now there is ambiguity in giving $(\gamma_1 p)^2$ a more general interpretation; shall it be $-p^2$, or $\tfrac{1}{6}\sum_{\alpha<\beta}\gamma_\alpha p\gamma_\beta p$? In fact, we shall use a specific linear combination of the two, so chosen that powers of momenta in ψ are held to the minimum.

The actual expression is

$$W = \frac{1}{2} \int \frac{(dp)}{(2\pi)^4} \, \eta(-p) \prod (\gamma_\alpha^0) \psi(p),$$

$$\psi(p) = \frac{1}{(2m)^3} \left[\frac{\prod (m - \gamma p)_\alpha}{p^2 + m^2 - i\epsilon} + 7m^2 - m \sum_\alpha \gamma_\alpha p \right.$$

$$\left. + p^2 + \frac{1}{3} \sum_{\alpha < \beta} \gamma_\alpha p \gamma_\beta p \right] \eta. \qquad (3\text{-}4.29)$$

The following indicates the options available to the $\gamma_\alpha p$:

$\gamma_1 p = \gamma_2 p = \gamma_3 p = \gamma_4 p$:

$$\psi = \frac{1}{\gamma_1 p + m - i\epsilon} \, \eta, \qquad [\tfrac{1}{4} \sum \gamma_\alpha p + m]\psi = \eta, \qquad (3\text{-}4.30)$$

$\gamma_1 p = \gamma_2 p = \gamma_3 p = -\gamma_4 p$:

$$\psi = \frac{1}{m}\left(1 - \frac{1}{2m}\gamma_1 p \right) \eta, \qquad [\tfrac{1}{4} \sum \gamma_\alpha p + m]\psi = \left(1 + \frac{1}{4m^2} p^2 \right) \eta, \quad (3\text{-}4.31)$$

$\gamma_1 p = \gamma_2 p = -\gamma_3 p = -\gamma_4 p$:

$$\psi = \frac{1}{m}\left(1 + \frac{1}{3m^2} p^2 \right) \eta, \qquad [\tfrac{1}{4} \sum \gamma_\alpha p + m]\psi = \left(1 + \frac{1}{3m^2} p^2 \right) \eta. \quad (3\text{-}4.32)$$

The particular structure adopted in (3–4.29) is designed to simplify the field, for the situation of (3–4.31), by eliminating a possible p^2 term. The three examples of field equations are synthesized in

$$[\tfrac{1}{4} \sum \gamma_\alpha p + m]\psi = \left[1 - \frac{1}{12m^2} \sum_{\alpha < \beta} (\tfrac{1}{2}(\gamma_\alpha p - \gamma_\beta p))^2 \right] \eta. \qquad (3\text{-}4.33)$$

Another unifying statement, with its obvious generalizations to other index arrangements, is

$$(\gamma_1 p - \gamma_4 p)\psi = \frac{1}{m}\left[1 - \frac{1}{4m}(\gamma_2 p + \gamma_3 p) \right.$$

$$\left. - \frac{1}{12m^2}(\gamma_2 p - \gamma_3 p)^2 \right](\gamma_1 p - \gamma_4 p)\eta. \qquad (3\text{-}4.34)$$

Using the definitions of antisymmetrical functions that are illustrated by

$$\psi_{[14]} = \frac{1}{24m^2}(\gamma_1 p - \gamma_4 p)\eta,$$

$$\psi_{[14][23]} = \frac{1}{72m^3}(\gamma_1 p - \gamma_4 p)(\gamma_2 p - \gamma_3 p)\eta, \qquad (3\text{-}4.35)$$

one then writes the system of first-order differential equations:

$$[\tfrac{1}{4}\sum \gamma_\alpha p + m]\psi + \sum_{\alpha<\beta} \tfrac{1}{2}(\gamma_\alpha p - \gamma_\beta p)\psi_{[\alpha\beta]} = \eta,$$

$$[\tfrac{1}{2}(\gamma_2 p + \gamma_3 p) - 2m]\psi_{[14]} + (\tfrac{1}{6})\tfrac{1}{2}(\gamma_1 p - \gamma_4 p)\psi + \tfrac{1}{2}(\gamma_2 p - \gamma_3 p)\psi_{[14][23]} = 0,$$

$$3m\psi_{[14][23]} = (\gamma_2 p - \gamma_3 p)\psi_{[14]}, \qquad (3\text{-}4.36)$$

where the last two are representative of sets of such equations.

The elimination of the auxiliary fields produces the single (multicomponent) field equation

$$\left\{ \tfrac{1}{4}\sum \gamma_\alpha p + m + \frac{1}{6}\left[\frac{(\tfrac{1}{2}(\gamma_1 p - \gamma_4 p))^2}{2m - \tfrac{1}{2}(\gamma_2 p + \gamma_3 p) - \dfrac{1}{6m}(\gamma_2 p - \gamma_3 p)^2} + \cdots \right] \right\} \psi = \eta,$$

$$(3\text{-}4.37)$$

where the summation over all pairs of indices is indicated by a representative term. The various alternatives are illustrated by

$$\gamma_1 p = \cdots \gamma_4 p:$$
$$(\gamma_1 p + m)\psi = \eta, \qquad (3\text{-}4.38)$$

$$\gamma_1 p = \cdots \gamma_3 p = -\gamma_4 p:$$
$$\left(\tfrac{1}{2}\gamma_1 p + m + \frac{(\tfrac{1}{2}\gamma_1 p)^2}{m - \tfrac{1}{2}\gamma_1 p} \right) \psi = \eta,$$

$$\gamma_1 p = \gamma_2 p = -\gamma_3 p = -\gamma_4 p:$$
$$\left(m - \frac{\tfrac{1}{3}p^2}{m + \dfrac{1}{3m}p^2} \right) \psi = \eta,$$

which restate the field expressions of Eqs. (3–4.30, 31, 32). Another form of the equation is

$$\left[p^2 + m^2 - \tfrac{4}{3}\{p^2 + (\tfrac{1}{4}\sum \gamma_\alpha p)^2\} + \frac{1}{m^2 + \tfrac{1}{3}p^2}\frac{1}{18}\sum (\tfrac{1}{2}(\gamma_\alpha p - \gamma_\beta p))^2 \right.$$
$$\left. \times (\tfrac{1}{2}(\gamma_{\alpha'} p - \gamma_{\beta'} p))^2 \right] \psi = [m - \tfrac{1}{4}\sum \gamma_\alpha p]\eta, \quad (3\text{-}4.39)$$

in which the last summation is extended over distinct pairs, $\alpha < \beta$, $\alpha' < \beta'$, $\alpha \neq \alpha'$, $\beta \neq \beta'$, with no repetitious counting. This is equivalent to a fourth-order differential equation.

Turning to fifth-rank multispinors, we first note that

$$\frac{(m - \gamma_1 p)^5}{p^2 + m^2 - i\epsilon} = \frac{(2m)^4}{\gamma_1 p + m - i\epsilon} - 15m^3 + 11m^2\gamma_1 p - 5m(\gamma_1 p)^2 + (\gamma_1 p)^3.$$

$$(3\text{-}4.40)$$

Ambiguities have been resolved in stating the field as

$$\psi = \frac{1}{(2m)^4}\left[\frac{\Pi(m-\gamma p)_\alpha}{p^2+m^2-i\epsilon} + 15m^3 - \tfrac{11}{5}m^2\sum\gamma_\alpha p + 5mp^2 + m\sum_{\alpha<\beta}\gamma_\alpha p\gamma_\beta p\right.$$
$$\left. - \tfrac{1}{5}p^2\sum\gamma_\alpha p - \tfrac{1}{5}\sum_{\alpha<\beta<\gamma}\gamma_\alpha p\gamma_\beta p\gamma_\gamma p\right]\eta. \quad (3\text{–}4.41)$$

The available options are indicated by

$$\gamma_1 p = \cdots\gamma_5 p:$$
$$\psi = \frac{1}{\gamma_1 p + m - i\epsilon}\,\eta, \qquad [\tfrac{1}{5}\sum\gamma_\alpha p + m]\psi = \eta, \qquad (3\text{–}4.42)$$

$$\gamma_1 p = \cdots\gamma_4 p = -\gamma_5 p:$$
$$\psi = \frac{1}{m}\left(1 - \frac{3}{5m}\gamma_1 p\right)\eta, \qquad [\tfrac{1}{5}\sum\gamma_\alpha p + m]\psi = \left(1+\left(\frac{3}{5m}\right)^2 p^2\right)\eta,$$
$$(3\text{–}4.43)$$

$$\gamma_1 p = \cdots\gamma_3 p = -\gamma_4 p = -\gamma_5 p:$$
$$\psi = \frac{1}{m}\left(1-\frac{1}{5m}\gamma_1 p\right)\left(1+\frac{1}{2m^2}p^2\right)\eta,$$
$$[\tfrac{1}{5}\sum\gamma_\alpha p + m]\psi = \left(1+\frac{1}{(5m)^2}p^2\right)\left(1+\frac{1}{2m^2}p^2\right)\eta$$
$$= \left(1+\frac{3}{2}\left(\frac{3}{5m}\right)^2 p^2 + \frac{1}{50m^4}(p^2)^2\right)\eta. \quad (3\text{–}4.44)$$

The following are generally valid statements:

$$[\tfrac{1}{5}\sum\gamma_\alpha p + m]\psi = \left[1-\left(\frac{3}{10m}\right)^2\sum_{\alpha<\beta}(\tfrac{1}{2}(\gamma_\alpha p - \gamma_\beta p))^2 + \left(\frac{1}{10m^2}\right)^2\frac{1}{3}\right.$$
$$\left.\times\sum(\tfrac{1}{2}(\gamma_\alpha p - \gamma_\beta p))^2(\tfrac{1}{2}(\gamma_{\alpha'} p - \gamma_{\beta'} p))^2\right]\eta, \quad (3\text{–}4.45)$$

where repetitious counting of double pairs is avoided, and

$$(\gamma_\alpha p - \gamma_\beta p)\psi = (\gamma_\alpha p - \gamma_\beta p)\left(1-\frac{1}{5m}\sum{}'\gamma_{\alpha'} p\right)$$
$$\times\left[1-\frac{1}{4m^2}\sum_{\alpha'<\beta'}{}'(\tfrac{1}{2}(\gamma_{\alpha'} p - \gamma_{\beta'} p))^2\right]\eta, \quad (3\text{–}4.46)$$

in which the summations marked $\sum'$ are extended over all index values other than α and β.

The auxiliary fields

$$\psi_{[\alpha\beta]} = \left(\frac{3}{10m}\right)^2\left[1-\frac{1}{(3m)^2}\frac{1}{6}\sum_{\alpha'<\beta'}{}'(\tfrac{1}{2}(\gamma_{\alpha'} p - \gamma_{\beta'} p))^2\right]\tfrac{1}{2}(\gamma_\alpha p - \gamma_\beta p)\eta$$
$$(3\text{–}4.47)$$

and, for example,

$$\psi_{[23][45]} = \frac{5}{(12m^2)^2}\,(m - \tfrac{1}{3}\gamma_1 p)\tfrac{1}{2}(\gamma_2 p - \gamma_3 p)\tfrac{1}{2}(\gamma_4 p - \gamma_5 p)\eta \quad (3\text{-}4.48)$$

then lead to the field equations

$$[\tfrac{1}{3}\textstyle\sum \gamma_\alpha p + m]\psi + \sum_{\alpha<\beta} \tfrac{1}{2}(\gamma_\alpha p - \gamma_\beta p)\psi_{[\alpha\beta]} = \eta,$$

$$[\tfrac{1}{3}\textstyle\sum' \gamma_{\alpha'} p + m]\psi_{[\alpha\beta]} + (\tfrac{3}{20})\tfrac{1}{2}(\gamma_\alpha p - \gamma_\beta p)\psi \qquad (3\text{-}4.49)$$
$$+ \sum_{\alpha'<\beta'} \tfrac{1}{2}(\gamma_{\alpha'} p - \gamma_{\beta'} p)\psi_{[\alpha\beta][\alpha'\beta']} = 0.$$

To complete the system we need a differential equation for the $\psi_{[\alpha\beta][\alpha'\beta']}$, relating them to the $\psi_{[\alpha\beta]}$. Now

$$(\gamma_1 p + 5m)\psi_{[23][45]} = \left(\frac{5}{12m}\right)^2\left(1 + \frac{1}{25m^2}p^2\right)\tfrac{1}{2}(\gamma_2 p - \gamma_3 p)\tfrac{1}{2}(\gamma_4 p - \gamma_5 p)\eta$$
$$(3\text{-}4.50)$$

while

$$\tfrac{1}{2}(\gamma_4 p - \gamma_5 p)\psi_{[23]} = \left(\frac{3}{10m}\right)^2\left(1 + \frac{1}{27m^2}p^2\right)\tfrac{1}{2}(\gamma_2 p - \gamma_3 p)\tfrac{1}{2}(\gamma_4 p - \gamma_5 p)\eta,$$
$$(3\text{-}4.51)$$

and thus no immediate connection exists, owing to the inescapable fact that $\frac{1}{25} \neq \frac{1}{27}$. It suggests, however, the introduction of another set of auxiliary fields, illustrated by

$$\chi_{[23][45]} = \frac{1}{(12m^2)^2}\,\gamma_1 p\tfrac{1}{2}(\gamma_2 p - \gamma_3 p)\tfrac{1}{2}(\gamma_4 p - \gamma_5 p)\eta, \quad (3\text{-}4.52)$$

which obey

$$(\gamma_1 p - 5m)\chi_{[\alpha\beta][\alpha'\beta']} + \gamma_1 p\psi_{[\alpha\beta][\alpha'\beta']} = 0, \qquad (3\text{-}4.53)$$

and enable one to write

$$(\gamma_1 p + 5m)\psi_{[\alpha\beta][\alpha'\beta']} + \tfrac{2}{27}\gamma_1 p\chi_{[\alpha\beta][\alpha'\beta']} - (\tfrac{25}{18})^2\tfrac{1}{2}(\gamma_{\alpha'} p - \gamma_{\beta'} p)\psi_{[\alpha\beta]} = 0.$$
$$(3\text{-}4.54)$$

The full system of first-order differential equations is given by Eqs. (3-4.49), (3-4.53), and (3-4.54). The auxiliary fields can be eliminated and a single equation for ψ constructed in various equivalent forms, but they have become too ponderous to be worth recording.

These procedures can be extended to higher-rank spinors. Without exhibiting the solution of the general problem, we do want to incorporate all available results into the larger framework appropriate to a multispinor of rank n. Generalized from Eqs. (3-4.6), (3-4.20), (3-4.36), and (3-4.49), the first two equations of the system are

$$\left[\frac{1}{n}\sum \gamma_\alpha p + m\right]\psi + \sum_{\alpha<\beta} \tfrac{1}{2}(\gamma_\alpha p - \gamma_\beta p)\psi_{[\alpha\beta]} = \eta,$$

$$\left[\frac{1}{n-2}\sideset{}{'}\sum \gamma_{\alpha'} p - \frac{n}{n-2} m\right]\psi_{[\alpha\beta]} + \frac{n-2}{n(n-1)} \tfrac{1}{2}(\gamma_\alpha p - \gamma_\beta p)\psi$$

$$+ \sideset{}{'}\sum_{\alpha'<\beta'} \tfrac{1}{2}(\gamma_{\alpha'} p - \gamma_{\beta'} p)\psi_{[\alpha\beta][\alpha'\beta']} = 0,$$

$$(3\text{–}4.55)$$

which imply

$$\gamma_1 p = \cdots \gamma_n p: \quad \psi = \frac{1}{\gamma_1 p + m - i\epsilon}\, \eta$$

$$\gamma_1 p = \cdots \gamma_{n-1} p = -\gamma_n p:$$

$$\psi = \frac{1}{m}\left(1 - \frac{n-2}{n}\frac{1}{m}\gamma_1 p\right)\eta.$$

$$(3\text{–}4.56)$$

The next in the sequence, which is as far as it shall be developed here, are

$$\left[\frac{1}{n-4}\sideset{}{'}\sum \gamma_{\alpha''} p + \frac{n}{n-4} m\right]\psi_{[\alpha\beta][\alpha'\beta']} + \frac{1}{2}\frac{(n-1)(n-4)^2}{(n-2)^3}\sideset{}{'}\sum \gamma_{\alpha''} p \chi_{[\alpha\beta][\alpha'\beta']}$$

$$-\frac{1}{2}\frac{(n-3)n^4}{(n-1)(n-2)^4(n-4)}\frac{1}{2}(\gamma_{\alpha'} p - \gamma_{\beta'} p)\psi_{[\alpha\beta]} + \cdots = 0,$$

$$(3\text{–}4.57)$$

$$\left[\frac{1}{n-4}\sideset{}{'}\sum \gamma_{\alpha''} p - \frac{n}{n-4} m\right]\chi_{[\alpha\beta][\alpha'\beta']} + \sideset{}{'}\sum \gamma_{\alpha''} p\psi_{[\alpha\beta][\alpha'\beta']} + \cdots = 0,$$

where the terms left unwritten are those referring to three index pairs. This set has the property that

$$\gamma_1 p = \cdots \gamma_{n-2} p = -\gamma_{n-1} p = -\gamma_n p:$$

$$\psi = \frac{1}{m}\left(1 - \frac{n-4}{n}\frac{1}{m}\gamma_1 p\right)\left(1 + \frac{n-3}{n-1}\frac{1}{m^2}p^2\right)\eta,$$

$$(3\text{–}4.58)$$

which is applicable even to $n = 3$.

The totally symmetric multispinor of rank $n = 1, 2, 3, \ldots$ gives a description for particles of spin $s = \tfrac{1}{2}n = \tfrac{1}{2}, 1, \tfrac{3}{2}, \ldots$. It has already been noted that the list is completed with $s = 0$ on considering $n = 2$ and replacing the symmetrical spinor by an antisymmetrical one. All the equations for $n = 2$ continue to apply with the changed symmetry. But this is a completely general remark. The system of equations we have been discussing involves operations on the fields that are entirely symmetrical among the spinor indices. Accordingly, the field ψ will acquire whatever symmetry characteristics the source η possesses. The significant description of the particle is given in the subspace $\gamma_1^0 = \cdots \gamma_n^0$, referring to the particle rest frame. The specific value of the particle spin will therefore be determined by the symmetry that is common to the source and the field. If there is total symmetry, the spin is $s = \frac{1}{2}n$; if there is one antisymmetrical index pair, and total symmetry among the rest, the spin is $s = \frac{1}{2}(n-2) = \frac{1}{2}n - 1$; and so forth. By appropriately choosing the symmetry, then, the third-rank spinor equations can be applied to $s = \frac{3}{2}$ or $\frac{1}{2}$; the fourth-rank equations to $s = 2, 1,$ or 0; and so on.

3-5 ACTION

A symbolic transcription of our discussion is contained in the following set of equations:

$$W = \tfrac{1}{2}\int S\gamma GS, \tag{3-5.1}$$

$$\delta W = \int \delta S\gamma\chi = \int \chi\gamma\,\delta S, \tag{3-5.2}$$

$$\chi = GS, \tag{3-5.3}$$

$$F\chi = S. \tag{3-5.4}$$

That is, beginning with the quadratic source expression for W, where γ is the appropriate representation of the metric, fields χ are defined through the consideration of an infinitesimal test source. The nonlocal space-time relation between field and source that is conveyed by G is then converted into a local differential one, which is symbolized by the operator F. Alternative expressions for W are

$$W = \tfrac{1}{2}\int S\gamma\chi = \tfrac{1}{2}\int \chi\gamma S \tag{3-5.5}$$

and

$$W = \tfrac{1}{2}\int \chi\gamma F\chi = \tfrac{1}{2}\int F\chi\gamma\chi. \tag{3-5.6}$$

Of particular importance is the linear combination

$$W = \int (S\gamma\chi - \tfrac{1}{2}\chi\gamma F\chi). \tag{3-5.7}$$

To this point, the field χ has been a derived quantity, a convenient shorthand for GS. It now acquires independent status, in the following sense. Forego the knowledge of any connection between χ and S, and subject them to independent variation in (3–5.7). This gives

$$\delta W = \int \delta S\gamma\chi + \int (\delta\chi\gamma S - \delta\chi\gamma F\chi). \tag{3-5.8}$$

But δW is $\int \delta S\gamma\chi$, and the additional term should vanish. Indeed it does, since $F\chi = S$. This means that, considered as a functional of χ for prescribed S, the expression (3-5.7) has vanishing first variations or is stationary at the unique field configuration selected by the field equations, in conjunction with boundary conditions. The quantity W is thereby invested with the attributes of *action*, producing the field equations through the principle of stationary action. We shall now illustrate these general remarks in the context of specific spin values.

Spin 0. The field equation is

$$(-\partial^2 + m^2)\phi(x) = K(x), \qquad (3\text{-}5.9)$$

and the action expression (3-5.7) reads

$$W = \int (dx)[K(x)\phi(x) - \tfrac{1}{2}\phi(x)(-\partial^2 + m^2)\phi(x)]. \qquad (3\text{-}5.10)$$

There is a more symmetrical form that contains only first derivatives of the field. No additional surface integral term is assigned to the partial integration. This can best be appreciated with the aid of the associated Euclidean description where fields decrease exponentially at large distances from the source, since that is the characteristic of $\Delta_E(x - x')$, $m \neq 0$. Even for massless particles the $(x - x')^{-2}$ behavior is sufficient to suppress infinitely remote surface integral contributions. Accordingly, we write

$$W = \int (dx)[K(x)\phi(x) + \mathfrak{L}(\phi(x))], \qquad (3\text{-}5.11)$$

where the field dependent quantity

$$\mathfrak{L}(\phi(x)) = -\tfrac{1}{2}[\partial^\mu \phi(x)\partial_\mu \phi(x) + m^2(\phi(x))^2] \qquad (3\text{-}5.12)$$

will be called the Lagrange function of the system.

We shall note here that the field equations can also be presented as the first-order set:

$$\partial_\mu \phi(x) - \phi_\mu(x) = K_\mu(x), \qquad -\partial_\mu \phi^\mu(x) + m^2\phi(x) = K(x). \qquad (3\text{-}5.13)$$

The elimination of the vector field ϕ_μ gives the second-order differential equation

$$(-\partial^2 + m^2)\phi(x) = K(x) - \partial_\mu K^\mu(x), \qquad (3\text{-}5.14)$$

which exhibits the same kind of effective scalar source that has already been encountered in Eq. (2-9.76). The corresponding action expression is

$$W = \int (dx)[K(x)\phi(x) + K^\mu(x)\phi_\mu(x) + \mathfrak{L}(\phi(x), \phi_\mu(x))], \qquad (3\text{-}5.15)$$

in which

$$\mathfrak{L} = -\phi^\mu \partial_\mu \phi + \tfrac{1}{2}(\phi^\mu \phi_\mu - m^2\phi^2). \qquad (3\text{-}5.16)$$

If the first equation of (3-5.13) is regarded as a definition, ϕ_μ loses its independent position and we recover the action expression (3-5.11), with the effective scalar source indicated in (3-5.14), and the additional contact term $-\int (dx)\tfrac{1}{2}K^\mu K_\mu$.

Spin 1. The field equation

$$(-\partial^2 + m^2)\phi_\mu(x) + \partial_\mu \partial^\nu \phi_\nu(x) = J_\mu(x) \qquad (3\text{-}5.17)$$

leads to the action expression

$$W = \int (dx)[J^\mu(x)\phi_\mu(x) + \mathcal{L}(\phi_\mu(x))], \qquad (3\text{-}5.18)$$

with the Lagrange function

$$\mathcal{L} = -\tfrac{1}{2}[\tfrac{1}{2}(\partial^\mu\phi^\nu - \partial^\nu\phi^\mu)(\partial_\mu\phi_\nu - \partial_\nu\phi_\mu) + m^2\phi^\mu\phi_\mu]. \qquad (3\text{-}5.19)$$

To obtain it one uses the rearrangements

$$\int (dx)\phi^\mu(-\partial^2)\phi_\mu = \int (dx)\partial^\nu\phi^\mu\partial_\nu\phi_\mu \qquad (3\text{-}5.20)$$

and

$$\int (dx)\phi^\mu\partial_\mu\partial^\nu\phi_\nu = -\int (dx)\partial^\nu\phi^\mu\partial_\mu\phi_\nu. \qquad (3\text{-}5.21)$$

But the latter could have been

$$\int (dx)\phi^\mu\partial_\mu\partial^\nu\phi_\nu = -\int (dx)\partial_\mu\phi^\mu\partial^\nu\phi_\nu, \qquad (3\text{-}5.22)$$

and this produces a different Lagrange function:

$$\mathcal{L} = -\tfrac{1}{2}[\partial^\nu\phi^\mu\partial_\nu\phi_\mu - (\partial_\mu\phi^\mu)^2 + m^2\phi^\mu\phi_\mu]; \qquad (3\text{-}5.23)$$

the restriction to first derivatives does not assure the uniqueness of the Lagrange function. Generally, two Lagrange functions that refer to the same system are connected by a relation of the form

$$\mathcal{L}_1 = \mathcal{L}_2 + \partial_\mu f^\mu, \qquad (3\text{-}5.24)$$

since the local divergence term does not contribute to the integrated action expression. In this situation,

$$f^\mu = \tfrac{1}{2}(\phi^\nu\partial_\nu\phi^\mu - \phi^\mu\partial_\nu\phi^\nu) \qquad (3\text{-}5.25)$$

converts (3-5.23) into (3-5.19).

A system of first-order equations is

$$\partial_\mu\phi_\nu - \partial_\nu\phi_\mu - G_{\mu\nu} = M_{\mu\nu} = -M_{\nu\mu}, \qquad \partial_\nu G^{\mu\nu} + m^2\phi^\mu = J^\mu. \qquad (3\text{-}5.26)$$

They imply the action

$$W = \int (dx)[J^\mu\phi_\mu + \tfrac{1}{2}M^{\mu\nu}G_{\mu\nu} + \mathcal{L}(\phi_\mu, G_{\mu\nu})], \qquad (3\text{-}5.27)$$

with the Lagrange function

$$\mathcal{L} = -\tfrac{1}{2}G^{\mu\nu}(\partial_\mu\phi_\nu - \partial_\nu\phi_\mu) + \tfrac{1}{2}(\tfrac{1}{2}G^{\mu\nu}G_{\mu\nu} - m^2\phi^\mu\phi_\mu). \qquad (3\text{-}5.28)$$

On regarding $G_{\mu\nu}$ as defined by the first equation of (3-5.26) we recover (3-5.19) with the effective vector source $J^\mu + \partial_\nu M^{\mu\nu}$, analogous to (2-9.78), and the added contact term $-\int (dx)\tfrac{1}{4}M^{\mu\nu}M_{\mu\nu}$.

Spin 2. The field equations are

$$(-\partial^2 + m^2)\phi_{\mu\nu} + \partial_\mu\partial^\lambda\phi_{\lambda\nu} + \partial_\nu\partial^\lambda\phi_{\mu\lambda} - \partial_\mu\partial_\nu\phi$$
$$- g_{\mu\nu}[(-\partial^2 + m^2)\phi + \partial_\kappa\partial_\lambda\phi^{\kappa\lambda}] = T_{\mu\nu}, \quad (3\text{-}5.29)$$

and one possibility for the Lagrange function, in the action

$$W = \int (dx)[T^{\mu\nu}\phi_{\mu\nu} + \mathcal{L}], \quad (3\text{-}5.30)$$

is

$$\mathcal{L} = -\tfrac{1}{2}[\partial^\lambda\phi^{\mu\nu}\partial_\lambda\phi_{\mu\nu} + m^2\phi^{\mu\nu}\phi_{\mu\nu} - \partial^\lambda\phi\partial_\lambda\phi - m^2\phi^2]$$
$$- \partial_\mu\phi^{\mu\nu}\partial_\nu\phi + \partial_\mu\phi^{\mu\nu}\partial^\lambda\phi_{\lambda\nu}, \quad (3\text{-}5.31)$$

with the option of replacing the last term by

$$\partial^\lambda\phi^{\mu\nu}\partial_\mu\phi_{\lambda\nu}, \quad (3\text{-}5.32)$$

or by any weighted average of the two.

First-order differential equations can be introduced in two different ways. One such set is associated with the action expression

$$W = \int (dx)[T^{\mu\nu}\phi_{\mu\nu} + \tfrac{1}{2}K^{\lambda\mu\nu}G_{\lambda\mu\nu} + \mathcal{L}], \quad (3\text{-}5.33)$$

$$\mathcal{L} = -\tfrac{1}{2}G^{\lambda\mu\nu}(\partial_\lambda\phi_{\mu\nu} - \partial_\nu\phi_{\lambda\mu}) - G^\lambda(\partial^\mu\phi_{\lambda\mu} - \partial_\lambda\phi)$$
$$+ \tfrac{1}{2}[G^{\lambda\mu\nu}G_{\lambda\nu\mu} - G^\lambda G_\lambda - m^2(\phi^{\mu\nu}\phi_{\mu\nu} - \phi^2)], \quad (3\text{-}5.34)$$

where $K_{\lambda\mu\nu}$ and $G_{\lambda\mu\nu}$ are antisymmetrical in λ and ν, and

$$G_\lambda = G_{\lambda\nu}{}^\nu. \quad (3\text{-}5.35)$$

These field equations are

$$\partial^\lambda G_{\mu\nu\lambda} - \partial_\nu G_\mu + m^2(\phi_{\mu\nu} + \tfrac{1}{2}g_{\mu\nu}\phi) = T_{\mu\nu} - \tfrac{1}{2}g_{\mu\nu}T, \quad (3\text{-}5.36)$$

with the left side understood to be symmetrized in μ and ν, and

$$\partial_\lambda\phi_{\mu\nu} - \partial_\nu\phi_{\lambda\mu} - G_{\lambda\mu\nu} = K'_{\lambda\mu\nu}. \quad (3\text{-}5.37)$$

The effective source of the latter equation:

$$K'_{\lambda\mu\nu} = \tfrac{1}{2}(K_{\lambda\mu\nu} + K_{\mu\lambda\nu} - K_{\mu\nu\lambda} - K_\lambda g_{\mu\nu} + K_\nu g_{\lambda\mu}),$$
$$K_\lambda = K_{\lambda\nu}{}^\nu, \quad (3\text{-}5.38)$$

indicates the rearrangements that are required to produce the form (3-5.37) from that yielded by the stationary action principle. It is also useful to remark, in connection with (3-5.36), that

$$\int (dx)T^{\mu\nu}\phi_{\mu\nu} = \int (dx)(T^{\mu\nu} - \tfrac{1}{2}g^{\mu\nu}T)(\phi_{\mu\nu} - \tfrac{1}{2}g_{\mu\nu}\phi). \quad (3\text{-}5.39)$$

If $G_{\lambda\mu\nu}$ is demoted to a derived quantity, one obtains a Lagrange function that is an equally weighted average of the two possibilities described in Eqs. (3-5.31, 32). Playing the role of source in this action is $T_{\mu\nu} + \partial^\lambda K_{\mu\nu\lambda}$, where symmetrization in μ and ν is understood, and there is an added contact term.

The second system of equations is represented by

$$W = \int (dx)[T^{\mu\nu}\phi_{\mu\nu} + L^{\mu\nu\lambda}H_{\mu\nu\lambda} + \mathcal{L}], \tag{3-5.40}$$

$$\begin{aligned}\mathcal{L} = H^{\mu\nu\lambda}\partial_\lambda\phi_{\mu\nu} - H^\nu\partial^\mu\phi_{\mu\nu} + \tfrac{1}{2}(H^\lambda - {}^\lambda H)\partial_\lambda\phi \\ - \tfrac{1}{2}[H^{\mu\nu\lambda}H_{\lambda\mu\nu} - {}^\lambda HH_\lambda + m^2(\phi^{\mu\nu}\phi_{\mu\nu} - \phi^2)],\end{aligned} \tag{3-5.41}$$

where $H_{\mu\nu\lambda}$ is symmetrical in μ and ν, while

$$^\lambda H = H_\nu{}^{\nu\lambda}, \qquad H^\lambda = H^{\lambda\nu}{}_\nu. \tag{3-5.42}$$

The field equations are

$$\partial^\lambda H_{\mu\nu\lambda} - \partial_\nu H_\mu + m^2(\phi_{\mu\nu} + \tfrac{1}{2}g_{\mu\nu}\phi) = T_{\mu\nu} - \tfrac{1}{2}g_{\mu\nu}T, \tag{3-5.43}$$

with the left side (specifically $\partial_\nu H_\mu$) symmetrized in μ and ν, and

$$\partial_\mu\phi_{\lambda\nu} + \partial_\nu\phi_{\lambda\mu} - \partial_\lambda\phi_{\mu\nu} - H_{\mu\nu\lambda} = L'_{\mu\nu\lambda}, \tag{3-5.44}$$

in which

$$\begin{aligned}L'_{\mu\nu\lambda} = L_{\mu\nu\lambda} - L_{\lambda\mu\nu} - L_{\lambda\nu\mu} + g_{\mu\nu}L_\lambda - \tfrac{1}{3}(g_{\lambda\nu}L_\mu + g_{\lambda\mu}L_\nu) \\ + \tfrac{1}{2}(g_{\lambda\nu\,\mu}L + g_{\lambda\mu\,\nu}L - g_{\mu\nu\,\lambda}L).\end{aligned} \tag{3-5.45}$$

When $H_{\mu\nu\lambda}$ is considered to be defined by (3–5.44), the Lagrange function turns into the one that uses the term given in (3–5.32). The effective source is

$$T_{\mu\nu} + \partial^\lambda L_{\mu\nu\lambda} - \partial^\lambda L_{\lambda\mu\nu} - \partial^\lambda L_{\lambda\nu\mu}, \tag{3-5.46}$$

and there are additional contact terms.

Spin 3. It has been seen that different versions of the Lagrange function can be introduced. We shall be content, however, to record only one in this relatively complicated situation. It is

$$W = \int (dx)[S^{\lambda\mu\nu}\phi_{\lambda\mu\nu} + \mathcal{L}], \tag{3-5.47}$$

$$\begin{aligned}\mathcal{L} = -\tfrac{1}{2}[\partial^\kappa\phi^{\lambda\mu\nu}\partial_\kappa\phi_{\lambda\mu\nu} + m^2\phi^{\lambda\mu\nu}\phi_{\lambda\mu\nu} - 3\partial_\kappa\phi^{\kappa\mu\nu}\partial^\lambda\phi_{\lambda\mu\nu} \\ + 6\partial_\mu\phi^{\lambda\mu\nu}\partial_\nu\phi_\lambda - 3\partial^\lambda\phi^\nu\partial_\lambda\phi_\nu - 3m^2\phi^\nu\phi_\nu - \tfrac{3}{2}(\partial_\nu\phi^\nu)^2] \\ + \tfrac{1}{2}m(\phi^\nu\partial_\nu\Phi - \Phi\partial_\nu\phi^\nu) + \partial^\nu\Phi\partial_\nu\Phi + 4m^2\Phi^2,\end{aligned} \tag{3-5.48}$$

where the auxiliary function Φ is to receive independent variation in the stationary action principle.

Spin $\tfrac{1}{2}$. Little need be said here. The field equation

$$[\gamma^\mu(1/i)\partial_\mu + m]\psi(x) = \eta(x) \tag{3-5.49}$$

implies the action

$$W = \int (dx)[\eta(x)\gamma^0\psi(x) + \mathcal{L}(\psi(x))] \tag{3-5.50}$$

with the Lagrange function

$$\mathcal{L} = -\tfrac{1}{2}\psi\gamma^0[\gamma^\mu(1/i)\partial_\mu + m]\psi. \tag{3–5.51}$$

There is an apparent arbitrariness, for the derivative could be transferred from the right-hand field to the left-hand field, with a minus sign. The two alternatives are identical, however,

$$\psi(x)\gamma^0\gamma^\mu\partial_\mu\psi(x) = -\partial_\mu\psi(x)\gamma^0\gamma^\mu\psi(x), \tag{3–5.52}$$

since the $\gamma^0\gamma^\mu$ are symmetrical matrices and $\psi(x)$ anticommutes with $\partial_\mu\psi(x)$.

Spin $\tfrac{3}{2}$. Vector-spinor field equations are

$$[\gamma^\lambda(1/i)\partial_\lambda + m]\psi_\mu - (1/i)\partial_\mu\gamma^\nu\psi_\nu - \gamma_\mu(1/i)\partial^\nu\psi_\nu + \gamma_\mu[m - \gamma^\lambda(1/i)\partial_\lambda]\gamma^\nu\psi_\nu = \eta_\mu, \tag{3–5.53}$$

and they are embodied in the action

$$W = \int (dx)[\eta^\mu\gamma^0\psi_\mu + \mathcal{L}] \tag{3–5.54}$$

with the Lagrange function

$$\begin{aligned}
\mathcal{L} = -\tfrac{1}{2}\{&\psi^\mu\gamma^0[\gamma^\lambda(1/i)\partial_\lambda + m]\psi_\mu - \psi^\mu\gamma^0(1/i)\partial_\mu\gamma^\nu\psi_\nu \\
&- \psi^\nu\gamma^0\gamma_\nu(1/i)\partial^\mu\psi_\mu + \psi^\mu\gamma^0\gamma_\mu[m - \gamma^\lambda(1/i)\partial_\lambda]\gamma^\nu\psi_\nu\}. \tag{3–5.55}
\end{aligned}$$

The second and third terms effect an explicit symmetrization between the application of the derivatives to the right and to the left,

$$-\partial_\mu\psi^\mu\gamma^0\gamma^\nu\psi_\nu = \psi^\nu\gamma^0\gamma_\nu\partial_\mu\psi^\mu. \tag{3–5.56}$$

This is automatic for the first and last terms of (3–5.55).

Spin $\tfrac{5}{2}$. The symmetrical tensor-spinor field $\psi_{\mu\nu}$ provides the following action description:

$$W = \int (dx)[\eta^{\mu\nu}\gamma^0\psi_{\mu\nu} + \mathcal{L}], \tag{3–5.57}$$

$$\begin{aligned}
\mathcal{L} = -\tfrac{1}{2}\{&\psi^{\mu\nu}\gamma^0[\gamma^\lambda(1/i)\partial_\lambda + m]\psi_{\mu\nu} - 2\psi^{\mu\nu}\gamma^0(1/i)\partial_\mu\gamma^\lambda\psi_{\lambda\nu} \\
&- 2\psi^{\mu\nu}\gamma^0\gamma_\mu(1/i)\partial^\lambda\psi_{\lambda\nu} + 2\psi^{\mu\nu}\gamma^0\gamma_\mu[m - \gamma^\kappa(1/i)\partial_\kappa]\gamma^\lambda\psi_{\lambda\nu} \\
&+ \psi\gamma^0\gamma^\mu(1/i)\partial^\nu\psi_{\mu\nu} + \psi^{\mu\nu}\gamma^0\gamma_\mu(1/i)\partial_\nu\psi - \tfrac{1}{2}\psi\gamma^0[\gamma^\lambda(1/i)\partial_\lambda + m]\psi\} \\
&+ \tfrac{1}{2}m(\psi\gamma^0\Psi - \Psi\gamma^0\psi) - \tfrac{2}{3}\Psi\gamma^0[\gamma^\lambda(1/i)\partial_\lambda - 3m]\Psi. \tag{3–5.58}
\end{aligned}$$

The auxiliary spinor Ψ is to be varied independently in applying the principle of stationary action.

Spins $0, \tfrac{1}{2}, 1, \tfrac{3}{2}, 2, \tfrac{5}{2}, \ldots$ Under this heading are collected the multispinor descriptions. Thus, the following is applicable to spins 0 or 1:

$$\begin{aligned}
W &= \int (dx)[\eta\gamma_1^0\gamma_2^0\psi + \mathcal{L}], \\
\mathcal{L} &= -\tfrac{1}{2}\psi\gamma_1^0\gamma_2^0[\tfrac{1}{2}(\gamma_1^\mu + \gamma_2^\mu)(1/i)\partial_\mu + m]\psi. \tag{3–5.59}
\end{aligned}$$

It should not be forgotten that these even-rank spinors are commutative quantities (B. E. statistics), matching the symmetry of $\gamma_1^0 \gamma_2^0$ and the antisymmetry of $\gamma_1^0 \gamma_2^0 (\gamma_1^\mu + \gamma_2^\mu)$. The second-order differential equation (3–4.26) can also be adopted as the basis of an action principle. Let us write it as

$$\left[-\left(\tfrac{1}{2} \sum_\alpha \gamma_\alpha^\mu (1/i)\partial_\mu \right)^2 + m^2 \right] \psi(x) = \varsigma(x), \qquad (3\text{--}5.60)$$

where

$$\varsigma(x) = \left[m - \tfrac{1}{2} \sum_\alpha \gamma_\alpha^\mu (1/i)\partial_\mu \right] \eta(x). \qquad (3\text{--}5.61)$$

Then we construct the action

$$W(\varsigma) = \frac{1}{2m} \int (dx)[\varsigma \gamma_1^0 \gamma_2^0 \psi + \mathcal{L}],$$
$$\mathcal{L} = -\tfrac{1}{2}\left[-\partial_\mu \psi \gamma_1^0 \gamma_2^0 \tfrac{1}{2} \sum_\alpha \gamma_\alpha^\mu \tfrac{1}{2} \sum_\beta \gamma_\beta^\nu \partial_\nu \psi + m^2 \psi \gamma_1^0 \gamma_2^0 \psi \right], \qquad (3\text{--}5.62)$$

where the factor $1/2m$ is supplied to make the two action expressions directly comparable; otherwise, it can be absorbed into a common scale factor for field and source. The value of the W that is implied by (3–5.62) can be presented as

$$W(\varsigma) = \tfrac{1}{2}(1/2m) \int (dx)\varsigma(x)\gamma_1^0 \gamma_2^0 \psi(x)$$
$$= \tfrac{1}{2}(1/2m) \int (dx)\eta(x)\gamma_1^0 \gamma_2^0 [m - \tfrac{1}{2} \sum \gamma_\alpha^\mu (1/i)\partial_\mu]\psi(x)$$
$$= \tfrac{1}{2} \int (dx)\eta(x)\gamma_1^0 \gamma_2^0 \psi(x) - (1/4m) \int (dx)\eta(x)\gamma_1^0 \gamma_2^0 \eta(x), \quad (3\text{--}5.63)$$

which asserts that the two actions differ in content only by a contact term. The latter removes the contact term that was added in Eq. (3–4.1). Another such remark is based on the commutativity of the symmetrical matrix $i\gamma_{51}i\gamma_{52}$ with $\gamma_1^0 \gamma_2^0$, as well as with the matrices $(\gamma^0 \gamma^\mu)_\alpha$. On decomposing ψ and ς into components with the aid of the projection matrices $\tfrac{1}{2}(1 \pm i\gamma_{51}i\gamma_{52})$, the action expression (3–5.62) completely separates into two independent parts. Thus, it is possible to use a reduced form of the action principle which contains only one of these field components and its associated source. The latter action should be multiplied by two in order to retain the same scale for sources and fields. That the above procedure only changes contact terms is verified by considering

$$W = \tfrac{1}{2}(1/2m) \int (dx)\varsigma \gamma_1^0 \gamma_2^0 (1 \pm i\gamma_{51}i\gamma_{52})\psi$$
$$= \tfrac{1}{2}(1/2m) \int (dx)\eta \gamma_1^0 \gamma_2^0 \{[m - \tfrac{1}{2} \sum \gamma_\alpha^\mu (1/i)\partial_\mu]\psi$$
$$\pm i\gamma_{51}i\gamma_{52}[m + \tfrac{1}{2} \sum \gamma_\alpha^\mu (1/i)\partial_\mu]\psi\}$$
$$= \tfrac{1}{2} \int (dx)\eta \gamma_1^0 \gamma_2^0 \psi - (1/4m) \int (dx)\eta \gamma_1^0 \gamma_2^0 (1 \mp i\gamma_{51}i\gamma_{52})\eta. \quad (3\text{--}5.64)$$

An analogous description can be introduced for spin $\frac{1}{2}$, incidentally, by considering the second-order differential equation

$$(-\partial^2 + m^2)\psi(x) = \zeta(x), \qquad \zeta(x) = [m - \gamma^\mu(1/i)\partial_\mu]\eta(x), \quad (3\text{-}5.65)$$

and the action

$$W(\zeta) = \frac{1}{2m}\int(dx)[\zeta(x)\gamma^0\psi(x) + \mathcal{L}(\psi(x))],$$

$$\mathcal{L} = -\tfrac{1}{2}[\partial^\mu\bar\psi\gamma^0\partial_\mu\psi + m^2\bar\psi\gamma^0\psi]. \qquad (3\text{-}5.66)$$

Again we observe that

$$W(\zeta) = \tfrac{1}{2}(1/2m)\int(dx)\zeta(x)\gamma^0\psi(x)$$

$$= \tfrac{1}{2}(1/2m)\int(dx)\eta(x)\gamma^0[m - \gamma^\mu(1/i)\partial_\mu]\psi(x)$$

$$= \tfrac{1}{2}\int(dx)\eta(x)\gamma^0\psi(x) - (1/4m)\int(dx)\eta(x)\gamma^0\eta(x) \qquad (3\text{-}5.67)$$

differs from the original action only by a contact term. Next, suppose that one of the two projection matrices

$$\tfrac{1}{2}(1 \pm i\gamma_5) = [\tfrac{1}{2}(1 \pm i\gamma_5)]^2, \qquad (3\text{-}5.68)$$

multiplied by the factor 2, is introduced into (3-5.66) so that it multiplies γ^0 everywhere on the right. Then, since

$$\gamma^0[\tfrac{1}{2}(1 \pm i\gamma_5)]^2 = [\tfrac{1}{2}(1 \pm i\gamma_5)]^T\gamma^0[\tfrac{1}{2}(1 \pm i\gamma_5)], \qquad (3\text{-}5.69)$$

a consistent projection of field and source onto a subspace has been brought about. The significance of this new action is given by

$$W = \tfrac{1}{2}(1/2m)\int(dx)\zeta(x)\gamma^0(1 \pm i\gamma_5)\psi(x)$$

$$= \tfrac{1}{2}(1/2m)\int(dx)\eta(x)\gamma^0\{[m - \gamma^\mu(1/i)\partial_\mu]\psi(x) \pm i\gamma_5[m + \gamma^\mu(1/i)\partial_\mu]\psi(x)\}$$

$$= \tfrac{1}{2}\int(dx)\eta(x)\gamma^0\psi(x) - (1/4m)\int(dx)\eta(x)\gamma^0(1 \mp i\gamma_5)\eta(x); \qquad (3\text{-}5.70)$$

only the contact term has been altered, and the same physical system is being described. But whether this second-order formulation of spin $\frac{1}{2}$ has any practical merit will not be discussed here. One remark is in order, however. The γ_5-dependent contact term in (3-5.70) is imaginary ($\gamma^0 i\gamma_5$ is antisymmetrical and real) and should be subtracted from the second-order action to preserve the detailed physical equivalence of the two descriptions. Since this subtractive term is given contact form through the use of the source η, and not ζ, it emphasizes that the second-order formulation could not be adopted as the fundamental spin $\frac{1}{2}$ description.

Spins $\frac{1}{2}$ or $\frac{3}{2}$ are represented by

$$W = \int (dx)[\eta\gamma\psi + \mathcal{L}], \tag{3-5.71}$$

where

$$\gamma = \prod \gamma_\alpha^0 = \gamma_1^0\gamma_2^0\gamma_3^0 \tag{3-5.72}$$

and

$$\mathcal{L} = -\tfrac{1}{2}\psi\gamma[\tfrac{1}{3}\sum \gamma_\alpha^\mu(1/i)\partial_\mu + m]\psi$$
$$- \tfrac{1}{2}\sum_{\alpha<\beta} [\psi\gamma\tfrac{1}{2}(\gamma_\alpha^\mu - \gamma_\beta^\mu)(1/i)\partial_\mu\psi_{[\alpha\beta]} + \psi_{[\alpha\beta]}\gamma\tfrac{1}{2}(\gamma_\alpha^\mu - \gamma_\beta^\mu)(1/i)\partial_\mu\psi]$$
$$- 3\psi_{[23]}\gamma[\gamma_1^\mu(1/i)\partial_\mu - 3m]\psi_{[23]} - \cdots . \tag{3-5.73}$$

In addition to ψ, the three auxiliary functions $\psi_{[23]}, \ldots$ are varied independently in the action principle. A second-order formulation is provided by

$$W(\zeta) = \frac{1}{2m} \int (dx)[\zeta\gamma\psi + \mathcal{L}],$$
$$\mathcal{L} = -\tfrac{1}{2}\left[-\tfrac{1}{8}\partial^\mu\psi\gamma\partial_\mu\psi - \tfrac{1}{8}\partial_\mu\psi\gamma\sum_{\alpha\beta}\gamma_\alpha^\mu\gamma_\beta^\nu\partial_\nu\psi + m^2\psi\gamma\psi\right], \tag{3-5.74}$$

in which

$$\zeta = [m - \tfrac{1}{3}\sum \gamma_\alpha^\mu(1/i)\partial_\mu]\eta. \tag{3-5.75}$$

For this situation, we have

$$W(\zeta) = \tfrac{1}{2}(1/2m)\int (dx)\zeta(x)\gamma\psi(x) = \tfrac{1}{2}(1/2m)\int (dx)\eta\gamma[m - \tfrac{1}{3}\sum \gamma_\alpha^\mu(1/i)\partial_\mu]\psi$$
$$= \tfrac{1}{2}\int (dx)\eta\gamma\psi - (1/4m)\int (dx)\eta\gamma\eta$$
$$+ (1/4m)\int (dx)\eta\gamma\sum_{\alpha<\beta}\tfrac{1}{2}(\gamma_\alpha^\mu - \gamma_\beta^\mu)(1/i)\partial_\mu\psi_{[\alpha\beta]}, \tag{3-5.76}$$

where the last term can also be written [cf. Eq. (3-4.19)]

$$9m\int (dx) \sum_{\alpha<\beta} \psi_{[\alpha\beta]}\gamma\psi_{[\alpha\beta]}. \tag{3-5.77}$$

It is of contact type since the $\psi_{[\alpha\beta]}$ are confined to the interior of the source. And only the contact terms are changed if γ in Eq. (3-5.74) is everywhere multiplied on the right by one of the matrices $1 \pm \prod(i\gamma_5)_\alpha$. This second-order formulation is physically equivalent to the first-order description.

We shall bypass the discussion of fourth- and fifth-rank spinors to present the action principle formulation for spinors of rank n, at least in its earlier stages. This incomplete Lagrange function, written without the symmetrization of derivatives that is used in (3–5.73), for example, is

$$\mathcal{L} = -\tfrac{1}{2}\psi\gamma\left[\frac{1}{n}\sum\gamma_\alpha^\mu\left(\frac{1}{i}\right)\partial_\mu + m\right]\psi - \sum\psi\gamma\tfrac{1}{2}(\gamma_\alpha^\mu - \gamma_\beta^\mu)\left(\frac{1}{i}\right)\partial_\mu\psi_{[\alpha\beta]}$$

$$- \frac{n(n-1)}{n-2}\frac{1}{2}\sum\psi_{[\alpha\beta]}\gamma\left[\frac{1}{n-2}\sum{}'\gamma_{\alpha'}^\mu\left(\frac{1}{i}\right)\partial_\mu - \frac{n}{n-2}m\right]\psi_{[\alpha\beta]}$$

$$- \frac{n(n-1)}{n-2}\sum\psi_{[\alpha\beta]}\gamma\tfrac{1}{2}(\gamma_{\alpha'}^\mu - \gamma_{\beta'}^\mu)\left(\frac{1}{i}\right)\partial_\mu\psi_{[\alpha\beta][\alpha'\beta']}$$

$$+ 2\frac{(n-1)^2(n-2)^3(n-4)}{n^3(n-3)}\frac{1}{2}\sum\psi_{[\alpha\beta][\alpha'\beta']}\gamma$$

$$\times\left[\frac{1}{n-4}\sum{}'\gamma_{\alpha''}^\mu\left(\frac{1}{i}\right)\partial_\mu + \frac{n}{n-4}m\right]\psi_{[\alpha\beta][\alpha'\beta']}$$

$$+ \frac{(n-1)^3(n-4)^3}{n^3(n-3)}\sum\psi_{[\alpha\beta][\alpha'\beta']}\gamma\sum{}'\gamma_{\alpha''}^\mu\left(\frac{1}{i}\right)\partial_\mu\chi_{[\alpha\beta][\alpha'\beta']}$$

$$+ \frac{(n-1)^3(n-4)^3}{n^3(n-3)}\frac{1}{2}\sum\chi_{[\alpha\beta][\alpha'\beta']}\gamma$$

$$\times\left[\frac{1}{n-4}\sum{}'\gamma_{\alpha''}^\mu\left(\frac{1}{i}\right)\partial_\mu - \frac{n}{n-4}m\right]\chi_{[\alpha\beta][\alpha'\beta']} + \cdots. \quad (3\text{–}5.78)$$

All repetitions are to be rejected in performing the summations over pairs of indices.

It is interesting to study the explicit connections between the multispinor action formulations and those employing tensors or tensor-spinors. Only the lowest, second-rank multispinors will be considered. Spinors of the second rank are usefully treated as matrices for this purpose. Some transcriptions of significant combinations are

$$\eta\gamma_1^0\gamma_2^0\psi = \eta_{\varsigma_1\varsigma_2}\gamma^0_{\varsigma_1\varsigma_1'}\gamma^0_{\varsigma_2\varsigma_2'}\psi_{\varsigma_1'\varsigma_2'} = \eta_{\varsigma_1\varsigma_2}\gamma^0_{\varsigma_2\varsigma_2'}\psi^T_{\varsigma_2'\varsigma_1'}(-\gamma^0)_{\varsigma_1'\varsigma_1}$$

$$= -\text{tr}\,(\eta\gamma^0\psi^T\gamma^0), \quad\quad (3\text{–}5.79)$$

$$\psi\gamma_1^0\gamma_2^0(\gamma_1^\mu + \gamma_2^\mu)\psi = \psi_{\varsigma_1\varsigma_2}[(\gamma^0\gamma^\mu)_{\varsigma_1\varsigma_1'}\gamma^0_{\varsigma_2\varsigma_2'} + \gamma^0_{\varsigma_1\varsigma_1'}(\gamma^0\gamma^\mu)_{\varsigma_2\varsigma_2'}]\psi_{\varsigma_1'\varsigma_2'}$$

$$= -\text{tr}\,(\psi\gamma^0[\gamma^\mu, \psi^T\gamma^0]),$$

and the action (3–5.59) is correspondingly rewritten as

$$W = \int(dx)\,\text{tr}\,(-\eta\gamma^0\psi^T\gamma^0 + \tfrac{1}{4}\psi\gamma^0(1/i)\partial_\mu[\gamma^\mu, \psi^T\gamma^0] + \tfrac{1}{2}m\psi\gamma^0\psi^T\gamma^0). \quad (3\text{–}5.80)$$

The symmetrical field and source spinors of unit spin are given general form by

$$2m^{-1/2}\psi = \gamma^\mu\gamma^0\phi_\mu + (1/2m)\sigma^{\mu\nu}\gamma^0 G_{\mu\nu},$$

$$2m^{1/2}\eta = \gamma^\mu\gamma^0 J_\mu - (m/2)\sigma^{\mu\nu}\gamma^0 M_{\mu\nu}, \quad\quad (3\text{–}5.81)$$

which has appeared before, in other notation, as the second line of Eq. (2–9.70). The commutators and trace evaluations stated in Section 2–9 apply here and give immediately:

$$W = \int (dx)[J^\mu \phi_\mu + \tfrac{1}{2} M^{\mu\nu} G_{\mu\nu} + \mathcal{L}],$$

$$\mathcal{L} = -\tfrac{1}{4} G^{\mu\nu}(\partial_\mu \phi_\nu - \partial_\nu \phi_\mu) - \tfrac{1}{2}\phi^\mu \partial^\nu G_{\mu\nu} + \tfrac{1}{4} G^{\mu\nu} G_{\mu\nu} - \tfrac{1}{2} m^2 \phi^\mu \phi_\mu.$$

(3–5.82)

This is the first-order form (3–5.27, 28), with the derivatives symmetrized in application to the vector and tensor fields. Similarly, the antisymmetrical spinors of zero spin are presented as

$$2m^{-1/2}\psi = i\gamma_5\gamma^0\phi - (1/m)\gamma^\mu\gamma_5\gamma^0\phi_\mu,$$

$$2m^{1/2}\eta = i\gamma_5\gamma^0 K + m\gamma^\mu\gamma_5\gamma^0 K_\mu,$$

(3–5.83)

and one obtains

$$W = \int (dx)[K\phi + K^\mu \phi_\mu + \mathcal{L}],$$

$$\mathcal{L} = -\tfrac{1}{2}\phi^\mu \partial_\mu \phi + \tfrac{1}{2}\phi \partial_\mu \phi^\mu + \tfrac{1}{2}\phi^\mu \phi_\mu - \tfrac{1}{2} m^2 \phi^2,$$

(3–5.84)

to be compared with Eqs. (3–5.15, 16).

The matrix transcription

$$i\gamma_{51} i\gamma_{52}\psi \rightarrow -i\gamma_5 \psi i\gamma_5$$

(3–5.85)

implies that a term of the matrix ψ that commutes (anticommutes) with $i\gamma_5$ is an eigenvector of $i\gamma_{51} i\gamma_{52}$ with the eigenvalue -1 ($+1$). Both eigenvalues are represented in the expressions (3–5.81) and (3–5.83). In applying the second-order action form (3–5.62), it is permissible to use projected versions of the field and source, as illustrated for spins 1 and 0, respectively, by

$$2m^{-1/2}\psi = \gamma^\mu\gamma^0\phi_\mu, \qquad 2m^{-1/2}\zeta = \gamma^\mu\gamma^0(J_\mu + \partial^\nu M_{\mu\nu}) \qquad (3–5.86)$$

and

$$2m^{-1/2}\psi = i\gamma_5\gamma^0\phi, \qquad 2m^{-1/2}\zeta = i\gamma_5\gamma^0(K - \partial_\mu K^\mu), \qquad (3–5.87)$$

where the ζ structures are the appropriate projections of (3–5.61). With these reduced fields and sources understood, the action (3–5.62) becomes (a factor of 2 is supplied)

$$W = (1/m)\int (dx)\ \mathrm{tr}\ (-\zeta\gamma^0\psi^T\gamma^0$$

$$+ \tfrac{1}{8}[\gamma^\mu, \partial_\mu\psi\gamma^0][\gamma^\nu, \partial_\nu\psi^T\gamma^0] + \tfrac{1}{2} m^2\psi\gamma^0\psi^T\gamma^0). \qquad (3–5.88)$$

The trace evaluations give directly the action expressions (3–5.18, 19) and (3–5.11, 12) with the effective vector and scalar sources, explicit in (3–5.86, 87), that have been stated previously. One can also make the opposite choices in these projections, and we record those action forms which are, for spin 1 and 0, respectively,

$$W = \int (dx) \left[\tfrac{1}{2} M^{\mu\nu} G_{\mu\nu} + \frac{1}{2m^2} \partial_\nu G^{\mu\nu} \partial^\lambda G_{\mu\lambda} + \tfrac{1}{4} G^{\mu\nu} G_{\mu\nu} \right] \qquad (3\text{-}5.89)$$

and

$$W = \int (dx) \left[K^\mu \phi_\mu + \frac{1}{2m^2} (\partial_\mu \phi^\mu)^2 + \tfrac{1}{2} \phi^\mu \phi_\mu \right]. \qquad (3\text{-}5.90)$$

The effective sources that appear here are $M_{\mu\nu} - (1/2m^2)(\partial_\mu J_\nu - \partial_\nu J_\mu)$ and $K_\mu - (1/m^2)\partial_\mu K$, respectively.

We shall close this section with a few varied comments. First we recall that, although we have not illustrated it here, the possibility exists of redefining sources and fields by linear transformations which change the detailed structure of the field equations and, therefore, of the Lagrange functions. Then it is noted that all discussion has been concerned with the vacuum amplitude $\langle 0_+ | 0_- \rangle^S$. The shift of attention to other transformation functions is conveyed by a change of boundary conditions in the action principle. Let us be explicit about the time cycle transformation function $\langle 0_- | 0_- \rangle^{S-, S+}$. Here the action separates into two analogous terms, with opposite signs, that are associated with the two senses of time development:

$$W(S_-, S_+) = \int (S_+ \gamma \chi_+ - \tfrac{1}{2} \chi_+ \gamma F \chi_+) - \int (S_- \gamma \chi_- - \tfrac{1}{2} \chi_- \gamma F \chi_-). \qquad (3\text{-}5.91)$$

The accompanying boundary conditions require negative frequencies for χ_+ and positive frequencies for χ_- at times before the sources come into operation, and the equality of χ_+ and χ_- after the sources have shut down. The latter has the following consequence. While the integrations in (3–5.91) can be regarded as extended over all space-time, there is complete cancellation of those regions that are subsequent to the functioning of the sources. Accordingly, the integration domain can be made semi-infinite, bounded by any space-like surface that has only source-free space in its future. The final comment is concerned with the relationship between the Lagrange functions of massive and massless particles.

It is known that, as $m \to 0$, the states of a particle with spin s decompose into those of various helicities, $\pm s, \pm(s-1), \dots$. Thus the description of all massless particles with helicities of magnitude $\leq s$, in integer steps, should be contained in the account of a massive particle with spin s. This has received some discussion in terms of sources. We want to trace the corresponding field decomposition in two important examples. Let the vector field and source of a unit spin particle with mass m be expressed by

$$\phi_\mu = A_\mu - (1/m)\partial_\mu \phi, \qquad \partial_\mu J^\mu = mK, \qquad (3\text{-}5.92)$$

where the latter is Eq. (2–3.44). Then the action (3–5.18, 19) becomes

$$W = \int (dx)[J^\mu A_\mu + K\phi - \tfrac{1}{4}(\partial^\mu A^\nu - \partial^\nu A^\mu)(\partial_\mu A_\nu - \partial_\nu A_\mu)$$
$$- \tfrac{1}{2}(mA^\mu - \partial^\mu \phi)(mA_\mu - \partial_\mu \phi)]. \qquad (3\text{-}5.93)$$

In the limit $m \to 0$, this action separates into two parts:

$$W(J) = \int (dx)[J^\mu A_\mu - \tfrac{1}{4}(\partial^\mu A^\nu - \partial^\nu A^\mu)(\partial_\mu A_\nu - \partial_\nu A_\mu)],$$

$$\partial_\mu J^\mu = 0, \tag{3-5.94}$$

which, as shall be discussed in greater detail later, describes the photon, and

$$W(K) = \int (dx)[K\phi - \tfrac{1}{2}\partial^\mu \phi \partial_\mu \phi], \tag{3-5.95}$$

referring to a massless particle of zero spin (helicity). Notice that the Lagrange function of the spinless particle comes entirely from the mass term of the original Lagrange function, and would have been overlooked had one merely set $m = 0$ in (3-5.19).

The other example is spin 2, where we express the tensor field and its source by

$$\phi_{\mu\nu} = h_{\mu\nu} - (2^{-1/2}/m)(\partial_\mu A_\nu + \partial_\nu A_\mu) + 6^{-1/2}[(2/m^2)\partial_\mu\partial_\nu\phi + g_{\mu\nu}\phi], \tag{3-5.96}$$

and

$$\partial_\nu T^{\mu\nu} = m2^{-1/2}J^\mu, \qquad \partial_\mu J^\mu = m(3^{1/2}K - 2^{-1/2}T), \tag{3-5.97}$$

the latter being Eq. (2-4.21). These structures are such that

$$\int (dx) T^{\mu\nu}\phi_{\mu\nu} = \int (dx)[T^{\mu\nu}h_{\mu\nu} + J^\mu A_\mu + K\phi]. \tag{3-5.98}$$

When (3-5.96) is inserted into the action (3-5.30, 31), and the limit $m \to 0$ performed, three independent parts are obtained. Two of them restate the unit and zero helicity actions, Eqs. (3-5.94) and (3-5.95). The third one is

$$W(T) = \int (dx)[T^{\mu\nu}h_{\mu\nu} - \tfrac{1}{2}(\partial^\lambda h^{\mu\nu}\partial_\lambda h^{\mu\nu} - \partial^\lambda h\partial_\lambda h) - \partial_\mu h^{\mu\nu}\partial_\nu h + \partial_\mu h^{\mu\nu}\partial^\lambda h_{\lambda\nu}],$$

$$\partial_\mu T^{\mu\nu} = 0. \tag{3-5.99}$$

As we shall also discuss later in more detail, it is the (or a) graviton action expression. This time the photon Lagrange function emerges completely from the mass term of the spin 2 particle, as indicated by

$$\int (dx)[-\tfrac{1}{4}(\partial^\mu A^\nu + \partial^\nu A^\mu)(\partial_\mu A_\nu + \partial_\nu A_\mu) + (\partial_\mu A^\mu)^2]$$

$$= \int (dx)(-\tfrac{1}{4})(\partial^\mu A^\nu - \partial^\nu A^\mu)(\partial_\mu A_\nu - \partial_\nu A_\mu), \tag{3-5.100}$$

but the scalar field action has contributions from both parts of the original Lagrange function. It is an interesting unification of the actions representing massless particles of various helicities to connect them with one action expression for a massive particle. Also implied are the relationships between different spins necessary to arrive at a common description for a given helicity.

3–6 INVARIANCE TRANSFORMATIONS AND FLUXES.   CHARGE

The vacuum amplitude $\langle 0_+|0_-\rangle^S$ is unaltered by a rigid translation or rotation of the sources; for charged particles it remains unchanged by a universal phase transformation of the sources. Physical information is obtained through the relative transformation of different parts of the source distribution. Relative translation gives information about energy-momentum, relative rotation about angular momentum, and relative phase displacement teaches about charge. When the source distribution is causally arranged, with the disjoint pieces treated as units, one acquires knowledge of integral physical quantities— total energy, total charge. At the next stage one considers transformations that vary arbitrarily in space-time, thereby supplying more localized data about the various physical quantities. Fields are the instrument for conveying these data.

To illustrate these remarks, let us consider spinless charged particles of mass m, as described by the action

$$W = \int (dx)[K(x)\phi(x) + \mathcal{L}], \qquad \mathcal{L} = -\tfrac{1}{2}[\partial^\mu\phi\partial_\mu\phi + m^2\phi^2], \qquad (3\text{–}6.1)$$

where $K(x)$ and $\phi(x)$ are now two-component objects in an appropriate Euclidean charge space. Invariance under the constant phase transformation of the source,

$$\overline{K}(x) = e^{iq\varphi}K(x), \qquad (3\text{–}6.2)$$

follows from the existence of the compensating field transformation

$$\bar{\phi}(x) = e^{iq\varphi}\phi(x), \qquad (3\text{–}6.3)$$

since all the Euclidean products in (3–6.1) are unchanged by a common rotation. Next, let φ become an arbitrary function of position. For simplicity we consider an infinitesimal phase transformation and write this generalization of (3–6.2) as

$$\delta K(x) = iq\,\delta\varphi(x)K(x). \qquad (3\text{–}6.4)$$

The associated action variation is

$$\delta W = \int (dx)\phi(x)\,\delta K(x) = \int (dx)\phi(x)iqK(x)\,\delta\varphi(x). \qquad (3\text{–}6.5)$$

But again we can introduce a compensating field transformation,

$$\delta\phi(x) = iq\,\delta\varphi(x)\phi(x), \qquad (3\text{–}6.6)$$

which leaves $K\phi$ and ϕ^2 unaltered, and fails to keep W invariant only because space-time derivatives now act upon $\delta\varphi(x)$:

$$\delta\big(\partial_\mu\phi(x)\big) = iq\,\delta\varphi(x)\partial_\mu\phi(x) + iq\phi(x)\partial_\mu\,\delta\varphi(x). \qquad (3\text{–}6.7)$$

Thus, through this method of calculation it is found that

$$\delta W = -\int (dx)j^\mu(x)\partial_\mu\,\delta\varphi(x), \qquad (3\text{–}6.8)$$

where

$$j^\mu(x) = \partial^\mu \phi(x) i q \phi(x). \tag{3-6.9}$$

The consistency of the two evaluations implies, with the aid of an integration by parts, that

$$\partial_\mu j^\mu(x) = \phi(x) i q K(x). \tag{3-6.10}$$

This is verified directly, on using the field equations. When the right-hand side is zero, which is true in source-free regions, we recognize the local statement of a conservation law. If the charge matrix q is diagonalized and complex sources introduced, the action expression and the properties of j^μ become

$$W = \int (dx)[K^*\phi + K\phi^* - \partial^\mu\phi^*\partial_\mu\phi - m^2\phi^*\phi], \tag{3-6.11}$$

and

$$j^\mu = i(\partial^\mu\phi^*\phi - \phi^*\partial^\mu\phi), \qquad \partial_\mu j^\mu = i(\phi^*K - K^*\phi), \tag{3-6.12}$$

where

$$K = 2^{-1/2}(K_{(1)} - iK_{(2)}), \qquad K^* = 2^{-1/2}(K_{(1)} + iK_{(2)})$$
$$\phi = 2^{-1/2}(\phi_{(1)} - i\phi_{(2)}), \qquad \phi^* = 2^{-1/2}(\phi_{(1)} + i\phi_{(2)}), \tag{3-6.13}$$

with the reminder that ϕ^* is not the complex conjugate of ϕ. Statements analogous to all these apply to the vacuum time cycle action with appropriate algebraic signs in δW to indicate the sense of time flow.

Using this more general framework, we now re-examine the causal situation with $K = K_1 + K_2$, where the phase of K_2 is changed by a constant and that of K_1 is held fixed. For the infinitesimal transformation being considered, we know that

$$\langle 0_+|0_-\rangle^K = \sum_{\{n\}} \langle 0_+|\{n\}\rangle^{K_1}\langle\{n\}|0_-\rangle^{K_2} \rightarrow \sum \langle 0_+|\{n\}\rangle^{K_1}[1 + i\,\delta\varphi Q]\langle\{n\}|0_-\rangle^{K_2},$$
$$\tag{3-6.14}$$

with

$$Q(\{n\}) = \sum_{pq} qn_{pq}, \tag{3-6.15}$$

or

$$\delta\langle 0_+|0_-\rangle^K = i\,\delta\varphi \sum \langle 0_+|\{n\}\rangle^{K_1} Q(\{n\})|0_-\rangle^{K_2}. \tag{3-6.16}$$

This weighted average of the charge values resembles an expectation value. Indeed it is one, if we consider the time cycle function, with the phase of $K_{(+)}$ displaced, after which the two sources are identified with K,

$$\delta\langle 0_-|0_-\rangle^{K_{(-)}, K_{(+)}} = i\,\delta\varphi \sum \langle 0_-|\{n\}\rangle^K Q(\{n\}|0_-\rangle^K$$
$$= i\,\delta\varphi \langle Q\rangle_0^K; \tag{3-6.17}$$

this is an infinitesimal version of the left-hand side of Eq. (2–2.123), for example. To apply (3–6.8) to this causal situation, we separate K_1 and K_2 by a space-like surface, which is otherwise arbitrary, and then make $\delta\varphi(x)$ vanish in the future of this surface and be the constant $\delta\varphi$ in its past. The step function derivative,

$-\partial_\mu \, \delta\varphi(x)$, confines integration to the surface and

$$\frac{\delta\langle 0_+|0_-\rangle^K}{\langle 0_+|0_-\rangle^K} = i \, \delta W = i \, \delta\varphi \int d\sigma_\mu j^\mu(x), \qquad (3\text{-}6.18)$$

which is the identification

$$\int d\sigma_\mu j^\mu = \frac{\sum \langle 0_+|\{n\}\rangle^{K_1} Q \langle\{n\}|0_-\rangle^{K_2}}{\langle 0_+|0_-\rangle^K}. \qquad (3\text{-}6.19)$$

In the time cycle description, integrations are extended up to a space-like surface, which follows the working of the sources. Then the choice $\delta\varphi_-(x) = 0$, $\delta\varphi_+(x) = \delta\varphi$ gives

$$\begin{aligned} \delta W &= -\int (dx) j^\mu(x)\partial_\mu \, \delta\varphi_+(x) + \int (dx) j^\mu(x)\partial_\mu \, \delta\varphi_-(x) \\ &= \delta\varphi \int d\sigma_\mu j^\mu, \end{aligned} \qquad (3\text{-}6.20)$$

which identifies

$$\int d\sigma_\mu j^\mu = \langle Q \rangle_0^K. \qquad (3\text{-}6.21)$$

In the latter situation $j^\mu(x)$ is computed from the real or mutually conjugate retarded fields, and is a real function.

It is evident that $j^\mu(x)$ provides a space-time account of the distribution and flow of charge—it is the charge flux vector or current vector. We shall evaluate it for a single-particle state. On referring to Eqs. (3-1.79–81), it is seen that the fields in the region between emission source K_2 and absorption source K_1, associated with a positively charged particle of momentum p, are

$$\begin{aligned} \phi(x) &= (d\omega_p)^{1/2} e^{ipx} iK_{2p+}, \\ \phi^*(x) &= (d\omega_p)^{1/2} e^{-ipx} iK_{1p+}^*, \end{aligned} \qquad (3\text{-}6.22)$$

and if only this excitation occurs,

$$j^\mu(x) = 2p^\mu \, d\omega_p(iK_{1p+}^*)(iK_{2p+}). \qquad (3\text{-}6.23)$$

The source factors identify the emission and absorption acts [compare Eq. (3-6.19)]; the current associated with the particle is $2p^\mu \, d\omega_p$. We can confirm this by verifying that the total charge is unity. But first it must be recognized that the uniform value of the current is an idealization, which applies in the interior of the particle beam but fails as one nears the edges. Of course, the momentum is not specified with arbitrary precision, as in (3-6.22), but within a cell of small but finite dimensions, having invariant measure $\Delta\omega_p$. Thus, the correct description is given by

$$(d\omega_p)^{1/2} e^{ipx} \rightarrow \frac{1}{(\Delta\omega_p)^{1/2}} \int_{\Delta\omega_p} d\omega_p e^{ipx}. \qquad (3\text{-}6.24)$$

To compute the total charge one can integrate the charge density $j^0(x)$ over all space at a given time:

$$Q(1_{p+}) = \frac{2p^0}{\Delta\omega_p} \int (dx) \left| \int_{\Delta\omega_p} d\omega_p \exp(i\mathbf{p} \cdot \mathbf{x}) \right|^2$$

$$= \frac{1}{\Delta\omega_p} \int_{\Delta\omega_p} d\omega_p (d\mathbf{p}')\, \delta(\mathbf{p} - \mathbf{p}') = 1. \qquad (3\text{-}6.25)$$

Similarly, for a negatively charged particle,

$$\phi^*(x) = (d\omega_p)^{1/2} e^{ipx} iK_{2p-}, \qquad \phi(x) = (d\omega_p)^{1/2} e^{-ipx} iK_{1p-}^*, \qquad (3\text{-}6.26)$$

and

$$j^\mu(x) = -2p^\mu\, d\omega_p (iK_{1p-}^*)(iK_{2p-}), \qquad (3\text{-}6.27)$$

with an analogous verification that the total charge of the particle is -1. The retarded fields of the time cycle description that are associated with a given momentum, and positive or negative charge, are [cf. Eq. (3-1.84)]

$$\phi(x) = (d\omega_p)^{1/2} e^{ipx} iK_{p+}, \qquad (3\text{-}6.28)$$

or

$$\phi^*(x) = (d\omega_p)^{1/2} e^{ipx} iK_{p-}, \qquad (3\text{-}6.29)$$

respectively, with their complex conjugates. The corresponding real currents are then given by

$$j^\mu(x) = \pm 2p^\mu\, d\omega_p |K_{p\pm}|^2. \qquad (3\text{-}6.30)$$

This is the contribution to the current expectation value attributed to a particular momentum and charge, being the current per particle multiplied by the expected number of particles of the given type emitted by the source. Incidentally, the consideration of a single momentum should not obscure the presence of interference terms, in these quadratic field structures, when several particles of different momenta are present. Such interference terms disappear on performing the spatial integrations that evaluate the total charge.

The non-conservation equations, (3-6.10) or (3-6.12), connect the particles and charges observed after the operation of a source to the activity in the interior of the source. Let us test this in the physical circumstance of the time cycle description, with $K_{(-)} = K_{(+)} = K$. Integration over a region that contains the source gives

$$\langle Q \rangle_0^K = \int d\sigma_\mu j^\mu = i \int (dx)(\phi_{\text{ret.}}^* K - K^* \phi_{\text{ret.}}), \qquad (3\text{-}6.31)$$

where the surface integral refers to any space-like surface that is subsequent to the source region. On any surface that precedes the source, the retarded fields and the current vanish. The explicit form of the right-hand side is

$$\langle Q \rangle_0^K = \int (dx)(dx') K^*(x) i [\Delta_{\text{ret.}}(x' - x) - \Delta_{\text{ret.}}(x - x')] K(x'). \qquad (3\text{-}6.32)$$

But according to Eqs. (3–1.27, 34),

$$(1/i)[\Delta_{\mathrm{ret.}}(x - x') - \Delta_{\mathrm{ret.}}(x' - x)] = (1/i)[\Delta_{\mathrm{ret.}}(x - x') - \Delta_{\mathrm{adv.}}(x - x')]$$
$$= \Delta^{(+)}(x - x') - \Delta^{(-)}(x - x'), \quad (3\text{–}6.33)$$

and

$$\langle Q \rangle_0^K = \int d\omega_p[|K(p)|^2 - |K^*(p)|^2]$$
$$= \sum_p [|K_{p+}|^2 - |K_{p-}|^2], \quad (3\text{–}6.34)$$

which is the expected result.

We have not yet remarked on the ambiguity in the current vector that is defined in general by (3–6.8). To a particular $j^\mu(x)$ vector can be added any expression of the form $\partial_\nu m^{\mu\nu}(x)$, where $m^{\mu\nu}$ is an antisymmetrical tensor, for

$$-\int (dx)\partial_\nu m^{\mu\nu}(x)\partial_\mu \, \delta\varphi(x) = \int (dx)\partial_\mu\partial_\nu m^{\mu\nu}(x) \, \delta\varphi(x) = 0. \quad (3\text{–}6.35)$$

This supplement to the charge density, $\boldsymbol{\nabla} \cdot \mathbf{n}(x)$, where $n_k = m^0{}_k$, adds a two-dimensional surface integral to the charge associated with a three-dimensional volume:

$$\int (dx)\boldsymbol{\nabla} \cdot \mathbf{n}(x) = \int d\mathbf{S} \cdot \mathbf{n}(x). \quad (3\text{–}6.36)$$

The calculation of total charge is not affected, therefore, nor is the value of the flux vector assigned to a uniform situation since this is also fixed by total charge considerations. Why can one not ignore the ambiguity and just accept the current expression that is naturally associated with the Lagrange function? One reason is that alternative Lagrange functions can produce different currents. This is illustrated by the unit spin situation.

The second-order Lagrange function (3–5.19) and the first-order Lagrange function (3–5.28) imply, respectively,

$$j^\mu(x) = (\partial^\mu\phi^\nu(x) - \partial^\nu\phi^\mu(x))iq\phi_\nu(x),$$
$$\partial_\mu j^\mu(x) = \phi^\mu(x)iqJ_\mu(x), \quad (3\text{–}6.37)$$

and

$$j^\mu(x) = G^{\mu\nu}(x)iq\phi_\nu(x),$$
$$\partial_\mu j^\mu(x) = \phi^\mu(x)iqJ_\nu(x) + \tfrac{1}{2}G^{\mu\nu}(x)iqM_{\mu\nu}(x). \quad (3\text{–}6.38)$$

In the absence of the source $M_{\mu\nu}$, these current expressions are equivalent. But, when we use the Lagrange function (3–5.23), there results

$$j^\mu = \partial^\mu\phi^\nu iq\phi_\nu + \phi^\mu iq\partial_\nu\phi^\nu$$
$$= (\partial^\mu\phi^\nu - \partial^\nu\phi^\mu)iq\phi_\nu + \partial_\nu(\phi^\mu iq\phi^\nu), \quad (3\text{–}6.39)$$

where $\phi^\mu iq\phi^\nu$ is indeed an antisymmetrical tensor. The alternative current expression is somewhat simpler, since $\partial_\nu\phi^\nu$ vanishes outside source regions, and the total charge can be calculated as

$$\int d\sigma_\mu j^\mu = \int d\sigma_\mu \partial^\mu \phi' iq\phi_\nu. \tag{3-6.40}$$

Let us apply this to the region between the two causally separated sources J_1^μ, J_2^μ, where the field is [Eq. (3-3.3)]

$$\begin{aligned}
\phi_\mu(x) &= i\int (dx') \left[\int d\omega_p e^{ip(x-x')} (g_{\mu\nu} + m^{-2} p_\mu p_\nu) \right] J_2^\nu(x') \\
&+ i\int (dx') \left[\int d\omega_p e^{-ip(x-x')} (g_{\mu\nu} + m^{-2} p_\mu p_\nu) \right] J_1^\nu(x') \\
&= \sum_{p\lambda q} (d\omega_p)^{1/2} e_{\mu p\lambda q} e^{ipx} iJ_{2p\lambda q} + \sum_{p\lambda q} (d\omega_p)^{1/2} e_{\mu p\lambda q}^* e^{-ipx} iJ_{1p\lambda q}^*. \tag{3-6.41}
\end{aligned}$$

The eigenvector properties

$$q e_{p\lambda q'}^\mu = q' e_{p\lambda q'}^\mu, \qquad e_{p\lambda q'}^{\mu*} q = e_{p\lambda q'}^{\mu*} q' \tag{3-6.42}$$

and the normalization

$$e_{p\lambda q}^{\mu*} e_{\mu p\lambda q} = 1 \tag{3-6.43}$$

ensure that the contribution to j^μ of a specific particle state is

$$j^\mu = q 2p^\mu \, d\omega_p (iJ_{1p\lambda q}^*)(iJ_{2p\lambda q}). \tag{3-6.44}$$

The flux per particle, $2p^\mu \, d\omega_p$, is clearly of universal applicability. As we have mentioned, it is fixed by the normalization condition, in the manner made precise by (3-6.25).

One might think that the ambiguity of current expressions is an aspect of second-order Lagrange functions, with their options in arranging two derivatives, and would disappear if first-order Lagrange functions were adopted. To dispel this illusion it suffices to consider spin 2 charged particles, where two first-order forms are available. From the Lagrange function (3-5.34) we obtain

$$\begin{aligned}
j^\lambda &= G^{\lambda\mu\nu} iq\phi_{\mu\nu} + G_\mu iq\phi^{\lambda\mu} - G^\lambda iq\phi \\
&\to \partial^\lambda \phi^{\mu\nu} iq\phi_{\mu\nu} - \partial^\nu(\phi^{\lambda\mu} iq\phi_{\mu\nu}), \tag{3-6.45}
\end{aligned}$$

with the latter form applicable in source-free regions, while (3-5.41) gives

$$\begin{aligned}
j^\lambda &= -H^{\mu\nu\lambda} iq\phi_{\mu\nu} + H_\mu iq\phi^{\lambda\mu} - \tfrac{1}{2}(H^\lambda - {}^\lambda H) iq\phi \\
&\to \partial^\lambda \phi^{\mu\nu} iq\phi_{\mu\nu} - 2\partial^\nu(\phi^{\lambda\mu} iq\phi_{\mu\nu}). \tag{3-6.46}
\end{aligned}$$

The application of the phase transformation procedure to the spin $\tfrac{1}{2}$ action (3-5.50, 51) produces

$$j^\mu(x) = \tfrac{1}{2}\psi(x)\gamma^0\gamma^\mu q\psi(x), \tag{3-6.47}$$

together with

$$\partial_\mu j^\mu(x) = \psi(x)\gamma^0 iq\eta(x), \tag{3-6.48}$$

which is also a consequence of the field equations. The field in the interval between the causally separated sources η_1 and η_2 is [Eq. (3–2.14)]

$$\psi(x) = \sum_{p\sigma q} [\psi_{p\sigma q}(x) i\eta_{2p\sigma q} + \psi_{p\sigma q}(x)^* i\eta_{1p\sigma q}^*], \qquad (3\text{–}6.49)$$

where, it is recalled,

$$\psi_{p\sigma q}(x) = (2m\, d\omega_p)^{1/2} u_{p\sigma q} e^{ipx}, \qquad (3\text{–}6.50)$$

$$u_{p\sigma q}^* \gamma^0 u_{p\sigma q} = 1.$$

The contribution to j^μ associated with a single-particle state is

$$j^\mu(x) = q\psi_{p\sigma q}(x)^* \gamma^0 \gamma^\mu \psi_{p\sigma q}(x)(i\eta_{1p\sigma q}^*)(i\eta_{2p\sigma q}), \qquad (3\text{–}6.51)$$

which identifies the flux per particle:

$$\psi_{p\sigma q}(x)^* \gamma^0 \gamma^\mu \psi_{p\sigma q}(x) = 2m\, d\omega_p u_{p\sigma q}^* \gamma^0 \gamma^\mu u_{p\sigma q}$$

$$= 2p^\mu\, d\omega_p. \qquad (3\text{–}6.52)$$

This expected result expresses the evaluation

$$u_{p\sigma q}^* \gamma^0 \gamma^\mu u_{p\sigma q} = p^\mu/m, \qquad (3\text{–}6.53)$$

which has the following derivation, based on the normalization of Eq. (3–6.50) [cf. Eq. (2–6.97)]:

$$0 = u_{p\sigma q}^* \gamma^0 \{\gamma p + m, \gamma^\mu\} u_{p\sigma q}$$

$$= -2p^\mu + 2m u_{p\sigma q}^* \gamma^0 \gamma^\mu u_{p\sigma q}. \qquad (3\text{–}6.54)$$

As an alternative to computing charges by surface integration of j^μ, one can consider the volume integral of $\partial_\mu j^\mu$, extended over the region occupied by η_2. The field associated with the latter source does not contribute to this calculation,

$$\int (dx)(dx')\eta_2(x) iq\gamma^0 G_+(x - x')\eta_2(x') = 0, \qquad (3\text{–}6.55)$$

since the antisymmetrical matrix q removes the match between the anticommutativity of the sources and the antisymmetry of the kernel $\gamma^0 G_+(x - x')$. Accordingly,

$$\int d\sigma_\mu j^\mu = \int (dx) \left[\sum_{p\sigma q'} \psi_{p\sigma q'}(x)^* i\eta_{1p\sigma q'}^* \right] \gamma^0 iq\eta_2(x)$$

$$= \sum_{p\sigma q} (i\eta_{1p\sigma q}^*) q (i\eta_{2p\sigma q}), \qquad (3\text{–}6.56)$$

which exhibits the charges $q = \pm 1$ assigned to the single-particle states. Although this interpretation is quite clear, it may be helpful to give a more formal discussion, based upon the analogue of the relation (3–6.19), or

$$\left[\int d\sigma_\mu j^\mu \right] \langle 0_+|0_-\rangle^\eta = \sum \langle 0_+|\{n\}\rangle^{\eta_1} Q \langle \{n\}|0_-\rangle^{\eta_2}. \qquad (3\text{–}6.57)$$

To analyze the left-hand side:

$$\sum \langle 0_+|\{n\}\rangle^{n_1}\left[\int d\sigma_\mu j^\mu\right]\langle\{n\}|0_-\rangle^{n_2}, \tag{3-6.58}$$

we have only to note that

$$\langle 0_+|\{n\}\rangle^{n_1}in_{1a}^*in_{2a}\langle\{n\}|0_-\rangle^{n_2} = \begin{cases} n_a = 1: & 0 \\ n_a = 0: & \langle 0_+|\{n+1_a\}\rangle^{n_1}\langle\{n+1_a\}|0_-\rangle^{n_2}, \end{cases} \tag{3-6.59}$$

and the charge values are identified as

$$Q(\{n\}) = \sum_{p\sigma q} qn_{p\sigma q}; \qquad n_{p\sigma q} = 0, 1. \tag{3-6.60}$$

It has been emphasized that the arbitrariness in assigning a charge flux vector is not just the possibility of adding any $\partial_\nu m^{\mu\nu}$ term, but is inherent in the existence of alternative Lagrange function descriptions of the given physical system. This is true of spin $\frac{1}{2}$ as well. A third-rank spinor that is antisymmetrical in a pair of indices gives a spin $\frac{1}{2}$ description. The current deduced from (3–5.73) is

$$j^\mu = \tfrac{1}{2}\psi\gamma(\tfrac{1}{3}\sum\gamma^\mu_\alpha)q\psi + \sum_{\alpha<\beta}\psi\gamma\tfrac{1}{2}(\gamma^\mu_\alpha - \gamma^\mu_\beta)q\psi_{[\alpha\beta]} + 3\psi_{[23]}\gamma\gamma^\mu_1 q\psi_{[23]} + \cdots. \tag{3-6.61}$$

Outside all the sources the auxiliary fields $\psi_{[\alpha\beta]}$ vanish and

$$(\gamma^\mu_\alpha - \gamma^\mu_\beta)\partial_\mu\psi = 0. \tag{3-6.62}$$

An appropriate expression for the third-rank spinor is written as

$$2^{3/2}\Psi_{\zeta_1\zeta_2\zeta_3} = (i\gamma_5\gamma^0)_{\zeta_1\zeta_2}\Psi_{\zeta_3} + (1/m)(\gamma_5\gamma^\mu\gamma^0)_{\zeta_1\zeta_2}\partial_\mu\Psi_{\zeta_3}. \tag{3-6.63}$$

It is antisymmetrical in ζ_1 and ζ_2, and the properties (3–6.62), as illustrated by

$$(i\gamma^\mu\gamma_5\gamma^0)_{\zeta_1\zeta_2}\partial_\mu\Psi + (1/m)(\gamma^\mu\gamma_5\gamma^\nu\gamma^0)_{\zeta_1\zeta_2}\partial_\mu\partial_\nu\Psi$$
$$= (i\gamma_5\gamma^0)_{\zeta_1\zeta_2}\gamma^\mu\partial_\mu\Psi + (1/m)(\gamma_5\gamma^\mu\gamma^0)_{\zeta_1\zeta_2}\partial_\mu\gamma^\nu\partial_\nu\Psi \tag{3-6.64}$$

with matrix notation restored for the third spinor index, are satisfied since

$$[\gamma^\mu(1/i)\partial_\mu + m]\Psi = 0. \tag{3-6.65}$$

The current that is derived from the first term of (3–6.61) by inserting (3–6.63) is

$$j^\mu = \frac{2}{3}\frac{1}{4m}\left(\Psi\gamma^0 q\left(\frac{1}{i}\right)\partial^\mu\Psi - \left(\frac{1}{i}\right)\partial^\mu\Psi\gamma^0 q\Psi\right)$$
$$+ \left(\frac{1}{3}\right)\frac{1}{2}\Psi\gamma^0\gamma^\mu q\Psi - \frac{1}{3}\frac{1}{4m^2}\partial^2\left(\frac{1}{2}\Psi\gamma^0\gamma^\mu q\Psi\right). \tag{3-6.66}$$

With the aid of the identity

$$0 = \Psi\gamma^0\gamma^\mu q[\gamma^\nu(1/i)\partial_\nu + m]\Psi + [-(1/i)\partial_\nu\Psi\gamma^0\gamma^\nu + m\Psi\gamma^0]\gamma^\mu q\Psi$$
$$= 2m\Psi\gamma^0\gamma^\mu q\Psi - (\Psi\gamma^0 q(1/i)\partial^\mu\Psi - (1/i)\partial^\mu\Psi\gamma^0 q\Psi) - \partial_\nu(\Psi\gamma^0\sigma^{\mu\nu}q\Psi),$$
$$(3\text{–}6.67)$$

it is presented as

$$j^\mu = \tfrac{1}{2}\Psi\gamma^0\gamma^\mu q\Psi + \partial_\nu m^{\mu\nu}, \qquad (3\text{–}6.68)$$

where

$$m_{\mu\nu} = -\frac{2}{3}\frac{1}{2m}\frac{1}{2}\Psi\gamma^0\sigma_{\mu\nu}q\Psi$$
$$+ \frac{1}{3}\frac{1}{(2m)^2}[\partial_\mu(\tfrac{1}{2}\Psi\gamma^0\gamma_\nu q\Psi) - \partial_\nu(\tfrac{1}{2}\Psi\gamma^0\gamma_\mu q\Psi)]. \qquad (3\text{–}6.69)$$

The $\sigma_{\mu\nu}$ term is the only such structure that does not involve coordinate derivatives. Analogous but more elaborate comparisons can be made between alternative descriptions for particles with spins $\tfrac{3}{2}, \tfrac{5}{2}, \ldots$.

The technique of variable phase transformation has been used to give a more detailed space-time description for the average charge distribution. It also supplies such information about charge fluctuations. We shall illustrate this for spinless particles, confining the discussion to the simplest measure of fluctuations. Consider, then, the time cycle vacuum amplitude with the sources

$$K_{(+)}(x) = e^{iq\varphi}K(x), \qquad K_{(-)}(x) = K(x), \qquad (3\text{–}6.70)$$

which is

$$\langle 0_-|0_-\rangle^{K(-),K(+)} = \sum \langle 0_-|\{n\}\rangle^K e^{iQ\varphi}\langle\{n\}|0_-\rangle^K \qquad (3\text{–}6.71)$$

or

$$\exp[iW(K_{(-)}, K_{(+)})] = \langle e^{iQ\varphi}\rangle_0^K. \qquad (3\text{–}6.72)$$

An infinitesimal variation of the phase constant φ gives

$$(\partial/\partial\varphi)W = \int d\sigma_\mu j^\mu_{(+)} = \langle Qe^{iQ\varphi}\rangle/\langle e^{iQ\varphi}\rangle. \qquad (3\text{–}6.73)$$

This equation, with $\varphi = 0$, has been discussed   Now let us differentiate once, before setting $\varphi = 0$, with the consequence that

$$\langle Q^2\rangle - \langle Q\rangle^2 = \int d\sigma_\mu j^\mu_{\text{fluct.}}, \qquad (3\text{–}6.74)$$

where

$$j^\mu_{\text{fluct.}}(x) = (\partial/\partial i\varphi)j^\mu_{(+)}(x)\Big]_{\varphi=0} \qquad (3\text{–}6.75)$$

is a charge fluctuation flux vector.

In evaluating this vector from

$$j^\mu_{(+)}(x) = \partial^\mu\phi_{(+)}(x)iq\phi_{(+)}(x), \qquad (3\text{–}6.76)$$

one of the $\phi_{(+)}$ fields is taken at $\varphi = 0$, and becomes $\phi_{\text{ret.}}(x)$. For the other we use Eq. (3–1.19), appropriate to the vacuum initial state, and obtain

$$(\partial/\partial i\varphi)\phi_{(+)}(x)\Big]_{\varphi=0} = q\phi(x) \qquad (3\text{–}6.77)$$

with

$$\phi(x) = \int (dx')\Delta_+(x - x')K(x'). \qquad (3\text{–}6.78)$$

In terms of these fields we have

$$j^{\mu}_{\text{fluct.}}(x) = \phi_{\text{ret.}}(x)(1/i)\partial^{\mu}\phi(x) - \phi(x)(1/i)\partial^{\mu}\phi_{\text{ret.}}(x), \qquad (3\text{–}6.79)$$

which uses the charge property

$$q^2 = 1. \qquad (3\text{–}6.80)$$

It is helpful to decompose $\phi(x)$ into real and imaginary parts:

$$\phi(x) = \phi_{\text{real}}(x) + i\phi_{\text{imag.}}(x), \qquad (3\text{–}6.81)$$

where, according to the relation

$$\Delta_+(x - x') + \Delta_-(x - x') = \Delta_{\text{ret.}}(x - x') + \Delta_{\text{adv.}}(x - x'), \qquad (3\text{–}6.82)$$

the real component is

$$\phi_{\text{real}}(x) = \tfrac{1}{2}[\phi_{\text{ret.}}(x) + \phi_{\text{adv.}}(x)]. \qquad (3\text{–}6.83)$$

In the circumstances to which (3–6.79) refers, which are made explicit by the appearance of the function $\phi_{\text{ret.}}(x)$, the advanced field vanishes. And the contribution to ϕ of $\tfrac{1}{2}\phi_{\text{ret.}}$ cancels, leaving the real form

$$j^{\mu}_{\text{fluct.}}(x) = \phi_{\text{ret.}}(x)\partial^{\mu}\phi_{\text{imag.}}(x) - \phi_{\text{imag.}}(x)\partial^{\mu}\phi_{\text{ret.}}(x), \qquad (3\text{–}6.84)$$

in which

$$\phi_{\text{imag.}}(x) = \int (dx')\tfrac{1}{2}[\Delta^{(+)}(x - x') + \Delta^{(-)}(x - x')]K(x'). \qquad (3\text{–}6.85)$$

Since the latter is a solution of the homogeneous field equation, one finds that

$$\partial_\mu j^{\mu}_{\text{fluct.}}(x) = \phi_{\text{imag.}}(x)K(x). \qquad (3\text{–}6.86)$$

A calculation of the total charge fluctuation can be performed by integrating over the source:

$$\int d\sigma_\mu j^{\mu}_{\text{fluct.}} = \int (dx)\partial_\mu j^{\mu}_{\text{fluct.}} = \int (dx)\phi_{\text{imag.}}(x)K(x)$$

$$= \int (dx)(dx')K(x)\Delta^{(+)}(x - x')K(x') = \sum_{pq}|K_{pq}|^2. \qquad (3\text{–}6.87)$$

The latter is the expected total number of particles emitted by the source, and this fluctuation formula

$$\langle (Q - \langle Q \rangle)^2 \rangle_0^K = \langle N_+ + N_- \rangle_0^K \qquad (3\text{–}6.88)$$

is contained in the more general statement (2–2.123). In effect, $j^{\mu}_{\text{fluct.}}$ is a particle flux vector.

3-7 INVARIANCE TRANSFORMATIONS AND FLUXES.
MECHANICAL PROPERTIES

The action for spinless particles is invariant under a rigid source translation,

$$\bar{K}(x) = K(x + X). \tag{3-7.1}$$

This is expressed, in (3–6.1), by the existence of the compensating field transformation

$$\bar{\phi}(x) = \phi(x + X). \tag{3-7.2}$$

The infinitesimal versions of these transformations are

$$\delta K(x) = \delta X^\nu \partial_\nu K(x), \qquad \delta\phi(x) = \delta X^\nu \partial_\nu \phi(x). \tag{3-7.3}$$

The following are the proposed generalizations of these expressions when δX^ν becomes an arbitrary function of position, $\delta x^\nu(x)$,

$$\delta K(x) = \partial_\nu[\delta x^\nu(x) K(x)], \qquad \delta\phi(x) = \delta x^\nu(x) \partial_\nu \phi(x). \tag{3-7.4}$$

The distinction between these forms is necessary to maintain the invariance of the source term under the compensating source and field variations:

$$\delta_{K,\phi} \int (dx) K(x)\phi(x) = \int (dx)\{\phi \partial_\nu[\delta x^\nu K] + \partial_\nu \phi \, \delta x^\nu K\}$$

$$= \int (dx)\partial_\nu[\phi \, \delta x^\nu K]$$

$$= 0. \tag{3-7.5}$$

The response of the Lagrange function to the field variation of (3–7.4) is

$$\delta \mathcal{L} = \delta x^\nu \partial_\nu \mathcal{L} - \partial^\mu \phi \partial^\nu \phi \partial_\mu \, \delta x_\nu; \tag{3-7.6}$$

only the first term on the right would appear for a rigid translation. An equivalent presentation is

$$\delta \mathcal{L} = \partial_\nu(\delta x^\nu \mathcal{L}) - t^{\mu\nu} \partial_\mu \, \delta x_\nu, \tag{3-7.7}$$

where

$$t^{\mu\nu}(x) = \partial^\mu \phi(x)\partial^\nu \phi(x) + g^{\mu\nu}\mathcal{L}(\phi(x)) = t^{\nu\mu}(x). \tag{3-7.8}$$

Let us also note the relation

$$t = g_{\mu\nu}t^{\mu\nu} = \partial^\mu \phi \partial_\mu \phi + 4\mathcal{L} = -\partial^\mu \phi \partial_\mu \phi - 2m^2\phi^2$$

$$= -\partial^2(\tfrac{1}{2}\phi^2) - m^2\phi^2 - K\phi, \tag{3-7.9}$$

where the final form involves the use of the field equation. The change induced in the action by the source variation of (3–7.4) is computed alternatively as

$$\delta W(K) = \int (dx) \, \delta K(x)\phi(x) = -\int (dx) K(x) \, \delta\phi(x)$$

$$= -\int (dx) K(x)\partial_\nu \phi(x) \, \delta x^\nu(x), \tag{3-7.10}$$

and

$$\delta W(K) = -\int (dx) t^{\mu\nu}(x) \partial_\mu \, \delta x_\nu(x). \qquad (3\text{-}7.11)$$

The comparison of the two versions implies that

$$\partial_\mu t^{\mu\nu}(x) = -K(x) \partial^\nu \phi(x), \qquad (3\text{-}7.12)$$

which can be verified directly. This is the local statement of a vectorial conservation law when the right-hand side is zero, which is true in source-free regions.

In the causal circumstance indicated by $K = K_1 + K_2$, a rigid infinitesimal displacement of K_2 induces

$$\langle 0_+|0_-\rangle^K = \sum \langle 0_+|\{n\}\rangle^{K_1} \langle \{n\}|0_-\rangle^{K_2}$$
$$\rightarrow \sum \langle 0_+|\{n\}\rangle^{K_1} [1 + i\,\delta X^\mu P_\mu] \langle \{n\}|0_-\rangle^{K_2}, \qquad (3\text{-}7.13)$$

where

$$P_\mu(\{n\}) = \sum_{pq} p_\mu n_{pq}, \qquad (3\text{-}7.14)$$

which can be written

$$\delta \langle 0_+|0_-\rangle^K = i\,\delta X^\mu \sum \langle 0_+|\{n\}\rangle^{K_1} P_\mu(\{n\}|0_-\rangle^{K_2}. \qquad (3\text{-}7.15)$$

Similarly, in the time cycle situation the displacement of $K_{(+)}$ and its subsequent identification with $K_{(-)} = K$ gives

$$\delta \langle 0_-|0_-\rangle^{K(-), K(+)} = i\,\delta X^\mu \sum \langle 0_-|\{n\}\rangle^K P_\mu \langle \{n\}|0_-\rangle^K$$
$$= i\,\delta X^\mu \langle P_\mu\rangle^K. \qquad (3\text{-}7.16)$$

To compare (3–7.15) with (3–7.11), we separate K_1 and K_2 by a space-like surface and let δx^ν vanish in the future of this surface and be the constant δX^ν in its past. That gives

$$\delta W = \delta X_\nu \int d\sigma_\mu t^{\mu\nu}, \qquad (3\text{-}7.17)$$

which is the identification

$$\int d\sigma_\mu t^{\mu\nu} = \frac{\sum \langle 0_+|\{n\}\rangle^{K_1} P^\nu \langle \{n\}|0_-\rangle^{K_2}}{\langle 0_+|0_-\rangle^K}. \qquad (3\text{-}7.18)$$

The analogous equation of the time cycle description is

$$\int d\sigma_\mu t^{\mu\nu} = \langle P^\nu\rangle_0^K, \qquad (3\text{-}7.19)$$

where $t^{\mu\nu}(x)$, which is computed from the real retarded fields, is also real.

The distribution and flow of energy and momentum is described by $t^{\mu\nu}(x)$. It is the energy-momentum flux vector, or stress tensor. Let us evaluate it for the state of a single particle, chosen to be neutral, for simplicity. In the region between the causally separated sources, the field that is associated with a particle of momentum p is

$$\phi(x) = (d\omega_p)^{1/2} e^{ipx} iK_{2p} + (d\omega_p)^{1/2} e^{-ipx} iK_{1p}^*. \qquad (3\text{-}7.20)$$

If only this excitation is considered,

$$t^{\mu\nu}(x) = 2p^{\mu}p^{\nu}\,d\omega_p(iK^*_{1p})(iK_{2p}) - d\omega_p(p^{\mu}p^{\nu} + m^2 g^{\mu\nu})$$
$$\times [e^{-2ipx}(iK^*_{1p})^2 + e^{2ipx}(iK_{2p})^2], \quad (3\text{-}7.21)$$

which obeys the conservation law $\partial_\mu t^{\mu\nu} = 0$. In any consideration involving time or space averages, where the corresponding components of p^μ set the scale, the oscillatory terms of (3-7.21) can be ignored. The first term exhibits the anticipated factors: the representation of the emission and absorption acts by iK_{2p}, iK^*_{1p}, the particle flux factor $2p^\mu\,d\omega_p$, and the measure of the quantity being transported—the energy-momentum vector p^ν. In the analogous discussion using retarded fields, we have

$$\phi_{\text{ret.}}(x) = (d\omega_p)^{1/2}e^{ipx}iK_p - (d\omega_p)^{1/2}e^{-ipx}iK^*_p \quad (3\text{-}7.22)$$

and

$$t^{\mu\nu}(x) = 2p^{\mu}p^{\nu}\,d\omega_p|K_p|^2 + d\omega_p(p^{\mu}p^{\nu} + m^2 g^{\mu\nu})[e^{-2ipx}(K^*_p)^2 + e^{2ipx}(K_p)^2]. \quad (3\text{-}7.23)$$

The disappearance of all interference terms through integration can be verified, in the causal situation for example, by using Eq. (3-7.12) to obtain

$$\int d\sigma_\mu t^{\mu\nu} = -\int (dx)K_2(x)\partial^\nu\phi(x), \quad (3\text{-}7.24)$$

where, in the region occupied by the emission source K_2,

$$\phi(x) = \int (dx')\Delta_+(x - x')K_2(x') + i\int (dx')\Delta^{(-)}(x - x')K_1(x'). \quad (3\text{-}7.25)$$

The first part of the field, associated with K_2 itself, does not contribute:

$$\int (dx)(dx')K_2(x)\partial_\nu\Delta_+(x - x')K_2(x') = 0, \quad (3\text{-}7.26)$$

since the gradient of the symmetrical function $\Delta_+(x - x')$ is an antisymmetrical function. Accordingly, with a slight rearrangement we get

$$\int d\sigma_\mu t^{\mu\nu} = \int (dx)(dx')iK_1(x)(1/i)\partial^\nu\Delta^{(+)}(x - x')iK_2(x')$$
$$= \sum_p (iK^*_{1p})p^\nu(iK_{2p}), \quad (3\text{-}7.27)$$

which exhibits the physical property p^ν that is carried by individual particles between the initial emission and the final absorption acts. Perhaps we should also note that

$$\langle 0_+|\{n\}\rangle^{K_1}iK^*_{1p}iK_{2p}\langle\{n\}|0_-\rangle^{K_2} = \langle 0_+|\{n + 1_p\}\rangle^{K_1}(n_p + 1)\langle\{n + 1_p\}|0_-\rangle^{K_2}, \quad (3\text{-}7.28)$$

which shows how (3-7.18) is used to produce the general energy-momentum evaluation (for neutral particles)

$$P^\nu(\{n\}) = \sum_p p^\nu n_p. \qquad (3\text{-}7.29)$$

The analogue of (3–7.27) for the time cycle description is the expectation value

$$\int d\sigma_\mu t^{\mu\nu} = \sum_p p^\nu |K_p|^2 = \langle P^\nu \rangle_0^K. \qquad (3\text{-}7.30)$$

Energy-momentum fluctuations can also be given space-time localization, in the manner discussed for charge, but we shall not discuss the details here.

Spinless particles have an alternative characterization in the action (3–5.15), which uses scalar and vector fields and sources. The derivatives of the field variation (3–7.4),

$$\delta(\partial_\mu \phi) = \delta x^\nu \partial_\nu(\partial_\mu \phi) + (\partial_\nu \phi)\partial_\mu \delta x^\nu, \qquad (3\text{-}7.31)$$

provide a model for the response of vector fields to coordinate-dependent displacements:

$$\delta \phi_\mu(x) = \delta x^\nu(x)\partial_\nu \phi_\mu(x) + \phi_\nu(x)\partial_\mu \delta x^\nu(x). \qquad (3\text{-}7.32)$$

The corresponding source variation is that required to maintain the invariance of $\int(dx)K^\mu \phi_\mu$, namely

$$\delta K^\mu(x) = \partial_\nu[\delta x^\nu(x)K^\mu(x)] - K^\nu(x)\partial_\nu \delta x^\mu(x). \qquad (3\text{-}7.33)$$

Notice that the distinction between the two forms disappears for a rigid rotation, where

$$\partial_\mu \delta x_\nu = -\partial_\nu \delta x_\mu = \delta\omega_{\mu\nu}. \qquad (3\text{-}7.34)$$

The response of the Lagrange function

$$\mathcal{L} = -\phi^\mu \partial_\mu \phi + \tfrac{1}{2}(\phi^\mu \phi_\mu - m^2 \phi^2) \qquad (3\text{-}7.35)$$

is

$$\begin{aligned}\delta\mathcal{L} &= \delta x^\nu \partial_\nu \mathcal{L} + (\phi^\mu \phi^\nu - \phi^\mu \partial^\nu \phi - \phi^\nu \partial^\mu \phi)\partial_\mu \delta x_\nu \\ &= \partial_\nu(\delta x^\nu \mathcal{L}) - t^{\mu\nu}\partial_\mu \delta x_\nu,\end{aligned} \qquad (3\text{-}7.36)$$

with

$$t^{\mu\nu} = \phi^\mu \partial^\nu \phi + \phi^\nu \partial^\mu \phi - \phi^\mu \phi^\nu + g^{\mu\nu}\mathcal{L} = t^{\nu\mu}, \qquad (3\text{-}7.37)$$

and

$$\delta W = -\int(dx)t^{\mu\nu}\partial_\mu \delta x_\nu. \qquad (3\text{-}7.38)$$

We also have

$$\partial_\mu t^{\mu\nu} = -K\partial^\nu \phi + (\partial_\mu K^\mu)\phi^\nu + K_\mu(\partial^\mu \phi^\nu - \partial^\nu \phi^\mu), \qquad (3\text{-}7.39)$$

which can be given other forms with the aid of the field equations. The only field-dependent term on the right-hand side can be written as $-\partial^\nu \phi$, multiplied by the effective source $K - \partial_\mu K^\mu$. In the absence of the vectorial source K_μ, the two versions of the stress tensor coincide.

The Lagrange functions (3–5.19, 28) for unit spin particles contain vector fields, and their derivatives in the curl combination. Using the latter as the model for an antisymmetrical tensor field $G_{\mu\nu}$, we infer by differentiation of (3–7.32) that

$$\delta G_{\mu\nu}(x) = \delta x^\lambda(x)\partial_\lambda G_{\mu\nu}(x) + G_{\lambda\nu}(x)\partial_\mu \, \delta x^\lambda(x) + G_{\mu\lambda}(x)\partial_\nu \, \delta x^\lambda(x). \quad (3\text{–}7.40)$$

When tensor field sources are introduced, their infinitesimal variation is

$$\delta M^{\mu\nu} = \partial_\lambda(\delta x^\lambda M^{\mu\nu}) - M^{\lambda\nu}\partial_\lambda \, \delta x^\mu - M^{\mu\lambda}\partial_\lambda \, \delta x^\nu. \quad (3\text{–}7.41)$$

But first we consider

$$\mathcal{L} = -\tfrac{1}{4}G^{\mu\nu}G_{\mu\nu} - \tfrac{1}{2}m^2\phi^\mu\phi_\mu, \qquad G_{\mu\nu} = \partial_\mu\phi_\nu - \partial_\nu\phi_\mu, \quad (3\text{–}7.42)$$

which gives

$$\begin{aligned}
\delta\mathcal{L} &= \delta x^\nu\partial_\nu\mathcal{L} - G^{\mu\lambda}G^\nu{}_\lambda\partial_\mu \, \delta x_\nu - m^2\phi^\mu\phi^\nu\partial_\mu \, \delta x_\nu \\
&= \partial_\nu(\delta x^\nu\mathcal{L}) - t^{\mu\nu}\partial_\mu \, \delta x_\nu
\end{aligned} \quad (3\text{–}7.43)$$

and

$$\delta W(J) = -\int(dx)t^{\mu\nu}\partial_\mu \, \delta x_\nu. \quad (3\text{–}7.44)$$

Here

$$\begin{aligned}
t^{\mu\nu} &= G^{\mu\lambda}G^\nu{}_\lambda + m^2\phi^\mu\phi^\nu + g^{\mu\nu}\mathcal{L} \\
&= t^{\nu\mu},
\end{aligned} \quad (3\text{–}7.45)$$

and we note that

$$\begin{aligned}
t &= G^{\mu\nu}G_{\mu\nu} + m^2\phi^\mu\phi_\mu + 4\mathcal{L} \\
&= -m^2\phi^\mu\phi_\mu.
\end{aligned} \quad (3\text{–}7.46)$$

The direct evaluation of the source variation effect is

$$\delta W(J) = \int(dx) \, \delta J^\mu\phi_\mu, \qquad \delta J^\mu = \partial_\nu(\delta x^\nu J^\mu) - J^\nu\partial_\nu \, \delta x^\mu, \quad (3\text{–}7.47)$$

and the comparison shows that

$$\partial_\mu t^{\mu\nu} = J_\mu G^{\mu\nu} + (\partial_\mu J^\mu)\phi^\nu, \quad (3\text{–}7.48)$$

or, more immediately,

$$\partial_\mu t^{\mu\nu} = -J^\mu\partial^\nu\phi_\mu + \partial_\mu(J^\mu\phi^\nu), \quad (3\text{–}7.49)$$

where the last term does not contribute to the volume integrations that evaluate the total energy and momentum emitted by the source.

The alternative Lagrange function (3–5.23) differs from (3–5.19) by the divergence of a vector [Eqs. (3–5.24, 25)], and that remains true of the variations $\delta\mathcal{L}$. Such additional terms do not contribute to δW. But let us note that the relation

$$\begin{aligned}
0 &= \int(dx)\partial_\lambda[f^{\lambda\mu\nu}(x)\partial_\mu \, \delta x_\nu] \\
&= \int(dx)\partial_\lambda f^{\lambda\mu\nu}(x)\partial_\mu \, \delta x_\nu + \int(dx)f^{\lambda\mu\nu}(x)\partial_\lambda\partial_\mu \, \delta x_\nu,
\end{aligned} \quad (3\text{–}7.50)$$

with the last term annulled by the restriction

$$f^{\lambda\mu\nu}(x) = -f^{\mu\lambda\nu}(x), \quad (3\text{–}7.51)$$

implies an arbitrariness in the stress tensor, which is indicated by the combination

$$t^{\mu\nu}(x) - \partial_\lambda f^{\lambda\mu\nu}(x). \tag{3–7.52}$$

This arbitrariness disappears, however, if we insist upon a property which, thus far, has emerged automatically—the symmetry of the tensor $t^{\mu\nu}$. In order that the symmetry property continue to apply in (3–7.52) without recourse to differential identities, we demand that

$$f^{\lambda\mu\nu}(x) = f^{\lambda\nu\mu}(x). \tag{3–7.53}$$

But the combination of (3–7.51) and (3–7.53) has the following consequence:

$$
\begin{aligned}
f^{\lambda\mu\nu} &= f^{\lambda\nu\mu} = -f^{\nu\lambda\mu} = -f^{\nu\mu\lambda} = f^{\mu\nu\lambda} \\
&= f^{\mu\lambda\nu} = -f^{\lambda\mu\nu} = 0.
\end{aligned} \tag{3–7.54}
$$

Physical implications of the symmetry of the stress tensor will be considered later, in connection with the discussion of angular momentum.

Now we turn to the Lagrange function

$$\mathcal{L} = -\tfrac{1}{2}G^{\mu\nu}(\partial_\mu\phi_\nu - \partial_\nu\phi_\mu) + \tfrac{1}{4}G^{\mu\nu}G_{\mu\nu} - \tfrac{1}{2}m^2\phi^\mu\phi_\mu \tag{3–7.55}$$

and observe that it gives the stress tensor

$$
\begin{aligned}
t^{\mu\nu} &= t^{\nu\mu} \\
&= (\partial^\mu\phi^\lambda - \partial^\lambda\phi^\mu)G^\nu{}_\lambda + G^\mu{}_\lambda(\partial^\nu\phi^\lambda - \partial^\lambda\phi^\nu) - G^{\mu\lambda}G^\nu{}_\lambda + m^2\phi^\mu\phi^\nu + g^{\mu\nu}\mathcal{L},
\end{aligned} \tag{3–7.56}
$$

which obeys

$$
\begin{aligned}
\partial_\mu t^{\mu\nu} &= J_\mu(\partial^\mu\phi^\nu - \partial^\nu\phi^\mu) + (\partial_\mu J^\mu)\phi^\nu \\
&\quad - \tfrac{1}{2}M_{\lambda\mu}(\partial^\lambda G^{\mu\nu} + \partial^\mu G^{\nu\lambda} + \partial^\nu G^{\lambda\mu}) - (\partial^\lambda M_{\lambda\mu})G^{\mu\nu}. \tag{3–7.57}
\end{aligned}
$$

It remains true, however, that

$$
\begin{aligned}
t &= 2G^{\mu\nu}(\partial_\mu\phi_\nu - \partial_\nu\phi_\mu) - G^{\mu\nu}G_{\mu\nu} + m^2\phi^\mu\phi_\mu + 4\mathcal{L} \\
&= -m^2\phi^\mu\phi_\mu. \tag{3–7.58}
\end{aligned}
$$

According to the field equation

$$\partial_\mu\phi_\nu - \partial_\nu\phi_\mu = G_{\mu\nu} + M_{\mu\nu}, \tag{3–7.59}$$

we have

$$\partial_\lambda(G_{\mu\nu} + M_{\mu\nu}) + \partial_\mu(G_{\nu\lambda} + M_{\nu\lambda}) + \partial_\nu(G_{\lambda\mu} + M_{\lambda\mu}) = 0, \tag{3–7.60}$$

and all the field-dependent terms on the right-hand side of (3–7.57) can be expressed in terms of the effective vector source $J_\mu + \partial^\nu M_{\mu\nu}$. The specialization $M_{\mu\nu} = 0$ identifies the two versions of the stress tensor.

Another question about uniqueness presents itself. The generalization to arbitrary displacements could have emulated more closely rigid rotation behavior, so that (3–7.32) would have been written as

$$\delta\phi_\mu = \delta x^\nu \partial_\nu\phi_\mu + \phi'\tfrac{1}{2}(\partial_\mu \, \delta x_\nu - \partial_\nu \, \delta x_\mu), \tag{3–7.61}$$

with the associated vector source variation

$$\delta J^\mu = \partial_\nu(\delta x^\nu J^\mu) + J^\nu \tfrac{1}{2}(\partial_\mu \, \delta x_\nu - \partial_\nu \, \delta x_\mu).\tag{3-7.62}$$

Indeed, the distinction between field and source variations disappears completely were we to adopt

$$\delta\phi_\mu = \tfrac{1}{2}\,\delta x^\nu \partial_\nu \phi_\mu + \tfrac{1}{2}\partial_\nu(\delta x^\nu \phi_\mu) + \phi^\nu \tfrac{1}{2}(\partial_\mu \, \delta x_\nu - \partial_\nu \, \delta x_\mu).\tag{3-7.63}$$

How does this freedom to choose the form of the displacement-induced variations affect the identification of the stress tensor, defined generally by

$$\delta W = -\int(dx)t^{\mu\nu}\partial_\mu \, \delta x_\nu,\tag{3-7.64}$$

and presumably made unique by the requirement of symmetry? The various choices for $\delta\phi_\mu$ differ by terms proportional to

$$\tfrac{1}{2}(\partial_\mu \, \delta x_\nu + \partial_\nu \, \delta x_\mu),\tag{3-7.65}$$

which vanish for rigid translations and rotations. This tensor is a measure of the dilation produced by the displacement, and includes the scalar measure $\partial_\mu \, \delta x^\mu$. The same dilation tensor appears in (3–7.64), in consequence of the symmetry of the stress tensor. Consider, for definiteness, the effect of the additional field variation

$$\delta_{\text{dil.}}\phi_\mu = -\phi^\nu \tfrac{1}{2}(\partial_\mu \, \delta x_\nu + \partial_\nu \, \delta x_\mu)\tag{3-7.66}$$

on the action expression (3–5.18, 19). Since

$$\begin{aligned}\delta_{\text{dil.}}(\partial_\mu\phi_\nu - \partial_\nu\phi_\mu) &= -\partial_\mu\phi^\lambda \tfrac{1}{2}(\partial_\nu \, \delta x_\lambda + \partial_\lambda \, \delta x_\nu)\\ &+ \partial_\nu\phi^\lambda \tfrac{1}{2}(\partial_\mu \, \delta x_\lambda + \partial_\lambda \, \delta x_\mu)\\ &- \tfrac{1}{2}\phi^\lambda(\partial_\lambda\partial_\mu \, \delta x_\nu - \partial_\lambda\partial_\nu \, \delta x_\mu)\end{aligned}\tag{3-7.67}$$

contains second derivatives of the coordinate displacements, a partial integration is required to attain the form of (3–7.64). This introduces second derivatives of the fields, and the field equations will be called upon to eliminate them. The initial choice of field variations was such as to obviate the need for any application of the field equations. The dilation dependent variations can be treated as a unit, and the stationary action principle invoked.

$$\begin{aligned}\delta_{\text{dil.}}W &= \int(dx)[-m^2\phi^\mu - \partial_\nu G^{\mu\nu}]\,\delta_{\text{dil.}}\phi_\mu\\ &= -\int(dx)J^\mu\delta_{\text{dil.}}\phi_\mu,\end{aligned}\tag{3-7.68}$$

which shows that the stress tensor is changed, by the addition of

$$-\tfrac{1}{2}(J^\mu\phi^\nu + J^\nu\phi^\mu).\tag{3-7.69}$$

This is also what is required in order to maintain consistency with the direct evaluation of the action variation:

$$\delta_{\text{dil}}.W(J) = \int (dx)\delta_{\text{dil}}.J^\mu \phi_\mu, \qquad \delta_{\text{dil}}.J^\mu = J_\nu \tfrac{1}{2}(\partial^\mu \, \delta x^\nu + \partial^\nu \, \delta x^\mu). \quad (3\text{--}7.70)$$

We conclude that stress tensors do have a degree of arbitrariness corresponding to the freedom in assigning the effect of general coordinate displacements, but this arbitrariness is confined to the interior of the sources. In the source-free regions where the stress tensor controls the flux of energy and momentum it would seem to be unique, according to the rules we have laid down for its evaluation.

Rearrangements that require the use of the field equations can also be accomplished by a different choice of field variation. We illustrate this in the spin 0 situation with

$$\delta\phi = \delta x^\nu \partial_\nu \phi + \tfrac{1}{4}\phi \partial_\nu \, \delta x^\nu, \qquad (3\text{--}7.71)$$

which gives

$$\delta\mathcal{L} = \delta x^\nu \partial_\nu \mathcal{L} - \partial^\mu \phi \partial^\nu \phi \partial_\mu \, \delta x_\nu - \tfrac{1}{4}\partial^\mu \phi \phi \partial_\mu \partial_\nu \, \delta x^\nu - \tfrac{1}{4}[\partial^\mu \phi \partial_\mu \phi + m^2 \phi^2]\partial_\nu \, \delta x^\nu, \qquad (3\text{--}7.72)$$

or

$$\delta\mathcal{L} = \partial_\nu [\delta x^\nu \mathcal{L} - \tfrac{1}{8}(\partial^\nu \phi^2)\partial_\lambda \, \delta x^\lambda] - t^{\mu\nu}\partial_\mu \, \delta x_\nu. \qquad (3\text{--}7.73)$$

This stress tensor is

$$t^{\mu\nu} = \partial^\mu \phi \partial^\nu \phi + g^{\mu\nu}[\tfrac{1}{2}\mathcal{L} - \tfrac{1}{8}\partial^2(\phi^2)]. \qquad (3\text{--}7.74)$$

The field equation asserts that

$$-\tfrac{1}{8}\partial^2(\phi^2) = \tfrac{1}{2}\mathcal{L} + \tfrac{1}{4}K\phi, \qquad (3\text{--}7.75)$$

and the original stress tensor reappears, with an added term that vanishes outside sources. But the direct use of (3–7.74) yields

$$\begin{aligned} t &= \partial^\mu \phi \partial_\mu \phi + 2\mathcal{L} - \tfrac{1}{2}\partial^2(\phi^2) \\ &= -m^2 \phi^2 - \tfrac{1}{2}\partial^2(\phi^2), \end{aligned} \qquad (3\text{--}7.76)$$

which, in source-free regions, is the result obtained in (3–7.9) through the use of the field equations.

The question of uniqueness intrudes again in this example. The rearrangement that connects (3–7.72) and (3–7.73) might have been handled differently:

$$-\tfrac{1}{8}\partial^\mu(\phi^2)\partial_\mu \partial_\nu \, \delta x^\nu = -\partial_\nu[\tfrac{1}{8}(\partial^\mu \phi^2)\partial_\mu \, \delta x^\nu] + \tfrac{1}{8}\partial^\mu \partial^\nu(\phi^2)\partial_\mu \, \delta x_\nu. \quad (3\text{--}7.77)$$

This would change the identification of the stress tensor by the added term

$$-\tfrac{1}{8}[\partial^\mu \partial^\nu(\phi^2) - g^{\mu\nu}\partial^2(\phi^2)]. \qquad (3\text{--}7.78)$$

Note that this symmetrical stress tensor contribution has a divergence that vanishes identically,

$$\partial_\mu(\partial^\mu \partial^\nu - g^{\mu\nu}\partial^2) = 0. \qquad (3\text{--}7.79)$$

Furthermore, it makes no contribution to the total energy and momentum, in virtue of the integration theorem (1-4.11),

$$\int d\sigma_\mu (\partial^\mu \partial^\nu - g^{\mu\nu}\partial^2)\phi^2 = \int (d\sigma_\mu \partial^\nu - d\sigma^\nu \partial_\mu)\partial^\mu \phi^2 = 0. \qquad (3\text{-}7.80)$$

This is true for angular momentum also. The added term is presented in the form of (3-7.52) by choosing

$$f^{\lambda\mu\nu} = \tfrac{1}{16}[g^{\lambda\mu}\partial^\nu + g^{\lambda\nu}\partial^\mu - 2g^{\mu\nu}\partial^\lambda]\phi^2. \qquad (3\text{-}7.81)$$

Observe that this expression is symmetrical in μ and ν, but does not have the antisymmetry that is stated in (3-7.51). The annulment of the last term in (3-7.50) comes about, instead, through the differential identity (it is (3-7.79) again)

$$\partial_\lambda \partial_\mu f^{\lambda\mu\nu} = 0. \qquad (3\text{-}7.82)$$

The rejection of stress tensor terms that involve differential identities is thus an essential aspect of the computation rules. But we are now going to see that stress tensors are quite analogous to charge flux vectors. Any current vector j^μ can be replaced by $j^\mu + \partial_\nu m^{\mu\nu}$, with arbitrary antisymmetrical $m^{\mu\nu}$. One might agree to reject such additional terms in studying a given Lagrange function, but the existence of different Lagrange functions for the same system, leading to current expressions that differ in just this way, shows that the arbitrariness is intrinsic.

The arbitrariness in symmetrical stress tensors is expressed by the possibility of replacing $t^{\mu\nu}(x)$ with

$$t^{\mu\nu}(x) + \partial_\kappa \partial_\lambda m^{\mu\nu,\kappa\lambda}(x), \qquad (3\text{-}7.83)$$

where

$$m^{\mu\nu,\kappa\lambda} = m^{\nu\mu,\kappa\lambda} = m^{\mu\nu,\lambda\kappa}, \qquad (3\text{-}7.84)$$

and the symmetry restriction

$$m^{\mu\nu,\kappa\lambda} + m^{\kappa\nu,\lambda\mu} + m^{\lambda\nu,\mu\kappa} = 0 \qquad (3\text{-}7.85)$$

assures that

$$\partial_\mu \partial_\kappa \partial_\lambda m^{\mu\nu,\kappa\lambda} = 0. \qquad (3\text{-}7.86)$$

It also guarantees that

$$\int d\sigma_\mu \partial_\kappa \partial_\lambda m^{\mu\nu,\kappa\lambda} = \tfrac{2}{3}\int (d\sigma_\mu \partial_\kappa - d\sigma_\kappa \partial_\mu)\partial_\lambda m^{\mu\nu,\kappa\lambda}$$

$$= 0 \qquad (3\text{-}7.87)$$

and it will be verified later that the total angular momentum is equally unaffected by the additional term. In the simple example provided by (3-7.78),

$$m^{\mu\nu,\kappa\lambda} = (g^{\mu\nu}g^{\kappa\lambda} - \tfrac{1}{2}g^{\kappa\nu}g^{\mu\lambda} - \tfrac{1}{2}g^{\kappa\mu}g^{\lambda\nu})\tfrac{1}{8}\phi^2. \qquad (3\text{-}7.88)$$

As in the discussion of currents, the existence of two sets of first-order field equations for spin 2 particles provides a valuable proving ground for uniqueness questions.

The tensor variation structure (3–7.40) is compatible with both antisymmetry and symmetry of the tensor. Accordingly, the symmetrical tensor field of spin 2 particles can be assigned the displacement variation

$$\delta\phi_{\mu\nu} = \delta x^\lambda \partial_\lambda \phi_{\mu\nu} + \phi_{\lambda\nu}\partial_\mu \, \delta x^\lambda + \phi_{\mu\lambda}\partial_\nu \, \delta x^\lambda. \tag{3-7.89}$$

Two significant derivative combinations of this tensor are given in (3–5.37) and (3–5.44). For the first, antisymmetrical combination we deduce

$$\delta G_{\lambda\mu\nu} = \delta x^\kappa \partial_\kappa G_{\lambda\mu\nu} + G_{\kappa\mu\nu}\partial_\lambda \, \delta x^\kappa + G_{\lambda\kappa\nu}\partial_\mu \, \delta x^\kappa + G_{\lambda\mu\kappa}\partial_\nu \, \delta x^\kappa$$
$$+ \phi_{\kappa\nu}\partial_\lambda\partial_\mu \, \delta x^\kappa - \phi_{\kappa\lambda}\partial_\mu\partial_\nu \, \delta x^\kappa, \tag{3-7.90}$$

while the symmetrical structure obeys

$$\delta H_{\mu\nu\lambda} = \delta x^\kappa \partial_\kappa H_{\mu\nu\lambda} + H_{\kappa\nu\lambda}\partial_\mu \, \delta x^\kappa + H_{\mu\kappa\lambda}\partial_\nu \, \delta x^\kappa + H_{\mu\nu\kappa}\partial_\lambda \, \delta x^\kappa + 2\phi_{\kappa\lambda}\partial_\mu\partial_\nu \, \delta x^\kappa. \tag{3-7.91}$$

Notice that second derivatives of the coordinate displacements appear here. First, we consider the action expression (3–5.33), simplified by setting the third-rank tensor source equal to zero, so that

$$G_{\lambda\mu\nu} = \partial_\lambda\phi_{\mu\nu} - \partial_\nu\phi_{\mu\lambda} \tag{3-7.92}$$

and

$$\mathcal{L} = -\tfrac{1}{4}G^{\lambda\mu\nu}G_{\lambda\mu\nu} + \tfrac{1}{2}G^\lambda G_\lambda - \tfrac{1}{2}m^2(\phi^{\mu\nu}\phi_{\mu\nu} - \phi^2). \tag{3-7.93}$$

Infinitesimal coordinate displacements induce

$$\delta\mathcal{L} = \partial_\nu(\delta x^\nu \mathcal{L}) - t^{\mu\nu'}\partial_\mu \, \delta x_\nu - F^{\lambda\mu\nu}\partial_\lambda\partial_\mu \, \delta x_\nu, \tag{3-7.94}$$

where

$$t^{\mu\nu'} = t^{\nu\mu'}$$
$$= G^{\mu\kappa\lambda}G^\nu{}_{\kappa\lambda} + \tfrac{1}{2}G^{\kappa\mu\lambda}G_\kappa{}^\nu{}_\lambda - G^\mu G^\nu - G_\lambda(G^{\lambda\mu\nu} + G^{\lambda\nu\mu})$$
$$+ 2m^2\phi^{\mu\lambda}\phi^\nu{}_\lambda - 2m^2\phi\phi^{\mu\nu} + g^{\mu\nu}\mathcal{L} \tag{3-7.95}$$

and

$$F^{\lambda\mu\nu} = G^{\lambda\mu\kappa}\phi^\nu{}_\kappa - G^\lambda\phi^{\mu\nu} + g^{\mu\lambda}G_\kappa\phi^{\kappa\nu}. \tag{3-7.96}$$

To extract the stress tensor we use the following identity:

$$F^{\lambda\mu\nu}\partial_\lambda\partial_\mu \, \delta x_\nu = \tfrac{1}{2}(F^{\lambda\mu\nu} + F^{\lambda\nu\mu})\partial_\lambda\partial_\mu \, \delta x_\nu + \tfrac{1}{2}(F^{\mu\lambda\nu} - F^{\mu\nu\lambda})\partial_\lambda(\partial_\mu \, \delta x_\nu + \partial_\nu \, \delta x_\mu). \tag{3-7.97}$$

A partial integration in the action variation shows that

$$t^{\mu\nu} = t^{\mu\nu'} - \partial_\lambda[F^{\lambda\mu\nu} + F^{\mu\lambda\nu} - F^{\mu\nu\lambda}]_{(\mu\nu)}, \tag{3-7.98}$$

where $(\mu\nu)$ indicates symmetrization in the indices μ and ν. According to the evaluation

$$\delta W(T) = \int (dx) \, \delta T^{\mu\nu}\phi_{\mu\nu} = -\int (dx) T^{\mu\nu} \, \delta\phi_{\mu\nu}, \tag{3-7.99}$$

this stress tensor must obey

$$\partial_\mu t^{\mu\nu} = -T^{\lambda\mu}\partial^\nu\phi_{\lambda\mu} + 2\partial_\lambda(T^\lambda{}_\mu\phi^{\mu\nu})$$
$$= T_{\lambda\mu}H^{\lambda\mu\nu} + 2(\partial_\lambda T^\lambda{}_\mu)\phi^{\mu\nu}. \tag{3-7.100}$$

When we confine our attention to source-free regions a considerable simplification occurs since all vector and scalar field combinations vanish in these circumstances:

$$t^{\mu\nu} = G^{\mu\kappa\lambda}G^\nu{}_{\kappa\lambda} + \tfrac{1}{2}G^{\kappa\mu\lambda}G_\kappa{}^\nu{}_\lambda + 2m^2\phi^{\mu\lambda}\phi^\nu{}_\lambda + g^{\mu\nu}\mathcal{L}$$
$$- \partial_\lambda[G^{\lambda\mu\kappa}\phi^\nu{}_\kappa + G^{\mu\lambda\kappa}\phi^\nu{}_\kappa - G^{\mu\nu\kappa}\phi^\lambda{}_\kappa]_{(\mu\nu)}, \tag{3-7.101}$$

and further reduction can be accomplished with the aid of the field equations

$$\partial_\lambda G^{\lambda\mu\nu} = m^2\phi^{\mu\nu}, \qquad \partial_\lambda G^{\mu\lambda\nu} = 0. \tag{3-7.102}$$

Thus,

$$\partial_\lambda(G^{\lambda\mu\kappa}\phi^\nu{}_\kappa) = m^2\phi^{\mu\kappa}\phi^\nu{}_\kappa + \tfrac{1}{2}G^{\lambda\mu\kappa}G_\lambda{}^\nu{}_\kappa \tag{3-7.103}$$

gives

$$t_G^{\mu\nu} = G^{\mu\kappa\lambda}G^\nu{}_{\kappa\lambda} + m^2\phi^{\mu\lambda}\phi^\nu{}_\lambda + g^{\mu\nu}\mathcal{L} - \partial_\lambda[G^{\mu\lambda\kappa}\phi^\nu{}_\kappa - G^{\mu\nu\kappa}\phi^\lambda{}_\kappa]_{(\mu\nu)}, \tag{3-7.104}$$

from which we obtain

$$t_G = -m^2\phi^{\mu\nu}\phi_{\mu\nu}. \tag{3-7.105}$$

In discussing the alternative action expression (3-5.40), we shall proceed directly to source-free conditions by omitting all scalar and vector fields in the Lagrange function. This is justified, despite the variation that is going to be performed, since all such fields occur in pairs, one factor of which continues to vanish after the infinitesimal coordinate displacement has been applied. Then, with

$$H_{\mu\nu\lambda} = \partial_\mu\phi_{\nu\lambda} + \partial_\nu\phi_{\mu\lambda} - \partial_\lambda\phi_{\mu\nu}, \tag{3-7.106}$$

we have

$$\mathcal{L} = \tfrac{1}{2}H^{\mu\nu\lambda}H_{\lambda\nu\mu} - \tfrac{1}{2}m^2\phi^{\mu\nu}\phi_{\mu\nu} \tag{3-7.107}$$

and

$$t_H^{\mu\nu} = -\tfrac{1}{2}(H^{\mu\kappa\lambda}H_{\kappa\lambda}{}^\nu + H^{\nu\kappa\lambda}H_{\kappa\lambda}{}^\mu) + g^{\mu\nu}\mathcal{L} + 2\partial_\lambda(H^{\kappa\lambda\mu}\phi_{\nu\kappa} - H^{\kappa\nu\mu}\phi_{\lambda\kappa})_{(\mu\nu)}, \tag{3-7.108}$$

in which we have also used the field equation

$$\partial_\lambda H^{\mu\nu\lambda} + m^2\phi^{\mu\nu} = 0. \tag{3-7.109}$$

The comparison of the two stress tensors now confirms that

$$t_H^{\mu\nu} - t_G^{\mu\nu} = \partial_\kappa\partial_\lambda m^{\mu\nu,\kappa\lambda}, \tag{3-7.110}$$

where

$$m_{\mu\nu,\kappa\lambda} = \tfrac{1}{2}[\phi_{\mu\kappa}\phi_{\nu\lambda} + \phi_{\mu\lambda}\phi_{\nu\kappa} - 2\phi_{\mu\nu}\phi_{\kappa\lambda}] + \tfrac{1}{2}[g_{\mu\nu}\phi_{\kappa\alpha}\phi_\lambda{}^\alpha + g_{\kappa\lambda}\phi_{\mu\alpha}\phi_\nu{}^\alpha]$$
$$- \tfrac{1}{4}[g_{\mu\kappa}\phi_{\nu\alpha}\phi_\lambda{}^\alpha + g_{\mu\lambda}\phi_{\nu\alpha}\phi_\kappa{}^\alpha + g_{\nu\kappa}\phi_{\mu\alpha}\phi_\lambda{}^\alpha + g_{\nu\lambda}\phi_{\mu\alpha}\phi_\kappa{}^\alpha]$$
$$= m_{\kappa\lambda,\mu\nu} \tag{3-7.111}$$

satisfies the symmetry requirements laid down for this tensor. A shorter derivation of the result is based on direct consideration of the difference in the Lagrange functions,

$$\mathcal{L}_H - \mathcal{L}_G = \tfrac{1}{2}\partial^\lambda\phi^{\mu\nu}\partial_\mu\phi_{\nu\lambda}. \tag{3-7.112}$$

Concerning the expression for the displacement induced change of source
and field, we note that the availability of additional terms that involve the
general dilation tensor (3–7.65) is strongly spin-dependent, if additional coordinate derivatives are eschewed. Matrices that act upon objects carrying
spin s are limited to the construction of tensors with rank $\leq 2s$. At least unit
spin is required in order to realize second-rank tensors, and only the scalar contraction of the dilation tensor is available to spins 0 and $\tfrac{1}{2}$. The spin 0 example
is illustrated in (3–7.71). We now discuss spin $\tfrac{1}{2}$. On referring to Eq. (2–5.11),
which describes the response of any source to coordinate transformations or the
equivalent rigid translation and rotation of the source, we are led, for the spin $\tfrac{1}{2}$
source $\eta(x)$, to the generalization

$$\delta\eta(x) = \partial_\nu(\delta x^\nu(x)\eta(x)) + i\tfrac{1}{4}\sigma^{\mu\nu}\eta(x)\partial_\mu\,\delta x_\nu(x). \tag{3-7.113}$$

The spin term only contains $\partial_\mu\,\delta x_\nu - \partial_\nu\,\delta x_\mu$. There is no symmetrical tensor
that can be devised from matrices, other than $g^{\mu\nu}$ multiplied by the unit matrix.
A specific multiple of $\partial_\nu\,\delta x^\nu$ appears, analogous to the scalar source response
of (3–7.4). The corresponding field variation, which is designed to leave intact
$\int(dx)\eta\tfrac{1}{2}\gamma^0\psi$, is

$$\delta\psi(x) = \delta x^\nu(x)\partial_\nu\psi(x) + i\tfrac{1}{4}\sigma^{\mu\nu}\psi(x)\partial_\mu\,\delta x_\nu(x). \tag{3-7.114}$$

The spin $\tfrac{1}{2}$ Lagrange function

$$\mathcal{L} = -\tfrac{1}{2}\psi\gamma^0[\gamma^\mu(1/i)\partial_\mu + m]\psi \tag{3-7.115}$$

has the following immediate response to the variation (3–7.114),

$$\begin{aligned}
\delta\mathcal{L} &= \delta x^\nu\partial_\nu\mathcal{L} + i\tfrac{1}{2}\psi\gamma^0\gamma^\mu\partial_\nu\psi\partial_\mu\,\delta x^\nu - \tfrac{1}{2}\psi\gamma^0\tfrac{1}{2}[\gamma^\lambda,\tfrac{1}{2}\sigma^{\mu\nu}]\partial_\lambda\psi\partial_\mu\,\delta x_\nu \\
&\quad - \tfrac{1}{2}\psi\gamma^0\gamma^\lambda\tfrac{1}{4}\sigma^{\mu\nu}\psi\partial_\lambda\partial_\mu\,\delta x_\nu.
\end{aligned}\tag{3-7.116}$$

But the symmetry of the second derivative picks out the matrix combination

$$\gamma^\lambda\sigma^{\mu\nu} + \gamma^\mu\sigma^{\lambda\nu} = i[\gamma^\lambda g^{\mu\nu} + \gamma^\mu g^{\lambda\nu} - 2\gamma^\nu g^{\mu\lambda}], \tag{3-7.117}$$

and, since the $\gamma^0\gamma^\mu$ are symmetrical matrices, the last term of $\delta\mathcal{L}$ vanishes.
Furthermore,

$$[\gamma^\lambda,\tfrac{1}{2}\sigma^{\mu\nu}] = i(g^{\lambda\nu}\gamma^\mu - g^{\lambda\mu}\gamma^\nu), \tag{3-7.118}$$

and we get

$$\delta\mathcal{L} = \partial_\nu(\delta x^\nu\mathcal{L}) - t^{\mu\nu}\partial_\mu\,\delta x_\nu, \tag{3-7.119}$$

with

$$t^{\mu\nu} = t^{\nu\mu} = \tfrac{1}{2}\psi\gamma^0\tfrac{1}{2}[\gamma^\mu(1/i)\partial^\nu + \gamma^\nu(1/i)\partial^\mu]\psi + g^{\mu\nu}\mathcal{L}. \tag{3-7.120}$$

The associated scalar is

$$t = -m\tfrac{1}{2}\psi\gamma^0\psi - \tfrac{3}{2}\eta\gamma^0\psi, \tag{3-7.121}$$

since the field equation implies that

$$\mathcal{L} = -\tfrac{1}{2}\psi\gamma^0\eta = -\tfrac{1}{2}\eta\gamma^0\psi, \tag{3-7.122}$$

and this stress tensor must obey the divergence equation

$$\partial_\mu t^{\mu\nu} = -\eta\gamma^0\partial^\nu\psi + i\tfrac{1}{4}\partial_\mu(\eta\gamma^0\sigma^{\mu\nu}\psi), \tag{3-7.123}$$

as one can verify directly. The various terms of $\delta\mathcal{L}$ show explicitly how the scalar nature of the Lagrange function leads to the cancellation of the rotational structure, leaving the dilational part and thereby producing a symmetrical stress tensor.

In the region between two causally separated sources, η_1 and η_2, the field associated with a particle of specified momentum, spin, and charge is [Eq. 3-6.49)]

$$\psi(x) = \psi_{p\sigma q}(x)i\eta_{2p\sigma q} + \psi_{p\sigma q}(x)^*i\eta_{1p\sigma q}^*. \tag{3-7.124}$$

The corresponding stress tensor contribution is given by

$$\begin{aligned} t^{\mu\nu} &= (i\eta_{1p\sigma q}^*)\psi_{p\sigma q}^*\gamma^0\tfrac{1}{2}(\gamma^\mu p^\nu + \gamma^\nu p^\mu)\psi_{p\sigma q}(i\eta_{2p\sigma q}) \\ &= (i\eta_{1p\sigma q}^*)(2p^\mu p^\nu\,d\omega_p)(i\eta_{2p\sigma q}), \end{aligned} \tag{3-7.125}$$

according to (3-6.53). The expected energy-momentum flux per particle is evident here.

When particles of spins 0 and 1 are described by spinors of the second rank, it is natural to follow the example of spin $\tfrac{1}{2}$ and write the displacement induced field variation as

$$\delta\psi = \delta x^\nu\partial_\nu\psi + i\tfrac{1}{4}(\sigma_1^{\mu\nu} + \sigma_2^{\mu\nu})\psi\partial_\mu\,\delta x_\nu. \tag{3-7.126}$$

Its effect on the Lagrange function

$$\mathcal{L} = -\tfrac{1}{2}\psi\gamma_1^0\gamma_2^0[\tfrac{1}{2}(\gamma_1^\mu + \gamma_2^\mu)(1/i)\partial_\mu + m]\psi \tag{3-7.127}$$

is quite analogous to the spin $\tfrac{1}{2}$ situation, apart from the $\partial_\lambda\partial_\mu\,\delta x_\nu$ term, for now $[\gamma = \gamma_1^0\gamma_2^0]$

$$\begin{aligned} \delta\mathcal{L} &= \delta x^\nu\partial_\nu\mathcal{L} + \tfrac{1}{2}\psi\gamma\tfrac{1}{4}i\sum(\gamma_\alpha^\mu\partial^\nu + \gamma_\alpha^\nu\partial^\mu)\psi\partial_\mu\,\delta x_\nu \\ &\quad - \tfrac{1}{2}\psi\gamma\tfrac{1}{2}(\gamma_1^\lambda\tfrac{1}{4}\sigma_2^{\mu\nu} + \gamma_2^\lambda\tfrac{1}{4}\sigma_1^{\mu\nu})\psi\partial_\lambda\partial_\mu\,\delta x_\nu. \end{aligned} \tag{3-7.128}$$

The identity (3-7.97) produces the symmetrical stress tensor

$$t^{\mu\nu} = \tfrac{1}{2}\psi\gamma\tfrac{1}{2}[\sum\gamma_\alpha^\mu(1/i)\partial^\nu]_{(\mu\nu)}\psi + g^{\mu\nu}\mathcal{L} - \partial_\lambda[\tfrac{1}{2}\psi\gamma\tfrac{1}{4}(\gamma_1^\mu\sigma_2^{\lambda\nu} + \gamma_2^\mu\sigma_1^{\lambda\nu})\psi]_{(\mu\nu)}, \tag{3-7.129}$$

to which the null terms involving $(\gamma^\mu\sigma^{\lambda\nu})_{1,2}$ can be added. This is useful for the matrix transcription of (3-7.129),

$$\begin{aligned} t^{\mu\nu} &= -\tfrac{1}{4}\operatorname{tr}\psi\gamma^0[\gamma^\mu, (1/i)\partial^\nu\psi^T\gamma^0]_{(\mu\nu)} + g^{\mu\nu}\mathcal{L} \\ &\quad - \tfrac{1}{8}\partial_\lambda\operatorname{tr}\psi\gamma^0[[\gamma^\mu, \psi^T\gamma^0], \sigma^{\lambda\nu}]_{(\mu\nu)}, \end{aligned} \tag{3-7.130}$$

where the double commutator can be replaced by the double anticommutator without affecting the value of this term.

The relation between this and earlier stress tensor forms will be examined, using the zero spin example. Inserting the matrix field (3–5.83) one finds

$$t^{\mu\nu} = \tfrac{1}{2}(\phi^\mu\partial^\nu\phi - \phi\partial^\nu\phi^\mu)_{(\mu\nu)} + g^{\mu\nu}[\mathcal{L} + \tfrac{1}{2}\partial_\lambda(\phi\phi^\lambda)] + \tfrac{1}{2}\partial^\nu(\phi\phi^\mu)_{(\mu\nu)} - g^{\mu\nu}\tfrac{1}{2}\partial_\lambda(\phi\phi^\lambda),$$

$$(3-7.131)$$

where the appropriate Lagrange function, given by (3–5.84), has been replaced by the Lagrange function of Eq. (3–5.16) and the necessary divergence term. Observe how the derivatives of the vector field cancel, leaving

$$t^{\mu\nu} = \tfrac{1}{2}(\phi^\mu\partial^\nu\phi + \phi^\nu\partial^\mu\phi) + g^{\mu\nu}\mathcal{L},$$

$$(3-7.132)$$

which agrees with (3–7.8) and (3–7.37) in regions where the vector source vanishes.

The generalization of the field variation (3–7.126) to an arbitrary multi-spinor is

$$\delta\psi = \delta x^\nu\partial_\nu\psi + i\tfrac{1}{4}\left(\sum_{\alpha=1}^{n}\sigma_\alpha^{\mu\nu}\right)\psi\partial_\mu\,\delta x_\nu,$$

$$(3-7.133)$$

which can also be applied to the various auxiliary fields $\psi_{[\alpha\beta]}$, $\psi_{[\alpha\beta][\alpha'\beta']}$, $\cdots$. In source-free space all auxiliary fields vanish, and this property is not affected by arbitrary displacements since the transformation law (3–7.133) refers only to the field under consideration. Accordingly, only the first term of the Lagrange function (3–5.78) contributes, and the result obtained for the stress tensor is an immediate generalization of (3–7.129), but with $\mathcal{L} = 0$ as befits the sourceless circumstances:

$$t^{\mu\nu} = \tfrac{1}{2}\psi\gamma\frac{1}{n}\left[\sum_\alpha\gamma_\alpha^\mu\left(\frac{1}{i}\right)\partial^\nu\right]_{(\mu\nu)}\psi - \partial_\lambda\left[\tfrac{1}{2}\psi\gamma\frac{1}{2n}\sum_{\alpha\neq\beta}\gamma_\alpha^\mu\sigma_\beta^{\lambda\nu}\psi\right]_{(\mu\nu)}.$$

$$(3-7.134)$$

We shall elaborate this for $n = 3$, using the spinor, antisymmetrical in a pair of indices, that gives an alternative spin $\tfrac{1}{2}$ description. The calculation resembles the one already performed for the current, and the result is

$$t^{\mu\nu} = t_D^{\mu\nu} + \frac{1}{6m}(\partial^\mu\partial^\nu - g^{\mu\nu}\partial^2)\tfrac{1}{2}\Psi\gamma^0\Psi + \frac{1}{12m^2}\partial^2 t_D^{\mu\nu},$$

$$(3-7.135)$$

where

$$t_D^{\mu\nu} = \tfrac{1}{2}\Psi\gamma^0\tfrac{1}{2}[\gamma^\mu(1/i)\partial^\nu + \gamma^\nu(1/i)\partial^\mu]\Psi$$

$$(3-7.136)$$

is the stress tensor of the simple spinor or Dirac equation for source-free conditions. The additional terms can be exhibited in the form of Eq. (3–7.83), with

$$m^{\mu\nu,\kappa\lambda} = -(g^{\mu\nu}g^{\kappa\lambda} - \tfrac{1}{2}g^{\kappa\nu}g^{\mu\lambda} - \tfrac{1}{2}g^{\kappa\mu}g^{\lambda\nu})\frac{1}{6m}\frac{1}{2}\Psi\gamma^0\Psi$$
$$+ \frac{1}{12m^2}[g^{\mu\nu}t_D^{\kappa\lambda} + g^{\kappa\lambda}t_D^{\mu\nu} - \tfrac{1}{2}(g^{\mu\kappa}t_D^{\nu\lambda} + g^{\mu\lambda}t_D^{\nu\kappa} + g^{\nu\kappa}t_D^{\mu\lambda} + g^{\nu\lambda}t_D^{\mu\kappa})].$$

$$(3-7.137)$$

The relative rotation of sources supplies angular momentum information. To discuss this, consider the displacement

$$\delta x^\nu = \delta\omega^{\lambda\nu}(x)x_\lambda, \qquad \delta\omega^{\mu\nu}(x) = -\delta\omega^{\nu\mu}(x), \qquad (3\text{-}7.138)$$

and the accompanying dilation

$$\partial_\mu \, \delta x_\nu + \partial_\nu \, \delta x_\mu = x^\lambda \partial_\mu \, \delta\omega_{\lambda\nu}(x) + x^\lambda \partial_\nu \, \delta\omega_{\lambda\mu}(x). \qquad (3\text{-}7.139)$$

The latter vanishes, of course, when $\delta\omega_{\lambda\nu}(x)$ is a constant, describing a rigid rotation. The implied action variation is

$$\delta W = -\int (dx) t^{\mu\nu} \partial_\mu \, \delta x_\nu = -\tfrac{1}{2}\int (dx) j^{\lambda\mu\nu}(x) \partial_\lambda \, \delta\omega_{\mu\nu}(x), \qquad (3\text{-}7.140)$$

where

$$j^{\lambda\mu\nu} = x^\mu t^{\lambda\nu} - x^\nu t^{\lambda\mu}. \qquad (3\text{-}7.141)$$

The alternative evaluation refers directly to the response of the source. In order to deal uniformly with all spins we use the multispinor description, for which

$$\delta W = \int (dx) \, \delta\eta\gamma\psi = -\int (dx)\eta\gamma \, \delta\psi$$
$$= -\int (dx)\eta\gamma[\delta x^\nu \partial_\nu \psi + i\tfrac{1}{4}(\textstyle\sum \sigma_\alpha^{\mu\nu})\psi \partial_\mu \, \delta x_\nu]. \qquad (3\text{-}7.142)$$

Comparison of the two computations gives

$$\partial_\lambda j^{\lambda\mu\nu} = -\eta\gamma(x^\mu \partial^\nu - x^\nu \partial^\mu + i\tfrac{1}{2}\textstyle\sum \sigma_\alpha^{\mu\nu})\psi + \partial_\lambda\left[\eta\gamma \, \frac{i}{4}\,(x^\mu \textstyle\sum \sigma_\alpha^{\lambda\nu} - x^\nu \textstyle\sum \sigma_\alpha^{\lambda\mu})\psi\right].$$
$$(3\text{-}7.143)$$

This result is not independent of the divergence equation for the stress tensor,

$$\partial_\mu t^{\mu\nu} = -\eta\gamma\partial^\nu\psi + \partial_\mu\left[\eta\gamma \, \frac{i}{4}\textstyle\sum \sigma_\alpha^{\mu\nu}\psi\right], \qquad (3\text{-}7.144)$$

since

$$\partial_\lambda j^{\lambda\mu\nu} = x^\mu \partial_\lambda t^{\lambda\nu} - x^\nu \partial_\lambda t^{\lambda\mu} + t^{\mu\nu} - t^{\nu\mu}. \qquad (3\text{-}7.145)$$

It is here that we recognize the significance of the symmetry property possessed by the stress tensor. The quantity $j^{\lambda\mu\nu}$, which is formed merely by taking first moments of the stress tensor, is conserved in source-free regions. As the rotational analogue of $t^{\mu\nu}$, $j^{\lambda\mu\nu}$ evidently provides a space-time account of the distribution and flux of angular momentum, in a four-dimensional sense.

As in earlier discussions, the rigid rotation of source η_2 relative to the causally detached source η_1 is introduced by letting $\delta\omega_{\mu\nu}(x)$ vanish in the future of an intermediate space-like surface, after being the constant $\delta\omega_{\mu\nu}$ in its past. This gives

$$\delta W = \tfrac{1}{2} \, \delta\omega_{\mu\nu}\int d\sigma_\lambda j^{\lambda\mu\nu}. \qquad (3\text{-}7.146)$$

An analogous equation applies to the time cycle description, where one has the simpler physical interpretation in terms of an expectation value

$$\int d\sigma_\lambda j^{\lambda\mu\nu} = \langle J^{\mu\nu}\rangle_0^\eta. \tag{3-7.147}$$

Surface integration is replaced by volume integration, with the aid of (3–7.143),

$$\int d\sigma_\lambda j^{\lambda\mu\nu} = -\int (dx)\eta\gamma(x^\mu\partial^\nu - x^\nu\partial^\mu + i\tfrac{1}{2}\sum\sigma_\alpha^{\mu\nu})\psi; \tag{3-7.148}$$

the combination of orbital and spin angular momenta is evident here. The equivalent result in other formulations is illustrated, for the vectorial source and field of unit spin particles, by [cf. Eq. (3–7.49)]

$$\int d\sigma_\lambda j^{\lambda\mu\nu} = -\int (dx)[J^\lambda(x^\mu\partial^\nu - x^\nu\partial^\mu)\phi_\lambda + J^\mu\phi^\nu - J^\nu\phi^\mu]. \tag{3-7.149}$$

We shall now confirm that the evaluation of total angular momentum is unaffected by the arbitrariness of the stress tensor. Accompanying the redefinition

$$t^{\mu\nu} \rightarrow t^{\mu\nu} + \partial_\alpha\partial_\beta m^{\mu\nu,\alpha\beta} \tag{3-7.150}$$

is

$$j^{\lambda\mu\nu} \rightarrow j^{\lambda\mu\nu} + x^\mu\partial_\alpha\partial_\beta m^{\lambda\nu,\alpha\beta} - x^\nu\partial_\alpha\partial_\beta m^{\lambda\mu,\alpha\beta}$$
$$= j^{\lambda\mu\nu} + \partial_\alpha\partial_\beta[x^\mu m^{\lambda\nu,\alpha\beta} - x^\nu m^{\lambda\mu,\alpha\beta}] - 2\partial_\alpha[m^{\lambda\nu,\alpha\mu} - m^{\lambda\mu,\alpha\nu}]. \tag{3-7.151}$$

The verification that the additional second-derivative term gives no contribution to $\int d\sigma_\lambda j^{\lambda\mu\nu}$ is identical to that of Eq. (3–7.87)—only the cyclic symmetry property (3–7.85) is invoked. The latter also implies this statement of antisymmetry in λ and α:

$$[m^{\lambda\nu,\alpha\mu} - m^{\lambda\mu,\alpha\nu}] + [m^{\alpha\nu,\lambda\mu} - m^{\alpha\mu,\lambda\nu}] = 0, \tag{3-7.152}$$

from which follows the vanishing of the surface integral for the last term of (3–7.151).

Particle states that are labeled by three-dimensional angular momentum quantum numbers, rather than by linear momentum, have been exhibited for spins 0, $\tfrac{1}{2}$, 1. Their use is sufficiently similar to the charge and energy-momentum discussions that we shall not enter into details here.

In contrast to the rigid translations and rotations of physical interest, dilations would seem to be only a device that assists in the identification of energy-momentum fluxes. There is, however, a subset of these transformations that plays a more physical role in the special circumstance of massless particles. It is the group of isotropic dilations that is characterized by

$$\partial_\mu \delta x_\nu(x) + \partial_\nu \delta x_\mu(x) = g_{\mu\nu} \delta\varphi(x). \tag{3-7.153}$$

These conditions are very restrictive. We first note the scalar relation

$$\partial_\nu \delta x^\nu = 2 \delta\varphi. \tag{3-7.154}$$

The divergence of the tensor equation then gives

$$\partial^2 \delta x_\mu = -\partial_\mu \delta\varphi, \tag{3-7.155}$$

and a further divergence asserts that

$$\partial^2 \, \delta\varphi(x) = 0. \tag{3-7.156}$$

But even more is obtained by applying the operator ∂^2 to (3–7.153), namely

$$\partial_\mu \partial_\nu \, \delta\varphi(x) = 0. \tag{3-7.157}$$

Thus $\delta\varphi(x)$ is limited to a linear function of the coordinates,

$$\delta\varphi(x) = 2 \, \delta a + 4 \, \delta b_\nu x^\nu. \tag{3-7.158}$$

The corresponding form of δx^μ, apart from infinitesimal translations and rotations, is the quadratic function

$$\delta x^\mu = \delta a x^\mu + \delta b_\nu (2x^\mu x^\nu - g^{\mu\nu} x^2). \tag{3-7.159}$$

Together with translations and rotations, these transformations form a group of 15 parameters. It has the structure of the rotation group in $4 + 2$ dimensions, in the sense that the homogeneous Lorentz group is the rotation group in $3 + 1$ dimensions. Perhaps the quickest way to recognize this is through the introduction of homogeneous coordinates,

$$x^\mu = y^\mu/(y_5 + y_6), \tag{3-7.160}$$

that are defined in the five-dimensional space of the null 'sphere'

$$y^2 + y_5^2 - y_6^2 = 0. \tag{3-7.161}$$

The ten-parameter space-time translation-rotation group now appears as the subgroup of homogeneous transformations on the null sphere that leaves $y_5 + y_6$ invariant:

$$\delta y^\nu = \delta\epsilon^\nu (y_5 + y_6) + \delta\omega^{\mu\nu} y_\mu, \qquad \delta y_5 = -\delta y_6 = -\delta\epsilon^\nu y_\nu; \tag{3-7.162}$$

the one-parameter δa transformations, uniform scale changes in space-time, are the 'rotations'

$$\delta y_5 = -\delta a y_6, \qquad \delta y_6 = -\delta a y_5, \qquad \delta y^\mu = 0; \tag{3-7.163}$$

and the transformations parametrized by δb^ν hold $y_5 - y_6$ fixed,

$$\delta y_5 = \delta y_6 = -\delta b_\nu y^\nu, \qquad \delta y^\nu = \delta b^\nu (y_5 - y_6). \tag{3-7.164}$$

The quadratic form of (3–7.161) also admits reflections, including $y_5 \to -y_5$, which has the following effect upon the x^μ coordinates:

$$x^\mu = \frac{y^\mu}{y_5 + y_6} \to \frac{y^\mu}{-y_5 + y_6} = \frac{y^\mu}{y^2} (y_5 + y_6) \tag{3-7.165}$$

or

$$x^\mu \to x^\mu/x^2, \tag{3-7.166}$$

an inversion in the origin. A sequence of two inversions, first performed at the origin and then at the point with coordinates δb^ν, produces the infinitesimal transformation of (3–7.159) with $\delta a = 0$.

The isotropic dilations are known as conformal transformations. Their physical relevance emerges on considering the part of the Lagrange function response to displacements that is given by

$$-t^{\mu\nu}\tfrac{1}{2}(\partial_\mu\,\delta x_\nu + \partial_\nu\,\delta x_\mu) = -\tfrac{1}{2}t\,\delta\varphi(x), \qquad (3\text{-}7.167)$$

which singles out the scalar t as the significant quantity. Inspection of the examples with spins 0, $\tfrac{1}{2}$, 1, 2, shows that the scalar t, evaluated at source-free points and with $m \to 0$, either vanishes or is a second-derivative structure of the form implied by the arbitrariness of the stress tensor,

$$t = \partial_\kappa\partial_\lambda n^{\kappa\lambda}. \qquad (3\text{-}7.168)$$

Since all physically interesting possibilities are included in these examples, we shall forego the luxury of a general proof. But what is involved should be clear. When $1/m$ is infinite there is no standard of length in the action. If the scale of all coordinates is changed uniformly,

$$\bar{x}^\mu = \lambda^{-1}x^\mu, \qquad (3\text{-}7.169)$$

the form of the action can be maintained by a corresponding scale change of sources and fields that is determined by the number of derivatives in the Lagrange function when the minimum number of fields is used. This is illustrated for spins 0 ánd $\tfrac{1}{2}$ by

$$\bar{\phi}(x) = \lambda\phi(\lambda x), \qquad \bar{\psi}(x) = \lambda^{3/2}\psi(\lambda x). \qquad (3\text{-}7.170)$$

Considered in infinitesimal form these scaling laws specify a definite multiple of $\partial_\nu\,\delta x^\nu$—an example is (3-7.71) for scalar fields. The invariance of the action, stating a local property of the Lagrange function, requires that the scalar t vanish everywhere or, more generally, be the divergence of a vector. As one can verify in the simpler examples, this vector employs the gradient vector for its construction and (3-7.168) emerges as a generally valid statement for a suitable choice of field variations. Then, since the general $\delta\varphi(x)$, Eq. (3-7.158), has vanishing second derivatives,

$$-(\partial_\kappa\partial_\lambda n^{\kappa\lambda})\,\delta\varphi = \partial_\kappa(n^{\kappa\lambda}\partial_\lambda\,\delta\varphi - \delta\varphi\partial_\lambda n^{\kappa\lambda}), \qquad (3\text{-}7.171)$$

and the action is invariant for the whole 15-parameter group that incorporates conformal transformations.

As always, invariance of the action implies conserved physical quantities with space-time distributions and fluxes. The procedure is standard; the constants δa and δb_ν are replaced with arbitrary coordinate-dependent functions. We shall assume, for simplicity, that $t = 0$. If, instead, (3-7.168) is applicable, additional terms appear in the various fluxes but nothing basic is altered. The response of the action to the generalizations of the conformal transformations is

$$\delta W = -\int (dx)[c^\mu(x)\partial_\mu\,\delta a(x) + c^{\mu\nu}(x)\partial_\mu\,\delta b_\nu(x)], \qquad (3\text{-}7.172)$$

with

$$c^{\mu} = t^{\mu\nu}x_{\nu}, \qquad c^{\mu\nu} = t^{\mu}{}_{\lambda}(2x^{\lambda}x^{\nu} - g^{\lambda\nu}x^2). \qquad (3\text{-}7.173)$$

The tensor $c^{\mu\nu}$ is not symmetrical:

$$c^{\mu\nu} - c^{\nu\mu} = -2x_{\lambda}j^{\lambda\mu\nu}, \qquad (3\text{-}7.174)$$

and the implied scalar is

$$c = g_{\mu\nu}c^{\mu\nu} = 2x^{\lambda}c_{\lambda}. \qquad (3\text{-}7.175)$$

In source-free regions, the local conservation laws are

$$\partial_{\mu}c^{\mu} = 0, \qquad \partial_{\mu}c^{\mu\nu} = 0, \qquad (3\text{-}7.176)$$

and the existence of conserved total quantities is indicated, for the time cycle description, by

$$\int d\sigma_{\mu}c^{\mu} = \langle C \rangle, \qquad \int d\sigma_{\mu}c^{\mu\nu} = \langle C^{\nu} \rangle. \qquad (3\text{-}7.177)$$

The physical content of these conservation statements will be discussed in the context of the most familiar massless particle.

3–8 THE ELECTROMAGNETIC FIELD.   MAGNETIC CHARGE

Although frequent reference has been made to the $m \to 0$ limit for unit spin particles, it is important to give an independent discussion of the field associated with the massless, unit helicity particle—the photon.   The starting point is Eq. (2–3.45), written as

$$W(J) = \tfrac{1}{2}\int (dx)(dx')J^{\mu}(x)D_{+}(x - x')J_{\mu}(x'),$$
$$\partial_{\mu}J^{\mu}(x) = 0. \qquad (3\text{-}8.1)$$

In defining a field $A_{\mu}(x)$ through the effect of a test source $\delta J^{\mu}(x)$,

$$\delta W(J) = \int (dx)\, \delta J^{\mu}(x)A_{\mu}(x), \qquad (3\text{-}8.2)$$

strict account must be taken of the source restriction, which demands that

$$\partial_{\mu}\, \delta J^{\mu}(x) = 0. \qquad (3\text{-}8.3)$$

Thus, when comparing (3–8.2) with

$$\delta W(J) = \int (dx)(dx')\, \delta J^{\mu}(x)D_{+}(x - x')J_{\mu}(x'), \qquad (3\text{-}8.4)$$

one should not identify the coefficients of $\delta J^{\mu}(x)$.   The correct conclusion is that they differ by any expression that leads to a vanishing integral in consequence of the restriction (3–8.3).   The general form of such an expression is

$$-\int (dx)\lambda(x)\partial_{\mu}\, \delta J^{\mu}(x) = 0, \qquad (3\text{-}8.5)$$

and therefore

$$A_\mu(x) = \int (dx') D_+(x - x') J_\mu(x') + \partial_\mu \lambda(x). \tag{3-8.6}$$

The aspect of $A_\mu(x)$ that is governed by the arbitrary scalar function $\lambda(x)$ is picked out by forming the divergence of (3–8.6). This gives

$$\partial_\nu A^\nu(x) = \partial^2 \lambda(x), \tag{3-8.7}$$

and the application of the differential operator $-\partial^2$ to (3–8.6) then provides us with the second-order differential equation

$$-\partial^2 A_\mu(x) + \partial_\mu \partial_\nu A^\nu(x) = J_\mu(x). \tag{3-8.8}$$

We recognize Eq. (3–3.6), with $m = 0$. Since the arbitrariness of $\lambda(x)$ persists in this differential field equation, the latter must be unaffected by any redefinition of A_μ in the form

$$A_\mu(x) \rightarrow A_\mu(x) + \partial_\mu \lambda(x), \tag{3-8.9}$$

which is known as a gauge transformation (more frequently, as a guage [sic] transformation). This gauge invariance is emphasized by writing the field equation in the equivalent form [compare Eqs. (3–3.7, 8)]

$$\partial_\nu F^{\mu\nu}(x) = J^\mu(x), \tag{3-8.10}$$

$$\partial_\mu A_\nu(x) - \partial_\nu A_\mu(x) = F_{\mu\nu}(x), \tag{3-8.11}$$

since the addition of a gradient term to a vector does not alter the curl of the vector. That the divergenceless nature of J^μ is built into the field equations is also emphasized, for

$$\partial_\mu J^\mu = \partial_\mu \partial_\nu F^{\mu\nu} \equiv 0, \tag{3-8.12}$$

owing to the antisymmetry of $F^{\mu\nu}$. The curl construction of $F_{\mu\nu}$ is given another form in the differential equations

$$\partial_\lambda F_{\mu\nu}(x) + \partial_\mu F_{\nu\lambda}(x) + \partial_\nu F_{\lambda\mu}(x) = 0. \tag{3-8.13}$$

The tensor dual to $F_{\mu\nu}$ is defined by [Eq. (2–3.60)]

$$*F^{\mu\nu} = \tfrac{1}{2} \epsilon^{\mu\nu\kappa\lambda} F_{\kappa\lambda}, \tag{3-8.14}$$

where the totally antisymmetrical tensor of the fourth rank is normalized to

$$\epsilon^{0123} = +1. \tag{3-8.15}$$

Using this concept, we express the differential equations (3–8.13) as

$$\partial_\nu *F^{\mu\nu}(x) = 0. \tag{3-8.16}$$

The pair of equations, (3–8.10) and (3–8.16), are Maxwell's equations for the tensor of electromagnetic field strengths, $F_{\mu\nu}$. We recall the identifications of electric and magnetic fields:

$$E_k = F^0{}_k, \qquad H_k = \tfrac{1}{2}\epsilon_{klm}F_{lm},$$

$$H_k = {}^*F^0{}_k, \qquad -E_k = \tfrac{1}{2}\epsilon_{klm}\,{}^*F_{lm}. \tag{3-8.17}$$

The alternative evaluations of $W(J)$ that lead to the action expression are

$$W(J) = \tfrac{1}{2}\int (dx)J^\mu(x)A_\mu(x) \tag{3-8.18}$$

and

$$W(J) = \tfrac{1}{2}\int (dx)(\partial_\nu F^{\mu\nu})A_\mu = \tfrac{1}{4}\int (dx)F^{\mu\nu}F_{\mu\nu}. \tag{3-8.19}$$

The electromagnetic action, foreshadowed in (3-5.94), is

$$W(J) = \int (dx)[J^\mu A_\mu + \mathcal{L}], \tag{3-8.20}$$

with the Lagrange function

$$\mathcal{L} = -\tfrac{1}{4}F^{\mu\nu}F_{\mu\nu} = \tfrac{1}{2}(\mathbf{E}^2 - \mathbf{H}^2). \tag{3-8.21}$$

The Lagrange function is explicitly gauge invariant, and so is the action because of the differential conservation property of J^μ. This time we have begun with a conservation law and inferred an invariance of the action.

In seeking expressions for the source and field variations that are associated with arbitrary coordinate displacements, it is natural to maintain the conservation property of J^μ and the gauge invariance of $F_{\mu\nu}$. The vector source transformation law (3-7.33, 47) has the required characteristics since

$$\delta(\partial_\mu J^\mu) = \partial_\nu(\delta x^\nu \partial_\mu J^\mu) \tag{3-8.22}$$

assures the continued vanishing of $\partial_\mu J^\mu$; the associated tensor transformation (3-7.40),

$$\delta F_{\mu\nu} = \delta x^\lambda \partial_\lambda F_{\mu\nu} + F_{\lambda\nu}\partial_\mu \,\delta x^\lambda + F_{\mu\lambda}\partial_\nu \,\delta x^\lambda, \tag{3-8.23}$$

involves only gauge invariant quantities. And the vector field prescription

$$\delta A_\mu = \delta x^\nu \partial_\nu A_\mu + A_\nu \partial_\mu \,\delta x^\nu = \partial_\mu(A_\nu \,\delta x^\nu) - F_{\mu\nu}\,\delta x^\nu, \tag{3-8.24}$$

which was based initially on the transformation properties of the gradient of a scalar function, is maintained under a gauge transformation. The direct evaluation of the induced action variation is

$$\delta W(J) = \int (dx)\,\delta J^\mu A_\mu = -\int (dx)J^\mu \,\delta A_\mu$$

$$= \int (dx)J^\mu F_{\mu\nu}\,\delta x^\nu, \tag{3-8.25}$$

while the Lagrange function response is

$$\delta \mathcal{L} = \delta x^\nu \partial_\nu \mathcal{L} - F^{\mu\lambda}F^\nu{}_\lambda \partial_\mu \,\delta x_\nu$$

$$= \partial_\nu(\delta x^\nu \mathcal{L}) - t^{\mu\nu}\partial_\mu \,\delta x_\nu, \tag{3-8.26}$$

with

$$t^{\mu\nu} = t^{\nu\mu} = F^{\mu\lambda}F^\nu{}_\lambda - g^{\mu\nu}\tfrac{1}{4}F^{\kappa\lambda}F_{\kappa\lambda} \tag{3-8.27}$$

and

$$t = 0. \tag{3-8.28}$$

The comparison of the two evaluations gives

$$\partial_\mu t^{\mu\nu} = J_\mu F^{\mu\nu} = -J^\mu \partial^\nu A_\mu + \partial_\mu(J^\mu A^\nu). \tag{3-8.29}$$

Some explicit stress tensor expressions are

$$t^{00} = \tfrac{1}{2}(\mathbf{E}^2 + \mathbf{H}^2), \qquad t^0{}_k = (\mathbf{E} \times \mathbf{H})_k. \tag{3-8.30}$$

A direct approach to the conformal conservation laws will now be made. Multiplication of (3–8.29) by x_ν gives $(t = 0)$

$$\partial_\mu c^\mu = J_\mu F^{\mu\nu} x_\nu, \tag{3-8.31}$$

and similarly

$$J_\mu F^\mu{}_\lambda (2x^\lambda x^\nu - g^{\lambda\nu}x^2) = (\partial_\mu t^\mu{}_\lambda)(2x^\lambda x^\nu - g^{\lambda\nu}x^2)$$
$$= \partial_\mu c^{\mu\nu}. \tag{3-8.32}$$

The nature of the corresponding integral conservation laws most closely resembles that of

$$\int d\sigma_\mu j^{\mu 0 k} = \int (d\mathbf{x}) j^{00k} = x^0 \int (d\mathbf{x}) t^{0k} - \int (d\mathbf{x}) x^k t^{00}, \tag{3-8.33}$$

which, being an explicit function of time, is a statement about how something moves; it is the centroid of the energy distribution in this instance. The strongest assertion of this type that is implied by conformal invariance is contained in

$$\int d\sigma_\mu(c^{\mu 0} - 2c^\mu x^0) = \int (d\mathbf{x})(c^{00} - 2c^0 x^0)$$
$$= \int (d\mathbf{x})((x^0)^2 - \mathbf{x}^2)t^{00}, \tag{3-8.34}$$

or

$$\int (d\mathbf{x})\mathbf{x}^2 t^{00} = (x^0)^2 \int (d\mathbf{x}) t^{00} + 2x^0 \int (d\mathbf{x}) c^0 - \int (d\mathbf{x}) c^{00}. \tag{3-8.35}$$

Thus, with the weighting factor provided by the energy density, the average value of $\mathbf{x}^2$ varies quadratically in time, with unit coefficient of $(x^0)^2$. The interpretation in terms of the motion of the particles that carry the energy is clear: photons move at the speed of light. The coefficient of x^0 and the constant term supply information about the initial correlation between position and velocity and the initial average value of $\mathbf{x}^2$. This view of $\int (d\mathbf{x}) c^0$ is consistent with its significance in terms of the momentum distribution:

$$\int (d\mathbf{x}) c^0 = \int (d\mathbf{x}) t^{0k} x_k - x^0 \int (d\mathbf{x}) t^{00}. \tag{3-8.36}$$

The field strengths $F_{\mu\nu}$ and the vector potential A_μ are placed on the same footing in the following action principle:

$$W = \int (dx)[J^\mu A_\mu + \tfrac{1}{2} M^{\mu\nu} F_{\mu\nu} + \mathfrak{L}(A, F)], \qquad (3\text{-}8.37)$$

where

$$\mathfrak{L} = -\tfrac{1}{2} F^{\mu\nu}(\partial_\mu A_\nu - \partial_\nu A_\mu) + \tfrac{1}{4} F^{\mu\nu} F_{\mu\nu} \qquad (3\text{-}8.38)$$

is explicitly gauge invariant. The field equations now read

$$\partial_\nu F^{\mu\nu} = J^\mu, \qquad \partial_\mu A_\nu - \partial_\nu A_\mu = F_{\mu\nu} + M_{\mu\nu}, \qquad (3\text{-}8.39)$$

or

$$\partial_\nu F^{\mu\nu} = J^\mu, \qquad \partial_\nu \, {}^*F^{\mu\nu} = {}^*J^\mu, \qquad (3\text{-}8.40)$$

where

$$ {}^*J^\mu = -\partial_\nu \, {}^*M^{\mu\nu}, \qquad \partial_\mu \, {}^*J^\mu \equiv 0, \qquad (3\text{-}8.41)$$

and $ {}^*M^{\mu\nu}$ is the tensor dual to $M^{\mu\nu}$. The stress tensor is symmetrical,

$$t^{\mu\nu} = (\partial^\mu A^\lambda - \partial^\lambda A^\mu)F^\nu{}_\lambda + F^\mu{}_\lambda(\partial^\nu A^\lambda - \partial^\lambda A^\nu) - F^{\mu\lambda}F^\nu{}_\lambda + g^{\mu\nu}\mathfrak{L}, \quad (3\text{-}8.42)$$

and

$$\partial_\nu t^{\mu\nu} = (J_\mu + \partial^\lambda M_{\mu\lambda})F^{\mu\nu} + J_\mu M^{\mu\nu} + {}^*J_\mu \, {}^*M^{\mu\nu}. \qquad (3\text{-}8.43)$$

On setting $M_{\mu\nu} = 0$ and identifying the field strength tensor with the curl of the vector potential, the previous results are recovered.

If the photon source function $J^\mu(x)$ has the interpretation of an electric current, according to the first set of the Maxwell equations (3–8.40), is $ {}^*J^\mu(x)$, as realized in (3–8.41), a magnetic current? The answer is negative. It is consistent with this, but hardly decisive, that the total value of the apparent magnetic charge is zero,

$$\int d\sigma_\mu \, {}^*J^\mu = -\tfrac{1}{2} \int (d\sigma_\mu \partial_\nu - d\sigma_\nu \partial_\mu) \, {}^*M^{\mu\nu} = 0, \qquad (3\text{-}8.44)$$

provided $M^{\mu\nu}$ has the kind of spatial localizability that attaches to the source concept. The essential remark is that, through a redefinition of the field strength, the magnetic current is transformed into an equivalent electric current. Indeed, the equations (3–8.40, 41) are also given by

$$\partial_\nu(F^{\mu\nu} + M^{\mu\nu}) = J^\mu + \partial_\nu M^{\mu\nu}, \qquad \partial_\nu \, {}^*(F^{\mu\nu} + M^{\mu\nu}) = 0, \quad (3\text{-}8.45)$$

which contains the effective electric current already exhibited in (3–8.43). But this short-lived possibility does raise a fundamental question concerning the existence of real magnetic charge, distributed and flowing in a manner that, explicitly or in context, differs from (3–8.41).

To study this question, we go back to the beginning, to the source. Is it possible to distinguish two fundamentally different kinds of photon sources? But the two kinds must also be closely related, for the structure of the Maxwell equations is retained under the substitution

$$J^\mu \rightarrow {}^*J^\mu, \qquad F^{\mu\nu} \rightarrow {}^*F^{\mu\nu}$$
$$^*J^\mu \rightarrow -J^\mu, \qquad {}^*F^{\mu\nu} \rightarrow -F^{\mu\nu} \tag{3-8.46}$$

or, more generally with arbitrary angle φ,

$$J^\mu \rightarrow J^\mu \cos\varphi + {}^*J^\mu \sin\varphi, \qquad F^{\mu\nu} \rightarrow F^{\mu\nu} \cos\varphi + {}^*F^{\mu\nu} \sin\varphi,$$
$$^*J^\mu \rightarrow -J^\mu \sin\varphi + {}^*J^\mu \cos\varphi, \qquad {}^*F^{\mu\nu} \rightarrow -F^{\mu\nu} \sin\varphi + {}^*F^{\mu\nu} \cos\varphi. \tag{3-8.47}$$

The consistency of the field strength substitutions involves the repetition property of the dual,

$$^{**}F^{\mu\nu} = -F^{\mu\nu}. \tag{3-8.48}$$

All this brings to mind the discussion of Section 2–3. It was recognized there that nothing intrinsic is altered if all photon polarization vectors are rotated through the angle $\pi/2$, thus replacing the initial polarization vector set $\mathbf{e}_{p\lambda}$ by ($p^0 = |\mathbf{p}|$)

$$^*\mathbf{e}_{p\lambda} = (\mathbf{p}/p^0) \times \mathbf{e}_{p\lambda}. \tag{3-8.49}$$

This suggests that the desired distinction and relation between two kinds of sources is realized if the measure of their effectiveness in emitting a given photon, labeled $p\lambda$, utilizes $\mathbf{e}_{p\lambda}$ for one kind of source and $^*\mathbf{e}_{p\lambda}$ for the other. That is indicated by (real polarization vectors are used to reduce the number of stars in our eyes)

$$J_{p\lambda} = (d\omega_p)^{1/2}[\mathbf{e}_{p\lambda} \cdot \mathbf{J}(p) + {}^*\mathbf{e}_{p\lambda} \cdot {}^*\mathbf{J}(p)]. \tag{3-8.50}$$

The equivalence of the descriptions that are connected by the source transformation of (3–8.47) then expresses the freedom to rotate both systems of polarization vectors through the common angle φ.

Again we consider a causal situation, but now with component sources J_1^μ, $^*J_1^\mu$ and J_2^μ, $^*J_2^\mu$. The coupling between the emission and absorption sources that a single photon mediates is conveyed by

$$\sum_{p\lambda} J_{1p\lambda}^* J_{2p\lambda} = \int d\omega_p \sum_\lambda [\mathbf{J}_1(-p) \cdot \mathbf{e}_{p\lambda}\mathbf{e}_{p\lambda} \cdot \mathbf{J}_2(p)$$
$$+ {}^*\mathbf{J}_1(-p) \cdot {}^*\mathbf{e}_{p\lambda}{}^*\mathbf{e}_{p\lambda} \cdot {}^*\mathbf{J}_2(p) + \mathbf{J}_1(-p) \cdot \mathbf{e}_{p\lambda}{}^*\mathbf{e}_{p\lambda} \cdot {}^*\mathbf{J}_2(p)$$
$$+ {}^*\mathbf{J}_1(-p) \cdot {}^*\mathbf{e}_{p\lambda}\mathbf{e}_{p\lambda} \cdot \mathbf{J}_2(p)]. \tag{3-8.51}$$

The intrinsic equivalence of the two sets of polarization vectors is contained in the dyadic relation

$$\sum_\lambda \mathbf{e}_{p\lambda}\mathbf{e}_{p\lambda} = \sum_\lambda {}^*\mathbf{e}_{p\lambda}{}^*\mathbf{e}_{p\lambda}, \tag{3-8.52}$$

and the corresponding terms of (3–8.51) can be given invariant form in the known way. But what of the coupling between different kinds of sources? We first note that the summation in

$$\sum_\lambda {}^*\mathbf{e}_{p\lambda}\mathbf{e}_{p\lambda} = (\mathbf{p}/p^0) \times \sum_\lambda \mathbf{e}_{p\lambda}\mathbf{e}_{p\lambda} \tag{3-8.53}$$

can be extended to include the third unit vector parallel to $\mathbf{p}$, thereby introducing the unit dyadic:

$$\sum_\lambda {}^*\mathbf{e}_{p\lambda}\mathbf{e}_{p\lambda} = -\sum_\lambda \mathbf{e}_{p\lambda} {}^*\mathbf{e}_{p\lambda} = (\mathbf{p}/p^0)\times. \tag{3-8.54}$$

The coupling illustrated by $\mathbf{J}_1(-p) \times {}^*\mathbf{J}_2(p) \cdot \mathbf{p}/p^0$ is very three-dimensional in appearance. Nevertheless, in the physical circumstances under examination, this is a Lorentz scalar. That can be directly verified. It is more rewarding, however, to write this term in an explicitly covariant form.

We begin by remarking that (causal subscripts are omitted)

$$\begin{aligned}\mathbf{J}(-p) \times {}^*\mathbf{J}(p) \cdot \mathbf{p}/p^0 &= \epsilon^{m0kl}J_m(-p)(1/p^0)p_k {}^*J_l(p)\\ &= \epsilon^{\mu\nu\kappa\lambda}J_\mu(-p)f_\nu(p)ip_\kappa {}^*J_\lambda(p),\end{aligned} \tag{3-8.55}$$

where $f_\nu(p)$ has only a time component such that

$$ip^0 f_0(p) = 1. \tag{3-8.56}$$

The decisive observation is that (3-8.55) remains true in its four-dimensional form with any vector $f_\nu(p)$ that obeys

$$ip^\nu f_\nu(p) = 1, \tag{3-8.57}$$

under the causal conditions that require the photon energy-momentum relation

$$p^2 = 0. \tag{3-8.58}$$

Also relevant, but holding without regard to causal arrangements, are the current conservation statements

$$p_\nu J^\nu(p) = 0, \qquad p_\nu {}^*J_\nu(p) = 0. \tag{3-8.59}$$

The following identity, which is valid for arbitrary p^μ, is based on these conservation laws:

$$\begin{aligned}\epsilon^{\mu\nu\kappa\lambda}J_\mu(-p)f_\nu(p)ip_\kappa {}^*J_\lambda(p) &= [ip^\nu f_\nu(p)]\mathbf{J}(-p) \times {}^*\mathbf{J}(p) \cdot \mathbf{p}/p^0\\ &\quad -i(p^2/p^0)\mathbf{f}(p) \cdot \mathbf{J}(-p) \times {}^*\mathbf{J}(p).\end{aligned} \tag{3-8.60}$$

Of the two terms on the right-hand side, the second vanishes when p^μ is a photon momentum obeying (3-8.58), and the first is independent of the specific choice of vector $f_\nu(p)$ that obeys the restriction (3-8.57). This is in fact the proof of covariance for causal circumstances. But some explanation is called for. A class of functions that obey (3-8.57) is given by

$$f_\nu(p) = n_\nu/ipn, \tag{3-8.61}$$

where n_ν is an arbitrary constant vector. If n_ν points along the time axis, for example, we have the situation of Eq. (3-8.56). That characterization of $f_\nu(p)$ is not covariant; after a Lorentz transformation is performed, n_ν will have nonvanishing spatial components, although it is still a time-like vector. It is here that the arbitrariness of n_ν enters, for we can replace the time-like vector by one with only a temporal component. It is through such coupling of the choice of n_ν

to the choice of coordinate system that any discrimination among coordinate systems is avoided and covariance achieved.

The space-time transcription of (3–8.57) is

$$\partial_\nu f^\nu(x - x') = \delta(x - x'). \tag{3–8.62}$$

If, as in (3–8.61), $f^\nu(x - x')$ is proportional to a constant vector n^ν,

$$f^\nu(x - x') = n^\nu f(x - x'), \tag{3–8.63}$$

the differential equation

$$(n\partial)f(x - x') = \delta(x - x') \tag{3–8.64}$$

is effectively one in a single variable. For the situation of (3–8.56), with only $n^0 \neq 0$, we get

$$\partial_0 f^0(x - x') = \delta(x^0 - x^{0\prime})\,\delta(\mathbf{x} - \mathbf{x}'). \tag{3–8.65}$$

The solution is not unique. Two alternative solutions that correspond to retarded and advanced boundary conditions are

$$f^0(x - x') = \delta(\mathbf{x} - \mathbf{x}')[\eta(x^0 - x^{0\prime}), -\eta(x^{0\prime} - x^0)], \tag{3–8.66}$$

where

$$\eta(x^0 - x^{0\prime}) = \begin{cases} x^0 > x^{0\prime}: & 1, \\ x^0 < x^{0\prime}: & 0 \end{cases} \tag{3–8.67}$$

represents the Heaviside step function (the capital of the Greek letter η is H, as the capital of δ is D, in the Chalcidian alphabet). Another choice aligns the vector n^ν with the third spatial axis, for example. Then

$$\partial_3 f_3(x - x') = \delta(x^0 - x^{0\prime})\,\delta(x_1 - x_1')\,\delta(x_2 - x_2')\,\delta(x_3 - x_3'), \tag{3–8.68}$$

and alternative solutions are

$$f_3(x - x') = \delta(x^0 - x^{0\prime})\,\delta(x_1 - x_1')\,\delta(x_2 - x_2')[\eta(x_3 - x_3'), -\eta(x_3' - x_3)]. \tag{3–8.69}$$

We specifically note the equally weighted linear combination

$$f_3(x - x') = \delta(x^0 - x^{0\prime})\,\delta(x_1 - x_1')\,\delta(x_2 - x_2')\tfrac{1}{2}\epsilon(x_3 - x_3'), \tag{3–8.70}$$

in which

$$\epsilon(x_3 - x_3') = -\epsilon(x_3' - x_3) = \begin{cases} x_3 > x_3': & 1, \\ x_3 < x_3': & -1. \end{cases} \tag{3–8.71}$$

More generally, it is compatible with the differential equation (3–8.62) to impose the symmetry restriction

$$-f^\nu(x' - x) = f^\nu(x - x'). \tag{3–8.72}$$

The four-dimensional replacement in (3–8.55) will be used to perform the space-time extrapolation of the source couplings inferred under causal circumstances. If one of the f^ν functions in (3–8.66) were adopted, an additional causal element, which is arbitrary and physically irrelevant, would be injected into the description. In contrast, the kind of function illustrated in (3–8.69) is temporally inert, and its arbitrary aspects are confined to spatial directions. Since causality is a fundamental guiding principle, we reject the use of functions such as those in (3–8.66). Without being committed to the specific examples of (3–8.69, 70), we do insist that $f^\nu(x - x')$ have a space-like direction and be localized in its time-like coordinate excursions.

The desired space-time extrapolation is given by

$$\langle 0_+|0_-\rangle^{J \ ^*J} = \exp[iW(J \ ^*J)] \tag{3–8.73}$$

with

$$W(J \ ^*J) = \tfrac{1}{2}\int (dx)(dx')J^\mu(x)D_+(x - x')J_\mu(x')$$

$$+ \tfrac{1}{2}\int (dx)(dx') \ ^*J^\mu(x)D_+(x - x') \ ^*J_\mu(x')$$

$$+ \int (dx)(dx')(dx'')\epsilon^{\mu\nu\kappa\lambda}J_\mu(x)D_+(x - x')f_\nu(x' - x'')\partial''_\kappa \ ^*J_\lambda(x''). \tag{3–8.74}$$

In verifying that this properly represents the initial causal situation, we encounter the Fourier transforms

$$\int (dx')(dx'')e^{-ipx'}f_\nu(x' - x'')\partial''_\kappa \ ^*J_\lambda(x'') = f_\nu(p)ip_\kappa \ ^*J_\lambda(p) \tag{3–8.75}$$

and

$$\int (dx')(dx'')e^{ipx'}f_\nu(x' - x'')\partial''_\kappa \ ^*J_\lambda(x'') = -f_\nu(-p)ip_\kappa \ ^*J_\lambda(-p), \tag{3–8.76}$$

which are involved in reproducing the last two terms of (3–8.51). The latter are interchanged by the substitution:

$$J_\mu(p) \to \ ^*J_\mu(p), \qquad \ ^*J_\mu(p) \to -J_\mu(p). \tag{3–8.77}$$

To test W for this symmetry property it is convenient to introduce four-dimensional momentum notation:

$$W(J \ ^*J) = \int \frac{(dp)}{(2\pi)^4} \frac{1}{p^2 - i\epsilon} [\tfrac{1}{2}J^\mu(-p)J_\mu(p) + \tfrac{1}{2} \ ^*J^\mu(-p) \ ^*J_\mu(p)$$

$$+ \epsilon^{\mu\nu\kappa\lambda}J_\mu(-p)f_\nu(p)ip_\kappa \ ^*J_\lambda(p)]. \tag{3–8.78}$$

The effect, on the last term, of the substitution (3–8.77), combined with $p_\mu \to -p_\mu$ and $\mu \leftrightarrow \lambda$, is

$$\epsilon^{\mu\nu\kappa\lambda}J_\mu(-p)f_\nu(p)ip_\kappa \ ^*J_\lambda(p) \to \epsilon^{\mu\nu\kappa\lambda}J_\mu(-p)(-)f_\nu(-p)ip_\kappa \ ^*J_\lambda(p), \tag{3–8.79}$$

and invariance requires that

$$-f_\nu(-p) = f_\nu(p). \tag{3–8.80}$$

In a causal situation, however, $f_\nu(p)$ occurs only in the combination $ip^\nu f_\nu(p) = 1$, and this additional symmetry property is not necessary. But if the relation between the two source types is to be maintained generally, the condition (3-8.80), which is also (3-8.72), must be imposed. It follows that $W(J *J)$ preserves its form under the general source transformation of (3-8.47).

Two kinds of test sources define two kinds of fields:

$$\delta W(J *J) = \int (dx)[\delta J^\mu(x) A_\mu(x) + \delta *J^\mu(x) *A_\mu(x)], \qquad (3\text{-}8.81)$$

where

$$\partial_\mu \, \delta J^\mu(x) = 0, \qquad \partial_\mu \, \delta *J^\mu(x) = 0. \qquad (3\text{-}8.82)$$

There are also two independent kinds of gauge arbitrariness, which are incorporated in the field expressions

$$
\begin{aligned}
A_\mu(x) = &\int (dx') D_+(x - x') J_\mu(x') \\
&+ \int (dx')(dx'') D_+(x - x') f^\nu(x' - x'') \, {}^*(\partial_\mu'' \, {}^*J_\nu(x'') - \partial_\nu'' \, {}^*J_\mu(x'')) \\
&+ \partial_\mu \lambda(x),
\end{aligned}
$$

$$
\begin{aligned}
{}^*A_\mu(x) = &\int (dx') D_+(x - x') \, {}^*J_\mu(x') \\
&- \int (dx')(dx'') D_+(x - x') f^\nu(x' - x'') \, {}^*(\partial_\mu'' J_\nu(x'') - \partial_\nu'' J_\mu(x'')) \\
&+ \partial_\mu \, {}^*\lambda(x),
\end{aligned}
\qquad (3\text{-}8.83)
$$

where $\epsilon^{\mu\nu\kappa\lambda}$ has been used to form dual tensors. The following identity should be noted [it is the content of Eq. (3-8.13)]

$$\partial_\mu \, {}^*(\partial^\mu V^\nu - \partial^\nu V^\mu) = \epsilon^{\mu\nu\kappa\lambda} \partial_\mu \partial_\kappa V_\lambda = 0. \qquad (3\text{-}8.84)$$

Accordingly,

$$\partial_\mu A^\mu(x) = \partial^2 \lambda(x), \qquad \partial_\mu \, {}^*A^\mu(x) = \partial^2 \, {}^*\lambda(x), \qquad (3\text{-}8.85)$$

and we now get

$$
\begin{aligned}
-\partial^2 A_\mu(x) &+ \partial_\mu \partial_\nu A^\nu(x) \\
&= J_\mu(x) + \int (dx') f^\nu(x - x') \, {}^*(\partial_\mu' \, {}^*J_\nu(x') - \partial_\nu' \, {}^*J_\mu(x')), \\
-\partial^2 \, {}^*A_\mu(x) &+ \partial_\mu \partial_\nu \, {}^*A^\nu(x) \\
&= {}^*J_\mu(x) - \int (dx') f^\nu(x - x') \, {}^*(\partial_\mu' J_\nu(x') - \partial_\nu' J_\mu(x')).
\end{aligned}
\qquad (3\text{-}8.86)
$$

Observe also that, for example,

$$
\begin{aligned}
\int (dx') f^\nu(x - x') \, {}^*&(\partial_\mu' J_\nu(x') - \partial_\nu' J_\mu(x')) \\
&= \partial^\kappa \int (dx') \epsilon_{\mu\nu\kappa\lambda} f^\nu(x - x') J^\lambda(x') \\
&= -\partial^\nu \int (dx') \, {}^*(f_\mu(x - x') J_\nu(x') - f_\nu(x - x') J_\mu(x')), \qquad (3\text{-}8.87)
\end{aligned}
$$

where the last form involves the interchange of the indices κ and ν. The two gauge invariant fields

$$F_{\mu\nu}(x) = \partial_\mu A_\nu(x) - \partial_\nu A_\mu(x)$$
$$+ \int (dx') \, ^*(f_\mu(x - x') \, ^*J_\nu(x') - f_\nu(x - x') \, ^*J_\mu(x')),$$

$$^*F_{\mu\nu}(x) = \partial_\mu \, ^*A_\nu(x) - \partial_\nu \, ^*A_\mu(x)$$
$$- \int (dx') \, ^*(f_\mu(x - x')J_\nu(x') - f_\nu(x - x')J_\mu(x'))$$

(3–8.88)

then obey

$$\partial_\nu F^{\mu\nu}(x) = J^\mu(x), \qquad \partial_\nu \, ^*F^{\mu\nu}(x) = \, ^*J^\mu(x). \tag{3–8.89}$$

But only if $^*F^{\mu\nu}(x)$ is the dual of $F^{\mu\nu}(x)$ can we proclaim these to be the general form of Maxwell's equations, with electric and magnetic currents.

A direct procedure for this purpose is to evaluate the curls of the two vector potentials $A_\mu(x)$, $^*A_\mu(x)$ and compare the results in (3–8.88). Here is another identity that is valid for any antisymmetrical tensor $G_{\mu\nu}$,

$$\partial_\mu \, ^*G_{\nu\lambda} + \partial_\nu \, ^*G_{\lambda\mu} + \partial_\lambda \, ^*G_{\mu\nu} = -\epsilon_{\mu\nu\lambda\kappa}\partial_\alpha G^{\kappa\alpha}. \tag{3–8.90}$$

In consequence,

$$\partial_\mu A_\nu(x) - \partial_\nu A_\mu(x) - \int (dx')D_+(x - x')(\partial'_\mu J_\nu(x') - \partial'_\nu J_\mu(x'))$$
$$= -\int (dx')(dx'')D_+(x - x')f^\lambda(x' - x'')\partial''_\lambda \, ^*(\partial''_\mu \, ^*J_\nu(x'') - \partial''_\nu \, ^*J_\mu(x''))$$
$$- \int (dx')(dx'')D_+(x - x')f^\lambda(x' - x'')\epsilon_{\mu\nu\lambda\kappa}(-\partial''^2) \, ^*J^\kappa(x''), \tag{3–8.91}$$

with a similar expression involving $^*A_\mu(x)$, and the use of the differential equations

$$\partial'_\lambda f^\lambda(x' - x'') = \delta(x' - x''), \qquad -\partial^2 D_+(x - x') = \delta(x - x') \tag{3–8.92}$$

gives

$$F_{\mu\nu}(x) = \int (dx')D_+(x - x')[\partial'_\mu J_\nu(x') - \partial'_\nu J_\mu(x') - \, ^*(\partial'_\mu \, ^*J_\nu(x') - \partial'_\nu \, ^*J_\mu(x'))],$$

(3–8.93)

$$^*F_{\mu\nu}(x) = \int (dx')D_+(x - x')[\partial'_\mu \, ^*J_\nu(x') - \partial'_\nu \, ^*J_\mu(x') + \, ^*(\partial'_\mu J_\nu(x') - \partial'_\nu J_\mu(x'))].$$

The necessary dual relationship is exhibited here. Notice that the gauge invariant field strengths are also independent of the arbitrary vector f_ν. It is evident that these tensors obey Maxwell's equations. The converse is also true; the solution of the Maxwell equations with outgoing wave boundary conditions is just (3–8.93). To verify this the identity (3–8.90) is applied, in the form

$$\partial_\mu F_{\nu\lambda} + \partial_\nu F_{\lambda\mu} + \partial_\lambda F_{\mu\nu} = \epsilon_{\mu\nu\lambda\kappa} \, ^*J^\kappa, \tag{3–8.94}$$

to

$$\partial_\mu J_\nu - \partial_\nu J_\mu = \partial^\lambda[\partial_\mu F_{\nu\lambda} + \partial_\nu F_{\lambda\mu}], \tag{3–8.95}$$

which produces

$$-\partial^2 F_{\mu\nu} = \partial_\mu J_\nu - \partial_\nu J_\mu - {}^*(\partial_\mu {}^*J_\nu - \partial_\nu {}^*J_\mu). \qquad (3\text{-}8.96)$$

The desired solution is that stated, with its dual, in Eq. (3–8.93).

Apart from the characteristic freedom of gauge transformations, the two vector potentials can be exhibited in terms of the field strengths. First, let us observe that

$$a_\mu(x) = \int (dx')(dx'')f^\nu(x - x') \, {}^*\!\left(f_\mu(x' - x'') \, {}^*J_\nu(x'') - f_\nu(x' - x'') \, {}^*J_\mu(x'')\right)$$

$$(3\text{-}8.97)$$

has the Fourier transform

$$a_\mu(p) = f^\nu(p) \, {}^*\!\left(f_\mu(p) \, {}^*J_\nu(p) - f_\nu(p) \, {}^*J_\mu(p)\right)$$
$$= \epsilon_{\mu\nu\kappa\lambda} f^\nu(p) f^\kappa(p) \, {}^*J^\lambda(p) = 0. \qquad (3\text{-}8.98)$$

The consequent vanishing of $a_\mu(x)$ is exploited to derive from Eq. (3–8.88) that

$$\int (dx')f^\nu(x - x')F_{\mu\nu}(x') = \int (dx')f^\nu(x - x')\left(\partial'_\mu A_\nu(x') - \partial'_\nu A_\mu(x')\right)$$

$$= \partial_\mu\left[\int (dx')f^\nu(x - x')A_\nu(x')\right] - A_\mu(x), \qquad (3\text{-}8.99)$$

which also uses the differential equation obeyed by $f^\nu(x - x')$. Thus,

$$A_\mu(x) = -\int (dx')f^\nu(x - x')F_{\mu\nu}(x') + \partial_\mu\lambda(x),$$

$$\lambda(x) = \int (dx')f^\nu(x - x')A_\nu(x'), \qquad (3\text{-}8.100)$$

and a gauge transformation on $A_\mu(x)$ changes $\lambda(x)$ appropriately to maintain these relations. We do not mean to suggest, by the way, that the $\lambda(x)$ of Eqs. (3–3.83) and (3–8.100) are the same function. A common symbol is used since both functions embody the characteristic arbitrariness of the vector potential. By introducing the compensating gauge transformation, the $\lambda(x)$ of Eq. (3–8.100) can always be reduced to zero. The result is a particular gauge in which

$$A_\mu(x) = -\int (dx')f^\nu(x - x')F_{\mu\nu}(x') \qquad (3\text{-}8.101)$$

and

$$\int (dx')f^\nu(x - x')A_\nu(x') = 0. \qquad (3\text{-}8.102)$$

Similarly,

$${}^*A_\mu(x) = -\int (dx')f^\nu(x - x')F_{\mu\nu}(x'), \qquad (3\text{-}8.103)$$

$$\int (dx')f^\nu(x - x') \, {}^*A_\nu(x') = 0, \qquad (3\text{-}8.104)$$

and we note that the gauge conditions (3–8.102, 104) are not independent statements, but are implied by the constructions of Eqs. (3–8.101) and (3–8.103).

3-9 CHARGE QUANTIZATION. MASS NORMALIZATION

Preparatory to exhibiting various action expressions, we note some integral identities that incorporate the field equations. Thus, from the Maxwell equations (3–8.89) we infer that

$$\int (dx)J^\mu A_\mu = \int (dx)\tfrac{1}{2}F^{\mu\nu}(\partial_\mu A_\nu - \partial_\nu A_\mu),$$

$$\int (dx)\,{}^*J^\mu\,{}^*A_\mu = \int (dx)\tfrac{1}{2}\,{}^*F^{\mu\nu}(\partial_\mu\,{}^*A_\nu - \partial_\nu\,{}^*A_\mu),$$

(3–9.1)

while the equations of (3–8.88) lead to

$$\int (dx)\tfrac{1}{2}F^{\mu\nu}(\partial_\mu A_\nu - \partial_\nu A_\mu) = \int (dx)\tfrac{1}{2}F^{\mu\nu}F_{\mu\nu} + \int (dx)\,{}^*J^\mu\,{}^*A_\mu,$$

$$\int (dx)\tfrac{1}{2}\,{}^*F^{\mu\nu}(\partial_\mu\,{}^*A_\nu - \partial_\nu\,{}^*A_\mu) = \int (dx)\tfrac{1}{2}\,{}^*F^{\mu\nu}\,{}^*F_{\mu\nu} + \int (dx)J^\mu A_\mu.$$

(3–9.2)

In arriving at the last expressions the following property of the dual is used:

$$ {}^*A^{\mu\nu}B_{\mu\nu} = \tfrac{1}{2}\epsilon^{\mu\nu\kappa\lambda}A_{\kappa\lambda}B_{\mu\nu} = A^{\mu\nu}\,{}^*B_{\mu\nu}, \qquad (3\text{–}9.3)$$

which also implies that

$$ {}^*F^{\mu\nu}\,{}^*F_{\mu\nu} = -F^{\mu\nu}F_{\mu\nu}. \qquad (3\text{–}9.4)$$

The quadratic $W(J\,{}^*J)$ expression can be presented as

$$ W = \tfrac{1}{2}\int (dx)(J^\mu A_\mu + {}^*J^\mu\,{}^*A_\mu) \qquad (3\text{–}9.5)$$

or in various equivalent forms, including

$$ W = \int (dx)[J^\mu A_\mu + {}^*J^\mu\,{}^*A_\mu - \tfrac{1}{2}F^{\mu\nu}(\partial_\mu A_\nu - \partial_\nu A_\mu) + \tfrac{1}{4}F^{\mu\nu}F_{\mu\nu}] \quad (3\text{–}9.6)$$

and

$$ W = \int (dx)[J^\mu A_\mu + {}^*J^\mu\,{}^*A_\mu - \tfrac{1}{2}\,{}^*F_{\mu\nu}(\partial_\mu\,{}^*A_\nu - \partial_\nu\,{}^*A_\mu) + \tfrac{1}{4}\,{}^*F^{\mu\nu}\,{}^*F_{\mu\nu}]. $$

(3–9.7)

The latter possess the action property. But one must appreciate the context, describing the independent field variables. In (3–9.6), for example, the fields A_μ, $F_{\mu\nu}$ are subject to independent variation, while the symbol ${}^*A_\mu$ stands for the functional of the field strength tensor stated in (3–8.103),

$$ {}^*A_\mu(x) = -\int (dx')f^\nu(x-x')\,{}^*F_{\mu\nu}(x'). \qquad (3\text{–}9.8)$$

This is verified by performing the indicated operations, which give

$$ \partial_\nu F^{\mu\nu}(x) = J^\mu(x), $$

$$ \partial_\mu A_\nu(x) - \partial_\nu A_\mu(x) = F_{\mu\nu}(x) $$

(3–9.9)

$$ - \int (dx')\,{}^*(f_\mu(x-x')\,{}^*J_\nu(x') - f_\nu(x-x')\,{}^*J_\mu(x')). $$

Proceeding from the dual to the last equation, the second set of Maxwell's equations,

$$\partial_\nu \,{}^*F^{\mu\nu}(x) = {}^*J^\mu(x), \tag{3-9.10}$$

is generated by differentiation. The construction of $A_\mu(x)$ follows as in Eqs. (3–8.99, 100). The use of (3–9.7) is analogous, with ${}^*A_\mu$ and ${}^*F_{\mu\nu}$ as independent fields while A_μ is defined as a functional of the ${}^*F_{\mu\nu}$ by (3–8.101),

$$A_\mu(x) = \int (dx')f'(x - x')\,{}^{**}F_{\mu\nu}(x'). \tag{3-9.11}$$

The asymmetry involved in employing either A_μ or ${}^*A_\mu$ as independent fields is overcome with yet a third action expression:

$$W = \int (dx)(dx')[-f'(x - x')\,(J^\mu(x)F_{\mu\nu}(x') + {}^*J^\mu(x)\,{}^*F_{\mu\nu}(x'))$$
$$- \tfrac{1}{2}\partial_\mu f_\nu(x - x')\,(F^{\mu\lambda}(x)F^\nu{}_\lambda(x') + {}^*F^{\mu\lambda}(x)\,{}^*F^\nu{}_\lambda(x'))], \tag{3-9.12}$$

which is explicitly invariant under the rotational transformation of (3–8.47), and uses the field strengths as independent variables. The equation produced by the stationary action principle can be presented as

$$f^\mu(p)K^\nu(p) - f^\nu(p)K^\mu(p) + {}^*(f^\mu(p)\,{}^*K^\nu(p) - f^\nu(p)\,{}^*K^\mu(p)) = 0, \tag{3-9.13}$$

using momentum space for compactness, where

$$K^\mu(x) = J^\mu(x) - \partial_\nu F^{\mu\nu}(x), \qquad {}^*K^\mu(x) = {}^*J^\mu(x) - \partial_\nu \,{}^*F^{\mu\nu}(x). \tag{3-9.14}$$

Multiplication of (3–9.13) by $f_\mu(p)$ gives

$$f^\mu(p)f_\mu(p)K^\nu(p) = f^\nu(p)f_\mu(p)K^\mu(p). \tag{3-9.15}$$

Then, since $K^\mu(x)$ is divergenceless,

$$p_\nu K^\nu(p) = 0, \tag{3-9.16}$$

we learn, successively, that

$$f_\mu(p)K^\mu(p) = 0 \tag{3-9.17}$$

and

$$K^\nu(p) = 0, \tag{3-9.18}$$

which also uses the positiveness of $-f^\mu(p)f_\mu(p)$ that expresses the choice of n^μ and thereby of $if^\mu(p)$ in (3–8.61) as a space-like vector. A similar treatment of the dual to Eq. (3–9.13) supplies

$$^*K^\nu(p) = 0, \tag{3-9.19}$$

and both sets of Maxwell's equations have been derived in a symmetrical way from the action expression (3–9.12).

By this time the hypothetical alert reader of limitless dedication, henceforth acronymically known as Harold, can no longer restrain himself. The following exchange ensues.

H.: You showed in the previous section that the apparent magnetic charge given in (3–8.41) could be transformed away. It was intimated that a different kind of magnetic current would be forthcoming. Yet the action principle of (3–9.6) and the field equations (3–9.9) are identical in form to (3–8.37) and (3–8.39), with

$$*M_{\mu\nu}(x) = \int (dx')\big(f_\mu(x - x')\,{}^*J_\nu(x') - f_\nu(x - x')\,{}^*J_\mu(x')\big), \quad (3\text{–}9.20)$$

and indeed

$$-\partial_\nu\,{}^*M^{\mu\nu}(x) = {}^*J^\mu(x). \quad (3\text{–}9.21)$$

How then can you claim that true magnetic charge is now being discussed?

S.: Mistake me not, good Sagredo, er, Harold. The function (3–9.20) does differ—in context—from the source function of (3–8.41), for it lacks that degree of localizability which is characteristic of sources. Consider, for example, the choice of $f_\nu(x - x')$ with only the spatial component

$$f_3(x - x') = \delta(x^0 - x^{0'})\,\delta(x_1 - x_1')\,\delta(x_2 - x_2')\eta(x_3 - x_3'), \quad (3\text{–}9.22)$$

giving the nonvanishing tensor component

$$*M_{03}(x) = \int_0^\infty ds\,{}^*J^0(x^0, x_1, x_2, x_3 - s). \quad (3\text{–}9.23)$$

Unlike the spatially limited magnetic charge distribution $*J^0(x)$, $*M_{03}(x)$ becomes independent of x_3 after one has passed through the charge distribution, moving positively along the third axis. This limiting value is

$$*M_{03}(x) \rightarrow \int_{-\infty}^\infty dx_3'\,{}^*J^0(x^0, x_1, x_2, x_3') \quad (3\text{–}9.24)$$

and the surface integral

$$-\int dx_1\,dx_2\,{}^*M^0{}_3(x) \rightarrow \int (d\mathbf{x})\,{}^*J^0(x), \quad (3\text{–}9.25)$$

which need not be zero. That is in contrast with the null value of (3–8.44), which was based on the spatial localizability of $*M_{\mu\nu}(x)$. Had we used the odd f^μ function of (3–8.70), the explicit form of $*M_{03}(x)$ would be different, but not the value of the surface integral that produces the total magnetic charge. Thus, it is through the special properties of the class of f^μ functions that we make the transition from mere semblance to the reality of magnetic charge. At the same time this poses a fundamental problem since the detailed description would seem to depend upon the arbitrary choice of the f^μ function, for which there is no physical basis. Surmounting that formidable difficulty is the task to which we now address ourselves.

Let us introduce into (3–8.81), which is the differential statement of the dependence of W on the source functions, those expressions for $\delta J^\mu(x)$ and $\delta\,{}^*J^\mu(x)$

that convey the effect of an arbitrary coordinate-dependent displacement:

$$\delta J^\mu(x) = \partial_\nu(\delta x^\nu(x)J^\mu(x)) - J^\nu(x)\partial_\nu\delta x^\mu(x)$$
$$= -\partial_\nu[\delta x^\mu(x)J^\nu(x) - \delta x^\nu(x)J^\mu(x)] \qquad (3\text{-}9.26)$$

and

$$\delta\,{}^*J^\mu(x) = -\partial_\nu[\delta x^\mu(x)\,{}^*J^\nu(x) - \delta x^\nu(x)\,{}^*J^\mu(x)]. \qquad (3\text{-}9.27)$$

The conservation requirements (3–8.82) are identically satisfied. This insertion gives

$$\delta W = \int(dx)[\delta x^\nu J^\mu(\partial_\mu A_\nu - \partial_\nu A_\mu) + \delta x^\nu\,{}^*J^\mu(\partial_\mu\,{}^*A_\nu - \partial_\nu\,{}^*A_\mu)]$$
$$= \int(dx)[\delta x^\nu J^\mu F_{\mu\nu} + \delta x^\nu\,{}^*J^\mu\,{}^*F_{\mu\nu}]$$
$$+ \int(dx)(dx')[{}^*(\delta x^\mu(x)J^\nu(x) - \delta x^\nu(x)J^\mu(x))f_\mu(x-x')\,{}^*J_\nu(x')$$
$$- {}^*(\delta x^\mu(x)\,{}^*J^\nu(x) - \delta x^\nu(x)\,{}^*J^\mu(x))f_\mu(x-x')J_\nu(x')]; \qquad (3\text{-}9.28)$$

alternative expressions for the two terms are exhibited in writing

$$\delta W = -\int(dx)[F^{\mu\lambda}F^\nu{}_\lambda - \tfrac14 g^{\mu\nu}F^{\alpha\beta}F_{\alpha\beta}]\partial_\mu\,\delta x_\nu$$
$$- \int(dx)(dx')\,\epsilon^{\mu\nu\kappa\lambda}(\delta x_\mu(x) - \delta x_\mu(x'))f_\nu(x-x')J_\kappa(x)\,{}^*J_\lambda(x'). \qquad (3\text{-}9.29)$$

Two elementary statements concerning explicit f^μ dependence emerge from these forms. If electric and magnetic currents are proportional with a universal constant, the f^μ term vanishes, as it should since this is a rotated version of pure electric charge; when electric currents are causally separated from magnetic currents the f^μ term vanishes, according to the restriction on the class of f^μ functions that confines it to space-like vectors connecting points in space-like relation.

It is the situation of electric and magnetic charge coexisting with different space-time distributions that poses the problem of nonphysical f^μ dependence. To make this very explicit, suppose $f^\mu(x - x')$ is chosen as in (3–8.70), a spatial vector of fixed direction with its support, its nonvanishing domain, confined to a line of that direction. Those points in the two source distributions that can be connected by this line contribute to δW. When the direction of the line is varied continuously, δW and W itself also vary, continuously, thereby denying to W any physical meaning. Is this the death knell of magnetic charge? No. There is a subtle possibility concealed here. It depends upon the precise fact that not W but $\exp[iW]$ is the physically significant quantity. If, in altering the direction of f^μ continuously, W were indeed to change, but change discontinuously—by multiples of 2π—the exponential would remain unaltered and the mathematical arbitrariness of f^μ should be without physical consequence. This is impossible, of course, when, as is assumed above, the sources are continuously distributed objects. Instead, they must have a granular structure, giving values of the f^μ

integral that differ by finite amounts according as the f^μ line does or does not penetrate the kernels of that structure. And, since the magnitude of the integral is also measured by the product of electric and magnetic charge, this combination cannot be arbitrary but must be restricted to certain discrete values. These are remarkable conclusions—charge is completely localized and quantized in magnitude. The sweeping nature of such inferences should be emphasized. We are encountering restrictions on the structure of photon sources that are required for the consistency of a theory of electric and magnetic charges. Sources are introduced as idealizations of realistic physical mechanisms, idealizations that dispense with individual characteristics but respect all general laws. In uncovering fundamental restrictions on sources, we are revealing general laws of nature. Such was the argument when the divergenceless nature of the vector photon source, demanded by the null photon mass, was interpreted as the assertion of a general conservation law, that of electric charge.

A realization of electric and magnetic currents in terms of the motion of point charges is given by

$$J^\mu(x) = \sum_a e_a \int_{-\infty}^{\infty} ds\, \frac{dx_a^\mu(s)}{ds}\, \delta(x - x_a(s)),$$

$$^*J^\mu(x) = \sum_a {}^*e_a \int_{-\infty}^{\infty} ds\, \frac{dx_a^\mu(s)}{ds}\, \delta(x - x_a(s))$$

(3-9.30)

(instead of $^*e$, symbols such as g are also used but we wish to emphasize the symmetry between electric and magnetic quantities). The causal motion of the points is conveyed by the restrictions

$$-\frac{dx_a^\mu(s)}{ds}\frac{dx_{a\mu}(s)}{ds} > 0, \qquad \frac{dx_a^0(s)}{ds} > 0.$$

(3-9.31)

The conservation properties hold individually, according to the calculation

$$\partial_\mu \int_{-\infty}^{\infty} ds\, \frac{dx^\mu(s)}{ds}\, \delta(x - x(s)) = -\int_{-\infty}^{\infty} ds\, \frac{dx^\mu(s)}{ds}\, \frac{\partial}{\partial x^\mu(s)}\, \delta(x - x(s)) = 0,$$

(3-9.32)

since the point $x^\mu(s)$ is infinitely remote from x^μ at the terminals of the integration. The evident identification of the e_a and $^*e_a$ as charges attached to the individual moving points is consistent with the evaluation of the total charges, as in

$$\int d\sigma_\mu J^\mu = \sum_a \int d\sigma_\mu \, dx^\mu(s)\, \delta(x - x(s)) = \sum e_a,$$

(3-9.33)

where the integration sweeps the whole four-dimensional domain with $d\sigma_\mu\, dx^\mu(s)$ acting as volume element.

We cannot simply insert these expressions into $W(J\,{}^*J)$, however. The latter was devised for continuously distributed sources and should not be applied to a collection of point charges without reexamination of the *physical* significance of W. But it is useful, and serves as an intermediate stage in the development, to modify the known results in a manner that is without effect for continuously distributed sources but makes the consideration of point charges *mathematically meaningful*. This is achieved by introducing an arbitrarily small space-like vector λ^μ and constructing the provisional action

$$W(\lambda) = \int (dx)[J^\mu(x)A_\mu(x \pm \lambda) + {}^*J^\mu(x) \, {}^*A_\mu(x \pm \lambda)$$
$$- \tfrac{1}{2}F^{\mu\nu}(x)(\partial_\mu A_\nu(x \pm \lambda) - \partial_\nu A_\mu(x \pm \lambda)) + \tfrac{1}{4}F^{\mu\nu}(x)F_{\mu\nu}(x \pm \lambda)], \tag{3-9.34}$$

where the appearance of $\pm\lambda$ signifies the procedure of equal averaging for expressions containing $+\lambda^\mu$ and $-\lambda^\mu$. This action continues to be stationary for field variations about the solutions of the Maxwell equations:

$$\delta W(\lambda) = \int (dx) \, \delta A_\mu(x \pm \lambda)[J^\mu(x) - \partial_\nu F^{\mu\nu}(x)]$$
$$- \int (dx)\tfrac{1}{2} \, \delta F^{\mu\nu}(x \pm \lambda)[\partial_\mu A_\nu(x) - \partial_\nu A_\mu(x) - F_{\mu\nu}(x)$$
$$+ \int (dx') \, {}^*(f_\mu(x - x') \, {}^*J_\nu(x') - f_\nu(x - x') \, {}^*J_\mu(x'))] = 0, \quad (3\text{-}9.35)$$

which uses the possibility of performing a displacement to transfer $\pm\lambda^\mu$ from one field factor to the other. To evaluate $W(\lambda)$ we use

$$W(\lambda) = \tfrac{1}{2}\int (dx)[J^\mu(x)A_\mu(x \pm \lambda) + {}^*J^\mu(x) \, {}^*A_\mu(x \pm \lambda)]. \tag{3-9.36}$$

The point charge construction of the currents gives

$$W(\lambda) = \tfrac{1}{2} \sum_{a \neq b} W_{ab}(\lambda) + \sum_a W_a(\lambda), \tag{3-9.37}$$

where

$$W_{ab}(\lambda) = (e_a e_b + {}^*e_a \, {}^*e_b)$$
$$\times \int ds \, ds' \frac{dx_a^\mu(s)}{ds} \frac{dx_{b\mu}(s')}{ds'} D_+(x_a(s) - x_b(s') \pm \lambda) + (e_a \, {}^*e_b - {}^*e_a e_b)$$
$$\times \int ds \, ds'(dx)\epsilon^{\mu\nu\kappa\lambda} \frac{dx_{a\mu}(s)}{ds} D_+(x_a(s) - x \pm \lambda)$$
$$\times \partial_\kappa f_\nu(x - x_b(s')) \frac{dx_{b\lambda}(s')}{ds'} \tag{3-9.38}$$

is symmetrical in a and b, and

$$W_a(\lambda) = \tfrac{1}{2}(e_a^2 + {}^*e_a^2)\int ds \, ds' \frac{dx_a^\mu(s)}{ds} \frac{dx_{a\mu}(s')}{ds'} D_+(x_a(s) - x_a(s') \pm \lambda). \tag{3-9.39}$$

The mathematical existence problem which the λ device is designed to overcome is concentrated in $W_a(\lambda)$. In the neighborhood of $s - s' \sim 0$, D_+ would be singular without the addition of the space-like λ^μ to its argument. This difficulty is restricted to the real part of D_+, however,

$$\text{Re } D_+(x - x') = (1/4\pi) \, \delta[(x - x')^2], \tag{3-9.40}$$

[cf. Eq. (2-1.64)], as contrasted with

$$\text{Im } D_+(x - x') = \text{Re} \int d\omega_p e^{ip(x-x')}. \tag{3-9.41}$$

Thus

$$\text{Im } W_a(\lambda) = \tfrac{1}{2}(e_a^2 + \,^*e_a^2)$$
$$\times \int d\omega_p \left[\int_{-\infty}^{\infty} ds \, \frac{dx_a^\mu(s)}{ds} \, e^{ipx_a(s)} \right] \left[\int_{-\infty}^{\infty} ds \, \frac{dx_{a\mu}(s)}{ds} \, e^{-ipx_a(s)} \right] \cos p\lambda \tag{3-9.42}$$

has natural upper frequency limits if the motion of the particle is without discontinuity, and the limit $\lambda^\mu \to 0$ can be introduced directly into (3-9.42). To discuss

$$w_a(\lambda) = \text{Re } W_a(\lambda)$$
$$= \frac{e_a^2 + \,^*e_a^2}{8\pi} \int ds \, ds' \, \frac{dx_a^\mu(s)}{ds} \, \frac{dx_{a\mu}(s')}{ds'} \, \delta[(x_a(s) - x_a(s') \pm \lambda)^2] \tag{3-9.43}$$

for sufficiently small λ^μ and s correspondingly close to s', it suffices to consider uniform motion. Let us use the rest frame, for simplicity, and identify ds with dx_a^0 in that frame of reference, while choosing λ^μ to be a spatial vector. Then, with $s - s' = \sigma$,

$$w_a(\lambda) \cong - \frac{e_a^2 + \,^*e_a^2}{8\pi} \int ds_a \int_{-\infty}^{\infty} d\sigma \, \delta(\sigma^2 - \lambda^2)$$
$$= - \frac{e_a^2 + \,^*e_a^2}{8\pi} \frac{1}{(\lambda^2)^{1/2}} \int ds_a. \tag{3-9.44}$$

Does $w_a(\lambda)$ have a physical significance? It does not. This quantity is associated with a single point charge or particle. Since the particles that comprise a source have prescribed motions they are being idealized as very massive particles, which are uninfluenced by the effects they produce. The description of their individual mechanical properties logically precedes the discussion of interactions. The nature of this description can be inferred from the results concerning stress tensors and their values in single-particle states: $t^{\mu\nu} = 2 \, d\omega_p p^\mu p^\nu$. As we have explained, this is a simplification valid in the interior of a beam where the variation of momentum and the associated finite spatial extension can be neglected. To reinstate these, we identify p^μ with the gradient of a phase function $\varphi(x)$, and introduce a variable weight function,

$$t^{\mu\nu}(x) = \rho(x)\partial^\mu\varphi(x)\partial^\nu\varphi(x); \tag{3-9.45}$$

the mass restriction

$$\partial^\mu\varphi\partial_\mu\varphi + m^2 = 0 \tag{3-9.46}$$

recalls the momentum significance of $\partial^\mu \varphi$. Note that

$$\partial_\mu t^{\mu\nu} = \partial_\mu(\rho\partial^\mu \varphi)\partial^\nu \varphi + \rho\tfrac{1}{2}\partial_\nu(\partial^\mu \varphi \partial_\mu \varphi) \tag{3-9.47}$$

and the local mechanical conservation laws are satisfied by the conservation of particle flux,

$$\partial_\mu(\rho\partial^\mu \varphi) = 0. \tag{3-9.48}$$

This interpretation also supplies the value of the integral:

$$\int d\sigma_\mu \rho\partial^\mu \varphi = 1. \tag{3-9.49}$$

Within this picture of prescribed motion it is consistent to take

$$\rho(x) = (1/m)\int_{-\infty}^{\infty} ds\, \delta(x - x(s)),$$

$$\partial^\nu \varphi(x(s)) = m\frac{dx^\nu(s)}{ds}. \tag{3-9.50}$$

Indeed, the conservation law is satisfied,

$$\partial_\mu(\rho\partial^\mu \varphi) = -\int_{-\infty}^{\infty} ds\, \frac{dx^\mu(s)}{ds}\frac{\partial}{\partial x^\mu(s)}\, \delta(x - x(s))$$

$$= -\int_{-\infty}^{\infty} ds\, \frac{d}{ds}\, \delta(x - x(s)) = 0, \tag{3-9.51}$$

and

$$\int d\sigma_\mu \rho\partial^\mu \varphi = \int (dx)\, dx^0(s)\, \delta(x - x(s)) = 1. \tag{3-9.52}$$

In transferring these results to the connection between action and stress tensor,

$$\delta W = -\int (dx)t^{\mu\nu}\partial_\mu\, \delta x_\nu, \tag{3-9.53}$$

one must not forget the meaning of $\delta x_\nu(x)$. It arose as a generalization of the rigid displacements given to sources, which were intended to simulate the displacement of a reference surface and are therefore in the opposite sense. Thus, when translating into the motion of point particles a minus sign must be affixed:

$$\delta W = m\int ds\, \frac{dx^\nu(s)}{ds}\frac{d\delta x_\nu(s)}{ds}. \tag{3-9.54}$$

It is now necessary to generalize the identification of ds with dx^0, in the rest frame, to the invariant proper time definition

$$-(ds)^2 = dx^\nu\, dx_\nu. \tag{3-9.55}$$

Its consequence for a varied motion,

$$-ds\, \delta ds = dx^\nu\, d\delta x_\nu, \tag{3-9.56}$$

converts (3-9.54) into

$$\delta W = -m\delta\int ds \tag{3-9.57}$$

and supplies the action expression for a single particle, labeled a, performing a prescribed motion,

$$W_a = -m_a \int ds_a. \tag{3-9.58}$$

The phenomenological orientation of source theory has the following corollary. Physical parameters identified under restricted physical circumstances do not change their meaning when a wider class of phenomena is considered. The mass parameter m_a is determined from the response of the particle to weak, slowly varying, prescribed forces as in beam deflection experiments. When electromagnetic interactions among several particles are considered, this parameter is not assigned a different value. It has already been fixed, normalized, by experiment. Thus the single-particle term (3-9.44) must not be added to (3-9.58), thereby changing the value of m_a. There is no question here of assigning some fraction of the total mass to an electromagnetic origin. What is at issue is the consistency between the various levels of dynamical description through which one passes in the course of the evolution of the theory. The prescribed forces of the most elementary level become assigned to the motion of particles at the next stage, but in neither one is there any reference to individual particle structure and the phenomenological parameter m_a must be common to both. The conclusion is that the real $w_a(\lambda)$ terms, which contribute neither to the vacuum persistence probability nor to the couplings among sources, must be struck out. Here, then, is the action to be associated with a point charge realization of photon sources:

$$W = \underset{\lambda \to 0}{\mathrm{Lim}} \left[W(\lambda) - \sum_a w_a(\lambda) \right]. \tag{3-9.59}$$

Consider again the effect of a source displacement, now pictured through the motion of point charges. We use (3-9.28), but with attention to the λ displacement and the minus sign required to translate $\delta x^\nu(x)$ into $\delta x_a^\nu(s)$:

$$\delta W = \sum_a \int ds\, \delta x_a^\mu(s) \frac{dx_a^\nu(s)}{ds} \left[e_a F_{\mu\nu}(x_a(s) \pm \lambda) + {}^*e_a\, {}^*F_{\mu\nu}(x_a(s) \pm \lambda) \right]$$

$$- \sum_{ab} (e_a\, {}^*e_b - {}^*e_a e_b) \int ds\, ds'\, {}^* \left(\delta x_a^\mu(s) \frac{dx_a^\nu(s)}{ds} - \delta x_a^\nu(s) \frac{dx_a^\mu(s)}{ds} \right)$$

$$\times f_\mu(x_a(s) - x_b(s') \pm \lambda) \frac{dx_{b\nu}(s')}{ds'} - \sum_a \delta w_a(\lambda) \Big|_{\lambda^\mu \to 0}. \tag{3-9.60}$$

The antisymmetrical product of two vector displacements defines a two-dimensional element of area,

$$\delta x_a^\mu\, dx_a^\nu - \delta x_a^\nu\, dx_a^\mu = d\sigma_a^{\mu\nu}, \tag{3-9.61}$$

and the antisymmetrical product of three displacements produces a three-dimensional volume element, or the equivalent directed surface element for the coordinates $x_a^\mu - x_b^\mu$,

$$d\, {}^*\sigma_a^{\mu\nu}\, dx_{b\nu} = d\sigma_{ab}^\mu. \tag{3-9.62}$$

The corresponding presentation of (3–9.60)

$$\delta W = \sum_a \int \tfrac{1}{2} d\sigma_a^{\mu\nu} [e_a F_{\mu\nu}(x_a \pm \lambda) + {}^* e_a \, {}^* F_{\mu\nu}(x_a \pm \lambda)]$$
$$- \sum_{ab} (e_a \, {}^* e_b - {}^* e_a e_b) \int d\sigma_{ab}^{\mu} f_\mu (x_a - x_b \pm \lambda) - \sum_a \delta w_a(\lambda) \Big|_{\lambda^\mu \to 0},$$
$$(3\text{–}9.63)$$

is no longer limited to infinitesimal displacements; the integrations extend over the geometrical domains defined by the initial and final particle trajectories.

Given the various three-dimensional surfaces that occur in (3–9.63), all the individual f^μ integrals can be made to vanish by appropriate choice of the f^μ support, which need not be restricted to straight lines. For any other selection of f^μ that gives nonvanishing values to one or more of the integrals, those values must be confined to integral multiples of 2π. Consider a pair of particles a and b, for which the three dimensional surface σ that is traced out by $x_a^\mu - x_b^\mu$ is effectively displaced by $\pm\lambda^\mu$. We designate these surfaces by $\sigma(\pm\lambda)$ and write the condition guaranteeing physical uniqueness as

$$(e_a \, {}^* e_b - {}^* e_a e_b) \tfrac{1}{2} \left[\int_{\sigma(\lambda)} + \int_{\sigma(-\lambda)} \right] d\sigma_\mu f^\mu(x) = 2\pi n, \qquad (3\text{–}9.64)$$

where n is an integer. In order to ensure that nonphysical elements do not intervene during the limiting process $\lambda^\mu \to 0$, we demand that this hold for almost all λ^μ. The scale of f^μ is fixed by the differential equation (3–8.62), or the equivalent integral statement

$$\int d\sigma_\mu f^\mu(x) = 1, \qquad (3\text{–}9.65)$$

referring to any surface that encloses the origin. The discreteness required by (3–9.64) implies that the support of f^μ on any such surface is confined to a finite number of points. And, in virtue of the symmetry property (3–8.72),

$$-f^\mu(-x) = f^\mu(x), \qquad (3\text{–}9.66)$$

that number must be an even integer, 2ν. We may visualize this number of filaments drawn out from the origin in a way that assigns to each filament its image in the origin. Let the contribution to the surface integral (3–9.65) that is associated with an individual point α, $\alpha = 1, \ldots, 2\nu$, be designated r_α so that

$$\sum_{\alpha=1}^{2\nu} r_\alpha = 1. \qquad (3\text{–}9.67)$$

The basic situation for (3–9.64) is that $\sigma(\lambda)$, for example, includes a single point α, while $\sigma(-\lambda)$ contains no support point of f^μ. Then

$$(e_a \, {}^* e_b - {}^* e_a e_b) \tfrac{1}{2} r_\alpha = 2\pi n_\alpha, \qquad (3\text{–}9.68)$$

and the addition of such expressions represents any other possibility. In particular, the summation over all $\alpha = 1, \ldots, 2\nu$ gives

$$\tfrac{1}{2}(e_a \,{}^*e_b - {}^*e_a e_b) = 2\pi \sum_{\alpha=1}^{2\nu} n_\alpha \tag{3-9.69}$$

or, making explicit that the points of support occur in pairs with equal values of r_α and n_α,

$$\sum_{\alpha=1}^{2\nu} n_\alpha = 2 \sum_{\alpha=1}^{\nu} n_\alpha = 2n_{ab}, \tag{3-9.70}$$

we get the charge quantization condition

$$\frac{1}{4\pi} (e_a \,{}^*e_b - e_b \,{}^*e_a) = 2n_{ab}. \tag{3-9.71}$$

Note that the weight factors r_α take the rational form

$$r_\alpha = n_\alpha/2n_{ab}. \tag{3-9.72}$$

If all 2ν points are equivalent, $r_\alpha = 1/(2\nu)$, and the integer n_{ab} is an integral multiple of ν. The simplest possibility, $\nu = 1$, is illustrated in the f^μ function of (3-8.70).

With the success in removing the arbitrary aspects of δW through the recognition that only $\exp[iW]$ is significant, we can present (3-9.63) effectively as

$$\delta W = \sum_a \int \tfrac{1}{2} d\sigma_a^{\mu\nu}[e_a F_{\mu\nu}(x_a \pm \lambda) + {}^*e_a \,{}^*F_{\mu\nu}(x_a \pm \lambda)] - \sum_a \delta w_a(\lambda)\Big|_{\lambda^\mu \to 0}. \tag{3-9.73}$$

This might seem to pose another problem, however. Although we retain the symbol δW, it is no longer the change of a quantity W and the question of uniqueness arises. Consider a continuous deformation of the trajectories that finally returns them to the initial configuration, thereby defining a surface enclosing a three-dimensional volume. As the covariant generalization of the three-dimensional relation

$$\int d\mathbf{S} \cdot \mathbf{H} = \int (d\mathbf{x}) \boldsymbol{\nabla} \cdot \mathbf{H} = \int (d\mathbf{x}) \,{}^*J^0, \tag{3-9.74}$$

we have

$$\int \tfrac{1}{2} d\sigma^{\mu\nu} F_{\mu\nu} = \int d\sigma_\mu \partial_\nu \,{}^*F^{\mu\nu} = \int d\sigma_\mu \,{}^*J^\mu, \tag{3-9.75}$$

and similarly

$$\int \tfrac{1}{2} d\sigma^{\mu\nu} \,{}^*F_{\mu\nu} = -\int d\sigma_\mu \partial_\nu F^{\mu\nu} = -\int d\sigma_\mu J^\mu. \tag{3-9.76}$$

The net change of W on completing this circuit is, therefore,

$$\Delta W = \sum_a \tfrac{1}{2} \left[\int_{\sigma_a(\lambda)} + \int_{\sigma_a(-\lambda)} \right] d\sigma_\mu (e_a \,{}^*J^\mu - {}^*e_a J^\mu), \tag{3-9.77}$$

where $\sigma_a(\pm\lambda)$ indicates the three-dimensional volume, associated with particle a, which is subjected to the alternative space-like displacements $\pm\lambda^\mu$. The integrals of (3-9.77) record the amounts of electric and magnetic charge within the various volumes. Here the basic situation occurs when particle b lies within the volume $\sigma_a(\lambda)$, for example, but is outside of $\sigma_a(-\lambda)$. The associated contribution to ΔW is $\frac{1}{2}(e_a \, {}^*e_b - {}^*e_a e_b)$, a multiple of 2π according to (3-9.69). This affirmation of the single-valuedness of $\exp[iW]$ was inevitable; it was only of some interest to see how the charge quantization condition brought it about.

The charge quantization demanded by magnetic charge provides a most satisfying explanation for one of the more striking empirical regularities in nature. Despite the widest variation in other properties possessed by particles, the magnitude of the unit of pure electric charge is universal. It is measured by the fine structure constant

$$\alpha = e^2/4\pi = 1/137.036. \tag{3-9.78}$$

If we assume that the smallest magnetic charge magnitude, ${}^*e_0$, corresponds to the smallest integer in (3-9.71), the latter becomes

$$\frac{1}{4\pi} e \, {}^*e_0 = 2, \tag{3-9.79}$$

and

$$ {}^*\alpha_0 = \frac{{}^*e_0^2}{4\pi} = 4(137.036). \tag{3-9.80}$$

This is very large indeed, being the equivalent of the electric charge $2(137)e$. However, one might think, if only for a moment, that this great asymmetry could be apparent since there is the freedom to redefine all electric and magnetic charges by the rotation of (3-8.47):

$$e_a' = e_a \cos\varphi + {}^*e_a \sin\varphi, \qquad {}^*e_a' = -e_a \sin\varphi + {}^*e_a \cos\varphi. \tag{3-9.81}$$

Of course, there are invariants of this rotation in the two-dimensional charge space, including

$$e_a^2 + {}^*e_a^2, \qquad e_a \, {}^*e_b - {}^*e_a e_b, \tag{3-9.82}$$

which correspond geometrically to lengths and angles between two-dimensional vectors. Also relevant is the inequality

$$(e_a \, {}^*e_b - {}^*e_a e_b)^2 \leq (e_a^2 + {}^*e_a^2)(e_b^2 + {}^*e_b^2). \tag{3-9.83}$$

Now consider the following invariant statement. For all known particles, $(e_a^2 + {}^*e_a^2)/4\pi$ is small compared to unity. Comparison of the inequality (3-9.83) with the charge quantization condition (3-9.71) then shows that the integers n_{ab} must all be zero. The corresponding points with coordinates e_a, ${}^*e_a$ are confined to a single line, which thus acquires an absolute significance. It is conventional to identify that line with the axis of pure electric charge. The complete reduction of the line to equally spaced points demands the existence of a second class of particles for which $(e_a^2 + {}^*e_a^2)/4\pi$ is large compared to unity. Among such particles there is no necessity for an absolute charge line although, if the integers of the charge quantization condition assume only moderate values, the charge points will cluster near a line, which is the conventional axis of pure magnetic charge.

It is remarkable that we have been led to the existence of two types of charged particles that are characterized internally by relatively weak and relatively strong forces, for this corresponds to the empirical distinction between leptons and hadrons, respectively. Certainly hadrons—mesons and baryons— are not magnetically charged particles, nor do their interactions possess a strength as great as (3-9.80). Rather, we view them as magnetically neutral composites of particles that carry both electric and magnetic charges, with the observed strong interactions of hadrons emerging as residuals of the considerably stronger magnetic forces, which thus far have successfully prevented the experimental recognition of free magnetic charge. It is essential for this explanation that a magnetically neutral composite appear as an ordinary electrical particle. If we have a group of particles with charges e_a, $*e_a$ such that

$$\sum_a {}^*e_a = 0, \qquad \sum_a e_a = e', \qquad (3\text{-}9.84)$$

the comparison with a reference particle of charges e_0, $*e_0$ gives

$$\frac{1}{4\pi} \sum_a (e_a {}^*e_0 - {}^*e_a e_0) = 2 \sum_a n_{a0} \qquad (3\text{-}9.85)$$

or

$$\frac{1}{4\pi} e' {}^*e_0 = 2n. \qquad (3\text{-}9.86)$$

This is the required charge relation,

$$e' = ne. \qquad (3\text{-}9.87)$$

The automatic appearance of conventional electrical behavior for a magnetically neutral composite is significant because the individual electric charges on particles that carry both types of charge—dual charged particles—can assume unconventional values. We make the specific assumption that the smallest magnetic charge, $*e_0$, is found on a dual charged particle with accompanying electric charge $e_0 \neq 0$ [the value of e_0 was irrelevant in (3-9.85)]. For any other set of dual charges, e_0', $*e_0'$, reference to the unit of pure electric charge shows that $*e_0'$ is a multiple of $*e_0$,

$$*e_0' = m *e_0, \qquad (3\text{-}9.88)$$

and the application of the charge quantization condition to the pair of dual charged particles gives

$$\frac{1}{4\pi} (e_0' {}^*e_0 - {}^*e_0' e_0) = \frac{1}{4\pi} (e_0' - me_0) {}^*e_0 = 2n \qquad (3\text{-}9.89)$$

or

$$e_0' = me_0 + ne, \qquad m, n = 0, \pm 1, \ldots . \qquad (3\text{-}9.90)$$

This exhibits e_0 and e as independent units in a two-dimensional lattice that produces all possible electric charges. Since m measures magnetic charge, in units of $*e_0$, we again recognize that a magnetically neutral composite is restricted to e as a charge unit. It also follows that electric charge differences, for a common value of magnetic charge, are confined to multiples of e.

The discussion of electrical particles and of dual charged particles naturally suggests the consideration of purely magnetic particles. The unit of pure magnetic charge, $*e$, must be an integral multiple of the smallest magnetic charge, as in (3-9.88). We write this specific relation in terms of an integer N,

$$*e = N \, *e_0. \tag{3-9.91}$$

The analogue of (3-9.79), connecting the unit of pure electric charge with the smallest magnetic charge, is the following connection between the unit of pure magnetic charge and the smallest electric charge:

$$\frac{1}{4\pi} e_0 \, *e = 2. \tag{3-9.92}$$

Its immediate consequence is

$$e = Ne_0. \tag{3-9.93}$$

From our various assumptions, which are grounded in the symmetry between electric and magnetic charge, we have inferred that the charge units on a dual charged particle are the same fraction, $1/N$, of the units of pure electric and magnetic charge. Among the possibilities, $2, 3, \ldots$, which value has nature selected for the integer N?

But first we must digress to discuss the relation between the statistics of composite particles and their constituents. One approach uses the spin-statistics connection. A composite formed of an odd number of particles with integer $+ \frac{1}{2}$ spin (F.D. statistics) has a resultant spin angular momentum that is also integer $+ \frac{1}{2}$. This composite particle obeys F.D. statistics. If there are an even number of constituent particles with integer $+ \frac{1}{2}$ spin, the composite possesses integral spin and is a B.E. particle. It is as though a F.D. (B.E.) particle carries a minus (plus) sign and these signs are multiplied to give the statistics of a composite structure. This is more than a mnemonic, for the plus and minus signs identify the algebraic properties of the individual sources that are multiplied to produce the effective source of the composite system. Now, as we have mentioned, there are two varieties of hadrons: mesons, which are B.E. particles, and baryons, which are F.D. particles. If both types of hadrons are to be constructed as magnetically neutral composites of dual charged particles, the latter cannot all be B.E. particles. The simplest assumption is that they are all F.D. particles; an even number of such constituents produces a B.E. particle, an odd number builds a F.D. particle.

Can the dual charged particles exhibit only one strength of magnetic charge? Note that both signs of the magnetic charge, linked to sign changes in electric charge, will occur. This is the antiparticle concept, with both charges involved in order to maintain the structure of the two sets of Maxwell's equations, which have the field strength tensor in common. If the only values of magnetic charge are $(1/N)\,{}^*e$ and $-(1/N)\,{}^*e$, they must be combined to produce a neutral composite, and such pairs of F.D. dual charged particles are B.E. particles; baryons cannot be manufactured in this way. Hence there must be at least two different charge magnitudes. According to the magnetic analogue of the electric lattice construction (3–9.90), the magnetic charges on dual charged particles with the same electric charge must differ by a multiple of ${}^*e$, the unit of pure magnetic charge. It would seem to be a reasonable characterization of dual charged particles to describe them as carrying charges that are smaller in magnitude than the units of pure charge. If that is granted, just two values of magnetic charge are admitted. With a conventional sign choice, they are $-(1/N)\,{}^*e$ and $[(N-1)/N]\,{}^*e$. The possible values of electric charge are analogous: $-(1/N)e$ and $[(N-1)/N]e$. Either electric charge can be assigned to either choice of magnetic charge, giving four dual charge combinations, although there may be duplications of these assignments.

In addition to neutralizing a magnetic charge by its negative, which builds a meson, we can now balance the magnetic charge $[(N-1)/N]\,{}^*e$ against $N-1$ units of the magnetic charge $-(1/N)\,{}^*e$. This is a composite of N F.D. particles, and $N = 2, 3, \ldots$ must be odd if the result is to be a F.D. baryon. The simplest possibility, which we adopt, is $N = 3$. Thus, baryons are viewed as composites of three entities that bear the magnetic charges, in ${}^*e$ units, of $\frac{2}{3}$, $-\frac{1}{3}$, $-\frac{1}{3}$. We learn, incidentally, from ${}^*e = 3\,{}^*e_0$, that

$$ {}^*\alpha = {}^*e^2/4\pi = 36(137.036). \tag{3–9.94} $$

It remains undecided whether the two magnetic charges of $-\frac{1}{3}$ refer to duplicates of the same particle, or to different particles with a common value of magnetic charge. To this we can only offer the observation that, without reference to antiparticles, the magnetic charge average over all distinct dual charged particles will not be zero in the first possibility, but does vanish in the second one where charge $-\frac{1}{3}$ has twice the multiplicity of charge $\frac{2}{3}$. We accept the situation of greater symmetry, and extend it to electric charge as well. Thus, whether we speak of electric charge in units of e or magnetic charge in units of ${}^*e$, there are three options with values $\frac{2}{3}$, $-\frac{1}{3}$, $-\frac{1}{3}$. It is natural to regard these nine possibilities as different states of a fundamental dual charged particle. To emphasize its basic dyadic character in regard to charge, this particle is called the dyon.

Although the hypothetical picture of magnetic charge as the basis of hadronic behavior is still quite incomplete, we have already far outrun our ability to test it, particularly since a quantitative phenomenological analysis of the properties of hadrons is not yet before us. We must turn away from these heady speculations and begin the study of ordinary electrical particles in dynamical contexts. However, Harold finds himself compelled to comment.

H.: You were quite persuasive concerning the importance of avoiding speculative assumptions about the structure of particles, and yet you have just entertained a very bold speculation indeed. Is this not inconsistent?

S.: The final goal of a phenomenological theory is to establish contact with an underlying fundamental theory. My injunction was against the confusion of phenomenological theory with fundamental theory. The organization and theoretical simplification of experimental data should not involve implicit structural assumptions. But, quite independently of that development, one may devise speculative candidates for eventual contact with the phenomenological theory. Ultimate success should be speeded through the logical separation of these two phases.

3-10 PRIMITIVE ELECTROMAGNETIC INTERACTIONS AND SOURCE MODELS

The conserved nature of the photon electric source $J^\mu(x)$ sets the pattern for any realization of such sources by an electric current vector associated with a specific type of particle. The electric currents that we have already considered for various spin choices fail to meet this standard since they are conserved only outside source regions. Let us repeat that discussion for spinless particles, using the slightly different procedure that is based on the action expression

$$W = \int (dx)[K\phi + K^\mu \phi_\mu + \mathcal{L}], \qquad \mathcal{L} = -\phi^\mu \partial_\mu \phi + \tfrac{1}{2}\phi^\mu \phi_\mu - \tfrac{1}{2}m^2\phi^2. \qquad (3\text{-}10.1)$$

The consideration of infinitesimal, variable phase transformations of the sources:

$$\delta K(x) = ieq\, \delta\varphi(x)K(x), \qquad \delta K^\mu(x) = ieq\, \delta\varphi(x)K^\mu(x) \qquad (3\text{-}10.2)$$

and of the compensating field transformations

$$\delta\phi(x) = ieq\, \delta\varphi(x)\phi(x), \qquad \delta\phi^\mu(x) = ieq\, \delta\varphi(x)\phi^\mu(x), \qquad (3\text{-}10.3)$$

gives directly

$$\delta W = \int (dx)[\phi(x)ieqK(x) + \phi^\mu(x)ieqK_\mu(x)]\, \delta\varphi(x) \qquad (3\text{-}10.4)$$

and indirectly

$$\delta W = -\int (dx)j^\mu(x)\partial_\mu\, \delta\varphi(x), \qquad (3\text{-}10.5)$$

with

$$j^\mu(x) = \phi^\mu(x)ieq\phi(x). \qquad (3\text{-}10.6)$$

The comparison of the two evaluations implies that

$$\partial_\mu j^\mu(x) = \phi(x)ieqK(x) + \phi^\mu(x)ieqK_\mu(x). \qquad (3\text{-}10.7)$$

Notice that we have written *eq* everywhere, replacing the charge matrix q of the earlier treatment, in order to measure charge in the physical unit e.

The observation that j^μ is not conserved in the interior of sources means only that the physical description begins with the creation of the charge-bearing particle and ignores the pre-existence of that charge, if not the particle, in the source. We must find a way to insert the fact that charge is transmitted, not created. It will be seen that this requires the introduction of an electromagnetic model of sources, which is simplified to the point of retaining only the charge conservation property, but still has arbitrary elements. One procedure creates a conserved electric current by amputating the nonconserved part, in a way that retains the original current in the regions that are causally separated from the emission and absorption acts, where the current is conserved. This is accomplished by the construction

$$j^\mu_{\text{cons.}}(x) = j^\mu(x) - \int (dx') f^\mu(x - x') \partial'_\nu j^\nu(x'), \qquad (3\text{-}10.8)$$

where

$$\partial_\mu f^\mu(x - x') = \delta(x - x') \qquad (3\text{-}10.9)$$

defines a not unfamiliar class of functions. When the support of $f^\mu(x - x')$ is restricted to space-like intervals, the subtracted term in (3-10.8) vanishes at any time for which the sources are causally inoperative. To keep uniformity of treatment between $j^\mu(x)$ and $J^\mu(x)$, we shall relate the conserved vector, now designated $J^\mu_{\text{cons.}}(x)$, to an arbitrary vectorial function $J^\mu(x)$ by

$$J^\mu_{\text{cons.}}(x) = J^\mu(x) - \int (dx') f^\mu(x - x') \partial'_\nu J^\nu(x'). \qquad (3\text{-}10.10)$$

The vector potential $A^\mu(x)$ must multiply the total current, in the action expression. That can be rearranged to give

$$\int (dx)(J^\mu_{\text{cons.}}(x) + j^\mu_{\text{cons.}}(x)) A_\mu(x) = \int (dx)(J^\mu(x) + j^\mu(x)) A^f_\mu(x), \quad (3\text{-}10.11)$$

in which

$$A^f_\mu(x) = A_\mu(x) - \partial_\mu \int (dx') f^\nu(x - x') A_\nu(x'), \qquad (3\text{-}10.12)$$

and where, for convenience, we have accepted the symmetry restriction

$$-f^\mu(x' - x) = f^\mu(x - x'), \qquad (3\text{-}10.13)$$

which has no apparent physical significance here. Note that the construction of $A^f_\mu(x)$ from $A_\mu(x)$ is a gauge transformation, such that the new vector potential is characterized by

$$\int (dx') f^\mu(x - x') A^f_\mu(x') = 0. \qquad (3\text{-}10.14)$$

This is a unique characterization, for, if the general gauge transformation

$$\overline{A}_\mu(x) = A_\mu(x) + \partial_\mu \lambda(x) \qquad (3\text{-}10.15)$$

is designed to make $\bar{A}_\mu(x)$ satisfy (3–10.14), we get

$$0 = \int (dx') f^\mu(x - x') A_\mu(x') + \lambda(x), \qquad (3\text{--}10.16)$$

which produces the construction of Eq. (3–10.12).

When two different kinds of particles are considered under physical conditions of noninteraction, the vacuum amplitudes are multiplied and the actions added. Thus, for noninteracting photons and spinless particles,

$$W_{\text{nonint.}} = \int (dx)[J^\mu_{\text{cons.}} A_\mu + K\phi + K^\mu\phi_\mu + \mathcal{L}_{\text{nonint.}}],$$

$$\mathcal{L}_{\text{nonint.}} = -\tfrac{1}{4}F^{\mu\nu}F_{\mu\nu} - \phi^\mu\partial_\mu\phi + \tfrac{1}{2}\phi^\mu\phi_\mu - \tfrac{1}{2}m^2\phi^2. \qquad (3\text{--}10.17)$$

An interaction between photons and charged particles is introduced by replacing $J^\mu_{\text{cons.}}$ with the total current. We call this interaction primitive because it is not the final statement of all interactions, but rather characterizes a first elementary stage, which implies and is supplemented by further, increasingly elaborate levels of description. Precisely in what sense it is the first of a series of dynamical steps will be discussed later. The action expression that characterizes this first stage is

$$W = \int (dx)[J^\mu A'_\mu + K\phi + K^\mu\phi_\mu + \mathcal{L}(\phi\phi^\mu, A'_\mu)],$$

$$\mathcal{L}(\phi\phi^\mu, A_\mu) = -\tfrac{1}{4}F^{\mu\nu}F_{\mu\nu} - \phi^\mu(\partial_\mu - ieqA_\mu)\phi + \tfrac{1}{2}\phi^\mu\phi_\mu - \tfrac{1}{2}m^2\phi^2, \qquad (3\text{--}10.18)$$

where we have chosen to incorporate into the Lagrange function the interaction term

$$j^\mu(x) A'_\mu(x) = \phi^\mu(x) ieq\phi(x) A_\mu(x). \qquad (3\text{--}10.19)$$

Although the Lagrange function here employs the vector potential of a specific gauge, it is a gauge invariant combination that remains unchanged under the unified gauge and phase transformation

$$A_\mu(x) \rightarrow A_\mu(x) + \partial_\mu\lambda(x),$$

$$\phi(x) \rightarrow e^{ieq\lambda(x)}\phi(x), \qquad \phi^\mu(x) \rightarrow e^{ieq\lambda(x)}\phi^\mu(x). \qquad (3\text{--}10.20)$$

This is a consequence of replacing $\partial_\mu\phi$, with transformation behavior

$$\partial_\mu\phi(x) \rightarrow e^{ieq\lambda(x)}[\partial_\mu\phi(x) + ieq\partial_\mu\lambda(x)\phi(x)], \qquad (3\text{--}10.21)$$

by the gauge covariant combination

$$(\partial_\mu - ieqA_\mu(x))\phi(x) \rightarrow e^{ieq\lambda(x)}(\partial_\mu - ieqA_\mu(x))\phi(x). \qquad (3\text{--}10.22)$$

The field equations deduced from the stationary action principle by varying ϕ^μ and ϕ are, respectively,

$$(\partial_\mu - ieqA'_\mu)\phi - \phi_\mu = K_\mu, \qquad -(\partial_\mu - ieqA'_\mu)\phi^\mu + m^2\phi = K, \quad (3\text{--}10.23)$$

where the gauge covariant combination stays intact since the sign reversal of the derivative on partial integration is matched by the antisymmetry of the charge matrix q. In performing the variation of A_μ^f we must not violate the gauge restriction on the vector potential,

$$\int (dx') f^\mu(x - x')\, \delta A_\mu^f(x') = 0. \tag{3-10.24}$$

Thus, the correct conclusion from

$$\delta_A W = \int (dx)[J^\mu + j^\mu - \partial_\nu F^{\mu\nu}]\, \delta A_\mu^f = 0 \tag{3-10.25}$$

is that

$$J^\mu(x) + j^\mu(x) - \partial_\nu F^{\mu\nu}(x) = \int (dx') f^\mu(x - x') \gamma(x'), \tag{3-10.26}$$

where $\gamma(x)$ is arbitrary as far as the action principle is concerned. But the divergence of this equation gives

$$\partial_\mu J^\mu(x) + \partial_\mu j^\mu(x) = \gamma(x), \tag{3-10.27}$$

and we recognize the Maxwell equation

$$\partial_\nu F^{\mu\nu}(x) = J^\mu_{\text{cons.}}(x) + j^\mu_{\text{cons.}}(x). \tag{3-10.28}$$

To connect the use of $f^\mu(x - x')$ in defining a specific gauge with the concept of electromagnetic source models, we perform the following phase transformation on ϕ and ϕ^μ, without the accompanying gauge transformation:

$$\phi(x) \to e^{-ieq\Lambda(x)}\phi(x), \qquad \phi^\mu(x) \to e^{-ieq\Lambda(x)}\phi^\mu(x), \tag{3-10.29}$$

where

$$\Lambda(x) = \int (dx') f^\mu(x - x') A_\mu(x') \tag{3-10.30}$$

and $A_\mu(x)$ is the vector potential in an arbitrary gauge. This transformation does two things. It replaces A_μ^f in $\mathcal{L}$ by

$$A_\mu^f(x) + \partial_\mu \Lambda(x) = A_\mu(x), \tag{3-10.31}$$

which is the inverse of the gauge transformation (3–10.12), and the transferal of the uncompensated phase factor to the sources replaces them by

$$K^A(x) = e^{ieq\Lambda(x)}K(x), \qquad K_\mu^A(x) = e^{ieq\Lambda(x)}K_\mu(x). \tag{3-10.32}$$

With the introduction of the arbitrary vector potential $A_\mu(x)$, we return to the use of $J^\mu_{\text{cons.}}(x)$. The additional label will be omitted, however, for one can understand from the context whether $J^\mu(x)$ is an arbitrary vector, since the vector potential is limited to a particular gauge, or is a conserved vector, since the vector potential admits gauge transformations. The new action expression is

$$W = \int (dx)[J^\mu A_\mu + K^A \phi + K_\mu^A \phi^\mu + \mathcal{L}(\phi\phi^\mu, A_\mu)]. \tag{3-10.33}$$

The gauge invariance of the Lagrange function is now matched by that of all the source terms, since $A_\mu \to A_\mu + \partial_\mu \lambda$ induces

$$\Lambda(x) \to \Lambda(x) + \lambda(x) \tag{3-10.34}$$

and

$$K^A(x) \to e^{ieq\lambda(x)}K^A(x), \qquad K_\mu^A(x) \to e^{ieq\lambda(x)}K_\mu^A(x). \tag{3-10.35}$$

While the charged particle field equations that are implied by the action (3-10.33) continue to be given by (3-10.23) with the sources K^A, K_μ^A, the electromagnetic field equation presents a different aspect. In contrast with the action of Eq. (3-10.18), δA_μ is arbitrary and the charged particle sources are functionals of the vector potential. The implication of the latter fact is indicated by

$$\delta_A \int (dx)\phi(x)K^A(x) = \int (dx)\phi(x)ieqK^A(x)\,\delta\Lambda(x)$$
$$= -\int (dx)(dx')\,\delta A_\mu(x)f^\mu(x-x')\phi(x')ieqK^A(x'). \tag{3-10.36}$$

Thus we now get

$$\partial_\nu F^{\mu\nu}(x) = J^\mu(x) + j^\mu(x) - \int (dx')f^\mu(x-x')[\phi(x')ieqK^A(x') + \phi^\nu(x')ieqK_\nu^A(x')]. \tag{3-10.37}$$

It is just the Maxwell equation of (3-10.28), since

$$\partial_\mu j^\mu(x) = \phi(x)ieqK^A(x) + \phi^\mu(x)ieqK_\mu^A(x), \tag{3-10.38}$$

but this time we have made explicit a contribution to the electric current that is associated directly with the charged particle source.

Consider the following fictitious source problem: A point charge e moves uniformly with four-vector velocity n^μ,

$$n^\mu n_\mu = -1, \tag{3-10.39}$$

until at a given location, which we adopt as the origin, it goes out of existence. What is the description of the photons emitted or absorbed by this act? The current vector is given by

$$J^\mu(x) = en^\mu \int_{-\infty}^0 ds\,\delta(x-ns); \tag{3-10.40}$$

it obeys the nonconservation equation

$$\partial_\mu J^\mu(x) = -e\int_{-\infty}^0 ds\,\frac{d}{ds}\delta(x-ns) = -e\,\delta(x). \tag{3-10.41}$$

Thus

$$J^\mu(x) = -ef^\mu(x), \tag{3-10.42}$$

where this f^μ function is time-like,

$$f^\mu(x) = -n^\mu \int_{-\infty}^0 ds\,\delta(x-ns) \tag{3-10.43}$$

and has the momentum representation [cf. Eq. (3–8.61)]

$$if^\mu(p) = -n^\mu i \int_{-\infty}^{0} dse^{-ispn} = n^\mu/pn, \qquad pn \neq 0. \qquad (3\text{-}10.44)$$

Recall the description of the emission and absorption of an arbitrary number of particles, here photons, by a given source distribution, $J^\mu(x)$. The factor in the vacuum amplitude that couples J^μ to the creation and detection sources, J_2^μ and J_1^μ, respectively, is

$$\exp\left[i\int (dx)(dx')J_1^\mu(x)D_+(x-x')J_\mu(x) + i\int(dx)(dx')J^\mu(x)D_+(x-x')J_{2\mu}(x')\right]$$
$$= \exp\left[i\int(dx)J^\mu(x)A_\mu(x)\right], \qquad (3\text{-}10.45)$$

where $A_\mu(x)$ combines the field associated with J_2^μ and the initial photons with that having analogous reference to the final photons. In view of the causal arrangement of sources, wherever $A_\mu(x)$ is of interest in (3–10.45), it is a solution of the source-free Maxwell equations or, in momentum space,

$$p^2 A_\mu(p) - p_\mu p_\nu A^\nu(p) = 0. \qquad (3\text{-}10.46)$$

If we insert the current of (3–10.42) into (3–10.45) it becomes

$$\exp\left[-ie\int (dx)f^\mu(x)A_\mu(x)\right] = \exp\left[-ie\int \frac{(dp)}{(2\pi)^4}f^\mu(-p)A_\mu(p)\right]. \qquad (3\text{-}10.47)$$

But, observe that

$$\frac{n_\mu}{pn} - \frac{p_\mu + n_\mu pn}{p^2 + (pn)^2} = \frac{p^2 n_\mu - p_\mu p_\nu n^\nu}{pn[p^2 + (pn)^2]} \qquad (3\text{-}10.48)$$

which shows the equivalence, for the purpose of evaluating (3–10.47), of the time-like f^μ function with the space-like

$$if^\mu(p) = \frac{p^\mu + n^\mu pn}{p^2 + (pn)^2}, \qquad n_\mu f^\mu(p) = 0. \qquad (3\text{-}10.49)$$

The latter is also an odd function of p without restriction, unlike (3–10.44) which mirrors the asymmetry of the coordinate function in (3–10.43). We recognize in (3–10.47) precisely the exponential factor that is associated with a single charged particle emission act, as in

$$\int (dx')\phi(x')K^A(x') = \int (dx')\phi(x')\exp\left[-ieq\int(dx)f^\mu(x-x')A_\mu(x)\right]K(x'), \qquad (3\text{-}10.50)$$

where x' serves as the reference point at which charge eq disappears in the source and emerges on the particle of interest.

The members of the class of f^μ functions given in (3–10.49) differ only in the choice of the time-like unit vector n^μ, which represents the motion of the charge in the source model. When the coordinate system identifies n^μ with the time axis, $f^\mu(p)$ has only spatial components that are independent of p^0,

$$if(p) = \mathbf{p}/\mathbf{p}^2, \qquad (3\text{-}10.51)$$

and

$$\mathbf{f}(x) = -\nabla \mathfrak{D}(\mathbf{x}) \, \delta(x^0), \qquad (3\text{-}10.52)$$

where

$$\mathfrak{D}(\mathbf{x}) = \int \frac{(d\mathbf{p})}{(2\pi)^3} \frac{e^{i\mathbf{p}\cdot\mathbf{x}}}{\mathbf{p}^2} = \frac{1}{4\pi} \frac{1}{|\mathbf{x}|} \qquad (3\text{-}10.53)$$

obeys

$$-\nabla^2 \mathfrak{D}(\mathbf{x}) = \delta(\mathbf{x}). \qquad (3\text{-}10.54)$$

There is one choice of f^μ that avoids the reference to an external unit vector by devising the latter from relevant physical parameters. It requires an extension of the structure of f^μ to include algebraic functions of derivatives that act upon the source function $K(x)$. We indicate this replacement in (3–10.50) and describe its meaning by writing

$$\exp\left[-ieq\int (dx)f^\mu(x - x', -i\partial_K)A_\mu(x)\right]K(x')$$

$$= \int \frac{(dP)}{(2\pi)^4} e^{iPx'} \exp\left[-ieq\int (dx)f^\mu(x - x', P)A_\mu(x)\right]K(P). \quad (3\text{-}10.55)$$

When $K(P)$ represents the emission or absorption of particles, the time-like vector P^μ can replace n^μ, apart from a scale factor. This gives

$$if_\mu(p, P) = \frac{p_\mu P^2 + P_\mu pP}{p^2 P^2 + (pP)^2} \rightarrow \frac{P_\mu}{pP} \qquad (3\text{-}10.56)$$

where the last form is the analogue of (3–10.44), one that is equivalent for the calculation of photon processes.

The discussion of spinless particles is particularly simple. A system without intrinsic angular momentum can only exhibit scalar properties in its rest frame. In the electromagnetic context this permits a monopole moment—charge—but forbids multipole moments. More generally, a particle of spin s, in its rest frame, can possess multipole moments to the maximum order $2s$. That is, a spin $\frac{1}{2}$ particle can have arbitrary dipole moments; a particle of unit spin can have arbitrary dipole and quadrupole moments; and so forth. A sufficiently general current expression for spin $\frac{1}{2}$ is

$$j^\mu(x) = e\tfrac{1}{2}\psi(x)\gamma^0\gamma^\mu q\psi(x) + \frac{e}{2m} (\tfrac{1}{2}g - 1)\partial_\nu[\tfrac{1}{2}\psi(x)\gamma^0\sigma^{\mu\nu}q\psi(x)]. \quad (3\text{-}10.57)$$

This way of writing the coefficient in the term that has the form $\partial_\nu m^{\mu\nu}$ anticipates the identification of g as the gyromagnetic ratio, the magnetic moment in the unit $\pm e/2m$ relative to the spin angular momentum [Eq. (1–2.44)]. That becomes clearer on using the identity (3–6.67), applicable in source-free regions, to rewrite (3–10.57) as

$$j^\mu(x) = \frac{e}{2m}\frac{1}{2}\left[\psi(x)\gamma^0 q\left(\frac{1}{i}\right)\partial^\mu\psi(x) - \left(\frac{1}{i}\right)\partial^\mu\psi(x)\gamma^0 q\psi(x)\right]$$

$$+ \frac{e}{2m}\frac{1}{2}g\partial_\nu[\tfrac{1}{2}\psi(x)\gamma^0\sigma^{\mu\nu}q\psi(x)]. \tag{3–10.58}$$

The magnetic moment of a system is

$$\boldsymbol{\mu} = \tfrac{1}{2}\int (d\mathbf{x})\mathbf{x}\times\mathbf{j}, \tag{3–10.59}$$

which here becomes

$$\boldsymbol{\mu} = \frac{e}{2m}\int(d\mathbf{x})\tfrac{1}{2}\psi\gamma^0 q\left[\mathbf{x}\times\left(\frac{1}{i}\right)\boldsymbol{\nabla}+g\tfrac{1}{2}\boldsymbol{\sigma}\right]\psi, \tag{3–10.60}$$

making explicit the roles of orbital angular momentum, spin angular momentum, and the g factor. The dipole moments permitted to a spin $\tfrac{1}{2}$ particle include an electric dipole moment. It would supplement the second term of (3–10.57) by a similar expression of arbitrary coefficient that uses the dual spin tensor

$$*\sigma^{\mu\nu} = \tfrac{1}{2}\epsilon^{\mu\nu\kappa\lambda}\sigma_{\kappa\lambda} = \sigma^{\mu\nu}\gamma_5. \tag{3–10.61}$$

No such property has yet been detected, however. Since the second term of the current is identically divergenceless, we still have [a factor of e is inserted compared to Eq. (3–6.48)]

$$\partial_\mu j^\mu(x) = \psi(x)\gamma^0 ieq\eta(x). \tag{3–10.62}$$

The current (3–10.57) is incorporated in the following action expression, analogous to (3–10.18),

$$W = \int(dx)[J^\mu A'_\mu + \eta\gamma^0\psi + \mathfrak{L}(\psi, A'_\mu)],$$

$$\mathfrak{L}(\psi, A_\mu) = -\tfrac{1}{4}F^{\mu\nu}F_{\mu\nu} - \tfrac{1}{2}\psi\gamma^0[\gamma^\mu(-i\partial_\mu - eqA_\mu) + m]\psi \tag{3–10.63}$$

$$+ \tfrac{1}{2}F^{\mu\nu}\frac{e}{2m}(\tfrac{1}{2}g - 1)\tfrac{1}{2}\psi\gamma^0\sigma_{\mu\nu}q\psi.$$

The omitted electric dipole interaction term is similar to the last one, with either of the antisymmetrical tensors replaced by its dual. The Lagrange function is invariant under the gauge transformation

$$A_\mu(x) \to A_\mu(x) + \partial_\mu\lambda(x), \qquad \psi(x) \to e^{ieq\lambda(x)}\psi(x), \tag{3–10.64}$$

and this property can be utilized, as in the spin 0 discussion, to remove reference to a specific gauge while introducing an electromagnetic model for the particle source:

$$\eta^A(x) = e^{ieq\Lambda(x)}\eta(x). \tag{3–10.65}$$

The field equations implied by the action (3–10.63) are

$$\left[\gamma^\mu(-i\partial_\mu - eqA'_\mu(x)) + m - \frac{eq}{2m}(\tfrac{1}{2}g - 1)\tfrac{1}{2}\sigma^{\mu\nu}F_{\mu\nu}(x)\right]\psi(x) = \eta(x),$$

$$(3\text{–}10.66)$$

together with the Maxwell equations employing the appropriate conserved currents.

The appearance of the gauge covariant derivative, $\partial_\mu - ieqA_\mu$, in the actions (3–10.18) and (3–10.63) is completely general. It expresses the identity between the two ways in which electric currents have been introduced. The first one considers the infinitesimal response to a variable phase transformation. For a typical particle field $\chi(x)$ this is

$$\delta\chi(x) = ieq\,\delta\varphi(x)\chi(x), \qquad (3\text{–}10.67)$$

giving

$$\delta W = -\int (dx)j^\mu(x)\partial_\mu\,\delta\varphi(x), \qquad (3\text{–}10.68)$$

where the factor of e in (3–10.67) provides the appropriate electromagnetic measure for the current. This kinematical definition is not unique. The dynamical definition of electric current imitates the role of the photon source. In particular, the response of the action to the field variation $\delta A_\mu = \partial_\mu\,\delta\lambda$ is required to be

$$\delta W = \int (dx)(J^\mu + j^\mu)\partial_\mu\,\delta\lambda. \qquad (3\text{–}10.69)$$

Thus, the identity of the two concepts is imposed by insisting that the action be invariant under the unified gauge-phase transformation with

$$\delta\varphi(x) = \delta\lambda(x). \qquad (3\text{–}10.70)$$

The replacement of derivatives on charge-bearing fields by gauge covariant derivatives accomplishes this for the whole group of gauge transformations, which is Abelian in structure. And the possibility of adding independently gauge invariant terms, as in (3–10.63), conveys the arbitrary aspects of the kinematical current definition. It is generally believed that there is something particularly simple and natural about the electromagnetic coupling produced by using only the gauge covariant substitution, and there is truth in this. But it must not be forgotten that alternative descriptions exist for the same spin value, and by following a common procedure we arrive at different electromagnetic properties. Thus, the third-rank spinor description of spin $\tfrac{1}{2}$, based on the Lagrange function (3–5.73) with gauge covariant derivatives, gives the current of Eq. (3–6.61), apart from the factor of e, and the corresponding g value, as contained in (3–6.68, 69), is $\tfrac{2}{3}$. If the very striking near-equality, $\tfrac{1}{2}g \cong 1$, that is observed for the electron and the muon has any single moral, it is the special relevance of the simple Dirac spinor equation for the description of these particles.

To illustrate the direct use of a gauge invariant Lagrange function for introducing primitive electromagnetic interactions, we shall discuss charged particles of unit spin. Such a Lagrange function, generalized from (3–5.28), is

$$\mathcal{L} = -\tfrac{1}{2}G^{\mu\nu}(D_\mu\phi_\nu - D_\nu\phi_\mu) + \tfrac{1}{4}G^{\mu\nu}G_{\mu\nu} - \tfrac{1}{2}m^2\phi^\mu\phi_\mu$$
$$+ \tfrac{1}{2}aF^{\mu\nu}\phi_\mu ieq\phi_\nu + \tfrac{1}{2}(b/m^2)F^{\mu\nu}G_\mu{}^\lambda ieqG_{\nu\lambda}, \qquad (3\text{–}10.71)$$

which finally uses an abbreviation for the gauge covariant derivative,

$$D_\mu = \partial_\mu - ieqA_\mu. \qquad (3\text{–}10.72)$$

Notice that we have devised two independently gauge invariant terms. The arbitrary coefficients a and b will be related to magnetic moment and electric quadrupole moment. We shall not consider the two additional couplings produced by replacing $F^{\mu\nu}$ with its dual. They would describe electric dipole and magnetic quadrupole moments. The particle field equations derived from the action principle are

$$D_\mu\phi_\nu - D_\nu\phi_\mu - G_{\mu\nu} - (b/m^2)(F_{\mu\lambda}ieqG^\lambda{}_\nu - F_{\nu\lambda}ieqG^\lambda{}_\mu) = M_{\mu\nu},$$
$$D_\nu G^{\mu\nu} + m^2\phi^\mu - aF^{\mu\nu}ieq\phi_\nu = J^\mu, \qquad (3\text{–}10.73)$$

and the electric current vector, in source-free regions, is

$$j^\mu = G^{\mu\nu}ieq\phi_\nu + \partial_\nu[a\phi^\mu ieq\phi^\nu + (b/m^2)G^{\mu\lambda}ieqG^\nu{}_\lambda]. \qquad (3\text{–}10.74)$$

If we are interested in the intrinsic electromagnetic properties of the particle, and not those induced by the electromagnetic field, it suffices to simplify (3–10.74) with the aid of the uncoupled particle field equations:

$$\partial_\mu\phi_\nu - \partial_\nu\phi_\mu = G_{\mu\nu}, \qquad \partial_\nu G^{\mu\nu} + m^2\phi^\mu = 0, \qquad \partial_\mu\phi^\mu = 0, \qquad (3\text{–}10.75)$$

the last of which is an important but not independent statement. This gives

$$j^\mu = \partial^\mu\phi^\nu ieq\phi_\nu + (b/2m^2)\partial^2(\partial^\mu\phi^\nu ieq\phi_\nu)$$
$$- (1 - a + b)\partial_\nu(\phi^\mu ieq\phi^\nu) + (b/2m^2)\partial^2\partial_\nu(\phi^\mu ieq\phi^\nu)$$
$$- (b/m^2)\partial_\nu\partial_\lambda(\partial^\mu\phi^\lambda ieq\phi^\nu - \partial^\nu\phi^\lambda ieq\phi^\mu), \qquad (3\text{–}10.76)$$

and the implied coupling with an electromagnetic potential in completely source-free regions is conveyed by

$$\int (dx)A_\mu j^\mu = \int (dx)[A_\mu(\partial^\mu\phi^\nu ieq\phi_\nu) - (1 - a + b)\tfrac{1}{2}F_{\mu\nu}(\phi^\mu ieq\phi^\nu)$$
$$+ (b/m^2)\partial_\lambda F_{\mu\nu}(\partial^\mu\phi^\lambda ieq\phi^\nu)]. \qquad (3\text{–}10.77)$$

The identity

$$(\partial^\mu\phi^\lambda ieq\phi^\nu) = (\partial^\mu\phi^\nu ieq\phi^\lambda) + \partial^\mu(\phi^\lambda ieq\phi^\nu) \qquad (3\text{–}10.78)$$

shows also that the field strength derivative in (3–10.77) should be symmetrized in the indices λ and ν.

For a slowly moving particle, of charge $\pm e$, the three field components ϕ_k dominate, and are conveniently combined in the vector $\boldsymbol{\phi}$. The spin matrix vector $\mathbf{s}$ is represented by the rotation

$$\mathbf{n} \cdot \mathbf{s}\boldsymbol{\phi} = i\mathbf{n} \times \boldsymbol{\phi}. \tag{3-10.79}$$

We use the spin matrices to present this specialization of (3–10.77) as

$$\int (dx) A_\mu j^\mu \rightarrow \int (dx)[-(\pm e)A^0 2m(\phi_1\phi_2) + (1 - a + b)(\pm e)\mathbf{H} \cdot (\phi_1\mathbf{s}\phi_2)$$
$$+ (b/m^2)(\pm e)\boldsymbol{\nabla}\mathbf{E} : 2m(\phi_1\mathbf{ss}\phi_2)], \tag{3-10.80}$$

where the dyadic $\boldsymbol{\nabla}\mathbf{E}$ is symmetrized, and we have also picked out the terms that describe the propagating particle in a causal arrangement. With the coupling of the scalar potential A^0 to the charge $\pm e$ serving as a reminder of the normalization, the linear coupling of the spin vector to the magnetic field identifies the g value:

$$g = 1 - a + b, \tag{3-10.81}$$

while the quadratic spin term gives the quadrupole moment Q, in the unit $(\pm e)/m^2$, as

$$Q = 2b. \tag{3-10.82}$$

The individual results obtained for g values when only the gauge covariant derivative is used ($s = \frac{1}{2}, g = 2, \frac{2}{3}; s = 1, g = 1$), are given uniformly by the general multispinor Lagrange function (3–5.78). The current that the latter implies in source and field-free space is

$$j^\mu(x) = e\tfrac{1}{2}\psi(x)\gamma \left(\frac{1}{n} \sum_\alpha \gamma^\mu_\alpha\right) q\psi(x). \tag{3-10.83}$$

The vanishing of all auxiliary fields under such circumstances, as expressed by

$$(\gamma^\mu_\alpha - \gamma^\mu_\beta)\partial_\mu\psi = 0, \tag{3-10.84}$$

implies the set of field equations

$$[\gamma^\mu_\alpha(1/i)\partial_\mu + m]\psi = 0, \qquad \alpha = 1, \ldots, n. \tag{3-10.85}$$

Hence, the rearrangement used for spin $\frac{1}{2}$ can be applied to each of the n terms that compose (3–10.83), giving

$$j^\mu(x) = \frac{e}{2m}\frac{1}{2}\left[\psi(x)\gamma q\left(\frac{1}{i}\right)\partial^\mu\psi(x) - \left(\frac{1}{i}\right)\partial^\mu\psi(x)\gamma q\psi(x)\right]$$
$$+ \frac{e}{2m}\partial_\nu\left[\tfrac{1}{2}\psi(x)\gamma\frac{1}{n}\sum\sigma^{\mu\nu}_\alpha q\psi(x)\right]. \tag{3-10.86}$$

Since the particle spin vector is

$$\mathbf{s} = \tfrac{1}{2}\sum\sigma_\alpha, \tag{3-10.87}$$

the g value is immediately identified as

$$g = 2/n, \qquad (3\text{–}10.88)$$

and all other multipole moments are zero. Note that the actual spin value enters only through the inequality $s \leq \tfrac{1}{2}n$, and

$$gs = 2s/n \leq 1, \qquad (3\text{–}10.89)$$

where the equality sign applies to totally symmetrical spinors. Incidentally, a very similar unified treatment applies to all spinor-symmetrical tensor fields used to describe integer $+ \tfrac{1}{2}$ spin values. As one can recognize from the examples of spin $\tfrac{3}{2}$ and $\tfrac{5}{2}$ Lagrange functions, Eqs. (3–5.55) and (3–5.58), the gauge covariant electromagnetic interaction implies a current vector that, in source and field-free space, is

$$j^{\mu}(x) = e\tfrac{1}{2}\psi^{\nu_1 \nu_2 \cdots}(x)\gamma^0 \gamma^{\mu} q \psi_{\nu_1 \nu_2 \ldots}(x). \qquad (3\text{–}10.90)$$

The same spin $\tfrac{1}{2}$ rearrangement, combined with projection of $\boldsymbol{\sigma}$ on the total spin and the observation that σ_3, for example, is unity when $s_3 = s$, gives directly

$$gs = 1. \qquad (3\text{–}10.91)$$

3–11 EXTENDED SOURCES. SOFT PHOTONS

Complementary to the principle of space-time uniformity is a principle of uniformity for phenomena that differ only in the values of energy-momentum that are engaged. The source concept was introduced as an idealization of collisions in which precisely the right balance of energy-momentum or, invariantly expressed, mass is transferred to create a specific particle. But the same laws of physics are operative when less mass, or more mass, is transferred. Long ago, in Section 2–3, we used an extrapolation to quasi-static source distributions, which are incapable of emitting particles, in order to connect the properties of photons with the Coulomb-Ampèrian laws of charge and current interactions. Perhaps in our recent preoccupation with the very familiar equations of Maxwell, we may have forgotten the initial logical basis for that contact. And now, through our concern with the electric currents that are associated with the operation of charged particle sources, we are moving in the opposite direction. The physical situation is quite simple. The creation of a charged particle generally involves the transfer of that charge from other particles having different states of motion. Accelerated charges radiate. Hence, unless precise control is exercised over the energy-momentum balance, the charged particle has a nonzero probability of being accompanied by photons. If we were to take too narrow a view of the source concept and decline to extend it to this multiparticle emission act, we would divorce the dynamical significance of charge from its kinematical aspects.

The emission or absorption of photons is not a localized process. The photon that accompanies the creation of a charged particle cannot be assigned to the agency of that particle, nor to the charges in the source, but involves interference between both effects. This is implicit in the additive construction of the electric current from contributions of the particles and the source. It is instructive to examine such phenomena in some detail. We begin by evaluating the probability amplitude for the emission of one photon of momentum k^μ and polarization λ, accompanying the creation of one spinless particle of momentum p^μ and charge $\pm e$. The physical context that underlies the use of the primitive interaction to compute this probability amplitude is that, between creation and detection, particle and photon propagate under conditions of noninteraction. Accordingly, it is useful to review the description of that situation when the two particles are produced by independent sources. This is contained in the vacuum amplitude

$$\langle 0_+|0_-\rangle^{JK} = \langle 0_+|0_-\rangle^J \langle 0_+|0_-\rangle^K \qquad (3\text{-}11.1)$$

as the term involving one emission and one absorption source of each kind (we place $K^\mu = 0$ in these considerations),

$$\langle 0_+|0_-\rangle^{JK} = 1 + \cdots + i \int (d\xi)(d\xi') J_1^\mu(\xi) D_+(\xi - \xi') J_{2\mu}(\xi')$$
$$\times \, i \int (dx)(dx') K_1(x) \Delta_+(x - x') K_2(x') + \cdots , \qquad (3\text{-}11.2)$$

or

$$\langle 0_+|0_-\rangle^{JK} = 1 + \cdots + i \int (d\xi) A_1^\mu(\xi) J_{2\mu}(\xi) i \int (dx) \phi_1(x) K_2(x) + \cdots ,$$
$$(3\text{-}11.3)$$

where we have used ξ^μ as an alternative to x^μ for assistance in distinguishing between the two kinds of particles. (And let us hope that no confusion results from speaking of particle, in the singular, when we mean charged particle.) The application of the primitive interaction will retain the noninteraction context but replace the independent sources $J_2^\mu(\xi)$, $K_2(x)$ by a joint source, $J_2^\mu(\xi) K_2(x)|_{\text{eff.}}$, which we now exhibit.

The restriction to the single action of a photon detection source can be introduced by considering

$$\delta_J W = \int (d\xi) \, \delta J^\mu(\xi) A_\mu(\xi) \qquad (3\text{-}11.4)$$

or

$$\delta_J \langle 0_+|0_-\rangle^{JK} = i \int (d\xi) \, \delta J^\mu(\xi) A_\mu(\xi) \langle 0_+|0_-\rangle^{JK}, \qquad (3\text{-}11.5)$$

for one can identify the probe source δJ^μ with J_1^μ. Since the field $A_\mu(\xi)$ is to be evaluated for $J_\mu(\xi) = 0$, it is given by

$$A^\mu(\xi) = \int (d\xi') D_+(\xi - \xi') j_{\text{cons.}}^\mu(\xi'), \qquad (3\text{-}11.6)$$

apart from an irrelevant gauge term. The process in which we are interested involves the causal coupling of three sources: J_1^μ, K_1, and K_2. The emission source K_2 is used to inject into the system the momentum P^μ that is realized as two particles,

$$P^\mu = p^\mu + k^\mu, \qquad (3\text{-}11.7)$$

where

$$-p^2 = m^2, \qquad k^2 = 0. \qquad (3\text{-}11.8)$$

Thus

$$-P^2 = m^2 - 2pk > m^2, \qquad (3\text{-}11.9)$$

since

$$-pk = p^0 k^0 [1 - (\mathbf{p}/p^0) \cdot (\mathbf{k}/k^0)] > 0. \qquad (3\text{-}11.10)$$

This source is operating in the extended sense, and we shall use the designation 'extended source' to distinguish its mode of action from that of K_1, which detects the particle by absorbing mass m. A source utilized in that way, performing only its initial mission, is a 'simple source.' Now, the current of Eq. (3-11.6) is a quadratic functional of the particle source and therefore contains a portion $j_{12}^\mu(\xi)$, that is bilinear in K_1 and K_2. This gives a factor on the right-hand side of (3-11.5) that already has the required three sources. All other terms describe different processes than the one of interest, which is displayed as

$$\langle 0_+|0_-\rangle^{JK} = 1 + \cdots + i \int (d\xi) A_1^\mu(\xi) j_{12\mu}(\xi) + \cdots . \qquad (3\text{-}11.11)$$

The relevant current structure, obtained from Eqs. (3-10.37) and (3-10.6), with $K_\mu = 0$, is

$$j_{12\mu}(\xi) = \partial_\mu \phi_1(\xi) ieq \phi_2(\xi) - \phi_1(\xi) ieq \partial_\mu \phi_2(\xi) - \int (d\xi') f_\mu(\xi - \xi') \phi_1(\xi') ieq K_2(\xi'). \qquad (3\text{-}11.12)$$

The omission of another term involving $K_1(\xi') ieq \phi_2(\xi')$ expresses the causal arrangement. The field $\phi_2(x)$ is related to its source by

$$(-\partial^2 + m^2) \phi_2(x) = K_2(x), \qquad (3\text{-}11.13)$$

or, in momentum space,

$$\phi_2(P) = \frac{1}{P^2 + m^2} K_2(P). \qquad (3\text{-}11.14)$$

The fact that $P^2 + m^2 \neq 0$ [Eq. (3-11.9)] means that the field $\phi_2(x)$ has no propagation characteristics, and is localized in the neighborhood of the source $K_2(x)$. Thus the field $\phi_2(x)$ will have no overlap with a sufficiently remote detection source $K_1(x)$, which is the assumed causal situation. The term 'virtual particle' is used to extrapolate ordinary particle concepts to such situations where the energy-momentum balance is not suitable to the creation of a 'real' particle. With our new terminology we can characterize the content

of (3–11.11, 12) by saying that the extended source may emit a virtual particle
which quickly is transformed or decays into a real particle and a (real) photon,
or it may emit both final particles in one act, although the photon originates
at a different point than the particle.

The precise meaning of these phrases is conveyed, on comparing (3–11.11, 12)
with (3–11.3), by

$$iJ_{2\mu}(\xi)K_2(x)\Big|_{\text{eff.}} = -\partial_\mu[\delta(x - \xi)ieq\phi_2(x)] - \delta(x - \xi)ieq\partial_\mu\phi_2(x)$$
$$- f_\mu(\xi - x)ieqK_2(x) \qquad (3\text{–}11.15)$$

where the first derivative refers to the x^μ coördinates. An equivalent momentum
version, which also introduces (3–11.14), is

$$iJ_{2\mu}(k)K_2(p)\Big|_{\text{eff.}} = \left[\frac{p_\mu + P_\mu}{P^2 + m^2} - if_\mu(k)\right]eqK_2(P). \qquad (3\text{–}11.16)$$

In this form it is easy to verify the conservation property

$$k^\mu J_{2\mu}(k)K_2(p)\Big|_{\text{eff.}} = 0, \qquad (3\text{–}11.17)$$

which is valid for $p^2 + m^2 = 0$ and arbitrary k^2:

$$\frac{kp + kP}{P^2 + m^2} - 1 = \frac{2kp + k^2}{2kp + k^2} - 1 = 0. \qquad (3\text{–}11.18)$$

An important simplification appears when one considers "soft" photons,
those for which energy and momentum are negligibly small compared to the
values associated with the particle. Then (3–11.16) can be written as

$$iJ_{2\mu}(k)K_2(p)\Big|_{\text{eff.}} \cong \left[\frac{p_\mu}{pk} - \frac{n_\mu}{nk}\right]eqK_2(p) \qquad (3\text{–}11.19)$$

in which we have also introduced the form (3–10.44) for $f_\mu(k)$. The interpretation is clear. From the viewpoint of the soft photon, the charge eq has made an
instantaneous transition from uniform motion with velocity n_μ to uniform
motion with velocity p_μ/m. This is expressed by the photon emission source

$$J_2^\mu(k) = -ieq\left[\frac{p^\mu}{pk} - \frac{n^\mu}{nk}\right], \qquad (3\text{–}11.20)$$

which is the transform of the conserved electric current

$$J^\mu(\xi) = eq\int_0^\infty ds(p^\mu/m)\,\delta(x - (p/m)s) + eq\int_{-\infty}^0 dsn^\mu\,\delta(x - ns). \qquad (3\text{–}11.21)$$

Notice how the two contributions, one associated with the particle source, the
other with the particle, are fitted together in an equivalent photon source. This
is an illustration of the self-consistency that is demanded of the source concept.
The source is introduced as an idealization of realistic dynamical processes.

The dynamical theory that is erected on this foundation must, under appropriate restrictions, validate its starting point. Thus we learn, not surprisingly, that the simple photon source description becomes applicable to a realistic system when there is negligible reaction associated with the emission or absorption process.

We should also recognize the physical significance of the covariant f_μ function given in (3-10.56), which we now write as

$$if_\mu(k, P) \rightarrow \frac{P_\mu}{kP} \simeq \frac{p_\mu}{kp}, \qquad (3\text{-}11.22)$$

where the latter version refers to soft photons. The effective photon source vanishes; the charge has not changed velocity and does not radiate. This is the most natural of source models, in which the emitted particle determines the velocity of the charge in the source and thereby suppresses the accompanying radiation. That suppression is not limited to soft photons, however. If we insert the unapproximated version of $if_\mu(k, P)$ in (3-11.16), it becomes ($k^2 = 0$):

$$iJ_{2\mu}(k)K_2(p)\Big|_{\text{eff.}} = \left[\frac{2p_\mu + k_\mu}{2kp} - \frac{p_\mu + k_\mu}{kp}\right]eqK_2(P)$$

$$= -\frac{1}{2}\left(\frac{k_\mu}{kp}\right)eqK_2(P). \qquad (3\text{-}11.23)$$

The probability amplitude for the emission of the two particles labelled $k\lambda$, pq requires, beyond (3-11.23), the additional factors $(d\omega_k)^{1/2}$ and $(d\omega_p)^{1/2}$, together with the explicit selection of charge $\pm e$ and the photon polarization λ. The latter is produced by scalar multiplication with the polarization vector $e_{k\lambda}^{\mu*}$, and

$$k_\mu e_{k\lambda}^\mu = 0. \qquad (3\text{-}11.24)$$

There is another point that can be illustrated by the effective source (3-11.16). Equivalent to a particle source model characterized by $f^\mu(k)$ is the assignment of electrical properties only to the particle, combined with the use of vector potentials in a specific gauge such that [Eq. (3-10.14)]

$$f^\mu(k)A_\mu^f(k) = 0. \qquad (3\text{-}11.25)$$

The vector potential that represents the emitted photon is proportional to the polarization vector $e_{k\lambda}^{\mu*}$, and the gauge condition (3-11.25) demands that

$$f_\mu(k)e_{k\lambda}^\mu = 0, \qquad (3\text{-}11.26)$$

which supplements (3-11.24). Thus, with the choice of $f_\mu(k)$ that is displayed in (3-10.49) we have

$$n_\mu e_{k\lambda}^\mu = 0, \qquad (3\text{-}11.27)$$

and this becomes $e_{k\lambda}^0 = 0$ in the appropriate coordinate frame. The significant observation is that, on multiplying (3-11.16) by one of these polarization vectors, the term associated with radiation from the source vanishes, as it should.

It is possible to remove the limitation to single photon emission, at least when attention is confined to soft photons. Since there is still only one particle detection source, we change tactics and use

$$\delta_K W = \int (dx)\ \delta K(x)\phi(x) \tag{3-11.28}$$

or

$$\delta_K \langle 0_+|0_-\rangle^{JK} = i\int (dx)\ \delta K(x)\phi(x)\langle 0_+|0_-\rangle^{JK}, \tag{3-11.29}$$

in which $\delta K(x) \to K_1(x)$ and $\phi(x)$ is related to the source $K_2(x)$ by the field equations

$$-(\partial_\mu - ieqA_\mu(x))\phi^\mu(x) + m^2\phi(x) = K_2^A(x), \qquad (\partial_\mu - ieA_\mu(x))\phi(x) = \phi_\mu(x). \tag{3-11.30}$$

The elimination of ϕ_μ gives the second-order differential equation

$$[-(\partial^\mu - ieqA^\mu(x))(\partial_\mu - ieqA_\mu(x)) + m^2]\phi(x)$$
$$= \exp\left[ieq\int (d\xi)f^\mu(x - \xi)A_\mu(\xi)\right]K_2(x). \tag{3-11.31}$$

Since both particle sources already appear in (3–11.29), the class of processes we wish to select are exhibited by

$$\langle 0_+|0_-\rangle^{JK} = \cdots + i\int (dx)K_1(x)\phi_2^{A_1}(x)\langle 0_+|0_-\rangle^{J_1} + \cdots, \tag{3-11.32}$$

where the notation emphasizes the dependence of the particle field $\phi_2(x)$ upon the vector potential $A_1^\mu(\xi)$ that represents the emitted photons in relation to their detection source $J_1^\mu(\xi)$.

Let us first recover the known single photon result in this new way. For this we need the part of $\phi_2^{A_1}(x)$ that is linear in the vector potential. The field equation (3–11.31) retains just that amount of information when it is simplified to

$$(-\partial^2 + m^2)\phi_2^A(x) = K_2(x) + (1/i)\partial^\mu[eqA_\mu(x)\phi_2(x)] + eqA_\mu(x)(1/i)\partial^\mu\phi_2(x)$$
$$+ ieq\int (d\xi)f^\mu(x - \xi)A_\mu(\xi)K_2(x). \tag{3-11.33}$$

We get what is required in (3–11.32) by multiplying this field equation by $\phi_1(x)$ and integrating:

$$\int (dx)K_1(x)\phi_2^{A_1}(x) = \int (dx)\phi_1(x)K_2(x) + \int (d\xi)A_1^\mu(\xi)j_{12\mu}(\xi). \tag{3-11.34}$$

The first term on the right represents the radiationless emission of the particle, and the second one reproduces (3–11.11). The nth term of the power series expansion of $\phi_2^A(x)$ in $A^\mu(\xi)$ describes n-photon emission processes. If we agree to consider only soft photons, all such processes can be combined in a compact formula which, as we would now expect, is equivalent to a photon source description.

The differential equation (3–11.31) is formally solved by

$$\phi_2^A(x) = \int (dx') \Delta_+^A(x, x') \exp\left[ieq \int (d\xi) f^\mu (x' - \xi) A_\mu(\xi)\right] K_2(x'), \quad (3\text{--}11.35)$$

where the Green's function $\Delta_+^A(x, x')$ obeys

$$[-(\partial - ieqA(x))^2 + m^2]\Delta_+^A(x, x') = \delta(x - x'). \quad (3\text{--}11.36)$$

We introduce the following transformation:

$$\Delta_+^A(x, x') = \exp\left[ieq \int_{x'}^x d\xi^\mu A_\mu(\xi)\right] \Delta_+^{A'}(x, x'), \quad (3\text{--}11.37)$$

in which the integration path is a straight line connecting x and x', as parametrized by

$$\xi^\mu = x^\mu \frac{1+\lambda}{2} + x^{\mu'} \frac{1-\lambda}{2}, \qquad 1 \geq \lambda \geq -1. \quad (3\text{--}11.38)$$

This transformation induces a gauge transformation on A_μ, replacing it with

$$A_\mu'(x) = A_\mu(x) - \partial_\mu \int_{x'}^x d\xi^\nu A_\nu(\xi), \quad (3\text{--}11.39)$$

and giving the new Green's function equation

$$[-(\partial - ieqA'(x))^2 + m^2]\Delta_+^{A'}(x, x') = \delta(x - x'). \quad (3\text{--}11.40)$$

The identity

$$A_\mu(x) - \partial_\mu \int_{x'}^x d\xi^\nu A_\nu(\xi) = \int_{-1}^1 d\lambda \frac{d}{d\lambda}\left[\frac{1+\lambda}{2} A_\mu(\xi)\right]$$
$$- \partial_\mu \int_{-1}^1 \tfrac{1}{2} d\lambda (x - x')^\nu A_\nu(\xi) \quad (3\text{--}11.41)$$

produces the gauge invariant construction

$$A_\mu'(x) = -\int_{-1}^1 \tfrac{1}{2} d\lambda \frac{1+\lambda}{2} (x - x')^\nu F_{\mu\nu}(\xi). \quad (3\text{--}11.42)$$

This vector potential has two other significant properties. In regions far from the electromagnetic source $J_\nu(\xi)$,

$$\partial^\mu A_\mu'(x) = \int_{-1}^1 \tfrac{1}{2} d\lambda \left(\frac{1+\lambda}{2}\right)^2 (x - x')^\nu J_\nu(\xi) = 0, \quad (3\text{--}11.43)$$

and, generally,

$$(x - x')^\mu A_\mu'(x) = 0. \quad (3\text{--}11.44)$$

Hence, if we were to begin a construction of $\Delta_+^{A'}(x, x')$ as a power series in A_μ', representing photon fields far from their detection source, the initial terms would be obtained from

$$(-\partial^2 + m^2)\Delta_+^{A'}(x, x') = \delta(x - x') + 2eqA_\mu'(x)(1/i)\partial^\mu\Delta_+(x - x') + \cdots .$$
$$(3-11.45)$$

But $\Delta_+(x - x')$, being an invariant function, depends only upon $(x - x')^2$, and its gradient is a multiple of the vector $(x - x')^\mu$. We learn that $\Delta_+^{A'}(x, x')$ has no term linear in A_μ'.

More can be said if the field strengths are treated as homogeneous, as is appropriate to soft photons, of negligible momenta. Then

$$A_\mu'(x) = -\tfrac{1}{2}F_{\mu\nu}(x - x')^\nu,$$
$$(3-11.46)$$

which implies the translational invariance of the Green's function,

$$\Delta_+^{A'}(x, x') = \Delta_+^{A'}(x - x'),$$
$$(3-11.47)$$

and the differential equation (3–11.36) becomes

$$[-\partial^2 + m^2 + ieq\tfrac{1}{2}F^{\mu\nu}(x_\mu\partial_\nu - x_\nu\partial_\mu) - \tfrac{1}{4}e^2 x^\mu F_{\mu\nu}^2 x^\nu]\Delta_+^{A'}(x) = \delta(x), \quad (3-11.48)$$

where

$$F_{\mu\nu}^2 = F_\mu{}^\lambda F_{\lambda\nu}.$$
$$(3-11.49)$$

The angular momentum structure of the linear field strength term assures its commutativity with ∂^2; it also commutes with the quadratic combination of coordinates:

$$[\tfrac{1}{2}F^{\mu\nu}(x_\mu\partial_\nu - x_\nu\partial_\mu), \tfrac{1}{2}x^\kappa F_{\kappa\lambda}^2 x^\lambda] = x^\mu F_{\mu\nu}^3 x^\nu$$
$$= 0,$$
$$(3-11.50)$$

since

$$F_{\mu\nu}^3 = F_{\mu\kappa}F^{\kappa\lambda}F_{\lambda\nu}$$
$$(3-11.51)$$

is an antisymmetrical function of μ and ν. All this, and the rotational invariance of $\delta(x)$, shows that the differential equation (3–11.48) can be simplified to

$$[-\partial^2 + m^2 - \tfrac{1}{4}e^2 x^\mu F_{\mu\nu}^2 x^\nu]\Delta_+^{A'}(x) = \delta(x).$$
$$(3-11.52)$$

We shall not stop now to solve the above equation. It suffices to know that $\Delta_+^{A'}(x - x')$ is an even function of field strengths, for this means that the field dependence of the latter function can be neglected relative to its partner in (3–11.37), since, compared to vector potentials, field strengths contain an additional photon momentum factor.

Introducing these soft photon simplifications, we arrive at

$$\int(dx)K_1(x)\phi_2^A(x) = \int(dx)(dx')K_1(x)\Delta_+(x - x') \exp\left[ieq\left(\int_{x'}^x d\xi^\mu A_\mu(\xi)\right.\right.$$
$$\left.\left. + \int(d\xi)f^\mu(x' - \xi)A_\mu(\xi)\right)\right]K_2(x').$$
$$(3-11.53)$$

The straight line integral that occurs here begins at x' and moves, in a direction determined by the vector $(x - x')^\mu$, toward an effectively infinitely distant point, since the photon emission processes are localized near the extended source K_2. And, if the coupling between the particle sources is to be appreciable, the geometrical displacement $(x - x')^\mu$ must coincide closely in direction with that of the momentum vector of the exchanged particle. Accordingly,

$$\exp\left[ieq\left(\int_{x'}^x d\xi^\mu A_\mu(\xi) + \int(d\xi)f^\mu(x' - \xi)A_\mu(\xi)\right)\right] = \exp\left[i\int(d\xi)J^\mu(\xi)A_\mu(\xi)\right],$$

(3–11.54)

where

$$J^\mu(\xi) = eq\int_0^\infty ds(p^\mu/m)\,\delta(\xi - x' - (p/m)s) + eq\int_{-\infty}^0 dsn^\mu\,\delta(\xi - x' - ns)$$

(3–11.55)

differs only inconsequentially from (3–11.21), through the explicit appearance of x' as the transition point; it is used implicitly as the origin in (3–11.21), since the variation of x' over K_2 is not significant in the soft photon context. This is the anticipated source description of multi-soft photon emission processes. Notice that the effective photon source characterizes the probability amplitudes for additional photon emission, relative to that of the radiationless process, which is supplied by the significance of K_2 when it acts as a simple particle emission source.

The time has come to face up to a characteristic feature of soft photons. With a continual diminution of the energy assigned to a soft photon in a given experimental arrangement, one eventually reaches a point where it is no longer possible to decide whether the photon has or has not been emitted. It is a somewhat complementary space-time observation that, with increasing wavelength, one eventually loses the possibility of isolating the soft photon emission process since the disposition of surrounding matter becomes relevant. Thus, more than the usual amount of detail concerning the experimental arrangement is required. This is emphasized by using the photon source (3–11.55) to compute the average number of photons emitted along with a given particle. That number is

$$\langle N \rangle = \sum_{k\lambda} |J_{k\lambda}|^2 = \int d\omega_k J^\mu(k)^* J_\mu(k) = e^2\int d\omega_k \left(\frac{p}{pk} - \frac{n}{nk}\right)^2. \quad (3\text{–}11.56)$$

To see the essence of the situation, it suffices to consider a slowly moving particle,

$$p^0 \simeq m + \tfrac{1}{2}mv^2, \qquad \mathbf{p} = m\mathbf{v}, \qquad |\mathbf{v}| \ll 1, \quad (3\text{–}11.57)$$

and a coordinate system in which n^μ has only a time component, $n^0 = 1$. Then the components of the vector combination in (3–11.56) are, approximately,

$$\frac{p^0}{pk} - \frac{n^0}{nk} \simeq \frac{1}{k^0}\left[1 - \frac{1}{1 - \mathbf{v}\cdot(\mathbf{k}/k^0)}\right] \simeq -\frac{1}{k^0}\mathbf{v}\cdot\frac{\mathbf{k}}{k^0},$$

$$\frac{\mathbf{p}}{pk} - \frac{\mathbf{n}}{nk} \simeq -\frac{1}{k^0}\mathbf{v}.$$

(3–11.58)

On writing

$$dw_k = \frac{1}{(2\pi)^3} \frac{(d\mathbf{k})}{2k^0} = \frac{d\Omega}{(2\pi)^3} \frac{1}{2} k^0 \, dk^0, \qquad (3\text{--}11.59)$$

where $d\Omega$ is the solid angle within which the photon moves, we get

$$\langle N \rangle = \left(\frac{e^2}{4\pi}\right) \frac{1}{\pi} \int \frac{dk^0}{k^0} \int \frac{d\Omega}{4\pi} \left[\mathbf{v}^2 - \left(\mathbf{v} \cdot \frac{\mathbf{k}}{k^0}\right)^2\right]$$

$$= \frac{2\alpha}{3\pi} \mathbf{v}^2 \int \frac{dk^0}{k^0}. \qquad (3\text{--}11.60)$$

This photon energy integral does not exist mathematically, diverging both at the upper and lower limits. But clearly there are physical restrictions at both ends. When one reaches energies at which the photon ceases to be soft, the evaluation (3–11.60) no longer applies, and a lower limit is set by the minimum detectable photon energy of the experimental arrangement. Once upon a time, the mathematical divergence at zero energy was taken literally, and this soft photon phenomenon became known as the 'infrared catastrophe.' As a catastrophe, it rates rather low on the scale. Consider the difference that is implied in the value of $\langle N \rangle$, depending upon whether the softest photon considered has a wavelength of visible light, $\sim 10^{-5}$ cm, or has a wavelength comparable to the nominal radius of the universe, $\sim 10^{28}$ cm. Since $\mathbf{v}^2 < 1$, that difference is

$$\Delta\langle N \rangle < \frac{2\alpha}{3\pi} \log 10^{33} \sim 0.1! \qquad (3\text{--}11.61)$$

If the radius of the universe is replaced by a typical laboratory dimension, this difference drops to $\sim 10^{-2}$.

To illustrate the discussion of spin values other than zero, we shall consider spin $\frac{1}{2}$, using the Lagrange function of (3–10.63). The current of (3–11.11) is replaced by

$$j^\mu_{12}(\xi) = e\psi_1(\xi)\gamma^0\gamma^\mu q\psi_2(\xi) + \frac{e}{2m} (\tfrac{1}{2}g - 1)\partial_\nu[\psi_1(\xi)\gamma^0\sigma^{\mu\nu}q\psi_2(\xi)]$$

$$- \int (dx)f^\mu(\xi - x)\psi_1(x)\gamma^0 ieq\eta_2(x), \qquad (3\text{--}11.62)$$

where

$$[\gamma(1/i)\partial + m]\psi_2(x) = \eta_2(x) \qquad (3\text{--}11.63)$$

is solved, in momentum space, by

$$\psi_2(P) = \frac{m - \gamma P}{P^2 + m^2} \eta_2(P). \qquad (3\text{--}11.64)$$

The comparison with the exchange of one particle and one photon under non-interaction conditions,

$$\langle 0_+|0_-\rangle^{J\eta} = 1 + \cdots + i\int (d\xi) A^\mu_1(\xi) J_{2\mu}(\xi) i \int (dx)\psi_1(x)\gamma^0\eta_2(x) + \cdots$$

$$(3\text{--}11.65)$$

supplies the effective two-particle source that represents the emission of the extended particle source:

$$iJ_2^\mu(\xi)\eta_2(x)\Big|_{\text{eff.}} = \delta(x-\xi)\gamma^\mu eq\psi_2(x) + \frac{e}{2m}(\tfrac{1}{2}g-1)\frac{\partial}{\partial\xi^\nu}\delta(x-\xi)\sigma^{\mu\nu}q\psi_2(x)$$
$$- f^\mu(\xi-x)ieq\eta_2(x). \tag{3-11.66}$$

The momentum space equivalent is

$$iJ_2^\mu(k)\eta_2(p)\Big|_{\text{eff.}} = \left[\gamma^\mu eq + \frac{e}{2m}(\tfrac{1}{2}g-1)\sigma^{\mu\nu}ik_\nu q\right]\psi_2(P) - f^\mu(k)ieq\eta_2(P)$$
$$= \left[\left(\gamma^\mu + \frac{1}{2m}(\tfrac{1}{2}g-1)\sigma^{\mu\nu}ik_\nu\right)\frac{m-\gamma P}{P^2+m^2} - if^\mu(k)\right]eq\eta_2(P). \tag{3-11.67}$$

Using the latter form, we observe that

$$ik_\mu J_2^\mu(k)\eta_2(p)\Big|_{\text{eff.}} = \left[\gamma k\frac{m-\gamma P}{P^2+m^2}-1\right]eq\eta_2(P) \tag{3-11.68}$$

and, on writing

$$\gamma k = \gamma P + m - (\gamma p + m), \tag{3-11.69}$$

we get

$$ik_\mu J_2^\mu(k)\eta_2(p)\Big|_{\text{eff.}} = -(\gamma p + m)\frac{m-\gamma P}{P^2+m^2}eq\eta_2(P) \tag{3-11.70}$$

or

$$\frac{\partial}{\partial\xi^\mu}J^\mu(\xi)\eta_2(x)\Big|_{\text{eff.}} = -\left[\gamma^\mu\frac{1}{i}\frac{\partial}{\partial x^\mu}+m\right]\delta(\xi-x)eq\psi_2(x). \tag{3-11.71}$$

But this is to be used in the context of Eq. (3–11.65) where the field $\psi_1(x)$ represents particles far from their detection source, and the Dirac differential operator in (3–11.71) produces the required null result. Alternatively, we can use the momentum form (3–11.70) and recall that

$$\psi_1(x) = \sum_{p\sigma q} i\eta^*_{1p\sigma q}(2m\,d\omega_p)^{1/2}e^{-ipx}u^*_{p\sigma q}, \tag{3-11.72}$$

where

$$u^*_{p\sigma q}\gamma^0(\gamma p + m) = 0. \tag{3-11.73}$$

Let us also note the photon analogue of (3–11.72),

$$A_1^\mu(\xi) = \sum_{k\lambda} iJ_{1k\lambda}(d\omega_k)^{1/2}e^{-ikx}e^{\mu*}_{k\lambda}, \tag{3-11.74}$$

with

$$e^{\mu*}_{k\lambda}k_\mu = 0, \tag{3-11.75}$$

since both factors are useful in producing a simplification of (3–11.67). A relevant algebraic property is

$$\gamma^\mu(m-\gamma P) = \gamma^\mu(m-\gamma p) - \gamma^\mu\gamma k$$
$$= 2p^\mu + \sigma^{\mu\nu}ik_\nu + [(\gamma p + m)\gamma^\mu + k^\mu], \tag{3-11.76}$$

where both terms in the square bracket can be omitted for our purposes. Similarly, we note that

$$-\sigma^{\mu\nu}ik_\nu\gamma k = (\gamma^\mu\gamma k + k^\mu)\gamma k = [k^\mu\gamma k], \qquad (3\text{-}11.77)$$

since $k^2 = 0$, and

$$\sigma^{\mu\nu}ik_\nu(m - \gamma p) = 2m\sigma^{\mu\nu}ik_\nu - 2kp\gamma^\mu + 2p^\mu\gamma k - [(\gamma p + m)\sigma^{\mu\nu}ik_\nu], \qquad (3\text{-}11.78)$$

where noncontributing terms have been isolated in brackets. The result is

$$iJ_2^\mu(k)\eta_2(p)\Big|_{\text{eff.}} = \left[\frac{p^\mu}{kp} - if^\mu(k) + \tfrac{1}{2}g\frac{\sigma^{\mu\nu}ik_\nu}{2kp}\right.$$
$$\left. + \frac{1}{2m}(\tfrac{1}{2}g - 1)\left(\frac{p^\mu}{kp}\gamma k - \gamma^\mu\right)\right]eq\eta_2(P), \qquad (3\text{-}11.79)$$

where one can also use the substitution

$$\frac{p^\mu}{kp}\gamma k - \gamma^\mu \rightarrow \frac{i}{m}\left[\sigma^{\mu\nu}p_\nu - \frac{p^\mu}{kp}\sigma^{\kappa\lambda}k_\kappa p_\lambda\right], \qquad (3\text{-}11.80)$$

which is derived from

$$m\gamma^\mu - p^\mu + \sigma^{\mu\nu}ip_\nu = [(\gamma p + m)\gamma^\mu]. \qquad (3\text{-}11.81)$$

It is evident that, in the limit of soft photons, there is an effective photon source which is identical with the one encountered for zero spin. This is to be expected, for every spin value. The successive multipole moment effects involve increasing powers of the photon momentum, and all become negligible compared to the charge acceleration radiation for sufficiently soft photons. But the particular choice of $f^\mu(k)$ that removes the acceleration radiation no longer suppresses photon emission completely, since the spin-dependent effects of magnetic dipole moments remain in (3-11.79), and no specialization of g can annul both terms.

We have illustrated the extended source concept in the context of emission. It can all be repeated when the extended source acts to absorb a particle and a photon. But these inverse processes are also related by the TCP operation, concerning which nothing has been said recently. The effect of the Euclidean based coordinate transformation

$$\bar{x}_\mu = -x_\mu \qquad (3\text{-}11.82)$$

on sources and fields is given by

$$\bar{J}_\mu(\bar{x}) = -J_\mu(x), \qquad \bar{A}_\mu(\bar{x}) = -A_\mu(x)$$
$$\bar{F}_{\mu\nu}(\bar{x}) = F_{\mu\nu}(x), \qquad (3\text{-}11.83)$$

and, for spin $\tfrac{1}{2}$,

$$\bar{\eta}(\bar{x}) = \gamma_5\eta(x), \qquad \bar{\psi}(\bar{x}) = \gamma_5\psi(x). \qquad (3\text{-}11.84)$$

The field-dependent source

$$\eta^A(x) = \exp\left[ieq\int(d\xi)f^\mu(x - \xi)A_\mu(\xi)\right]\eta(x) \qquad (3\text{-}11.85)$$

has the same transformation behavior as $\eta(x)$ if

$$-f^\mu(-x) = f^\mu(x), \qquad (3\text{-}11.86)$$

which finally provides a physical basis for the symmetry property that, thus far, has been adopted for convenience. The purely electromagnetic part of the action retains its form under this transformation,

$$\int (dx)[J^\mu(x)A_\mu(x) - \tfrac{1}{4}F^{\mu\nu}(x)F_{\mu\nu}(x)] = \int (dx)[\overline{J}^\mu(x)\overline{A}_\mu(x) - \tfrac{1}{4}\overline{F}^{\mu\nu}(x)\overline{F}_{\mu\nu}(x)],$$

$$(3\text{-}11.87)$$

while the particle contribution, including the interaction term, reverses sign:

$$\int (dx) \left\{ \eta^A(x)\gamma^0\psi(x) - \tfrac{1}{2}\psi(x)\gamma^0 \left[\gamma^\mu(-i\partial_\mu - eqA_\mu(x)) + m \right.\right.$$
$$\left.\left. - \frac{e}{2m}(\tfrac{1}{2}g - 1)\tfrac{1}{2}\sigma^{\mu\nu}F_{\mu\nu}(x) \right]\psi(x) \right\}$$
$$= -\int (dx) \left\{ \overline{\eta}^A(x)\gamma^0\overline{\psi}(x) - \tfrac{1}{2}\overline{\psi}(x)\gamma^0 \left[\gamma^\mu(-i\partial_\mu - eq\overline{A}_\mu(x)) + m \right.\right.$$
$$\left.\left. - \frac{e}{2m}(\tfrac{1}{2}g - 1)\tfrac{1}{2}\sigma^{\mu\nu}\overline{F}_{\mu\nu}(x) \right]\overline{\psi}(x) \right\}. \qquad (3\text{-}11.88)$$

But the complete statement of the *TCP* operation includes the reversal of all factors. The anticommutativity of the sources and fields associated with the spin $\tfrac{1}{2}$, F. D. particle provides the additional minus sign needed to produce the anticipated invariance of the action under the *TCP* transformation.

The *TCP* operation inverts the causal order, and interchanges emission and absorption processes. On applying the transformation to (3-11.66), one quickly verifies that the whole structure is maintained, and it is therefore only necessary to change the causal labels. The same remark applies to the momentum version (3-11.67), of course, except that we follow the practice of reversing the signs of all momenta when absorption processes are being described, which the transformation automatically supplies. What has been shown in the spin $\tfrac{1}{2}$ framework is of general validity.

3-12 INTERACTION SKELETON.   SCATTERING CROSS SECTIONS

A given primitive interaction implies a set of coupled field equations. Here is the example of the photon and the charged spin $\tfrac{1}{2}$ particle, written, for simplicity, with $\tfrac{1}{2}g = 1$:

$$[\gamma(-i\partial - eqA(x)) + m]\psi(x) = \eta^A(x),$$
$$-\partial^2 A^\mu(x) + \partial^\mu \partial A(x) = J^\mu(x) + \tfrac{1}{2}\psi(x)\gamma^0\gamma^\mu eq\psi(x)$$
$$- \int (dx')f^\mu(x - x')\psi(x')\gamma^0 ieq\eta^A(x'). \qquad (3\text{-}12.1)$$

In view of the nonlinearity of this system, the construction of the fields in terms of the sources will be given by doubly infinite power series. That is also the nature of the action when the fields are eliminated and W is expressed as a functional of the sources. The successive terms of this series, $W_{n\nu}$, with n particle and ν photon sources, represent increasingly complicated physical processes which are thus acknowledged to occur, but will not be given their final description at this first level of dynamical evolution. That is the meaning of an interaction skeleton. At later stages of the dynamical development, processes already present in skeletal form are provided with more complete descriptions, and some additional processes are recognized. In this section, we propose to carry the discussion of the simplest terms in the interaction skeleton to the point of displaying their observational implications.

There are two asymmetrical ways to eliminate the fields. In the first, one introduces the formal solution of the particle field equation:

$$\psi^A(x) = \int (dx')G_+^A(x, x')\eta^A(x'),$$

$$[\gamma(-i\partial - eqA(x)) + m]G_+^A(x, x') = \delta(x - x'),$$

$$\text{(3-12.2)}$$

which gives the partial action expression

$$W = \int (dx)[J^\mu A_\mu - \tfrac{1}{4}F^{\mu\nu}F_{\mu\nu}] + \tfrac{1}{2}\int (dx)(dx')\eta^A(x)\gamma^0 G_+^A(x, x')\eta^A(x'). \quad \text{(3-12.3)}$$

The stationary requirement on variations of A_μ recovers the Maxwell equation of (3-12.1), with $\psi(x)$ given by (3-12.2)—a highly nonlinear equation for the vector potential. One can still exercise the option of removing A_μ from the particle source by adopting the special gauge of the A'_μ potentials.

The latter procedure is particularly recommended if we follow the second course and eliminate the vector potential, replacing it with

$$A'_\mu(x) = \int (dx')D_+(x - x')[J^\mu(x) + j^\mu_{\text{cons.}}(x)] + \partial_\mu\lambda(x), \quad \text{(3-12.4)}$$

where

$$j^\mu_{\text{cons.}}(x) = j^\mu(x) - \int (dx')f^\mu(x - x')\partial'_\nu j^\nu(x') \quad \text{(3-12.5)}$$

and the gauge condition determines $\lambda(x)$ as

$$\lambda(x) = -\int (dx')(dx'')f_\mu(x - x')D_+(x' - x'')[J^\mu(x'') + j^\mu_{\text{cons.}}(x'')]. \quad \text{(3-12.6)}$$

Another way of presenting this potential is [J' is now an arbitrary vector]

$$A'_\mu(x) = \int (dx')D'_+(x - x')_{\mu\nu}[J^\nu(x') + j^\nu(x')], \quad \text{(3-12.7)}$$

where, written in momentum space for convenience,

$$D'_+(k)_{\mu\nu} = (g_{\mu\kappa} - ik_\mu f_\kappa(k))g^{\kappa\lambda}D_+(k)(g_{\lambda\nu} - f_\lambda(k)ik_\nu)$$
$$= (g_{\mu\nu} - ik_\mu f_\nu(k) - f_\mu(k)ik_\nu - k_\mu k_\nu f^\lambda(k)f_\lambda(k))D_+(k) \quad \text{(3-12.8)}$$

is the Green's function of the second-order Maxwell equation that satisfies the gauge condition

$$f^\mu(k)D^f_+(k)_{\mu\nu} = 0. \tag{3-12.9}$$

The second partial action expression can be written as

$$W = \int(dx)[\eta\gamma^0\psi - \tfrac{1}{2}\psi\gamma^0(-i\gamma\partial + m)\psi]$$
$$+ \tfrac{1}{2}\int(dx)(dx')(J^\mu(x) + j^\mu_{\text{cons.}}(x))D_+(x - x')(J_\mu(x') + j_{\mu\,\text{cons.}}(x')), \tag{3-12.10}$$

or in the equivalent form that uses the nonconserved currents and $D^f_+(x - x')_{\mu\nu}$. The nonlinear field equation for ψ that is derived from this action is that of (3-12.1), with A_μ replaced by (3-12.4) or (3-12.7).

Which of these asymmetric forms it is most convenient to consider depends upon the process of interest. Suppose, for example, that no photons are in evidence. Then one can set $J^\mu = 0$ in (3-12.10) and examine the nonlinear properties of the particle field. If the causal situation is such that interactions occur far from the particle emission and detection sources, which is part of the arrangement of a scattering experiment, the f^μ term in $j^\mu_{\text{cons.}}$—causally tied to the source—can be ignored. The interaction term of (3-12.10) contains four particle fields and thereby at least four source factors. When we consider processes that involve only four sources, as in particle-particle scattering, the stationary aspects of the action principle permit us to identify ψ with the field of noninteracting particles,

$$\psi(x) = \int(dx')G_+(x - x')\eta(x').^7 \tag{3-12.11}$$

In omitting further terms of a more complete solution of the effective field equation,

$$(-i\gamma\partial + m)\psi(x) - eq\gamma^\mu\psi(x)\int(dx')D_+(x - x')\tfrac{1}{2}\psi(x')\gamma^0\gamma_\mu eq\psi(x') = \eta(x), \tag{3-12.12}$$

which are at least cubic in the source, what is thereby lacking in W has no less than six powers of the source since first-order effects of the field change are annulled through the stationary action property. Thus we have identified

$$W_{40} = \tfrac{1}{2}\int(dx)(dx')j^\mu(x)D_+(x - x')j_\mu(x'), \tag{3-12.13}$$

where

$$j^\mu(x) = \tfrac{1}{2}\psi(x)\gamma^0\gamma^\mu eq\psi(x) \tag{3-12.14}$$

and $\psi(x)$ is the field given in (3-12.11). Analogous results hold for any other spin value. With spinless particles, for example,

$$j^\mu(x) = i\partial^\mu\phi(x)eq\phi(x), \tag{3-12.15}$$

and

$$\phi(x) = \int (dx')\Delta_+(x - x')K(x').$$
(3–12.16)

Processes that involve only two particle sources but any number of photon sources will be described most conveniently by the action (3–12.3). The stationary action principle permits the identification of A_μ with the field of the photon source J_μ, the omission of j^μ, which is at least quadratic in the particle source, changing those terms in W that contain no less than four particle sources. Thus the whole set of skeletal interaction terms $W_{2\nu}$ is given by

$$W_{2\dots} = \tfrac{1}{2} \int (dx)(dx')\eta(x)\gamma^0 G_+^A(x, x')\eta(x')$$
(3–12.17)

with

$$A^\mu(x) = \int (dx')D_+(x - x')J^\mu(x').$$
(3–12.18)

The reference to the vector potential in the particle source has been dropped, with the understanding that (3–12.17) will be applied to processes in which $\eta(x)$ is used as a simple particle source, all particle-photon interactions occurring far from any of the sources. To exhibit the individual $W_{2\nu}$, we must expand in power series the A^μ dependence of $G_+^A(x, x')$ and extract the term containing ν vector potentials. For this purpose it is useful to rewrite the Green's function equation of (3–12.2) as

$$(-\gamma i\partial + m)G_+^A(x, x') = \delta(x - x') + eq\gamma A(x)G_+^A(x, x'),$$
(3–12.19)

which is converted to an integral equation by the formal solution

$$G_+^A(x, x') = G_+(x - x') + \int (d\xi)G_+(x - \xi)eq\gamma A(\xi)G_+^A(\xi, x').$$
(3–12.20)

The desired power series expansion can now be constructed by successive substitution in this equation. Such manipulations are facilitated, however, by adopting a matrix notation in which the coordinates x and x' join the discrete spinor and charge indices as continuous row and column labels. Thus, we transcribe (3–12.20) into

$$G_+^A = G_+ + G_+ eq\gamma A G_+^A$$
(3–12.21)

and write the formal solution of this matrix equation as

$$G_+^A = (1 - G_+ eq\gamma A)^{-1}G_+.$$
(3–12.22)

A compact statement of the expansion is, therefore,

$$G_+^A = \sum_{\nu=0} (G_+ eq\gamma A)^\nu G_+ = G_+ + G_+ eq\gamma A G_+ + G_+ eq\gamma A G_+ eq\gamma A G_+ + \cdots.$$
(3–12.23)

We can now return to (3–12.17) and write out the successive $W_{2\nu}$. In doing this one recognizes that each particle source is multiplied by a propagation function

G_+ to form the field ψ of (3–12.11):

$$W_{21} = \tfrac{1}{2}\int (dx)\psi(x)\gamma^0 eq\gamma A(x)\psi(x),$$

$$W_{22} = \tfrac{1}{2}\int (dx)(dx')\psi(x)\gamma^0 eq\gamma A(x)G_+(x - x')eq\gamma A(x')\psi(x'),$$

$$W_{23} = \tfrac{1}{2}\int (dx)(dx')(dx'')\psi(x)\gamma^0 eq\gamma A(x)G_+(x - x')eq\gamma A(x') \tag{3–12.24}$$

$$\times\, G_+(x' - x'')eq\gamma A(x'')\psi(x'').$$

The spin 0 analogue of Eq. (3–12.17) is

$$W_{2...} = \tfrac{1}{2}\int (dx)(dx')K(x)\Delta_+^A(x, x')K(x'), \tag{3–12.25}$$

where the Green's function differential equation (3–11.36) is presented as

$$(-\partial^2 + m^2)\Delta_+^A(x, x') = \delta(x - x') + (1/i)\partial^\mu[eqA_\mu(x)\Delta_+^A(x, x')]$$
$$+ eqA_\mu(x)(1/i)\partial^\mu\Delta_+^A(x, x') - e^2 A^\mu(x)A_\mu(x)\Delta_+^A(x, x'). \tag{3–12.26}$$

The equivalent integral equation is of the following symbolic appearance,

$$\Delta_+^A = \Delta_+ + \Delta_+[eq(pA + Ap) - e^2 A^2]\Delta_+^A, \tag{3–12.27}$$

which has the formal solution

$$\Delta_+^A = [1 - \Delta_+(eq(pA + Ap) - e^2 A^2)]^{-1}\Delta_+$$
$$= \Delta_+ + \Delta_+(eq(pA + Ap) - e^2 A^2)\Delta_+$$
$$+ \Delta_+(eq(pA + Ap) - e^2 A^2)\Delta_+(eq(pA + Ap) - e^2 A^2)\Delta_+ + \cdots. \tag{3–12.28}$$

The successive powers of A^μ are not presented quite so neatly as with spin $\tfrac{1}{2}$. The first two terms of the series $W_{2\nu}$ are

$$W_{21} = \tfrac{1}{2}\int (dx)\phi(x)eq(pA(x) + A(x)p)\phi(x),$$

$$W_{22} = \tfrac{1}{2}\int (dx)(dx')\phi(x)eq(pA(x) + A(x)p)\Delta_+(x - x') \tag{3–12.29}$$

$$\times\, eq(pA(x') + A(x')p)\phi(x') - \tfrac{1}{2}\int (dx)\phi(x)e^2 A^2(x)\phi(x),$$

in which it has been expedient to retain the symbol

$$p_\mu = -i\partial_\mu. \tag{3–12.30}$$

Note that

$$pA(x) - A(x)p = -i\partial_\mu A^\mu(x), \tag{3–12.31}$$

which means that the careful ordering of factors can be ignored if the vector potential has a vanishing four-dimensional divergence, as is the situation for (3–12.18). Potentials having this property are said to be in the Lorentz gauge.

The immediate applications of the interaction skeleton for which we have been preparing refer to scattering processes. Let us therefore review the general connection between the source description and the transition probabilities that describe the effects of interactions among particles. The causal situation is this. Emission sources, generally referring to different kinds of particles, act to produce a multiparticle state of particles in a physically noninteracting condition, owing to their initial spatial separation. After a sufficient time lapse, some of these particles approach each other, interact, and then separate to be eventually annihilated along with their noninteracting companions by suitable detection sources. The causal analysis of the arrangement is given by

$$\langle 0_+|0_-\rangle^{S_1+S_2} = \sum_{\{n\}\{n'\}} \langle 0_+|\{n\}\rangle^{S_1}\langle\{n\}|\{n'\}\rangle\langle\{n'\}|0_-\rangle^{S_2}, \quad (3\text{-}12.32)$$

where the individual probability amplitudes $\langle\{n\}|\{n'\}\rangle$ describe the transitions induced by the particle interactions, and

$$\langle\{n\}|0_-\rangle^S = \langle 0_+|0_-\rangle^S \prod_a \frac{(iS_a)^{n_a}}{(n_a!)^{1/2}},$$
$$\langle 0_+|\{n\}\rangle^S = \prod_a^T \frac{(iS_a^*)^{n_a}}{(n_a!)^{1/2}} \langle 0_+|0_-\rangle^S \quad (3\text{-}12.33)$$

represent the noninteracting multiparticle states. The latter are labeled by the numbers of particles in the various single-particle modes, and the products also range over all the different kinds of particles, of both statistics. As a generating function of the probability amplitudes, (3–12.32) is more usefully presented in this version,

$$\frac{\langle 0_+|0_-\rangle^{S_1+S_2}}{\langle 0_+|0_-\rangle^{S_1}\langle 0_+|0_-\rangle^{S_2}} = \sum \frac{\langle 0_+|\{n\}\rangle^{S_1}}{\langle 0_+|0_-\rangle^{S_1}} \langle\{n\}|\{n'\}\rangle \frac{\langle\{n'\}|0_-\rangle^{S_2}}{\langle 0_+|0_-\rangle^{S_2}}, \quad (3\text{-}12.34)$$

since the various powers of the emission and detection sources serve to directly identify initial and final states.

The vacuum probability amplitude is determined by the action

$$W(S) = \tfrac{1}{2}\int S\gamma GS + W'(S), \quad (3\text{-}12.35)$$

in which we have specifically exhibited in symbolic form the quadratic structure that represents noninteracting particles. All relevant types of particles are included, so that S is being used as a supersource. By removing from both sides of (3–12.34) the expression that refers to noninteracting particles we arrive at

$$\exp\left(i\int S_1\gamma GS_2\right)[\exp(iW'(S_1, S_2)) - 1]$$
$$= \sum\left[\prod_a^T \frac{(iS_{1a}^*)^{n_a}}{(n_a!)^{1/2}}\right][\langle\{n\}|\{n'\}\rangle - \delta(\{n\}, \{n'\})]\left[\prod \frac{(iS_{2a})^{n_a'}}{(n_a'!)^{1/2}}\right], \quad (3\text{-}12.36)$$

where
$$W'(S_1, S_2) = W'(S_1 + S_2) - W'(S_1) - W'(S_2). \quad (3\text{-}12.37)$$

The factor $\exp[i\int S_1 \gamma G S_2]$ represents the exchange of those particles that happen not to interact. And higher powers in the expansion of $\exp[iW']$ indicate the possibility of repeating independently in disjoint space-time regions all configurations of interacting particles. Thus, the irreducible interaction processes, those that do not contain additional noninteracting particles and cannot be analyzed into two or more disconnected processes, are obtained from

$$iW'(S_1, S_2) = \sum \left[\prod^T \frac{(iS_{1a}^*)^{n_a}}{(n_a!)^{1/2}}\right] \langle\{n\}|\{n'\}\rangle_{\text{irred.}} \left[\prod \frac{(iS_{2a})^{n_a'}}{(n_a'!)^{1/2}}\right]. \quad (3\text{-}12.38)$$

Invariances of the action imply selection rules for the transition probabilities. Rigid translations or constant phase transformations of all sources, for example, which do not change $W'(S_1, S_2)$, must leave the right-hand side of (3–12.38) unaltered. The emission and absorption source products are multiplied by reciprocal phase constants, related to momentum and charge in these examples. The individual transition probabilities must vanish if the phase constants do not cancel, expressing the necessary conservation of momentum or charge in the interaction process. The factor that imposes momentum conservation,

$$\sum_a n_a p_a^\mu = \sum_a n_a' p_a^\mu, \quad (3\text{-}12.39)$$

will emerge from a space-time integration over the interaction region. We make this explicit by writing

$$\langle\{n\}|\{n'\}\rangle_{\text{irred.}} = \left[\int(dx)\exp\left(i\sum_a (n_a' - n_a)p_a x\right)\right]i\langle\{n\}|T|\{n'\}\rangle, \quad (3\text{-}12.40)$$

thereby defining the elements of the transition matrix T. The integral is not a four-dimensional delta function since the integration domain is not infinite. To appreciate this we must recall that the precise specification of individual momenta used here is an idealization that holds well within a particle beam, but fails near the boundaries. Where the initial and final beams overlap to give causal definition to the interaction region, (3–12.40) is applicable, and limiting the integration to that finite volume is an approximate way of recognizing the realities of the situation. It is probability that is physically significant, and we are actually concerned with

$$\left|\int(dx)\exp\left[i\sum (n_a' - n_a)p_a x\right]\right|^2 = \int(dx)(dx')\exp\left[i\sum(n_a' - n_a)p_a(x-x')\right]$$
$$= \int(dX)(d\xi)\exp\left[i\sum(n_a' - n_a)p_a\xi\right] \quad (3\text{-}12.41)$$

where
$$X^\mu = \tfrac{1}{2}(x^\mu + x^{\mu'}), \qquad \xi^\mu = x^\mu - x^{\mu'}. \quad (3\text{-}12.42)$$

The ξ integration can now be identified as a delta function, and the X integral measures the total interaction volume V, within the uncertainties attached to the boundary layers. The proportionality of the transition probability to the volume of the four-dimensional interaction region indicates that the important quantity is the coefficient of proportionality, the transition probability per unit four-dimensional volume, or, per unit time in a unit three-dimensional volume. This ratio is

$$|\langle\{n\}|\{n'\}\rangle_{\text{irred.}}|^2 V^{-1} = (2\pi)^4 \, \delta[\sum (n_a - n_a')p_a]|\langle\{n\}|T|\{n'\}\rangle|^2, \quad (3\text{-}12.43)$$

which supplies the physical interpretation of the transition matrix.

Let us begin the specific discussions of skeletal interactions with the scattering of spinless particles, as described by (3-12.13, 15, 16). The field $\phi(x)$ is required in the interaction region, which is causally intermediate between the emission source $K_2(x)$ and the detection source $K_1(x)$. The total field is the superposition of parts related to these sources,

$$\phi(x) = \phi_1(x) + \phi_2(x), \quad (3\text{-}12.44)$$

where

$$\phi_1(x) = i\int (dx')\Delta^{(-)}(x - x')K_1(x'),$$
$$\phi_2(x) = i\int (dx')\Delta^{(+)}(x - x')K_2(x'), \quad (3\text{-}12.45)$$

and the particular forms of $\Delta_+(x - x')$ disclose the causal situation. The process we are concerned with involves the action of two emission sources and two absorption sources. Thus, when (3-12.13) is considered, with the current

$$j^\mu(x) = i\partial^\mu\phi_1(x)eq\phi_1(x) + i\partial^\mu\phi_2(x)eq\phi_2(x) + i\partial^\mu\phi_1(x)eq\phi_2(x) - \phi_1(x)eqi\partial^\mu\phi_2(x)$$
$$= j^\mu_{11}(x) + j^\mu_{22}(x) + j^\mu_{12}(x), \quad (3\text{-}12.46)$$

we must retain only those contributions having the required overall characteristics, as conveyed by the causal indices. Those terms are

$$W_{40} \rightarrow \tfrac{1}{2}\int (dx)(dx')j^\mu_{12}(x)D_+(x - x')j_{12\mu}(x)$$
$$+ \int (dx)(dx')j^\mu_{11}(x)D_+(x - x')j_{22\mu}(x'), \quad (3\text{-}12.47)$$

all others having too many or too few emission or detection indices.

In earlier discussions of charged spinless particles we have worked with complex sources. But experience with spin $\tfrac{1}{2}$, for example, has shown the greater convenience of retaining real multicomponent sources and making the appropriate complex projections for specific charge values. Henceforth we shall write

$$K_{pq} = (d\omega_p)^{1/2}\varphi_q^* K(p), \quad (3\text{-}12.48)$$

where the two complex charge eigenvectors are

$$\varphi_+^* = 2^{-1/2}(1, -i), \qquad \varphi_-^* = 2^{-1/2}(1, i). \quad (3\text{-}12.49)$$

These vectors have the properties of orthonormality,

$$\varphi_q^* \varphi_{q'} = \delta_{qq'}, \tag{3-12.50}$$

completeness,

$$\sum \varphi_q \varphi_q^* = 1, \tag{3-12.51}$$

and, relative to the charge matrix

$$q = \begin{pmatrix} 0 & -i \\ i & 0 \end{pmatrix}, \tag{3-12.52}$$

they obey

$$\varphi_{q'}^* q = \varphi_{q'}^* q', \qquad q \varphi_{q'} = q' \varphi_{q'}. \tag{3-12.53}$$

The effect of complex conjugation is given by

$$\varphi_q^* = \varphi_{-q}. \tag{3-12.54}$$

Using this notation, we present the fields of (3-12.45) as

$$\phi_1(x) = \sum_{pq} iK_{1pq}^* \phi_{pq}(x)^*$$

$$\phi_2(x) = \sum_{pq} \phi_{pq}(x) iK_{2pq}, \tag{3-12.55}$$

where

$$\phi_{pq}(x) = (d\omega_p)^{1/2} \varphi_q e^{ipx} \tag{3-12.56}$$

is the field associated with the specific particle labeled pq, which enters the interaction region after its creation by the source K_{2pq}. Similarly, $\phi_{pq}(x)^*$ is the field of the particle labeled pq which, after leaving the interaction region, is annihilated by the detection source K_{1pq}^*.

The charge structure of the various partial currents that compose (3-12.46) is of importance. In $j_{22}^\mu(x)$, for example, the charge factor associated with two incident particles of charges q' and q'' is

$$\varphi_{q'} q \varphi_{q''} = \varphi_{-q'}^* q \varphi_{q''} = q'' \delta_{-q',q''}. \tag{3-12.57}$$

As we see, it vanishes unless $q' + q'' = 0$; only zero charge is brought into the interaction region. A similar restriction to opposite charges applies to $j_{11}^\mu(x)$. When we consider $j_{12}^\mu(x)$, the charge factor associated with charge q'' entering the interaction region and charge q' leaving it is

$$\varphi_{q'}^* q \varphi_{q''} = q'' \delta_{q',q''}, \tag{3-12.58}$$

and the necessary equality of q' and q'' implies that no charge accumulates in the interaction region. These are different ways of satisfying charge conservation in the scattering process. The second term of Eq. (3-12.47) does not contribute to the scattering of particles with like charge and we examine that process first.

The form of the current $j_{12}^\mu(x)$ is

$$j_{12}^\mu(x) = \sum_{p_1 p_2 q} iK_{1p_1q}^* eq(p_1^\mu + p_2^\mu)(d\omega_{p_1}\, d\omega_{p_2})^{1/2} e^{i(p_2-p_1)x} iK_{2p_2q}, \quad (3\text{-}12.59)$$

in which we can recognize the current, $eq2p^\mu\, d\omega_p$, that is associated with a single undeflected particle. When only contributions from incident particles of the same charge are retained,

$$W_{40} = \tfrac{1}{2}e^2 \sum iK_{1p_1q}^* iK_{1p_1'q}^* (d\omega_{p_1}\, d\omega_{p_1'}\, d\omega_{p_2}\, d\omega_{p_2'})^{1/2}(p_1 + p_2)(p_1' + p_2')$$

$$\times \left[\int (dx)(dx')e^{i(p_2-p_1)x} D_+(x - x')e^{i(p_2'-p_1')x'} \right] iK_{2p_2q} iK_{2p_2'q}, \quad (3\text{-}12.60)$$

where

$$\int (dx)(dx')e^{i(p_2-p_1)x} D_+(x - x')e^{i(p_2'-p_1')x'} = \frac{1}{(p_1 - p_2)^2} \int (dx)e^{i(p_2+p_2'-p_1-p_1')x}$$

$$(3\text{-}12.61)$$

exhibits the momentum form of the photon propagation function, $D_+(k) = (k^2)^{-1}$, and produces the space-time integration that enforces energy-momentum conservation. In picking out the desired T matrix element we must take into account that the source factors identifying a particular pair of incident particles, $iK_{2p_2q}iK_{2p_2'q}$, and a particular pair of scattered particles, $iK_{1p_1q}^* iK_{1p_1'q}^*$, can each be produced in two ways corresponding to the symmetry of these products in p_2, p_2' and in p_1, p_1'. Thus the transition matrix element will have those symmetries, which is a statement of B. E. statistics. The matrix element is

$$\langle 1_{p_1q} 1_{p_1'q} | T | 1_{p_2q} 1_{p_2'q} \rangle = (d\omega_{p_1}\, d\omega_{p_1'}\, d\omega_{p_2}\, d\omega_{p_2'})^{1/2}$$

$$\times e^2 \left[\frac{(p_1 + p_2)(p_1' + p_2')}{(p_1 - p_2)^2} + \frac{(p_1 + p_2')(p_1' + p_2)}{(p_1 - p_2')^2} \right],$$

$$(3\text{-}12.62)$$

which is explicitly symmetrical in p_2, p_2' and also has the required p_1, p_1' symmetry since overall momentum conservation implies that

$$p_1' - p_2 = -(p_1 - p_2'), \qquad p_1' - p_2' = -(p_1 - p_2). \quad (3\text{-}12.63)$$

The experimental measure of the effectiveness of a given scattering act, as observed in beam scattering arrangements, is an area or cross section. It expresses the rate at which the designated process occurs per unit time and per unit spatial volume, relative to the incident particle flux and the density of the scatterers, in the usual situation of a fixed target. The controllable factors referring to the initial particles can be given a general form that permits the cross section concept to be applied to colliding beams as well as stationary targets. Let $s_{a,b}^\mu$ be the particle flux vectors of two such beams. An invariant measure of their relative flux is suggested by the requirement that it must vanish when the vectors are proportional, and the beams run with the same velocity. This definition is

$$F = [(s_a s_b)^2 - s_a^2 s_b^2]^{1/2}, \qquad (3\text{–}12.64)$$

which does produce a real positive quantity since the flux vectors are time-like. If we write this out in terms of particle density s^0 and particle velocity $\mathbf{v} = \mathbf{s}/s^0$, the flux definition becomes

$$\begin{aligned} F &= s_a^0 s_b^0 [(1 - \mathbf{v}_a \cdot \mathbf{v}_b)^2 - (1 - \mathbf{v}_a^2)(1 - \mathbf{v}_b^2)]^{1/2}, \\ &= s_a^0 s_b^0 [(\mathbf{v}_a - \mathbf{v}_b)^2 - (\mathbf{v}_a \times \mathbf{v}_b)^2]^{1/2}, \end{aligned} \qquad (3\text{–}12.65)$$

and when one of the beams is a stationary target ($\mathbf{v}_b = 0$) it reduces to

$$F = |\mathbf{s}_a| s_b^0, \qquad (3\text{–}12.66)$$

the magnitude of the incident flux multiplied by the target density.

Since the particle flux associated with a single particle in a small momentum cell is

$$s^\mu = 2p^\mu \, d\omega_p, \qquad (3\text{–}12.67)$$

the version in which we shall apply (3–12.64) is

$$F = d\omega_a \, d\omega_b 4[(p_a p_b)^2 - m_a^2 m_b^2]^{1/2}, \qquad (3\text{–}12.68)$$

which introduces the masses of the particles. Other forms can be used, particularly one involving the total mass M, the invariant measure of the total momentum,

$$M^2 = -(p_a + p_b)^2 = m_a^2 + m_b^2 - 2p_a p_b, \qquad (3\text{–}12.69)$$

namely

$$F = d\omega_a \, d\omega_b 2[M^2 - (m_a + m_b)^2]^{1/2}[M^2 - (m_a - m_b)^2]^{1/2}. \qquad (3\text{–}12.70)$$

The following ratio, probability of a transition per unit four-dimensional volume [(3–12.43)] divided by invariant flux [(3–12.70)], defines a differential cross section. It is differential since the final particles are specified within small ranges of momenta, as limited by momentum conservation. Integrations over these differential elements supply various differential cross sections of lesser degrees of specification, leading finally to a total cross section, although the latter may not exist if very slight deflections carry a disproportionate weight. We shall use the symbol $d\sigma$ generally for all differential cross sections, relying on the explicitly stated differentials to indicate its precise nature.

Energy-momentum conservation in a two-particle scattering process fixes the energies of the scattered particles and leaves free only two parameters that give the direction of the line along which both particles move, in the rest frame of the total momentum. We may as well integrate immediately over the distributions of those variables that assume precise values. Let us consider any pair of particles and evaluate

$$\int d\omega_a\, d\omega_b (2\pi)^4\, \delta(p_a + p_b - P)$$

$$= \frac{1}{16\pi^2} \int \frac{(d\mathbf{p}_a)}{p_a^0} \frac{(d\mathbf{p}_b)}{p_b^0}\, \delta(\mathbf{p}_a + \mathbf{p}_b)\, \delta(p_a^0 + p_b^0 - M), \qquad (3\text{-}12.71)$$

where the second version refers to the rest frame of P^μ, in which $P^0 = M$. The magnitude of the relative momentum

$$\mathbf{p} = \mathbf{p}_a = -\mathbf{p}_b \qquad (3\text{-}12.72)$$

is given by

$$|\mathbf{p}| = \frac{1}{2M}\,[M^2 - (m_a + m_b)^2]^{1/2}[M^2 - (m_a - m_b)^2]^{1/2}. \quad (3\text{-}12.73)$$

In carrying out the energy integration that selects this value one must write

$$(d\mathbf{p}_a) = (d\mathbf{p}) = |\mathbf{p}|^2\, d|\mathbf{p}|\, d\Omega = |\mathbf{p}|\, \frac{p_a^0 p_b^0}{M}\, d(p_a^0 + p_b^0)\, d\Omega, \quad (3\text{-}12.74)$$

where $d\Omega$ is the element of solid angle for the relative momentum. The immediate result is

$$\int d\omega_a\, d\omega_b (2\pi)^4\, \delta(p_a + p_b - P)$$

$$= \frac{1}{32\pi^2}\, \frac{1}{M^2}\,[M^2 - (m_a + m_b)^2]^{1/2}[M^2 - (m_a - m_b)^2]^{1/2}\, d\Omega, \quad (3\text{-}12.75)$$

which reduces the differential aspect to the angles that specify the direction of the relative momentum. We note that the same square-root kinematical mass factors occur in the final state integration (3–12.75) and in the incident flux (3–12.70). These relatively complicated factors will cancel for a purely elastic scattering process where initial and final particles are the same.

The transition matrix element (3–12.62) provides a simple application of the cross section definition, giving directly

$$d\sigma = \int d\omega_{p_1}\, d\omega_{p_1'}(2\pi)^4\, \delta(p_1 + p_1' - p_2 - p_2')\, \frac{1}{2M(M^2 - 4m^2)^{1/2}}\,(4\pi\alpha)^2$$

$$\times \left[\frac{(p_1 + p_2)(p_1' + p_2')}{(p_1 - p_2)^2} + \frac{(p_1 + p_2')(p_1' + p_2)}{(p_1 - p_2')^2}\right]^2, \qquad (3\text{-}12.76)$$

and

$$\frac{d\sigma}{d\Omega} = \frac{1}{4}\,\frac{\alpha^2}{M^2}\left[\frac{(p_1 + p_2)(p_1' + p_2')}{(p_1 - p_2)^2} + \frac{(p_1 + p_2')(p_1' + p_2)}{(p_1 - p_2')^2}\right]^2$$

$$= \frac{\alpha^2}{M^2}\left[\frac{M^2 - 2m^2}{M^2 - 4m^2}\left(\frac{1}{\sin^2(\theta/2)} + \frac{1}{\cos^2(\theta/2)}\right) - 1\right]^2. \quad (3\text{-}12.77)$$

In the latter form, θ is the deflection angle, and the full equivalence of the angles

θ and $\pi - \theta$ expresses the indistinguishability of the B. E. particles. The reduction is performed in the rest frame by noting that each of the four particle energies equals $\frac{1}{2}M$ and this gives, for example,

$$- \frac{(p_1 + p_2)(p_1' + p_2')}{(p_1 - p_2)^2} = \frac{2M^2 - 4m^2}{(p_1 - p_2)^2} - 1 = 2 \frac{M^2 - 2m^2}{M^2 - 4m^2} \frac{1}{\sin^2 (\theta/2)} - 1.$$

$$(3\text{-}12.78)$$

Of particular interest are the very high and very low energy limits:

$M \gg 2m$:

$$\frac{d\sigma}{d\Omega} = \frac{\alpha^2}{M^2} \left(\frac{1}{\sin^2 (\theta/2)} + \frac{1}{\cos^2 (\theta/2)} - 1 \right)^2,$$

$M \sim 2m$:

$$(3\text{-}12.79)$$

$$\frac{d\sigma}{d\Omega} = \frac{1}{16} \frac{\alpha^2}{(M - 2m)^2} \left(\frac{1}{\sin^2 (\theta/2)} + \frac{1}{\cos^2 (\theta/2)} \right)^2.$$

Note that at small scattering angles the latter reduces to the Rutherford differential cross section for the scattering of distinguishable particles,

$$\left(\frac{d\sigma}{d\Omega} \right)_{\text{Ruth.}} = \left(\frac{\alpha}{2\mu v^2} \right)^2 \frac{1}{\sin^4 (\theta/2)},$$

$$(3\text{-}12.80)$$

where

$$\tfrac{1}{2}\mu v^2 = M - 2m$$

$$(3\text{-}12.81)$$

is the relative kinetic energy of the particles.

When particles of opposite charge scatter, they are distinguishable by their charges and only one kind of term emerges from the analogue of (3–12.60) that has the source factors replaced by $iK^*_{1p_1q}iK^*_{1p_1' -q}$ and $iK_{2p_2q}iK_{2p_2' -q}$. An additional minus sign is also needed. Of course, each process appears twice owing to the combined symmetry: $p_1 \leftrightarrow p_1'$, $p_2 \leftrightarrow p_2'$, $q \to -q$. But now the second term of (3–12.47) comes into play, with

$$j^\mu_{11}(x) = \sum_{p_1 p_1' q} iK^*_{1p_1q}iK^*_{1p_1' -q}eq\tfrac{1}{2}(p_1^\mu - p_1^{\mu\prime})(d\omega_{p_1} d\omega_{p_1'})^{1/2}e^{-i(p_1+p_1')x},$$

$$(3\text{-}12.82)$$

$$j^\mu_{22}(x) = \sum_{p_2 p_2' q} eq\tfrac{1}{2}(p_2^\mu - p_2^{\mu\prime})(d\omega_{p_2} d\omega_{p_2'})^{1/2}e^{i(p_2+p_2')x}iK_{2p_2q}iK_{2p_2' -q},$$

where each current contributes two equal terms to a given process, corresponding to the symmetries expressed by $p_1 \leftrightarrow p_1'$, $q \to -q$ and $p_2 \leftrightarrow p_2'$, $q \to -q$. The implied transition matrix element is

$$\langle 1_{p_1q}1_{p_1' -q}|T|1_{p_2q}1_{p_2' -q} \rangle$$
$$= (d\omega_{p_1} d\omega_{p_1'} d\omega_{p_2} d\omega_{p_2'})^{1/2}$$
$$\times e^2 \left[- \frac{(p_1 + p_2)(p_1' + p_2')}{(p_1 - p_2)^2} + \frac{(p_1 - p_1')(p_2 - p_2')}{(p_1 + p_1')^2} \right].$$

$$(3\text{-}12.83)$$

Notice the simple connection between the matrix elements (3–12.62) and (3–12.83); they are interchanged by either of the substitutions

$$p_1 \leftrightarrow -p_2, \qquad p_1' \leftrightarrow -p_2'. \tag{3–12.84}$$

Correspondences of this type between different transitions have become known as crossing relations. Their origin is not far to seek. Emission and absorption processes are united in the field $\phi(x)$. The formal substitution $p^\mu \to -p^\mu$ interchanges the physical effects that identify emission and absorption acts. And the numerical characterizations provided by the individual fields $\phi_{pq}(x)$ respond appropriately:

$$\phi_{pq}(x) \leftrightarrow \phi_{p-q}(x)^*. \tag{3–12.85}$$

Given the transition matrix element for one process, the substitution generates another one in which an initial particle of properties p, q is replaced by a final particle of properties p, $-q$, or conversely. Of course, this must be done at both ends of the reaction if one is to retain a scattering process. When particles of opposite charge are present, the outcome can be a symmetry of a given matrix element, as illustrated by the invariance of (3–12.83) under either of the substitutions

$$p_1 \leftrightarrow -p_2', \qquad p_1' \leftrightarrow -p_2. \tag{3–12.86}$$

Note that we are considering individual applications of a transformation that, used wholesale, is the *TCP* operation.

The square bracket factor of (3–12.83) has the following evaluation:

$$2\frac{M^2 - 2m^2}{M^2 - 4m^2}\frac{1}{\sin^2(\theta/2)} - 1 - \frac{M^2 - 4m^2}{M^2}\cos\theta$$

$$= \cos^2\frac{\theta}{2} 2\left[\frac{M^2 - 2m^2}{M^2 - 4m^2}\left(\frac{1}{\sin^2(\theta/2)} + \frac{1}{\cos^2(\theta/2)}\right) - 1\right] + \frac{4m^2}{M^2}\cos\theta. \tag{3–12.87}$$

The second term, $(4m^2/M^2)\cos\theta$, is relatively negligible both at high energies and at low energies. This provides a simple connection between the cross sections for unlike and like charges, one that becomes accurate asymptotically at both extremes of the mass scale and constitutes a reasonable interpolation between these limits:

$$\left(\frac{d\sigma}{d\Omega}\right)_{+-} \simeq \cos^4\frac{\theta}{2}\left(\frac{d\sigma}{d\Omega}\right)_{++,--}. \tag{3–12.88}$$

The scattering of photons by spinless charged particles is contained in (3–12.29), along with other processes. The part we want is extracted by writing, as in (3–12.44),

$$\phi(x) = \phi_1(x) + \phi_2(x), \tag{3–12.89}$$

together with its photon analogue

$$A^\mu(x) = A_1^\mu(x) + A_2^\mu(x), \tag{3-12.90}$$

where

$$A_1^\mu(x) = i\int (dx') g^{\mu\nu} D^{(-)}(x - x') J_{1\nu}(x'),$$

$$A_2^\mu(x) = i\int (dx') g^{\mu\nu} D^{(+)}(x - x') J_{2\nu}(x'), \tag{3-12.91}$$

and then retaining those terms that have one photon and one particle emission source along with one photon and one particle detection source. They are

$$W_{22} \to \int (dx)(dx')\phi_1(x)[eq2pA_1(x)\Delta_+(x - x')eq2pA_2(x')$$
$$+ eq2pA_2(x)\Delta_+(x - x')eq2pA_1(x')]\phi_2(x')$$
$$- \int (dx)\phi_1(x)2e^2 A_1(x)A_2(x)\phi_2(x), \tag{3-12.92}$$

in which we have adopted the simplification that expresses the use of the Lorentz gauge for the vector potential. Let us recall that

$$g^{\mu\nu} = \sum_\lambda e_{k\lambda}^\mu e_{k\lambda}^{\nu*} + \frac{k^\mu \bar{k}^\nu + k^\nu \bar{k}^\mu}{k\bar{k}}. \tag{3-12.93}$$

Of the two terms that do not refer to polarization vectors, one vanishes because the source is divergenceless and the other, a gradient in coordinate space, can be removed by a gauge transformation. Thus, it is in a special class of gauges that we write

$$A_1^\mu(x) = \sum_{k\lambda} iJ_{1k\lambda}^* A_{k\lambda}^\mu(x)^*, \qquad A_2^\mu(x) = \sum_{k\lambda} A_{k\lambda}^\mu(x)iJ_{2k\lambda}, \tag{3-12.94}$$

where

$$A_{k\lambda}^\mu(x) = (d\omega_k)^{1/2} e_{k\lambda}^\mu e^{ikx}, \tag{3-12.95}$$

and they are Lorentz gauges, since

$$\partial_\mu A_{k\lambda}^\mu(x) = (d\omega_k)^{1/2} ik_\mu e_{k\lambda}^\mu e^{ikx} = 0. \tag{3-12.96}$$

Now select the coefficient of $iJ_{k_1\lambda_1}^* iK_{p_1q}^*$ and $iJ_{k_2\lambda_2} iK_{p_2q}$ (we have finally omitted causal labels on the sources since they are abundantly evident in the other indices). The resulting space-time integrals produce the Fourier transforms that convey the momentum specification of the scattering process. Here is an example:

$$\int (dx)(dx') e^{-ip_1 x} p^\mu e^{-ik_1 x} \Delta_+(x - x') e^{ik_2 x'} p^\nu e^{ip_2 x'}$$
$$= p_1^\mu p_2^\nu \int (dx) e^{-i(p_1 + k_1)(x - x')} \Delta_+(x - x') \int (dx') e^{i(k_2 + p_2 - k_1 - p_1)x'}$$
$$= \frac{p_1^\mu p_2^\nu}{(p_1 + k_1)^2 + m^2} \int (dx) e^{i(k_2 + p_2 - k_1 - p_1)x}, \tag{3-12.97}$$

which also uses the fact that the differential operators p^μ and p^ν act directly upon momentum eigenfunctions. The transition matrix element is obtained as

$$\langle 1_{k_1\lambda_1} 1_{p_1 q} | T | 1_{k_2\lambda_2} 1_{p_2 q}\rangle = (d\omega_{k_1}\, d\omega_{p_1}\, d\omega_{k_2}\, d\omega_{p_2})^{1/2} 2e^2 e^{\mu*}_{k_1\lambda_1} V_{\mu\nu} e^{\nu}_{k_2\lambda_2}, \quad (3\text{-}12.98)$$

with

$$V_{\mu\nu} = \frac{p_{1\mu}p_{2\nu}}{p_1 k_1} - \frac{p_{1\nu}p_{2\mu}}{p_1 k_2} - g_{\mu\nu}, \quad (3\text{-}12.99)$$

which uses the kinematical simplifications

$$(p_1 + k_1)^2 + m^2 = [p_1^2 + m^2 + k_1^2] + 2p_1 k_1 = 2p_1 k_1,$$
$$(p_1 - k_2)^2 + m^2 = [p_1^2 + m^2 + k_2^2] - 2p_1 k_2 = -2p_1 k_2. \quad (3\text{-}12.100)$$

Other aspects of the kinematics are these. The total momentum is

$$P = p_1 + k_1 = p_2 + k_2, \quad (3\text{-}12.101)$$

and therefore

$$-p_1 k_1 = -p_2 k_2 = \tfrac{1}{2}(M^2 - m^2) \quad (3\text{-}12.102)$$

while

$$-p_1 k_2 = -p_2 k_1 = \tfrac{1}{2}(M^2 - m^2) + k_1 k_2. \quad (3\text{-}12.103)$$

Invariant expressions for the particle energies in the center of mass system, the rest frame of P^μ, are

$$-\frac{1}{M} P p_1 = -\frac{1}{M} P p_2 = \frac{1}{2M}(M^2 + m^2),$$
$$-\frac{1}{M} P k_1 = -\frac{1}{M} P k_2 = \frac{1}{2M}(M^2 - m^2). \quad (3\text{-}12.104)$$

Written in terms of the center of mass scattering angle θ, we also have

$$-k_1 k_2 = \left(\frac{M^2 - m^2}{2M}\right)^2 (1 - \cos\theta) \quad (3\text{-}12.105)$$

and

$$-p_1 k_2 = -p_2 k_1 = \frac{M^2 - m^2}{2}\left[\frac{M^2 + m^2}{2M^2} + \frac{M^2 - m^2}{2M^2}\cos\theta\right]. \quad (3\text{-}12.106)$$

We should also note the crossing symmetries exhibited by the transition matrix element (3-12.98, 99). Since the sign of q is irrelevant, there is invariance, specifically of $V_{\mu\nu}$, under the substitution

$$p_1 \leftrightarrow -p_2, \quad (3\text{-}12.107)$$

for which the equality of $k_1 p_1$ with $k_2 p_2$, and of $k_1 p_2$ with $k_2 p_1$, is decisive. Concerning the photons, the use of linear polarizations with real polarization vectors implies the transformation

$$k^\mu \to -k^\mu : \qquad A^\mu_{k\lambda}(x) \leftrightarrow A^\mu_{k\lambda}(x)^*, \quad (3\text{-}12.108)$$

and the transition matrix element should be invariant under the interchange

$$k_1 \leftrightarrow -k_2, \qquad \lambda_1 \leftrightarrow \lambda_2. \qquad (3\text{-}12.109)$$

This induces the exchange of μ and ν in $V_{\mu\nu}$, which indeed shows the required invariance.

The tensor $V_{\mu\nu}$ also has the following important properties:

$$\begin{aligned} k_1^{\mu} V_{\mu\nu} &= (p_2 - p_1 - k_1)_{\nu} = -k_{2\nu}, \\ V_{\mu\nu} k_2^{\nu} &= (p_1 - p_2 - k_2)_{\mu} = -k_{1\mu}. \end{aligned} \qquad (3\text{-}12.110)$$

They bring about the necessary conservation of the effective sources for the emission of the final photon and the absorption of the incident one:

$$k_1^{\mu} V_{\mu\nu} e_{k_2\lambda_2}^{\nu} = -k_{2\nu} e_{k_2\lambda_2}^{\nu} = 0, \qquad e_{k\lambda}^{\mu*} V_{\mu\nu} k_2^{\nu} = -e_{k_1\lambda_1}^{\mu*} k_{1\mu} = 0. \quad (3\text{-}12.111)$$

The summation of the transition probability over both polarizations of the scattered photon can now be performed with the aid of (3-12.93),

$$\begin{aligned} \sum_{\lambda_1} |e_{k_1\lambda_1}^{\mu*} V_{\mu\nu} e_{k_2\lambda_2}^{\nu}|^2 &= e_{k_2\lambda_2}^{\alpha*} V_{\mu\alpha} \sum_{\lambda_1} e_{k_1\lambda_1}^{\mu} e_{k_1\lambda_1}^{\nu*} V_{\nu\beta} e_{k_2\lambda_2}^{\beta} \\ &= e_{k_2\lambda_2}^{\alpha*} V_{\alpha}^{\mu} V_{\mu\beta} e_{k_2\lambda_2}^{\beta}. \end{aligned} \qquad (3\text{-}12.112)$$

If the incident photon beam is unpolarized, both polarizations appearing with equal probability, the necessary average can also be performed by means of (3-12.93):

$$\begin{aligned} \tfrac{1}{2} \sum_{\lambda_1\lambda_2} |e_{k_1\lambda_1}^{\mu*} V_{\mu\nu} e_{k_2\lambda_2}^{\nu}|^2 &= \tfrac{1}{2} V_{\alpha}^{\mu} V_{\mu\beta} \sum_{\lambda_2} e_{k_2\lambda_2}^{\alpha*} e_{k_2\lambda_2}^{\beta} \\ &= \tfrac{1}{2} V_{\alpha}^{\mu} V_{\mu\beta} \left(g^{\alpha\beta} - \frac{k_2^{\alpha} \bar{k}_2^{\beta} + k_2^{\beta} \bar{k}_2^{\alpha}}{k_2 \bar{k}_2} \right) \\ &= \tfrac{1}{2} (V^{\mu\nu} V_{\mu\nu} - 2), \qquad (3\text{-}12.113) \end{aligned}$$

since

$$V_{\alpha}^{\mu} V_{\mu\beta} k_2^{\alpha} = -k_1^{\mu} V_{\mu\beta} = k_{2\beta}. \qquad (3\text{-}12.114)$$

A straightforward algebraic reduction gives

$$V^{\mu\nu} V_{\mu\nu} - 2 = 1 + \left[1 + m^2 \left(\frac{1}{p_1 k_2} - \frac{1}{p_1 k_1} \right) \right]^2. \qquad (3\text{-}12.115)$$

Let us again emphasize the relative simplicity of the kinematical factors in the cross section for elastic scattering, even for particles of unequal mass. The ratio of the integral in (3-12.75) to the invariant flux of (3-12.70) produces the factor

$$\frac{d\Omega}{(8\pi)^2} \frac{1}{M^2}, \qquad (3\text{-}12.116)$$

which supplies the unit for a differential scattering cross section. Then, since

$2e^2 = 8\pi\alpha$, we get directly

$$\frac{d\sigma}{d\Omega} = \frac{\alpha^2}{M^2}\frac{1}{2}\left[1 + \left(\frac{\dfrac{M^2 - m^2}{M^2 + m^2} + \cos\theta}{1 + \dfrac{M^2 - m^2}{M^2 + m^2}\cos\theta}\right)^2\right], \tag{3-12.117}$$

which uses the center of mass scattering angle evaluation for $p_1 k_2$. The differential and total cross sections for the extreme energetic limits are

$$M \gg m: \qquad \frac{d\sigma}{d\Omega} = \frac{\alpha^2}{M^2}, \qquad \sigma = 4\pi\frac{\alpha^2}{M^2},$$

$$\tag{3-12.118}$$

$$M \simeq m: \qquad \frac{d\sigma}{d\Omega} = \frac{\alpha^2}{m^2}\frac{1}{2}(1 + \cos^2\theta), \qquad \sigma = \frac{8\pi}{3}\frac{\alpha^2}{m^2},$$

the latter being the Thomson cross sections.

The conserved nature of the effective sources that emit the final photon and absorb the initial photon implies the gauge invariance of the transition probabilities. This permits one to exploit whatever simplifications can be introduced by special gauge choices. The question of gauge in connection with the polarization vectors $e^\mu_{k\lambda}$ is implicit in the choice of $\bar{k}^\mu$ which, in some coordinate frame, has its spatial components reversed relative to those of k^μ. This is expressed with the aid of a unit time-like vector n^μ as

$$\bar{k}^\mu = -(k^\mu + 2n^\mu nk), \tag{3-12.119}$$

and the polarization vectors have the two orthogonality properties,

$$k_\mu e^\mu_{k\lambda} = 0, \qquad n_\mu e^\mu_{k\lambda} = 0 \tag{3-12.120}$$

which are incorporated in the summation

$$\sum_\lambda e^\mu_{k\lambda}e^{\nu*}_{k\lambda} = g^{\mu\nu} - \frac{k^\mu k^\nu + (k^\mu n^\nu + k^\nu n^\mu)nk}{(nk)^2}. \tag{3-12.121}$$

We now return to (3-12.98, 99) and observe that the identification of n^μ with either p_1^μ/m or p_2^μ/m produces the simplification

$$e^{\mu*}_{k_1\lambda_1}V_{\mu\nu}e^\nu_{k_2\lambda_2} = -e^{\mu*}_{k_1\lambda_1}e_{\mu k_2\lambda_2}. \tag{3-12.122}$$

One can verify directly that the same result for the summation and average over polarizations is obtained in this way. Applied to the final photons, (3-12.121) gives

$$\sum_{\lambda_1}|e^*_{k_1\lambda_1}e_{k_2\lambda_2}|^2 = 1 - \frac{|k_1 e_{k_2\lambda_2}|^2}{(nk_1)^2}, \tag{3-12.123}$$

and then

$$\tfrac{1}{2}\sum_{\lambda_1\lambda_2}|e^*_{k_1\lambda_1}e_{k_2\lambda_2}|^2 = 1 + \frac{(k_1k_2)^2 + 2k_1k_2(nk_1)(nk_2)}{2(nk_1)^2(nk_2)^2}$$

$$= \tfrac{1}{2}\left[1 + \left(1 + \frac{k_1k_2}{(nk_1)(nk_2)}\right)^2\right]. \tag{3-12.124}$$

If we set $n = p_1/m$ or p_2/m and insert the relation

$$k_1k_2 = p_1k_1 - p_1k_2, \tag{3-12.125}$$

we regain (3–12.113, 115).

The gauges we have just described are particularly useful when the particle is at rest initially, or finally. Another choice of the vector n is P/M, and this is most convenient in the center of mass frame. In all these examples the coordinate system is chosen to identify n^μ with the time axis. Let us use the center of mass description to study the polarization dependence of the scattering cross section. The trajectories of the incident and scattered particles define a plane. We first choose linear polarization vectors that are either perpendicular to the plane, or lie in it at right angles to the appropriate photon momentum. The differential cross sections for the various polarization assignments can be read off from (3–12.98, 99), as simplified by the special choices of the several vectors, including the center of mass momentum relation

$$0 = \mathbf{p}_1 + \mathbf{k}_1 = \mathbf{p}_2 + \mathbf{k}_2, \tag{3-12.126}$$

which gives

$$-e^{\mu*}_{k_1\lambda_1}V_{\mu\nu}e^{\nu}_{k_2\lambda_2} = \frac{\mathbf{e}_{k_1\lambda_1}\cdot\mathbf{k}_2\mathbf{k}_1\cdot\mathbf{e}_{k_2\lambda_2}}{p_1k_2} + \mathbf{e}_{k_1\lambda_1}\cdot\mathbf{e}_{k_2\lambda_2}. \tag{3-12.127}$$

The cross section vanishes if one polarization vector lies in the scattering plane while the other is perpendicular to the plane. When both vectors are perpendicular to the scattering plane,

$$\left(\frac{d\sigma}{d\Omega}\right)_\perp = \frac{\alpha^2}{M^2}, \tag{3-12.128}$$

and when they are both in the scattering plane,

$$\left(\frac{d\sigma}{d\Omega}\right)_\| = \frac{\alpha^2}{M^2}\left[\frac{M^2 - m^2}{M^2 + m^2 + (M^2 - m^2)\cos\theta}\sin^2\theta + \cos\theta\right]^2$$

$$= \frac{\alpha^2}{M^2}\left[\frac{\dfrac{M^2 - m^2}{M^2 + m^2} + \cos\theta}{1 + \dfrac{M^2 - m^2}{M^2 + m^2}\cos\theta}\right]^2. \tag{3-12.129}$$

The average of the two, appropriate to an initially unpolarized beam, is (3–12.117), which gives the latter structure greater physical meaning. It is interesting to observe that photons scattered through the angle determined by

$$\cos \theta = - \frac{M^2 - m^2}{M^2 + m^2} \qquad (3\text{–}12.130)$$

would be completely polarized perpendicular to the scattering plane.

The differential cross sections referring to circular polarization or helicity states can be produced from the linear polarization results. The circular polarization vectors are linear combinations of those parallel and perpendicular to the scattering plane, relatively shifted in phase by $\frac{1}{2}\pi$ [(2–3.29)]. Since the complex conjugate polarization vector represents the outgoing photon, the probability amplitude with the same helicity initially and finally equals half the sum of the two linear polarization amplitudes, and that for opposite helicities is half the difference. The differential cross sections corresponding to no change in helicity, or to a helicity reversal, are therefore, respectively,

$$\left(\frac{d\sigma}{d\Omega}\right)_{\text{no}} = \frac{\alpha^2}{M^2} \frac{\cos^4 \frac{1}{2}\theta}{(\cos^2 \frac{1}{2}\theta + (m^2/M^2) \sin^2 \frac{1}{2}\theta)^2},$$

$$\left(\frac{d\sigma}{d\Omega}\right)_{\text{yes}} = \frac{\alpha^2}{M^2} \frac{(m/M)^4 \sin^4 \frac{1}{2}\theta}{(\cos^2 \frac{1}{2}\theta + (m^2/M^2) \sin^2 \frac{1}{2}\theta)^2}. \qquad (3\text{–}12.131)$$

The geometrical factors that appear here, $\cos^4 \frac{1}{2}\theta$ and $\sin^4 \frac{1}{2}\theta$, are familiar as probabilities, for unit angular momentum with magnetic quantum number $+1$ in a given direction, that a measurement made in a direction at the relative angle θ will yield magnetic quantum numbers $+1$ and -1, respectively. There is also a dynamical weighting factor that is unity at low energies, $M \simeq m$, and suppresses helicity changes at very high energies. The total differential cross section, which is independent of the initial helicity, is the sum of the partial cross sections in (3–12.131):

$$\frac{d\sigma}{d\Omega} = \frac{\alpha^2}{M^2} \frac{\cos^4 \frac{1}{2}\theta + (m/M)^4 \sin^4 \frac{1}{2}\theta}{(\cos^2 \frac{1}{2}\theta + (m^2/M^2) \sin^2 \frac{1}{2}\theta)^2}. \qquad (3\text{–}12.132)$$

It is equivalent to (3–12.117). At the scattering angle determined by

$$\tan \tfrac{1}{2}\theta = M/m, \qquad (3\text{–}12.133)$$

the two partial cross sections are equal, leading to zero average angular momentum in the direction of the scattered photon. This angle is, of course, the same as the one given in (3–12.130) at which the scattered photon is linearly polarized.

Other processes involving two particles and two photons are contained in W_{22}. When we select terms with two particle emission sources and two photon detection sources we are considering particle-antiparticle annihilation into two photons; two photon emission sources together with two particle detection sources indicate the inverse process, the creation of a particle-antiparticle pair through collision of two photons. The latter, for example, is described by

$$W_{22} \rightarrow \tfrac{1}{2} \int (dx)(dx') \phi_1(x) eq2pA_2(x) \Delta_+(x - x') eq2pA_2(x') \phi_1(x')$$
$$- \int (dx) \phi_1(x) e^2 (A_2(x))^2 \phi_1(x), \tag{3-12.134}$$

and we extract the coefficient of $iK^*_{p_1 q} iK^*_{p'_1 -q}$ and $iJ_{k_2 \lambda_2} iJ_{k'_2 \lambda'_2}$ to get

$$\langle 1_{p_1 q} 1_{p'_1 -q} | T | 1_{k_2 \lambda_2} 1_{k'_2 \lambda'_2} \rangle = (d\omega_{p_1} \, d\omega_{p'_1} \, d\omega_{k_2} \, d\omega_{k'_2})^{1/2} 2e^2 e^\mu_{k_2 \lambda_2} V'_{\mu\nu} e^\nu_{k'_2 \lambda'_2} \tag{3-12.135}$$

where

$$V'_{\mu\nu} = \frac{p_{1\mu} p'_{1\nu}}{p_1 k_2} + \frac{p_{1\nu} p'_{1\mu}}{p_1 k'_2} - g_{\mu\nu}. \tag{3-12.136}$$

Notice that the symmetry $k_2 \lambda_2 \leftrightarrow k'_2 \lambda'_2$ is explicit, as is $p_1 q \leftrightarrow p'_1 -q$ when one uses the kinematical relations

$$-p_1 k_2 = -p'_1 k'_2 = \tfrac{1}{2}M[\tfrac{1}{2}M - (\tfrac{1}{4}M^2 - m^2)^{1/2} \cos \theta],$$
$$-p_1 k'_2 = -p'_1 k_2 = \tfrac{1}{2}M[\tfrac{1}{2}M + (\tfrac{1}{4}M^2 - m^2)^{1/2} \cos \theta], \tag{3-12.137}$$

where θ is the angle between particle and photon relative momenta. The location of the threshold for the reaction at $M = 2m$ is apparent in the square root that gives the magnitude of the center of mass particle momentum, all particle and photon energies being equal to $\tfrac{1}{2}M$. Using real polarization vectors, the matrix element (3-12.135, 136) is obtained from (3-12.98, 99) by the crossing transformation

$$p_2 \rightarrow -p'_1, \qquad k_1 \lambda_1 \rightarrow -k'_2 \lambda'_2. \tag{3-12.138}$$

Since the final particles now differ from the initial ones, the ratio of the kinematical square root factors appears explicitly in the differential cross section. We shall give it for various polarization assignments, first using the linear polarizations that are parallel or perpendicular to the plane of the reaction:

$$\left(\frac{d\sigma}{d\Omega}\right)_\perp = \frac{\alpha^2}{M^2} \left(1 - \frac{4m^2}{M^2}\right)^{1/2}$$

$$\left(\frac{d\sigma}{d\Omega}\right)_\parallel = \frac{\alpha^2}{M^2} \left(1 - \frac{4m^2}{M^2}\right)^{1/2} \left[2 \frac{(\tfrac{1}{4}M^2 - m^2) \sin^2 \theta}{\tfrac{1}{4}M^2 - (\tfrac{1}{4}M^2 - m^2) \cos^2 \theta} - 1\right]^2$$

$$= \frac{\alpha^2}{M^2} \left(1 - \frac{4m^2}{M^2}\right)^{1/2} \left[\frac{1 - \left(\frac{M^2}{4m^2} - 1\right) \sin^2 \theta}{1 + \left(\frac{M^2}{4m^2} - 1\right) \sin^2 \theta}\right]^2 ; \tag{3-12.139}$$

the cross section vanishes when the two polarization vectors are at right angles.

Provided the threshold energy is exceeded by at least the factor $2^{1/2}$, $M^2 > 8m^2$, there is an angle at which the differential cross section for parallel polarizations vanishes,

$$\sin^2 \theta = \left(\frac{M^2}{4m^2} - 1\right)^{-1}. \tag{3-12.140}$$

The cross section appropriate to unpolarized photons, an average of the four possibilities, is

$$\frac{d\sigma}{d\Omega} = \frac{\alpha^2}{M^2}\left(1 - \frac{4m^2}{M^2}\right)^{1/2} \frac{1}{2} \frac{1 + \left(\frac{M^2}{4m^2} - 1\right)^2 \sin^4 \theta}{\left[1 + \left(\frac{M^2}{4m^2} - 1\right)\sin^2 \theta\right]^2} \cdot \tag{3-12.141}$$

The behavior near the threshold, and at high energies, is given by

$$M \simeq 2m: \qquad \frac{d\sigma}{d\Omega} = \frac{\sigma}{4\pi} = \frac{\alpha^2}{8m^2}\left(\frac{M - 2m}{m}\right)^{1/2};$$

$$\tag{3-12.142}$$

$$M \gg 2m: \qquad \frac{d\sigma}{d\Omega} = \frac{\sigma}{4\pi} = \frac{\alpha^2}{2M^2}.$$

The transition amplitudes for circularly polarized photons are again simple linear combinations of those referring to linear polarizations. The photons are oppositely directed in the center of mass frame and assigning them the same helicity, for example, means that the photons have opposite angular momentum along the common line of motion. It follows that the transition amplitude for equal helicities is half the sum of the two linear polarization amplitudes, and that for opposite helicities is half their difference:

$$\left(\frac{d\sigma}{d\Omega}\right)_{\lambda_2 = \lambda_2'} = \frac{\alpha^2}{M^2}\left(1 - \frac{4m^2}{M^2}\right)^{1/2} \frac{1}{\left[1 + \left(\frac{M^2}{4m^2} - 1\right)\sin^2 \theta\right]^2},$$

$$\tag{3-12.143}$$

$$\left(\frac{d\sigma}{d\Omega}\right)_{\lambda_2 = -\lambda_2'} = \frac{\alpha^2}{M^2}\left(1 - \frac{4m^2}{M^2}\right)^{1/2} \frac{\left(\frac{M^2}{4m^2} - 1\right)^2 \sin^4 \theta}{\left[1 + \left(\frac{M^2}{4m^2} - 1\right)\sin^2 \theta\right]^2}.$$

The dominant reaction thus shifts from equal helicities near the threshold to opposite helicities at very high energies. The crossing point occurs at

$$M^2 = 4m^2\left(1 + \frac{1}{\sin^2 \theta}\right) \geq 8m^2, \tag{3-12.144}$$

which is a restatement of (3–12.140). Note that the geometrical factors, 1 and $(\frac{3}{8}) \sin^4 \theta$, also refer to angular momentum. The first affirms the equivalence of all directions in a state of zero angular momentum, and the second gives the probability, for angular momentum quantum number 2, that connects magnetic quantum number ± 2 in one direction with magnetic quantum zero in another direction inclined at the angle θ.

The transition matrix element for the inverse process of particle-antiparticle annihilation into two photons has the same appearance as (3–12.135, 136), with the causal labels reversed and complex conjugate polarization vectors substituted. Since the roles of initial and final particles have been reversed, the kinematical square root factor becomes inverted, but all else is the same. With helicity labeling of the final photons,

$$\left(\frac{d\sigma}{d\Omega}\right)_{\lambda_1 = \lambda_1'} = \frac{\alpha^2}{M^2}\left(1 - \frac{4m^2}{M^2}\right)^{-1/2} \frac{1}{\left[1 + \left(\frac{M^2}{4m^2} - 1\right)\sin^2\theta\right]^2},$$

$$\left(\frac{d\sigma}{d\Omega}\right)_{\lambda_1 = -\lambda_1'} = \frac{\alpha^2}{M^2}\left(1 - \frac{4m^2}{M^2}\right)^{-1/2} \frac{\left(\frac{M^2}{4m^2} - 1\right)^2 \sin^4\theta}{\left[1 + \left(\frac{M^2}{4m^2} - 1\right)\sin^2\theta\right]^2},$$

(3–12.145)

and again the predominant helicity relationship of the photons changes in going from low to high energies. In the annihilation of slow particles, $M \simeq 2m$, there is no relative angular momentum for the photons to carry away and equal helicities for the oppositely moving photons dominates. At very high energies the photons sustain the maximum angular momentum along their common line of motion. Nevertheless both differential cross sections are isotropic, and the total cross sections are

$$M - 2m = \tfrac{1}{4}mv^2, \quad v \ll 1: \quad \sigma = 2\pi \frac{\alpha^2}{m^2} \frac{1}{v}\ ;$$

$$M \gg 2m: \quad \sigma = 4\pi \frac{\alpha^2}{M^2}.$$

(3–12.146)

The variation of the cross section with inverse relative speed v, when $v \ll 1$, means only that the rate of the annihilation process per unit volume is proportional to the product of the beam densities. The computation of the total cross section for reactions in which the final state contains two identical particles, such as the B. E. photons in this annihilation process, needs one note of caution. In summing individually over the final states of both particles, which is here the summation over helicities and the integration over all directions of motion, every physically distinct state of the two particles is counted twice. That trap has been avoided in stating the cross sections of (3–12.146).

3–13 SPIN $\frac{1}{2}$ PROCESSES

Let us begin with the scattering of spin $\frac{1}{2}$ particles that have like charges. As in the spin 0 discussion, the relevant part of W_{40}, without reference to specific charge values, is

$$W_{40} \rightarrow \tfrac{1}{2}\int (dx)(dx')j^\mu_{12}(x)D_+(x-x')j_{12\mu}(x')$$
$$+ \int (dx)(dx')j^\mu_{11}(x)D_+(x-x')j_{22\mu}(x'), \qquad (3\text{-}13.1)$$

but only the first term applies to particles of the same charge, with

$$j^\mu_{12}(x) = \psi_1(x)\gamma^0\gamma^\mu eq\psi_2(x). \qquad (3\text{-}13.2)$$

The causally labeled fields are

$$\psi_1(x) = \sum_{p_1\sigma_1 q} i\eta^*_{p_1\sigma_1 q}\psi_{p_1\sigma_1 q}(x)^*, \qquad \psi_2(x) = \sum_{p_2\sigma_2 q} \psi_{p_2\sigma_2 q}(x)i\eta_{p_2\sigma_2 q}, \qquad (3\text{-}13.3)$$

and

$$\psi_{p\sigma q} = (2m\,d\omega_p)^{1/2}u_{p\sigma q}e^{ipx}. \qquad (3\text{-}13.4)$$

The form obtained for W_{40}, analogous to (3–12.60), is

$$W_{40} = \tfrac{1}{2}e^2 \sum i\eta^*_{p'_1\sigma'_1 q}i\eta^*_{p_1\sigma_1 q}(2m)^2(d\omega_{p_1}\,d\omega_{p'_1}\,d\omega_{p_2}\,d\omega_{p'_2})^{1/2}$$
$$\times (u^*_{p_1\sigma_1}\gamma^0\gamma^\mu u_{p_2\sigma_2})(u^*_{p'_1\sigma'_1}\gamma^0\gamma_\mu u_{p'_2\sigma'_2})$$
$$\times \left[\int (dx)(dx')e^{i(p_2-p_1)x}D_+(x-x')e^{i(p'_2-p'_1)x'}\right]i\eta_{p_2\sigma_2 q}i\eta_{p'_2\sigma'_2 q}. \qquad (3\text{-}13.5)$$

The q label, which is common to all spinors, has been omitted. Note also that the initial order of the totally anticommutative source factors has been rearranged without the intervention of minus signs:

$$i\eta_{p_1\sigma_1 q}i\eta_{p_2\sigma_2 q}i\eta_{p'_1\sigma'_1 q}i\eta_{p'_2\sigma'_2 q}. \qquad (3\text{-}13.6)$$

The transition matrix element will be defined relative to the order of sources that appears explicitly in (3–13.5). But one must, of course, take into account that a particular product of detection sources occurs twice in the summation with a relative minus sign, owing to the F. D. antisymmetry under the exchange $p_1\sigma_1 \leftrightarrow p'_1\sigma'_1$. A similar remark applies to the emission sources and the permutation $p_2\sigma_2 \leftrightarrow p'_2\sigma'_2$. Thus, the required matrix element, which shows the antisymmetries that are characteristic of F. D. states, is

$$\langle 1_{p_1\sigma_1 q}1_{p'_1\sigma'_1 q}|T|1_{p_2\sigma_2 q}1_{p'_2\sigma'_2 q}\rangle = (2m)^2(d\omega_{p_1}\,d\omega_{p'_1}\,d\omega_{p_2}\,d\omega_{p'_2})^{1/2}$$
$$\times e^2\left[\frac{(u^*_{p_1\sigma_1}\gamma^0\gamma^\mu u_{p_2\sigma_2})(u^*_{p'_1\sigma'_1}\gamma^0\gamma_\mu u_{p'_2\sigma'_2})}{(p_1-p_2)^2}\right.$$
$$\left. -\frac{(u^*_{p_1\sigma_1}\gamma^0\gamma^\mu u_{p'_2\sigma'_2})(u^*_{p'_1\sigma'_1}\gamma^0\gamma_\mu u_{p_2\sigma_2})}{(p_1-p'_2)^2}\right]. \qquad (3\text{-}13.7)$$

Even when one is not interested in specific spin values in the initial and final states, perhaps the simplest general procedure is to evaluate the differential cross sections for the various helicity assignments. That is already suggested by the photon polarization considerations of the preceding section, where the outcome of polarization summations and averages required some algebraic reduction to attain the result that was produced directly by considering the various helicities (or linear polarizations). This simplification was particularly marked in high energy photon scattering where the helicity strongly preferred not to change. The same tendency appears in the present situation, which we may as well call electron-electron scattering since that is the outstanding realization. The general construction of the spinor $u_{p\sigma}$ was given in (2–6.90) as

$$u_{p\sigma} = \left[\left(\frac{p^0 + m}{2m}\right)^{1/2} + \left(\frac{p^0 - m}{2m}\right)^{1/2} i\gamma_5 \frac{\boldsymbol{\sigma} \cdot \mathbf{p}}{|\mathbf{p}|}\right] v_\sigma, \qquad (3\text{–}13.8)$$

where the v_σ are γ^0 eigenvectors with eigenvalue $+1$. When they are chosen to be eigenvectors of $\boldsymbol{\sigma} \cdot \mathbf{p}/|\mathbf{p}|$ as well, identifying σ with the helicity, we get

$$u_{p\sigma} = \left[\left(\frac{p^0 + m}{2m}\right)^{1/2} + \left(\frac{p^0 - m}{2m}\right)^{1/2} i\gamma_5\sigma\right] v_\sigma$$

$$\rightarrow \left(\frac{p^0}{2m}\right)^{1/2} (1 + i\gamma_5\sigma)v_\sigma, \qquad (3\text{–}13.9)$$

where the latter is the high energy limit in which helicity becomes linked to the eigenvalue of $i\gamma_5$. In the center of mass frame where all particle energies equal $\frac{1}{2}M$, consider the following high energy evaluations:

$$(u^*_{p_1\sigma_1} u_{p_2\sigma_2}) = (M/4m)v^*_{\sigma_1}(1 + i\gamma_5\sigma_1)(1 + i\gamma_5\sigma_2)v_{\sigma_2}$$

$$= (M/4m)(1 + \sigma_1\sigma_2)v^*_{\sigma_1}v_{\sigma_2}, \qquad (3\text{–}13.10)$$

since $i\gamma_5$ has no diagonal matrix elements in the $\gamma^{0\prime} = +1$ subspace, and, using the relation

$$\gamma^0\boldsymbol{\gamma} = i\gamma_5\boldsymbol{\sigma}, \qquad (3\text{–}13.11)$$

$$(u^*_{p_1\sigma_1}\gamma^0\boldsymbol{\gamma} u_{p_2\sigma_2}) = (M/4m)v^*_{\sigma_1}(1 + i\gamma_5\sigma_1)i\gamma_5\boldsymbol{\sigma}(1 + i\gamma_5\sigma_2)v_{\sigma_2}$$

$$= (M/4m)(\sigma_1 + \sigma_2)v^*_{\sigma_1}\boldsymbol{\sigma} v_{\sigma_2}. \qquad (3\text{–}13.12)$$

We see that $(u^*_{p_1\sigma_1}\gamma^0\gamma^\mu u_{p_2\sigma_2})$ vanishes if $\sigma_1 = -\sigma_2$. The helicity does not change in these products at high energy because $\gamma^0\gamma^\mu$ commutes with γ_5.

Accepting the restriction $\sigma_1 = \sigma_2$, $\sigma_1' = \sigma_2'$, we find that the product appearing in the first term of (3–13.7) is

$$(u^*_{p_1\sigma_1}\gamma^0\gamma^\mu u_{p_2\sigma_2})(u^*_{p_1'\sigma_1'}\gamma^0\gamma_\mu u_{p_2'\sigma_2'})$$

$$= (M/2m)^2[-(v^*_{\sigma_1}v_{\sigma_2})(v^*_{\sigma_1'}v_{\sigma_2'}) + \sigma_2\sigma_2'(v^*_{\sigma_1}\boldsymbol{\sigma} v_{\sigma_2}) \cdot (v^*_{\sigma_1'}\boldsymbol{\sigma} v_{\sigma_2'})]. \qquad (3\text{–}13.13)$$

There is a basic identity, expressing the completeness of the four matrices, 1, $\boldsymbol{\sigma}$, which is presented in Eq. (2–5.58). When the general matrices X and Y are appropriately chosen as dyadic products, it tells us that

$$-(v^*_{\sigma_1}v_{\sigma_2})(v^*_{\sigma'_1}v_{\sigma'_2}) + (v^*_{\sigma_1}\boldsymbol{\sigma}v_{\sigma_2}) \cdot (v^*_{\sigma'_1}\boldsymbol{\sigma}v_{\sigma'_2})$$
$$= -2[(v^*_{\sigma_1}v_{\sigma_2})(v^*_{\sigma'_1}v_{\sigma'_2}) - (v^*_{\sigma_1}v_{\sigma'_2})(v^*_{\sigma'_1}v_{\sigma_2})], \quad (3\text{-}13.14)$$

which is antisymmetrical in the indices σ_2, σ'_2 and in σ_1, σ'_1. But a word of caution about notation is called for here. Although we have written v_{σ_2}, say, one must not forget the implicit reference to the direction of the momentum $\mathbf{p}_2$ along which the spin is projected to give the component σ_2. What we have just referred to as antisymmetry in σ_2, σ'_2 is, properly speaking, antisymmetry in $\mathbf{p}_2\sigma_2$ and $\mathbf{p}'_2\sigma'_2$. The combination (3-13.14) does not vanish when the helicities σ_2 and σ'_2 are equal.

We consider first the situation of equal initial helicities. Then precisely the combination that is evaluated by the identity (3-13.14) appears in (3-13.13) and, owing to the antisymmetry just mentioned, the two terms of (3-13.7) are combined into one, with the factor

$$\frac{1}{(\mathbf{p}_1 - \mathbf{p}_2)^2} + \frac{1}{(\mathbf{p}_1 - \mathbf{p}'_2)^2} = \frac{1}{M^2}\left(\frac{1}{\sin^2 \tfrac{1}{2}\theta} + \frac{1}{\cos^2 \tfrac{1}{2}\theta}\right). \quad (3\text{-}13.15)$$

To evaluate the spinor products that appear in (3-3.14) we choose $\mathbf{p}_2 = -\mathbf{p}'_2$ as reference direction—the z-axis—and express the choice of equal helicities, or opposite spin projections along this direction, by standard two component spinors:

$$v_{\sigma_2} = v_+ = \begin{pmatrix} 1 \\ 0 \end{pmatrix}, \qquad v_{\sigma'_2} = v_- = \begin{pmatrix} 0 \\ 1 \end{pmatrix}. \quad (3\text{-}13.16)$$

The spinors v_{σ_1} and $v_{\sigma'_1}$ are in the same relation, but with respect to the direction of $\mathbf{p}_1 = -\mathbf{p}'_1$, which is rotated by the angle θ about the y-axis, for example. That is expressed by

$$v^*_{\sigma_1} = v^*_+ e^{(1/2)i\theta\sigma_y} = \overbrace{\cos \tfrac{1}{2}\theta \sin \tfrac{1}{2}\theta},$$
$$\qquad\qquad\qquad\qquad\qquad\qquad\qquad\qquad (3\text{-}13.17)$$
$$v^*_{\sigma'_1} = v^*_- e^{(1/2)i\theta\sigma_y} = \underbrace{-\sin \tfrac{1}{2}\theta \cos \tfrac{1}{2}\theta},$$

and

$$(v^*_{\sigma_1}v_{\sigma_2})(v^*_{\sigma'_1}v_{\sigma'_2}) - (v^*_{\sigma_1}v_{\sigma'_2})(v^*_{\sigma'_1}v_{\sigma_2}) = \cos^2 \tfrac{1}{2}\theta + \sin^2 \tfrac{1}{2}\theta = 1, \quad (3\text{-}13.18)$$

which, as a combination of matrix elements, is also the determinant of the unimodular rotation matrix. Since the various factors of $2m$ and M cancel in the matrix element, leaving $2e^2$, the result is immediate:

$$\left(\frac{d\sigma}{d\Omega}\right)_{\sigma_2=\sigma'_2} = \frac{\alpha^2}{M^2}\left(\frac{1}{\sin^2 \tfrac{1}{2}\theta} + \frac{1}{\cos^2 \tfrac{1}{2}\theta}\right)^2. \quad (3\text{-}13.19)$$

For the situation of opposite initial helicities, the combination that appears in (3–3.13), apart from a minus sign, is

$$(v^*_{\sigma_1}v_{\sigma_2})(v^*_{\sigma'_1}v_{\sigma'_2}) + (v^*_{\sigma_1}\boldsymbol{\sigma}v_{\sigma_2})\cdot(v^*_{\sigma'}{}'\boldsymbol{\sigma}v_{\sigma'_2}) = 2(v^*_{\sigma_1}v_{\sigma'_2})(v^*_{\sigma'_1}v_{\sigma_2}). \quad (3\text{–}13.20)$$

The two contributions of (3–13.7) are now associated with different final states, which do not interfere in differential cross sections. The helicity labels are, respectively, $\sigma_1 = \sigma_2$, $\sigma'_1 = \sigma'_2 = -\sigma_2$ and $\sigma_1 = \sigma'_2 = -\sigma_2$, $\sigma'_1 = \sigma_2$. Alternatively expressed, unit angular momentum along the initial direction of motion can lead to either of the magnetic quantum numbers, $+1$ or -1, along the common direction taken by the scattered particles. Leaving aside the multiplier of 2, the factors contributed by (3–13.20) in the two situations are

$\sigma_1 = \sigma_2, \sigma'_1 = \sigma'_2:$

$$(v^*_+ e^{(1/2)i\theta\sigma_\nu}v_+)(v^*_+ e^{(1/2)i\theta\sigma_\nu}v_+) = \cos^2\tfrac{1}{2}\theta,$$

$\sigma_1 = \sigma'_2, \sigma'_1 = \sigma_2:$

$$(v^*_- e^{(1/2)i\theta\sigma_\nu}v_+)(v^*_- e^{(1/2)i\theta\sigma_\nu}v_+) = \sin^2\tfrac{1}{2}\theta,$$

$$(3\text{–}13.21)$$

if one is careful in translating the helicity specifications into spin projections along the two relevant directions. The differential cross section produced by adding the noninterfering contributions is

$$\left(\frac{d\sigma}{d\Omega}\right)_{\sigma_2=-\sigma'_2} = \frac{\alpha^2}{M^2}\left[\frac{\cos^4\tfrac{1}{2}\theta}{\sin^4\tfrac{1}{2}\theta} + \frac{\sin^4\tfrac{1}{2}\theta}{\cos^4\tfrac{1}{2}\theta}\right]. \quad (3\text{–}13.22)$$

The trigonometric factor in square brackets can also be written as

$$\left(\frac{\cos^2\tfrac{1}{2}\theta}{\sin^2\tfrac{1}{2}\theta} + \frac{\sin^2\tfrac{1}{2}\theta}{\cos^2\tfrac{1}{2}\theta}\right)^2 - 2 = \left(\frac{1}{\sin^2\tfrac{1}{2}\theta} + \frac{1}{\cos^2\tfrac{1}{2}\theta} - 2\right)^2 - 2$$

$$= \left(\frac{1}{\sin^2\tfrac{1}{2}\theta} + \frac{1}{\cos^2\tfrac{1}{2}\theta}\right)^2$$

$$- 4\left(\frac{1}{\sin^2\tfrac{1}{2}\theta} + \frac{1}{\cos^2\tfrac{1}{2}\theta}\right) + 2. \quad (3\text{–}13.23)$$

The differential cross section appropriate to initially unpolarized electrons (one unpolarized beam will do) is the equally weighted average of the more specific cross sections given in (3–13.19) and (3–13.22, 23). That is

$$\frac{d\sigma}{d\Omega} = \frac{\alpha^2}{M^2}\left(\frac{1}{\sin^2\tfrac{1}{2}\theta} + \frac{1}{\cos^2\tfrac{1}{2}\theta} - 1\right)^2, \quad (3\text{–}13.24)$$

which has two remarkable features. It is a perfect square, as though only a single process contributed, and it is identical with the spin 0 high energy differential cross section of (3–12.79).

This interesting equivalence of different spin values is restricted to very high energies. That is most evident in the scattering of low energy particles, $M \simeq 2m$.

Then the spinors $u_{p\sigma}$ essentially reduce to the $\gamma^{0\prime} = +1$ eigenvectors v_σ. Helicity ceases to be the most useful description, the referral of all spins to a common direction in space taking its place. Indeed, with that choice,

$$(u^*_{p_1\sigma_1} u_{p_2\sigma_2}) \simeq \delta_{\sigma_1\sigma_2}, \qquad (u^*_{p_1\sigma_1} i\gamma_5 \sigma u_{p_2\sigma_2}) \simeq 0, \qquad (3\text{-}13.25)$$

and the individual scattering processes take place without change of magnetic quantum number. In this low energy limit, spin and orbital motion are dynamically independent, in contrast to very high energy conditions where they are tightly linked. When the initial, and therefore the final, magnetic quantum numbers are equal, both terms of (3-13.7) contribute, with reversed sign; for opposite initial magnetic quantum numbers one or the other of the two terms is effective depending upon the assignment of opposite magnetic quantum numbers in the final state. The spin-averaged differential cross section, is, therefore,

$$\begin{aligned}
\frac{d\sigma}{d\Omega} &= \frac{1}{16} \frac{\alpha^2}{(M-2m)^2} \frac{1}{2}\left[\left(\frac{1}{\sin^2 \tfrac{1}{2}\theta} - \frac{1}{\cos^2 \tfrac{1}{2}\theta}\right)^2 + \frac{1}{\sin^4 \tfrac{1}{2}\theta} + \frac{1}{\cos^4 \tfrac{1}{2}\theta}\right] \\
&= \frac{1}{16} \frac{\alpha^2}{(M-2m)^2} \left[\frac{3}{4}\left(\frac{1}{\sin^2 \tfrac{1}{2}\theta} - \frac{1}{\cos^2 \tfrac{1}{2}\theta}\right)^2 + \frac{1}{4}\left(\frac{1}{\sin^2 \tfrac{1}{2}\theta} + \frac{1}{\cos^2 \tfrac{1}{2}\theta}\right)^2\right].
\end{aligned}$$
$$(3\text{-}13.26)$$

The latter form corresponds to the alternative of averaging over the three symmetrical spin states, which are antisymmetrical in spatial coordinates, and the single antisymmetrical spin state, which is symmetrical in spatial coordinates. This F. D. result differs, of course, from that of (3-12.79), where only the symmetrical spatial combination appears.

States classified by the total spin of the particles are useful at high energy too, provided the reference direction of the magnetic quantum number differs for the initial and the final particles in conformity with the altered direction of motion. For unit angular momentum there is also a separation of the states with unit magnitude of the magnetic quantum number from the one of zero magnetic quantum number, reminiscent of that for a unit spin particle as its mass vanishes, effectively a high energy limit. We have already noted that the transitions of the initial unit magnetic quantum state take it only into final states of magnetic quantum number ± 1. Indeed the weight factors, $\cos^4 \tfrac{1}{2}\theta$ and $\sin^4 \tfrac{1}{2}\theta$, that appear in (3-13.22) are just the probabilities for unit angular momentum that connect magnetic quantum number $+1$ in the initial direction with $+1$ and -1 in the final direction. They occurred within a photon context in Eq. (3-12.131). The unit spin state of zero magnetic quantum number is the symmetrical combination of the two ways that realize equal helicity, $\sigma = \sigma' = \pm 1$. The zero spin state is the corresponding antisymmetrical combination. Since the helicities are maintained in individual scattering acts, and reversing the sign of all helicities is without effect, there are no transitions between states of different spin and the differential cross section (3–13.19) applies to either spin state of zero magnetic quantum number.

Not much more effort is required to obtain scattering cross sections for arbitrary energy, using the helicity classification of states. Now helicity changes in scattering do occur, as exhibited in the general evaluation of

$$
(u^*_{p_1\sigma_1} u_{p_2\sigma_2}) = v^*_{\sigma_1}\left(\frac{\tfrac{1}{2}M + m}{2m} + \frac{\tfrac{1}{2}M - m}{2m}\,\sigma_1\sigma_2\right)v_{\sigma_2}
$$

$$
= \begin{cases} \sigma_1 = \sigma_2: & (M/2m)v^*_{\sigma_1}v_{\sigma_2}, \\ \sigma_1 = -\sigma_2: & v^*_{\sigma_1}v_{\sigma_2}. \end{cases} \qquad (3\text{–}13.27)
$$

However, the vector structure

$$
(u^*_{p_1\sigma_1} i\gamma_5 \boldsymbol{\sigma} u_{p_2\sigma_2}) = \left(\frac{M^2}{4m^2} - 1\right)^{1/2}\tfrac{1}{2}(\sigma_1 + \sigma_2)v^*_{\sigma_1}\boldsymbol{\sigma}v_{\sigma_2} \qquad (3\text{–}13.28)
$$

still requires equality of the helicities. We shall now merely list relative contributions of the various processes that appear additively in the spin-averaged differential cross section. They are classified by initial and final magnetic quantum numbers that refer to corresponding directions of motion, and according to whether helicity transitions have occurred,

$$0 \to 0,\ \text{no:} \qquad \left[\left(\frac{M^2}{4m^2} - \frac{1}{2}\right)\left(\frac{1}{\sin^2\tfrac{1}{2}\theta} + \frac{1}{\cos^2\tfrac{1}{2}\theta}\right) - 1\right]^2,$$

$$0 \to 0,\ \text{yes:} \qquad 1,$$

$$0 \to \pm 1,\ \text{yes:} \qquad \frac{M^2}{4m^2}\left(\frac{\cos\tfrac{1}{2}\theta}{\sin\tfrac{1}{2}\theta} - \frac{\sin\tfrac{1}{2}\theta}{\cos\tfrac{1}{2}\theta}\right)^2, \qquad (3\text{–}13.29)$$

$$1 \to 1,\ \text{both:} \qquad \left[\left(\frac{M^2}{4m^2} - \frac{1}{2}\right)\frac{\cos^2\tfrac{1}{2}\theta}{\sin^2\tfrac{1}{2}\theta} - \frac{1}{2}\right]^2,$$

$$1 \to -1,\ \text{both:} \qquad \left[\left(\frac{M^2}{4m^2} - \frac{1}{2}\right)\frac{\sin^2\tfrac{1}{2}\theta}{\cos^2\tfrac{1}{2}\theta} - \frac{1}{2}\right]^2.$$

For the last two processes, the constant $-\tfrac{1}{2}$ is the contribution associated with helicity changes. Adding these terms and supplying the appropriate factor gives

$$
\frac{d\sigma}{d\Omega} = \frac{\alpha^2}{M^2}\left[\left(\frac{M^2 - 2m^2}{M^2 - 4m^2}\right)^2\left(\frac{1}{\sin^2\tfrac{1}{2}\theta} + \frac{1}{\cos^2\tfrac{1}{2}\theta} - 1\right)^2\right.
$$
$$
\left. - \frac{4m^2(M^2 - 3m^2)}{(M^2 - 4m^2)^2}\left(\frac{1}{\sin^2\tfrac{1}{2}\theta} + \frac{1}{\cos^2\tfrac{1}{2}\theta} + 1\right)\right], \qquad (3\text{–}13.30)
$$

which interpolates between the high and low energy forms, (3–13.24) and (3–13.26) respectively. While resembling the zero spin result in (3–12.77), it differs in detail, except at high energies.

To discuss electron-positron scattering we return to (3–13.1) and consider both terms, with

$$j_{11}^{\mu}(x) = \tfrac{1}{2}\psi_1(x)\gamma^0\gamma^\mu eq\psi_1(x)$$

$$= \sum_{p_1\sigma_1 p_1'\sigma_1'} i\eta^*_{p_1'\sigma_1'\,-q} i\eta^*_{p_1\sigma_1 q}(-eq)\psi^*_{p_1'\sigma_1'\,-q}(x)\gamma^0\gamma^\mu\psi^*_{p_1\sigma_1 q}(x) \quad (3\text{–}13.31)$$

and

$$j_{22}^{\mu}(x) = \tfrac{1}{2}\psi_2(x)\gamma^0\gamma^\mu eq\psi_2(x)$$

$$= \sum_{p_2\sigma_2 p_2'\sigma_2'} (-eq)\psi_{p_2\sigma_2 q}(x)\gamma^0\gamma^\mu\psi_{p_2'\sigma_2'\,-q}(x) i\eta_{p_2\sigma_2 q} i\eta_{p_2'\sigma_2'\,-q}, \quad (3\text{–}13.32)$$

where the two equal contributions that refer to a specific pair of oppositely charged particles have already been collected. The transition matrix element that is defined by the coefficient of

$$i\eta^*_{p_1'\sigma_1'\,-q} i\eta^*_{p_1\sigma_1 q} i\eta_{p_2\sigma_2 q} i\eta_{p_2'\sigma_2'\,-q} \quad (3\text{–}13.33)$$

is

$$\langle 1_{p_1\sigma_1 q} 1_{p_1'\sigma_1'\,-q}|T|1_{p_2\sigma_2 q} 1_{p_2'\sigma_2'\,-q}\rangle = (2m)^2(d\omega_{p_1}\,d\omega_{p_1'}\,d\omega_{p_2}\,d\omega_{p_2'})^{1/2}$$

$$\times\, e^2\left[-\frac{(u^*_{p_1\sigma_1}\gamma^0\gamma^\mu u_{p_2\sigma_2})(u^*_{p_1'\sigma_1'}\gamma^0\gamma_\mu u_{p_2'\sigma_2'})}{(p_1-p_2)^2} \right.$$

$$\left. +\frac{(u^*_{p_1'\sigma_1'\,-q}\gamma^0\gamma^\mu u^*_{p_1\sigma_1 q})(u_{p_2\sigma_2 q}\gamma^0\gamma_\mu u_{p_2'\sigma_2'\,-q})}{(p_1+p_1')^2} \right].$$

$$(3\text{–}13.34)$$

The crossing relation between this matrix element and the one for like charge scattering again follows from the unification of emission and absorption processes in the field $\psi(x)$, as conveyed by the formal substitutions

$$\eta_{2p\sigma q} \leftrightarrow \eta^*_{1p\,-\sigma\,-q}, \qquad \psi_{p\sigma q}(x) \leftrightarrow \psi^*_{p\,-\sigma\,-q}(x). \quad (3\text{–}13.35)$$

When applied to (3–13.7), the substitutions

$$p_2' \leftrightarrow -p_1', \qquad u_{p_2'\sigma_2' q} \to u^*_{p_1'\sigma_1'\,-q}, \qquad u^*_{p_1'\sigma_1' q} \to u_{p_2'\sigma_2'\,-q} \quad (3\text{–}13.36)$$

produce (3–13.34), with the additional minus sign coming from the rearrangements necessary to realize (3–13.33), the standard multiplication order of the sources [i^4 is omitted here]:

$$\eta^*_{p_1'\sigma_1' q}\eta^*_{p_1\sigma_1 q}\eta_{p_2\sigma_2 q}\eta_{p_2'\sigma_2' q} \to \eta_{p_2'\sigma_2'\,-q}\eta^*_{p_1\sigma_1 q}\eta_{p_2\sigma_2 q}\eta^*_{p_1'\sigma_1'\,-q}$$

$$= -\eta^*_{p_1'\sigma_1'\,-q}\eta^*_{p_1\sigma_1 q}\eta_{p_2\sigma_2 q}\eta_{p_2'\sigma_2'\,-q}, \quad (3\text{–}13.37)$$

since the two sources associated with charge $-q$ must reverse their relative position.

In order to treat both terms of (3–13.34) in the same way, we use the relation of (2–6.134),

$$u^*_{p\,-\sigma\,-q} = i\sigma\gamma_5 u_{p\sigma q}, \tag{3-13.38}$$

which gives

$$(u^*_{p'_1\sigma'_1} {}_{-q}\gamma^0\gamma^\mu u^*_{p_1\sigma_1 q})(u_{p_2\sigma_2 q}\gamma^0\gamma_\mu u_{p'_2\sigma'_2\,-q})$$
$$= \sigma_1\sigma_2(u^*_{p'_1\sigma'_1}\gamma^0\gamma^\mu i\gamma_5 u_{p_1\,-\sigma_1})(u^*_{p_2\,-\sigma_2}\gamma^0\gamma_\mu i\gamma_5 u_{p'_2\sigma'_2}), \quad (3-13.39)$$

where the now matching charge labels have been omitted. The effect of the additional γ_5 factor is illustrated by

$$u^*_{p_2\,-\sigma_2} i\gamma_5 u_{p'_2\sigma'_2} = \left(\frac{M^2}{4m^2} - 1\right)^{1/2}\tfrac{1}{2}(\sigma'_2 - \sigma_2)v^*_{-\sigma_2}v_{\sigma'_2} = 0,$$

$$u^*_{p_2\,-\sigma_2}\sigma u_{p'_2\sigma'_2} = \begin{cases}\sigma_2 = -\sigma'_2: & (M/2m)v^*_{-\sigma_2}\sigma v_{\sigma'_2}, \\ \sigma_2 = \sigma'_2: & v^*_{-\sigma_2}\sigma v_{\sigma'_2}.\end{cases} \tag{3-13.40}$$

The first statement depends upon the opposite motion of the two particles in the center of mass frame. There is an irreconcilable conflict between the numerical factor, demanding the equality of the helicities $-\sigma_2$ and σ'_2, and the spinor product, which requires equal magnetic quantum numbers and therefore opposite values of the helicities $-\sigma_2$ and σ'_2. The situation is analogous to that for a spin 1 particle, where the time component of the vector field vanishes in the rest frame.

As we recognize from (3–13.40), high energy collisions with the same initial helicities ($\sigma_2 = \sigma'_2$) are dominated by the first term of (3–13.34), which differs from its analogue in like particle scattering only in sign; therefore [(3–13.19)]

$$\left(\frac{d\sigma}{d\Omega}\right)_{\sigma_2=\sigma'_2} = \frac{\alpha^2}{M^2}\frac{1}{\sin^4\frac{1}{2}\theta}$$
$$= \frac{\alpha^2}{M^2}\cos^4\tfrac{1}{2}\theta\left(\frac{1}{\sin^2\frac{1}{2}\theta} + \frac{1}{\cos^2\frac{1}{2}\theta}\right)^2. \tag{3-13.41}$$

The high energy evaluation of (3–13.39), in the principal circumstance $\sigma_2 = -\sigma'_2$, $\sigma_1 = -\sigma'_1$, is

$$(u^*_{p'_1\sigma'_1} {}_{-q}\gamma^0\gamma^\mu u^*_{p_1\sigma_1 q})(u_{p_2\sigma_2 q}\gamma^0\gamma_\mu u_{p'_2\sigma'_2\,-q}) = (M/2m)^2\sigma_1\sigma_2(v^*_{\sigma'_1}\sigma v_{-\sigma_1})\cdot(v^*_{-\sigma_2}\sigma v_{\sigma'_2})$$
$$= (M/2m)^2\sigma_1\sigma_2 2(v^*_{\sigma'_1}v_{\sigma'_2})(v^*_{-\sigma_2}v_{-\sigma_1}), \tag{3-13.42}$$

according to (3–13.20) and the null property noted in (3–13.40). The first term of (3–13.34) contributes only to the process with $\sigma_1 = \sigma_2$, and, recalling that

$$(p_1 + p'_1)^2 = -M^2, \tag{3-13.43}$$

we get

$$\left(\frac{d\sigma}{d\Omega}\right)_{\sigma_2=-\sigma_2'} = \frac{\alpha^2}{M^2}\left[\left(\frac{\cos^2\frac{1}{2}\theta}{\sin^2\frac{1}{2}\theta} - \cos^2\frac{1}{2}\theta\right)^2 + (\sin^2\frac{1}{2}\theta)^2\right]$$

$$= \frac{\alpha^2}{M^2}\cos^4\frac{1}{2}\theta\left[\left(\frac{1}{\sin^2\frac{1}{2}\theta} + \frac{1}{\cos^2\frac{1}{2}\theta} - 2\right)^2 - 2\right]. \quad (3\text{-}13.44)$$

The average of (3-13.41) and (3-13.44), the differential cross section for un-polarized particles, is

$$\frac{d\sigma}{d\Omega} = \frac{\alpha^2}{M^2}\cos^4\frac{1}{2}\theta\left(\frac{1}{\sin^2\frac{1}{2}\theta} + \frac{1}{\cos^2\frac{1}{2}\theta} - 1\right)^2. \quad (3\text{-}13.45)$$

This, too, is identical with the high energy spin 0 differential cross section. As in (3-12.88), the simple factor $\cos^4\frac{1}{2}\theta$ relates high energy electron-positron scattering to high energy electron-electron scattering.

Owing to the disparity of the denominators in (3-13.34) at low energies, only the first term is significant for $M \sim 2m$ and

$$\frac{d\sigma}{d\Omega} = \frac{1}{16}\frac{\alpha^2}{(M-2m)^2}\frac{1}{\sin^4\frac{1}{2}\theta}. \quad (3\text{-}13.46)$$

A general formula, incorporating both limiting cross sections, (3-13.45) and (3-13.46), can be derived, as with electron-electron scattering, by considering all the helicity transitions that are possible at intermediate energies. But the crossing relations make it unnecessary to do this. The connections between individual transition amplitudes also apply to the helicity sums of the squares of these amplitudes. Thus one can begin with the electron-electron scattering result exhibited in (3-13.30) and derive the required electron-positron form. The crossing transformation $p_1' \leftrightarrow -p_2'$ implies

$$(p_1 - p_2')^2 \leftrightarrow (p_1 + p_1')^2 \quad (3\text{-}13.47)$$

or, translated into the parameters M and θ,

$$(M^2 - 4m^2)\cos^2\frac{1}{2}\theta \leftrightarrow -M^2 \quad (3\text{-}13.48)$$

while

$$(p_1 - p_2)^2 = (M^2 - 4m^2)\sin^2\frac{1}{2}\theta \quad (3\text{-}13.49)$$

retains its meaning. As a quick illustration of the procedure let us make these substitutions in the high energy limit of electron-electron scattering, as indicated by

$$\left(\frac{M^2}{M^2\sin^2\frac{1}{2}\theta} + \frac{M^2}{M^2\cos^2\frac{1}{2}\theta} - 1\right)^2 \rightarrow \left(\frac{-M^2\cos^2\frac{1}{2}\theta}{M^2\sin^2\frac{1}{2}\theta} + \frac{-M^2\cos^2\frac{1}{2}\theta}{-M^2} - 1\right)^2$$

$$= \cos^4\frac{1}{2}\theta\left(\frac{1}{\sin^2\frac{1}{2}\theta} + \frac{1}{\cos^2\frac{1}{2}\theta} - 1\right)^2. \quad (3\text{-}13.50)$$

The relation between the two high energy differential cross sections has become quite transparent.

When the substitutions (3–13.48) are introduced in (3–13.30), the general electron-electron differential cross section for unpolarized beams, the result is the corresponding electron-positron differential cross section (the kinematical factor $1/M^2$ does not take part in these transformations, of course):

$$\frac{d\sigma}{d\Omega} = \frac{\alpha^2}{M^2} \left\{ \left[\frac{2m^2}{M^2 - 4m^2} \frac{1}{\sin^2 \frac{1}{2}\theta} + \cos^2 \frac{1}{2}\theta \left(\frac{1}{\sin^2 \frac{1}{2}\theta} + \frac{1}{\cos^2 \frac{1}{2}\theta} - 1 \right) \right. \right.$$
$$\left. + \frac{2m^2}{M^2} \cos\theta \right]^2 - \frac{5m^2}{M^2 - 4m^2} \frac{4m^2}{M^2} \frac{1}{\sin^2 \frac{1}{2}\theta} - \frac{8m^2}{M^2} \frac{\cos^2 \frac{1}{2}\theta}{\sin^2 \frac{1}{2}\theta} \right\} . \qquad (3\text{--}13.51)$$

Although written in a slightly different way that exhibits the dominant low energy and high energy behavior, the first, square bracket, term is identical with the spin 0 differential cross section for scattering of opposite charges. The latter was only stated implicitly in (3–12.87), one half of which is the entry in the square bracket of (3–13.51). The two additional terms in this equation are relatively negligible at both low and high energies, but they can be of quantitative significance at intermediate energies.

In order to illustrate the scattering of different kinds of charged particles, we shall also consider the interaction between a spin 0 particle and a spin ½ particle. The appropriate electric current vector is the sum of those associated with the two types of particles and the interaction term in W_{40} is

$$W_{40} \rightarrow \int (dx)(dx') j^\mu(x)_{\text{spin } \frac{1}{2}} D_+(x - x') j_\mu(x')_{\text{spin } 0}. \qquad (3\text{--}13.52)$$

We can write the transition matrix element directly:

$$\langle 1_{p_1\sigma_1} 1_{p_1'q'} | T | 1_{p_2\sigma_2} 1_{p_2'q'} \rangle = 2m (d\omega_{p_1} \, d\omega_{p_1'} \, d\omega_{p_2} \, d\omega_{p_2'})^{1/2}$$
$$\times e^2 qq' \frac{(u^*_{p_1\sigma_1} \gamma^0 \gamma^\mu u_{p_2\sigma_2})(p_1' + p_2')_\mu}{(p_1 - p_2)^2}, \qquad (3\text{--}13.53)$$

where all primed quantities refer to the spin 0 particle. This will also extend to the masses, m and m', of the spin ½ and spin 0 particle, respectively. A simplification can be introduced with the aid of the total momentum,

$$P = p_1 + p_1' = p_2 + p_2', \qquad (3\text{--}13.54)$$

for

$$\gamma(p_1' + p_2') = 2\gamma P - \gamma p_1 - \gamma p_2 \qquad (3\text{--}13.55)$$

and

$$u^*_{p_1\sigma_1} \gamma^0 \gamma(p_1' + p_2') u_{p_2\sigma_2} = 2 u^*_{p_1\sigma_1} \gamma^0 (\gamma P + m) u_{p_2\sigma_2}$$
$$= -2 u^*_{p_1\sigma_1} (M - m\gamma^0) u_{p_2\sigma_2}, \qquad (3\text{--}13.56)$$

in which the last form refers to the center of mass frame. The introduction of the helicity construction (3–13.9) gives

$$u^*_{p_1\sigma_1}\gamma^0\gamma(p_1'+p_2')u_{p_2\sigma_2}$$

$$= -2\left[(M-m)\frac{p^0+m}{2m}+(M+m)\frac{p^0-m}{2m}\sigma_1\sigma_2\right]v^*_{\sigma_1}v_{\sigma_2},\quad (3\text{–}13.57)$$

where the kinematics of the situation state that the electron energy and momentum magnitude, which remain unaltered by the collision in the center of mass frame, are

$$p^0 = \frac{M^2+m^2-m'^2}{2M},$$

$$|\mathbf{p}| = \frac{1}{2M}[M^2-(m-m')^2]^{1/2}[M^2-(m+m')^2]^{1/2}.\qquad (3\text{–}13.58)$$

At high energies, where the individual particle masses are negligible, the electron helicity is maintained in scattering,

$$\sigma_1=\sigma_2:\quad 2mu^*_{p_1\sigma_1}\gamma^0\gamma(p_1'+p_2')u_{p_2\sigma_2} = -2M^2\cos\tfrac{1}{2}\theta,\qquad (3\text{–}13.59)$$

and

$$M \gg m, m':\qquad \frac{d\sigma}{d\Omega} = \frac{\alpha^2}{M^2}\frac{\cos^2\tfrac{1}{2}\theta}{\sin^4\tfrac{1}{2}\theta}.\qquad (3\text{–}13.60)$$

When, at low energies, the electron spin is referred to a fixed direction no change in magnetic quantum number occurs on scattering, and

$$M \simeq m+m':\quad \frac{d\sigma}{d\Omega} = \frac{1}{16}\frac{\alpha^2}{(M-m-m')^2}\frac{1}{\sin^4\tfrac{1}{2}\theta}.\qquad (3\text{–}13.61)$$

The general result, summed over final spins and averaged over the initial spins (the latter process is unnecessary here) is

$$\frac{d\sigma}{d\Omega} = \frac{\alpha^2}{[M^2-(m-m')^2]^2[M^2-(m+m')^2]^2}[M^2(M^2-m^2-m'^2)^2\cos^2\tfrac{1}{2}\theta$$

$$+ m^2(M^2-m^2+m'^2)^2\sin^2\tfrac{1}{2}\theta]\frac{1}{\sin^4\tfrac{1}{2}\theta}.\qquad (3\text{–}13.62)$$

It is interesting to consider the two limiting situations in which the mass of one particle becomes very large, not only compared to the other mass, but to the total energy of the second particle. If m' is that large mass, it is more convenient to introduce

$$M - m' \to p^0,\qquad p^0/m' \to 0,\qquad (3\text{–}13.63)$$

where the electron energy in the center of mass frame is indistinguishable from the energy in the coordinate system where the heavy particle remains at rest.

The limiting process gives

spin ½:

$$\frac{d\sigma}{d\Omega} = \tfrac{1}{4}\alpha^2 \left[\frac{p^0}{(p^0)^2 - m^2}\right]^2 \left(\cos^2 \tfrac{1}{2}\theta + \frac{m^2}{(p^0)^2}\sin^2 \tfrac{1}{2}\theta\right)\frac{1}{\sin^4 \tfrac{1}{2}\theta}$$

$$= \begin{cases} p^0 \gg m: & \dfrac{1}{4}\dfrac{\alpha^2}{(p^0)^2}\dfrac{\cos^2 \tfrac{1}{2}\theta}{\sin^4 \tfrac{1}{2}\theta}\,, \\[3mm] p^0 \simeq m: & \dfrac{1}{16}\dfrac{\alpha^2}{(p^0-m)^2}\dfrac{1}{\sin^4 \tfrac{1}{2}\theta}\,. \end{cases} \qquad (3\text{-}13.64)$$

The analogous limit in which it is the spin ½ particle that has become very heavy has the same form as (3–13.64), without the trigonometric factor in the numerator:

spin 0:
$$\frac{d\sigma}{d\Omega} = \tfrac{1}{4}\alpha^2 \left[\frac{p^0}{(p^0)^2 - m^2}\right]^2 \frac{1}{\sin^4 \tfrac{1}{2}\theta}$$

$$= \begin{cases} p^0 \gg m: & \dfrac{1}{4}\dfrac{\alpha^2}{(p^0)^2}\dfrac{1}{\sin^4 \tfrac{1}{2}\theta}\,, \\[3mm] p^0 \simeq m: & \dfrac{1}{16}\dfrac{\alpha^2}{(p^0-m)^2}\dfrac{1}{\sin^4 \tfrac{1}{2}\theta}\,; \end{cases} \qquad (3\text{-}13.65)$$

the now superfluous prime on m has been omitted. The two differential cross sections have been identified with the moving particle, the very massive one acting only as a stationary charge. The possibility of applying a source description to this circumstance will be developed in the next section.

But first let us examine some processes involving photons and spin ½ particles, as contained in the expression (3–12.24) for W_{22}. Electron-positron annihilation into two photons is described by

$$W_{22} \to \tfrac{1}{2}\int (dx)(dx')\psi_2(x)\gamma^0 eq\gamma A_1(x) G_+(x - x')eq\gamma A_1(x')\psi_2(x'), \quad (3\text{-}13.66)$$

and the coefficient of $iJ^*_{k_1\lambda_1}iJ^*_{k'_1\lambda'_1}i\eta_{p'_2\sigma'_2} {-}q^i\eta_{p_2\sigma_2 q}$ gives

$$\langle 1_{k_1\lambda_1}1_{k'_1\lambda'_1}|T|1_{p'_2\sigma'_2} {-}q1_{p_2\sigma_2 q}\rangle = 2m(d\omega_{k_1}\,d\omega_{k'_1}\,d\omega_{p_2}\,d\omega_{p'_2})^{1/2}$$

$$\times\, e^2 u_{p'_2\sigma'_2-q}\gamma^0 \left[\gamma e^*_{k'_1\lambda'_1}\frac{1}{\gamma(p_2 - k_1) + m}\gamma e^*_{k_1\lambda_1}\right.$$

$$\left. + \gamma e^*_{k_1\lambda_1}\frac{1}{\gamma(p_2 - k'_1) + m}\gamma e^*_{k'_1\lambda'_1}\right]u_{p_2\sigma_2 q}. \qquad (3\text{-}13.67)$$

The B. E. symmetry in $k_1\lambda_1$, $k'_1\lambda'_1$ is evident, and the F. D. antisymmetry in $p_2\sigma_2 q$, $p'_2\sigma'_2 {-}q$ can be verified (recall that $\gamma^T_\mu = -\gamma^0\gamma_\mu\gamma^0$) with the aid of the kinematical relations

$$-(p_2 - k'_1) = p'_2 - k_1, \qquad -(p_2 - k_1) = p'_2 - k'_1. \qquad (3\text{-}13.68)$$

We shall find it more convenient to write the dynamical factor of (3–13.67) as

$$e^2\sigma_2' u_{p_2'}^* {}_{-\sigma_2'}\gamma^0 i\gamma_5\left[\gamma e_{k_1'\lambda_1'}^* \frac{m-\gamma(p_2-k_1)}{-2p_2k_1}\gamma e_{k_1\lambda_1}^* \right.$$
$$\left. +\gamma e_{k_1\lambda_1}^* \frac{m-\gamma(p_2-k_1')}{-2p_2k_1'}\gamma e_{k_1'\lambda_1'}^* \right]u_{p_2\sigma_2}. \quad (3\text{–}13.69)$$

In the center of mass frame the energies of all electrons and photons equal $\tfrac{1}{2}M$, and

$$-2p_2k_1 = \tfrac{1}{2}M[M-(M^2-4m^2)^{1/2}\cos\theta],$$
$$-2p_2k_1' = \tfrac{1}{2}M[M+(M^2-4m^2)^{1/2}\cos\theta]. \quad (3\text{–}13.70)$$

Particularly simple is the annihilation of slowly moving particles, $M\simeq 2m$, for which (3–13.69) becomes

$$e^2\sigma_2' v_{-\sigma_2'}^* i\gamma_5\left[\gamma\cdot e_{k_1'\lambda_1'}^* \frac{m+\gamma\cdot k_1}{2m^2}\gamma\cdot e_{k_1\lambda_1}^* +\gamma\cdot e_{k_1\lambda_1}^* \frac{m-\gamma\cdot k_1}{2m^2}\gamma\cdot e_{k_1'\lambda_1'}^* \right]v_{\sigma_2}. \quad (3\text{–}13.71)$$

Multiplied by $2m$, this reduces to $(\gamma^0\gamma = i\gamma_5\sigma)$

$$(e^2/m)\sigma' v_{-\sigma'}^*[\sigma\cdot e^*, \sigma\cdot e'^*]\sigma\cdot kv_\sigma = 2e^2 i\sigma'\delta_{\sigma\sigma'}e^*\times e'^*\cdot k/m, \quad (3\text{–}13.72)$$

where the unneeded causal labels have been dropped. Only terms with an even number of γ_5 factors survive here, and we have used the fact that $e\times e'$ must be directed along k. The equality of the helicities states that the two magnetic quantum numbers are opposite, in the antisymmetrical way implied by the factor σ'. Accordingly, only the singlet state of zero total spin can undergo two-photon annihilation, for slowly moving particles. The corresponding zero angular momentum state of the two photons, a linear combination of the two equal helicity states, is identified by the perpendicular polarization vectors of the two photons. When we recall that

$$M-2m = \tfrac{1}{4}mv^2, \; v\ll 1: \quad \left(1-\frac{4m^2}{M^2}\right)^{-1/2}=\frac{2}{v}, \quad (3\text{–}13.73)$$

the differential cross section per unit solid angle for a given pair of perpendicularly polarized photons, with the particles in the singlet state, is obtained as $(\alpha^2/m^2)(1/v)$. To compute the total annihilation cross section for unpolarized particles we must supply the following additional factors: 2, for the number of polarizations available to one photon, the other polarization being fixed by the requirement of perpendicularity; $\tfrac{1}{4}$, the statistical weight of the singlet state; 2π, the total solid angle accessible to either photon without duplication of the final states. This gives

$$v \ll 1: \qquad\qquad \sigma = \pi \frac{\alpha^2}{m^2} \frac{1}{v}, \qquad\qquad (3\text{-}13.74)$$

which is half the analogous spin 0 annihilation cross section.

It is interesting to observe that the following effective interaction term,

$$W_{22\ \mathrm{eff.}} = \frac{e^2}{2m^3} \int (dx) \tfrac{1}{2}\psi(x)\gamma^0\gamma_5\psi(x)(-\tfrac{1}{4}) \,{}^*F^{\mu\nu}(x)F_{\mu\nu}(x), \quad (3\text{-}13.75)$$

where

$$(-\tfrac{1}{4})\,{}^*F^{\mu\nu}(x)F_{\mu\nu}(x) = \mathbf{E}(x) \cdot \mathbf{H}(x), \qquad\qquad (3\text{-}13.76)$$

will directly produce the transition matrix element expressed in (3–13.72), when evaluated for the same low energy collision. Its space-time locality, in contrast with the nonlocal structure of (3–13.66), is a specific reflection of those limited energetic circumstances which prevent any more detailed space-time characterization of the process.

In a high energy evaluation of (3–13.69) the mass m in the numerator would be neglected. We shall see that this is justified if one excludes very small forward or backward scattering angles,

$$\sin\theta \gg m/M. \qquad\qquad (3\text{-}13.77)$$

Then, since the resulting matrix commutes with γ_5, the helicities are maintained, $-\sigma_2' = \sigma_2$, and annihilation occurs only in unit magnetic quantum number states. One must be careful not to confuse the latter statement, which refers to the spinors $u_{p_2'\sigma_2'\,-q}$ and $u_{p_2\sigma_2 q}$, with the properties of the spinors $u^*_{p_2'\,-\sigma_2'}$ and $u_{p_2\sigma_2}$ where the magnetic quantum numbers are opposite, since the helicities $-\sigma_2'$ and σ_2 are equal. Written in a simplified notation and multiplied by $2m/e^2$, the high energy version of (3–13.69) is $(-\sigma' = \sigma)$:

$$\frac{1}{M}\,\sigma v^*_{-\sigma'} \left[\frac{\boldsymbol{\sigma}\cdot\mathbf{e}'^*\boldsymbol{\sigma}\cdot(\mathbf{p}-\mathbf{k})\boldsymbol{\sigma}\cdot\mathbf{e}^*}{\sin^2 \tfrac{1}{2}\theta} + \frac{\boldsymbol{\sigma}\cdot\mathbf{e}^*\boldsymbol{\sigma}\cdot(\mathbf{p}+\mathbf{k})\boldsymbol{\sigma}\cdot\mathbf{e}'^*}{\cos^2 \tfrac{1}{2}\theta} \right] v_\sigma. \quad (3\text{-}13.78)$$

It is convenient to use the photon momentum $\mathbf{k}$ as the spin reference direction—the z-axis. Then the orthogonal particle spinors describing magnetic quantum numbers of $\pm\tfrac{1}{2}$ in the $\mathbf{p}$ direction can be written as

$$v_\sigma = v_+ \cos\tfrac{1}{2}\theta + v_- \sin\tfrac{1}{2}\theta, \qquad v^*_{-\sigma'} = -v^*_+ \sin\tfrac{1}{2}\theta + v^*_- \cos\tfrac{1}{2}\theta. \quad (3\text{-}13.79)$$

The photon polarization vectors appear in the combinations

$$\begin{aligned}
\boldsymbol{\sigma}\cdot\mathbf{e}^*_+ = \boldsymbol{\sigma}\cdot\mathbf{e}'^*_- &= 2^{-1/2}(\sigma_x - i\sigma_y), \\
\boldsymbol{\sigma}\cdot\mathbf{e}^*_- = \boldsymbol{\sigma}\cdot\mathbf{e}'^*_+ &= 2^{-1/2}(\sigma_x + i\sigma_y),
\end{aligned} \qquad (3\text{-}13.80)$$

which have the effect of raising and lowering particle magnetic quantum numbers by unity:

$$\tfrac{1}{2}(\sigma_x + i\sigma_y)v_- = v_+, \qquad v_-^*\tfrac{1}{2}(\sigma_x - i\sigma_y) = v_+^*, \tag{3-13.81}$$
$$\tfrac{1}{2}(\sigma_x - i\sigma_y)v_+ = v_-, \qquad v_+^*\tfrac{1}{2}(\sigma_x + i\sigma_y) = v_-^*,$$

all other combinations being zero. Thus we can easily work out the values of (3-13.78) for any choice of photon helicities. With $\lambda = -\lambda' = +1$, for example, we get

$$\lambda = -\lambda' = +1: \quad \cos^2 \tfrac{1}{2}\theta \left(\frac{2}{M}\right) v_+^* \boldsymbol{\sigma} \cdot \mathbf{p}v_- \left(\frac{1}{\sin^2 \tfrac{1}{2}\theta} + \frac{1}{\cos^2 \tfrac{1}{2}\theta}\right)$$

$$= \cos^2 \tfrac{1}{2}\theta \sin \theta \left(\frac{4}{\sin^2 \theta}\right) = 2 \cot \tfrac{1}{2}\theta, \tag{3-13.82}$$

and similarly

$$\lambda = -\lambda' = -1: \quad \sin^2 \tfrac{1}{2}\theta \sin \theta \left(\frac{4}{\sin^2 \theta}\right) = 2 \tan \tfrac{1}{2}\theta, \tag{3-13.83}$$

while null results are obtained for equal helicities, $\lambda = \lambda' = \pm 1$. As in the spin 0 discussion, high energy annihilation photons carry only the maximum angular momentum, ± 2, along their line of motion. Again there is an elementary interpretation for the geometrical factors of (3-13.82, 83) which appear in transition probabilities as $\sin^2 \theta \cos^4 \tfrac{1}{2}\theta$ and $\sin^2 \theta \sin^4 \tfrac{1}{2}\theta$. These are the spin 2 probabilities that connect magnetic quantum number $+1$ in a given direction with magnetic quantum numbers $+2$ and -2 in another direction at the relative angle θ. The transition amplitude factor $1/\sin^2 \theta$ also appears for spin 0, in conjunction with the geometrical factor $\sin^2 \theta$, which produces an isotropic differential cross section. Now, however, the spin averaged differential cross section is

$$\frac{d\sigma}{d\Omega} = \frac{1}{2}\frac{\alpha^2}{M^2}\left(\frac{\cos^2 \tfrac{1}{2}\theta}{\sin^2 \tfrac{1}{2}\theta} + \frac{\sin^2 \tfrac{1}{2}\theta}{\cos^2 \tfrac{1}{2}\theta}\right)$$

$$= \frac{\alpha^2}{M^2}\left(\frac{2}{\sin^2 \theta} - 1\right), \tag{3-13.84}$$

and this alteration in angular distribution is attributable entirely to the change of the initial state from zero to unit magnetic quantum number.

It is the singularity of this differential cross section at angles $\theta = 0$ and π that denies universal validity in angle to the high energy evaluation. These singularities are spurious, and trace back to the failure of the high energy approximation

$$1 - \left(1 - \frac{4m^2}{M^2}\right)^{1/2} \cos \theta \to 2 \sin^2 \tfrac{1}{2}\theta,$$

$$1 + \left(1 - \frac{4m^2}{M^2}\right)^{1/2} \cos \theta \to 2 \cos^2 \tfrac{1}{2}\theta \tag{3-13.85}$$

at $\theta = 0$ and π, respectively. A sufficiently more accurate version is

$$1 - \left(1 - \frac{4m^2}{M^2}\right)^{1/2} \cos\theta \rightarrow 2\left(\sin^2 \tfrac{1}{2}\theta + \frac{m^2}{M^2}\right),$$

$$1 + \left(1 - \frac{4m^2}{M^2}\right)^{1/2} \cos\theta \rightarrow 2\left(\cos^2 \tfrac{1}{2}\theta + \frac{m^2}{M^2}\right),$$

$$\text{(3–13.86)}$$

which is also useful in the form of the product

$$1 - \left(1 - \frac{4m^2}{M^2}\right)\cos^2\theta \rightarrow \sin^2\theta + \frac{4m^2}{M^2}, \qquad \text{(3–13.87)}$$

making explicit the origin of the angle restriction (3–13.77). It would seem that one had only to replace (3–13.84) with

$$\frac{d\sigma}{d\Omega} = \frac{\alpha^2}{M^2}\left(\frac{2}{\sin^2\theta + (4m^2/M^2)} - 1\right), \qquad \text{(3–13.88)}$$

leading to the total annihilation cross section

$$\sigma = 2\pi \frac{\alpha^2}{M^2}\left(\log\frac{M^2}{m^2} - 1\right), \qquad \text{(3–13.89)}$$

and this is correct. But there is more here than meets the eye.

When the improvements of (3–13.86) are introduced in the denominators of (3–13.78) and thereby in (3–13.82, 83), the result is

$$\frac{d\sigma}{d\Omega} = \frac{\alpha^2}{M^2}\,\frac{2\sin^2\theta(1 - \tfrac{1}{2}\sin^2\theta)}{(\sin^2\theta + (4m^2/M^2))^2}, \qquad \text{(3–13.90)}$$

which is not the same as (3–13.88). In fact, something has been omitted and that is the contribution of the m-terms in (3–13.69), which are not negligible for $\sin\theta \sim m/M$. These terms anticommute with γ_5; the helicity reverses or $\sigma_2' = \sigma_2$, and only initial states of zero magnetic quantum number are significant. Multiplied by $2m/e^2$, this contribution to (3–13.69) is ($\sigma' = \sigma$)

$$-\frac{m}{M}\,v_{-\sigma'}^*\left[\frac{\boldsymbol{\sigma}\cdot\mathbf{e}'^*\boldsymbol{\sigma}\cdot\mathbf{e}^*}{\sin^2 \tfrac{1}{2}\theta + (m^2/M^2)} + \frac{\boldsymbol{\sigma}\cdot\mathbf{e}^*\boldsymbol{\sigma}\cdot\mathbf{e}'^*}{\cos^2 \tfrac{1}{2}\theta + (m^2/M^2)}\right]v_\sigma. \quad \text{(3–13.91)}$$

Since this mechanism is unimportant under high energy conditions, except for small values of $\sin\theta$, we need not distinguish between the directions associated with the vectors $\pm\mathbf{p}$ and $\pm\mathbf{k}$. A particular choice of spinors is

$$v_\sigma = v_+, \qquad v_{-\sigma'}^* = v_+^* \qquad \text{(3–13.92)}$$

and only photons with the same helicity can be emitted:

$$\lambda = \lambda' = +1: \qquad -\frac{2m}{M}\frac{1}{\sin^2 \frac{1}{2}\theta + (m^2/M^2)}$$

$$\lambda = \lambda' = -1: \qquad -\frac{2m}{M}\frac{1}{\cos^2 \frac{1}{2}\theta + (m^2/M^2)} \cdot \qquad (3\text{–}13.93)$$

That gives the following supplement to the spin averaged differential cross section,

$$\frac{1}{2}\frac{\alpha^2}{M^2}\left(\frac{m}{M}\right)^2\left[\frac{1}{(\sin^2 \frac{1}{2}\theta + (m^2/M^2))^2} + \frac{1}{(\cos^2 \frac{1}{2}\theta + (m^2/M^2))^2}\right]$$

$$= \frac{\alpha^2}{M^2} 2 \frac{4m^2/M^2}{(\sin^2 \theta + (4m^2/M^2))^2}, \qquad (3\text{–}13.94)$$

and its addition to (3–13.90) effectively produces (3–13.88). Note that the differential cross section for forward and backward emission comes entirely from this last process. The value of that cross section per unit solid angle, $\frac{1}{2}(\alpha^2/m^2)$, differs only by a factor of 2 from the result of the low energy calculation when the kinematical factor $2/v$ is replaced by its high energy value of unity [Eq. (3–13.73)].

The general evaluation of (3–13.69) involves little that has not been encountered at high energies, apart from the frequent appearance of the parameter

$$\kappa = \left(1 - \frac{4m^2}{M^2}\right)^{1/2}. \qquad (3\text{–}13.95)$$

The helicity construction of the spinors in (3–13.69), together with the factor $2m/e^2$, gives it as

$$-\frac{2}{M}\left(\frac{\sigma' - \sigma}{2} + \frac{2m}{M}\frac{\sigma + \sigma'}{2}\right)$$

$$\times v_{-\sigma'}^*\left[\frac{\sigma \cdot e'^*\sigma \cdot (\mathbf{p} - \mathbf{k})\sigma \cdot e^*}{1 - \kappa \cos\theta} + \frac{\sigma \cdot e^*\sigma \cdot (\mathbf{p} + \mathbf{k})\sigma \cdot e'^*}{1 + \kappa \cos\theta}\right]v_\sigma$$

$$-\frac{2m}{M}\kappa\frac{\sigma + \sigma'}{2}v_{-\sigma'}^*\left[\frac{\sigma \cdot e'^*\sigma \cdot e^*}{1 - \kappa \cos\theta} + \frac{\sigma \cdot e^*\sigma \cdot e'^*}{1 + \kappa \cos\theta}\right]v_\sigma. \qquad (3\text{–}13.96)$$

Here is a list indicating the various possibilities:

$$\sigma = -\sigma' = +1, \ \lambda = -\lambda' = +1: \quad \cos^2 \tfrac{1}{2}\theta \sin\theta\kappa\frac{4}{1 - \kappa^2 \cos^2\theta},$$

$$\lambda = -\lambda' = -1: \quad \sin^2 \tfrac{1}{2}\theta \sin\theta\kappa\frac{4}{1 - \kappa^2 \cos^2\theta},$$

$$\lambda = \lambda' = \pm 1: \quad 0, \qquad (3\text{–}13.97)$$

and

$$\sigma = \sigma' = +1, \quad \lambda = -\lambda' = \pm 1:$$

$$-\frac{4m}{M} \kappa \frac{\sin^2 \theta}{1 - \kappa^2 \cos^2 \theta},$$

$$\lambda = \lambda' = +1:$$

$$-\frac{4m}{M}(1 + \kappa) \frac{1}{1 - \kappa^2 \cos^2 \theta}, \tag{3-13.98}$$

$$\lambda = \lambda' = -1:$$

$$\frac{4m}{M}(1 - \kappa) \frac{1}{1 - \kappa^2 \cos^2 \theta}.$$

The only transition not considered in the high energy discussion is the one with zero initial magnetic quantum and final magnetic quantum number of ± 2. It has the anticipated geometrical factor, $\sin^2 \theta$. The immediately obtained form of the differential cross section for unpolarized particles is

$$\frac{d\sigma}{d\Omega} = \frac{\alpha^2}{M^2 \kappa}\left[\kappa^2 \sin^2 \theta(2 - \sin^2 \theta) + \frac{4m^2}{M^2}\kappa^2 \sin^4 \theta + \frac{4m^2}{M^2}(1 + \kappa^2)\right]$$

$$\times \frac{1}{(1 - \kappa^2 \cos^2 \theta)^2}. \tag{3-13.99}$$

Explicit here is the contribution of the only processes that occur at $\sin \theta = 0$, those with zero initial and final magnetic quantum numbers:

$$\frac{d\sigma}{d\Omega}(\sin \theta = 0) = \frac{1}{4}\frac{\alpha^2}{m^2 \kappa}(1 + \kappa^2). \tag{3-13.100}$$

They are also the only ones that survive at low energy. It is the function $1 + \kappa^2$ that produces the variation by a factor of 2 in proceeding from low energy ($\kappa = 0$) to high energy ($\kappa = 1$). Another presentation of the differential cross section is

$$\frac{d\sigma}{d\Omega} = \frac{\alpha^2}{M^2 \kappa}\left[\frac{2(2 - \kappa^2)}{1 - \kappa^2 \cos^2 \theta} - 1 - \frac{2(1 - \kappa^2)^2}{(1 - \kappa^2 \cos^2 \theta)^2}\right]. \tag{3-13.101}$$

The last term can be neglected at high energies, and we recover (3–13.88). The total annihilation cross section is evaluated as

$$\sigma = \pi \frac{\alpha^2}{M^2 \kappa}\left[(3 - \kappa^4)\frac{1}{\kappa}\log\frac{1 + \kappa}{1 - \kappa} - 4 + 2\kappa^2\right], \tag{3-13.102}$$

which reduces to the limiting forms (3–13.74) and (3–13.89) in the appropriate circumstances.

Apart from changing the kinematical factor κ^{-1} into κ, the same differential cross section (3–13.101) applies to the inverse process of electron-positron creation in the collision of two photons. Factors associated with the summation over final polarizations and the averaging of initial ones do not change since both particles, electron and photon, have two possible polarizations. But there is a difference in the evaluation of the total cross section, for electron and positron are distinct particles and the full solid angle of 4π applies. This gives the total pair creation cross section

$$\sigma = 2\pi \frac{\alpha^2}{M^2} \kappa \left[(3 - \kappa^4) \frac{1}{\kappa} \log \frac{1 + \kappa}{1 - \kappa} - 4 + 2\kappa^2 \right]$$

$$= \begin{cases} v \ll 1: & \frac{1}{2}\pi \dfrac{\alpha^2}{m^2} v, \\ \\ M \gg m: & 4\pi \dfrac{\alpha^2}{M^2} \left(\log \dfrac{M^2}{m^2} - 1 \right). \end{cases} \qquad (3\text{–}13.103)$$

The differential cross section for electron-photon scattering can be derived from the electron-positron annihilation differential cross section by means of the crossing substitutions

$$p_2' \rightarrow -p_1, \qquad k_1' \rightarrow -k_2. \qquad (3\text{–}13.104)$$

The corresponding transformations on the parameters M and θ are indicated by the combinations

$$-2p_2 k_1 = \frac{M^2}{2} (1 - \kappa \cos \theta) \rightarrow -2p_2 k_1$$

$$= \frac{M^2 - m^2}{M^2} \left(\frac{M^2 + m^2}{2} + \frac{M^2 - m^2}{2} \cos \theta \right), \qquad (3\text{–}13.105)$$

$$-2p_2 k_1' = \frac{M^2}{2} (1 + \kappa \cos \theta) \rightarrow 2p_2 k_2 = -(M^2 - m^2),$$

from which we derive

$$M^2 \rightarrow -\frac{(M^2 - m^2)^2}{M^2} \sin^2 \tfrac{1}{2}\theta \qquad (3\text{–}13.106)$$

and

$$M^2(1 - \kappa^2 \cos^2 \theta) \rightarrow 4(M^2 \cot^2 \tfrac{1}{2}\theta + m^2). \qquad (3\text{–}13.107)$$

In the limit of high energies these correspondences simplify to

$$M^2 \rightarrow -M^2 \sin^2 \tfrac{1}{2}\theta, \qquad M^2 \sin^2 \tfrac{1}{2}\theta \rightarrow M^2 \cos^2 \tfrac{1}{2}\theta, \qquad M^2 \cos^2 \tfrac{1}{2}\theta \rightarrow -M^2. \qquad (3\text{–}13.108)$$

If we apply them to the first version of (3–13.84), we get (the kinematical factor $1/M^2$ is not involved)

$$\frac{M^2 \cos^2 \tfrac{1}{2}\theta}{M^2 \sin^2 \tfrac{1}{2}\theta} + \frac{M^2 \sin^2 \tfrac{1}{2}\theta}{M^2 \cos^2 \tfrac{1}{2}\theta} \rightarrow -\left(\frac{1}{\cos^2 \tfrac{1}{2}\theta} + \cos^2 \tfrac{1}{2}\theta \right). \qquad (3\text{–}13.109)$$

What is the significance of the minus sign?

Consider the individual transition matrix elements, which are multiples of cot $\frac{1}{2}\theta$ and tan $\frac{1}{2}\theta$. They change from real to imaginary values in response to the substitutions (3-13.108). Since it is the absolute square of the matrix elements that give probabilities, the additional factors of i are without effect, but $i^2 = -1$ makes a spurious appearance when the crossing substitutions are applied directly to the differential cross section. The generality of this effect, for crossing substitutions involving a single spin $\frac{1}{2}$ particle, can be recognized with the aid of a technique that we have not used thus far—the evaluation of polarization summations of transition probabilities by means of the spinor completeness property

$$\sum_\sigma u_{p\sigma}u_{p\sigma}^*\gamma^0 = \frac{m - \gamma p}{2m}. \tag{3-13.110}$$

The crossing substitution on spinors, $u_{p\sigma} \leftrightarrow u_{p\,-\sigma}^*$, which is effectively produced by the negative complex conjugate of (3-13.110), gives

$$\sum_\sigma u_{p\sigma}^* u_{p\sigma}\gamma^0 = \frac{-m - \gamma p}{2m}, \tag{3-13.111}$$

whereas the formal replacement of p with $-p$ in (3-13.110) produces the negative of this result. We did not encounter this phenomenon in relating electron-electron scattering to electron-positron scattering since two spin $\frac{1}{2}$ substitutions are used there.

The high energy limit of the electron-photon differential cross section for unpolarized particles is, therefore,

$$\frac{d\sigma}{d\Omega} = \frac{1}{2}\frac{\alpha^2}{M^2}\left(\frac{1}{\cos^2\frac{1}{2}\theta} + \cos^2\frac{1}{2}\theta\right), \tag{3-13.112}$$

where the two terms correspond to collisions with equal and with opposite signs of the initial helicities, respectively; the electron and photon helicities are maintained in scattering. The apparent singularity at $\theta = \pi$ is removed if we use (3-13.88) and the high energy correspondence

$$M^2 \sin^2\theta \to 4M^2 \cot^2\tfrac{1}{2}\theta. \tag{3-13.113}$$

The result can be presented as

$$\frac{d\sigma}{d\Omega} = \frac{1}{2}\frac{\alpha^2}{M^2}\left(\frac{1}{\cos^2\frac{1}{2}\theta + (m^2/M^2)} + \cos^2\tfrac{1}{2}\theta\right), \tag{3-13.114}$$

and the corresponding total cross section is

$$\sigma = 2\pi\frac{\alpha^2}{M^2}\left(\log\frac{M^2}{m^2} + \frac{1}{2}\right). \tag{3-13.115}$$

To get the electron-photon differential cross section at arbitrary energies we make the appropriate substitutions in (3-13.101) and remove the kinematical factor $1/\kappa$ since now initial and final particles are identical. This gives directly

$$\frac{d\sigma}{d\Omega} = \frac{1}{2}\frac{\alpha^2}{M^2}\left[\frac{\dfrac{(M^2-m^2)^2}{M^2}\sin^2\tfrac{1}{2}\theta - 4m^2}{M^2\cot^2\tfrac{1}{2}\theta + m^2} + 2 + \left(\frac{2m^2}{M^2\cot^2\tfrac{1}{2}\theta + m^2}\right)^2\right],$$

(3–13.116)

which can be rearranged as

$$\frac{d\sigma}{d\Omega} = \frac{1}{2}\frac{\alpha^2}{M^2}\left[\frac{1 + \left(\cos^2\tfrac{1}{2}\theta - \dfrac{m^2}{M^2}\sin^2\tfrac{1}{2}\theta\right)^2}{\cos^2\tfrac{1}{2}\theta + \dfrac{m^2}{M^2}\sin^2\tfrac{1}{2}\theta}\right.$$

$$\left. - \frac{4m^2}{M^2}\left(1 - \frac{m^2}{M^2}\right)\frac{\sin^4\tfrac{1}{2}\theta\cos^2\tfrac{1}{2}\theta}{\left(\cos^2\tfrac{1}{2}\theta + \dfrac{m^2}{M^2}\sin^2\tfrac{1}{2}\theta\right)^2}\right].$$

(3–13.117)

The last term does not contribute at high energies, where (3–13.114) is regained, nor at low energies where the Thomson cross section emerges,

$$M \simeq m: \qquad \frac{d\sigma}{d\Omega} = \frac{\alpha^2}{m^2}\frac{1}{2}(1 + \cos^2\theta).$$

(3–13.118)

The total cross section for electron-photon scattering is

$$\sigma = 2\pi\alpha^2\left[\frac{1}{M^2-m^2}\log\frac{M^2}{m^2} + \frac{M^2+m^2}{2M^4}\right.$$

$$\left. - \frac{4m^2}{(M^2-m^2)^2}\left(\frac{M^2+m^2}{M^2-m^2}\log\frac{M^2}{m^2} - 2\right)\right]$$

$$= \begin{cases} M \gg m: \ 2\pi\dfrac{\alpha^2}{M^2}\left(\log\dfrac{M^2}{m^2} + \dfrac{1}{2}\right), \\[2mm] M \simeq m: \ \dfrac{8\pi}{3}\dfrac{\alpha^2}{m^2}. \end{cases}$$

(3–13.119)

3–14 SOURCES AS SCATTERERS

The photon sources that appear with increasing powers in the interactions $W_{21}, W_{22}, \ldots$ can also be used in the extended sense to give an idealized description of charged particles. As we have already suggested, this simplified treatment is appropriate for a particle that is sufficiently heavy to be uninfluenced by its scattering partner. Consider then a point charge of strength Ze that is stationed at the origin,

$$J^0(\mathbf{x}) = Ze\,\delta(\mathbf{x}), \qquad \mathbf{J}(\mathbf{x}) = 0,$$

(3–14.1)

for which the potentials are

$$A^0(\mathbf{x}) = \frac{Ze}{4\pi} \frac{1}{|\mathbf{x}|}, \qquad \mathbf{A}(\mathbf{x}) = 0. \tag{3-14.2}$$

Beginning with spinless particles and the interaction W_{21} we have, representing a scattering process,

$$
\begin{aligned}
W_{21} &\rightarrow \int (dx)\phi_1(x) 2eq A^0(\mathbf{x})(-i\partial_0)\phi_2(x)\\
&= \sum iK^*_{p_1q}(d\omega_{p_1}\, d\omega_{p_2})^{1/2}(-2eqp_2^0)\left[\int (dx)e^{i(p_2-p_1)x}A^0(\mathbf{x})\right]iK_{p_2q},
\end{aligned}
\tag{3-14.3}
$$

where

$$
\begin{aligned}
\int (dx)e^{i(p_2-p_1)x}A^0(\mathbf{x}) &= \int dx^0 e^{i(p_1^0-p_2^0)x^0} \int (d\mathbf{x})e^{i(\mathbf{p}_2-\mathbf{p}_1)\cdot\mathbf{x}}A^0(\mathbf{x})\\
&= \left[\int dx^0 e^{i(p_1^0-p_2^0)x^0}\right]\frac{Ze}{(\mathbf{p}_1-\mathbf{p}_2)^2}.
\end{aligned}
\tag{3-14.4}
$$

When dealing with an immobile scatterer, all reference to momentum conservation is lost, but energy conservation survives. The definition of the transition matrix is a correspondingly simplified version of the general definition (3–12.40), containing only the time integral factor, and (3–12.43) similarly degenerates to a statement of the transition probability per unit time that displays on the right-hand side a single factor of 2π and the one delta function that establishes energy conservation. In the present situation, then, the transition matrix is

$$\langle 1_{p_1q}|T|1_{p_2q}\rangle = (d\omega_{p_1}\, d\omega_{p_2})^{1/2}\left[-\frac{2Ze^2qp^0}{(\mathbf{p}_1-\mathbf{p}_2)^2}\right], \tag{3-14.5}$$

giving the transition probability per unit time as

$$d\omega_{p_1}\, d\omega_{p_2}\, 2\pi\, \delta(p_1^0-p_2^0)\left[\frac{2Ze^2p^0}{(\mathbf{p}_1-\mathbf{p}_2)^2}\right]^2. \tag{3-14.6}$$

A differential cross section in angle is produced on dividing this by the incident particle flux, $2|\mathbf{p}|\, d\omega_{p_2}$, and integrating over the well-defined final energy, $p_1^0 = p_2^0 = p^0$:

$$
\begin{aligned}
\int d\omega_{p_1} 2\pi\, \delta(p_1^0-p_2^0) &= \frac{1}{8\pi^2}\int \frac{(d\mathbf{p}_1)}{p^0}\, \delta(p_1^0-p^0)\\
&= \frac{1}{8\pi^2}\,|\mathbf{p}|\, d\Omega.
\end{aligned}
\tag{3-14.7}
$$

The immediate result is

$$d\sigma = \frac{1}{(4\pi)^2}\, d\Omega\left[\frac{2Ze^2p^0}{(\mathbf{p}_1-\mathbf{p}_2)^2}\right]^2 \tag{3-14.8}$$

or

$$\frac{d\sigma}{d\Omega} = \tfrac{1}{4}Z^2\alpha^2\left[\frac{p^0}{(p^0)^2 - m^2}\right]^2\frac{1}{\sin^4\tfrac{1}{2}\theta}, \qquad (3\text{-}14.9)$$

which does indeed agree with (3-13.65), apart from the use of Ze as the charge of the stationary scatterer.

The similar consideration for spin $\tfrac{1}{2}$ begins with

$$W_{21} \rightarrow \int (dx)\psi_1(x)(-eq)A^0(\mathbf{x})\psi_2(x)$$
$$= \sum i\eta_{p_1\sigma_1q}^* 2m(d\omega_{p_1}\,d\omega_{p_2})^{1/2}(-eq)(u_{p_1\sigma_1}^* u_{p_2\sigma_2})$$
$$\times\left[\int (dx)e^{i(p_2-p_1)x}A^0(\mathbf{x})\right]i\eta_{p_2\sigma_2q}, \qquad (3\text{-}14.10)$$

which is expressed by

$$\langle 1_{p_1\sigma_1q}|T|1_{p_2\sigma_2q}\rangle = 2m(d\omega_{p_1}\,d\omega_{p_2})^{1/2}\left[-\frac{Ze^2q}{(\mathbf{p}_1 - \mathbf{p}_2)^2}\right](u_{p_1\sigma_1}^* u_{p_2\sigma_2}), \qquad (3\text{-}14.11)$$

and the transition probability per unit time is

$$d\omega_{p_1}\,d\omega_{p_2}2\pi\,\delta(p_1^0 - p_2^0)\left[\frac{2mZe^2}{(\mathbf{p}_1 - \mathbf{p}_2)^2}\right]^2|u_{p_1\sigma_1}^* u_{p_2\sigma_2}|^2. \qquad (3\text{-}14.12)$$

When helicity states are used,

$$(u_{p_1\sigma_1}^* u_{p_2\sigma_2}) = \left(\frac{p^0 + m}{2m} + \frac{p^0 - m}{2m}\sigma_1\sigma_2\right)(v_{\sigma_1}^* v_{\sigma_2})$$

$$= \begin{cases}\sigma_1 = \sigma_2: & (p^0/m)\cos\tfrac{1}{2}\theta, \\ \sigma_1 = -\sigma_2: & \sigma_1\sin\tfrac{1}{2}\theta,\end{cases} \qquad (3\text{-}14.13)$$

where the factor σ_1 in the second entry reproduces the algebraic signs that are exhibited in (3-13.17). For either choice of σ_2 the summation over σ_1 gives the differential cross section

$$\frac{d\sigma}{d\Omega} = \tfrac{1}{4}Z^2\alpha^2\left[\frac{p^0}{(p^0)^2 - m^2}\right]^2\left(\cos^2\tfrac{1}{2}\theta + \frac{m^2}{(p^0)^2}\sin^2\tfrac{1}{2}\theta\right)\frac{1}{\sin^4\tfrac{1}{2}\theta}, \qquad (3\text{-}14.14)$$

as contained in (3-13.64). It is quite clear, in (3-14.13), that the electron retains its helicity at high energy while the spin remains inert in space at low energy.

If W_{21} describes scattering by the fixed charge, what do W_{22}, W_{23}, ... represent? Consider, for example,

$$W_{22} = \tfrac{1}{2}e^2\int (dx)(dx')\psi(x)A^0(\mathbf{x})G_+(x - x')\gamma^0A^0(\mathbf{x}')\psi(x'). \qquad (3\text{-}14.15)$$

Only the field $\psi(x)$ refers to propagating particles and therefore W_{22} also describes an electron scattering process, as do all the other $W_{2\nu}$. Thus, the expansion in powers of the static potential A^0 is no longer a classification into successively more complicated processes, but represents successive approximations to the complete treatment of the motion of the particle in the Coulomb field of the point source. The interaction skeleton here acquires more substance, and thereby indicates one aspect of the dynamical scheme that is generally lacking at the first dynamical level, namely, the possibility for unlimited repetition of a particular interaction mechanism.

Since all powers of A^0 contribute to the scattering process one should like to avoid that power series expansion and work directly with the appropriate Green's function, $\Delta_+^A(x, x')$ or $G_+^A(x, x')$. Unfortunately, the ability to solve the Green's function equations in a reasonably closed form is restricted to the nonrelativistic limit, in the physically interesting situation of a point source and the Coulomb potential. The latter has a simple connection with the differential equation (3–11.36) for Δ_+^A, which here assumes the three-dimensional form

$$-[\nabla^2 + (p^0)^2 - m^2 - 2p^0 eqA^0(\mathbf{x}) + e^2 A^0(\mathbf{x})^2]\,\Delta_+^A(\mathbf{x}\mathbf{x}'p^0) = \delta(\mathbf{x} - \mathbf{x}') \tag{3–14.16}$$

when one introduces the time Fourier transform in this time translationally invariant situation:

$$\Delta^A(\mathbf{x}\mathbf{x}'p^0) = \int_{-\infty}^{\infty} dx^0 e^{ip^0(x^0 - x^{0'})}\,\Delta_+^A(\mathbf{x}x^0, \mathbf{x}'x^{0'}). \tag{3–14.17}$$

The transition to the nonrelativistic limit is conveyed by the altered meaning of energy,

$$(p^0)^2 - m^2 \to 2mE, \qquad p^0 \to m, \tag{3–14.18}$$

and the term quadratic in the scalar potential is neglected. It is this omission that produces the essential difference between the two regimes, which otherwise are connected by the reciprocal correlation of (3–14.18). Thus, if we begin with the nonrelativistic form of the differential cross section and introduce the inverse of the substitutions (3–14.18), we get

$$\frac{d\sigma}{d\Omega} = \tfrac{1}{4}Z^2\alpha^2 \left(\frac{m}{2mE}\right)^2 \frac{1}{\sin^4 \tfrac{1}{2}\theta} \to \tfrac{1}{4}Z^2\alpha^2 \left[\frac{p^0}{(p^0)^2 - m^2}\right]^2 \frac{1}{\sin^4 \tfrac{1}{2}\theta}, \tag{3–14.19}$$

in agreement with (3–14.9). This is, furthermore, the exact consequence of (3–14.16), with the $(A^0)^2$ term struck out, since the nonrelativistic solution has the special property that all higher powers of A^0, or Z, lead only to a multiplicative phase that disappears on forming the transition probability. Accordingly, the first significant deviation from (3–14.9) arises from the last, quadratic term in W_{22}, Eq. (3–12.29). It produces the following modification in the transition matrix:

$$\delta\langle 1_{p_1 q}|T|1_{p_2 q}\rangle = (d\omega_{p_1}\, d\omega_{p_2})^{1/2}(Z\alpha)^2 \int (d\mathbf{x})\, \exp[i(\mathbf{p}_2 - \mathbf{p}_1)\cdot\mathbf{x}]\frac{1}{|\mathbf{x}|^2}, \tag{3–14.20}$$

where

$$\int (d\mathbf{x})\, \exp[i(\mathbf{p}_2 - \mathbf{p}_1)\cdot\mathbf{x}]\frac{1}{|\mathbf{x}|^2} = 4\pi \int_0^{\infty} d|\mathbf{x}|\,\frac{\sin|\mathbf{p}_1 - \mathbf{p}_2|\,|\mathbf{x}|}{|\mathbf{p}_1 - \mathbf{p}_2|\,|\mathbf{x}|} = \frac{2\pi^2}{|\mathbf{p}_1 - \mathbf{p}_2|}. \tag{3–14.21}$$

The correction to the transformation matrix is indicated by the substitution

$$\frac{1}{(\mathbf{p}_1 - \mathbf{p}_2)^2} \rightarrow \frac{1}{(\mathbf{p}_1 - \mathbf{p}_2)^2}\left[1 - \frac{1}{4}\pi Z\alpha q\, \frac{|\mathbf{p}_1 - \mathbf{p}_2|}{p^0}\right], \qquad (3\text{-}14.22)$$

which changes the differential cross section for spin 0 scattering into

$$\frac{d\sigma}{d\Omega} = \frac{1}{4}Z^2\alpha^2\left[\frac{p^0}{(p^0)^2 - m^2}\right]^2 \frac{1}{\sin^4 \frac{1}{2}\theta}\left[1 - \pi Z\alpha q\left(1 - \frac{m^2}{(p^0)^2}\right)^{1/2}\sin\frac{1}{2}\theta\right].$$

$$(3\text{-}14.23)$$

The factor Zq implies that the correction diminishes the cross section for particles of like charge and increases it for the scattering of opposite charges. In dealing only once with the effect of the quadratic interaction term and ignoring the phase factors that represent the consequence of repeated linear interactions, we have obtained only the first term of a multiplicative power series in $Z\alpha q$. As this first term displays, there must also be at least one factor of the particle speed, since the correction is a relativistic phenomenon.

The corresponding discussion of spin $\frac{1}{2}$ scattering proceeds somewhat more indirectly since both the desired relativistic correction and the repetition of the effective nonrelativistic interaction are combined in W_{22}, Eq. (3-14.15). Taken as it stands, the latter implies the following transition matrix modification,

$$'\delta'\langle 1_{p_1\sigma_1 q}|T|1_{p_2\sigma_2 q}\rangle = 2m(d\omega_{p_1}\, d\omega_{p_2})^{1/2}(Z\alpha)^2 u^*_{p_1\sigma_1}\left[\int (dx)(dx')e^{-i\mathbf{p}_1\cdot\mathbf{x}}\right.$$

$$\left. \times \frac{1}{|\mathbf{x}|}\, G_+(\mathbf{x} - \mathbf{x}', p^0)\gamma^0\, \frac{1}{|\mathbf{x}'|}\, e^{i\mathbf{p}_2\cdot\mathbf{x}'}\right]u_{p_2\sigma_2}, \qquad (3\text{-}14.24)$$

where the time integration has introduced the Fourier transform of the electron Green's function, which appears as

$$G_+(\mathbf{x} - \mathbf{x}', p^0)\gamma^0 = \int_{-\infty}^{\infty} dx^0 e^{ip^0(x^0 - x^{0\prime})}G_+(\mathbf{x} - \mathbf{x}', x^0 - x^{0\prime})\gamma^0$$

$$= \int \frac{(d\mathbf{p})}{(2\pi)^3}\, e^{i\mathbf{p}\cdot(\mathbf{x} - \mathbf{x}')}\, \frac{m\gamma^0 + p^0 + \gamma^0\boldsymbol{\gamma}\cdot\mathbf{p}}{\mathbf{p}^2 + m^2 - (p^0)^2 - i\epsilon}. \qquad (3\text{-}14.25)$$

In the nonrelativistic limit, $\gamma^0 \rightarrow 1$, $\gamma^0\boldsymbol{\gamma} = i\gamma_5\boldsymbol{\sigma}$ is neglected, $p^0 \rightarrow m$, and $(p^0)^2 - m^2 \rightarrow 2mE$. This indicates the structure of the terms that are to be regarded as already included in the phase factor. The nonrelativistic presence of $2m$ in the Green's function implies that $2p^0$ is its relativistic counterpart. Indeed, the addition of the relations

$$u^*_{p_1\sigma_1}\gamma^0(m - \gamma^0 p^0 + \boldsymbol{\gamma}\cdot\mathbf{p}_1)u_{p_2\sigma_2} = 0, \qquad u^*_{p_1\sigma_1}\gamma^0(m - \gamma p^0 + \boldsymbol{\gamma}\cdot\mathbf{p}_2)u_{p_2\sigma_2} = 0,$$

$$(3\text{-}14.26)$$

supplies the replacement

$$m\gamma^0 \to p^0 - \gamma^0\boldsymbol{\gamma} \cdot \tfrac{1}{2}(\mathbf{p}_1 + \mathbf{p}_2), \tag{3-14.27}$$

within the context of the spin matrix element used in (3-14.24). Accordingly, the actual correction contained in (3-14.24) is (but see below)

$$\delta\langle 1_{p_1\sigma_1 q}|T|1_{p_2\sigma_2 q}\rangle = 2m(d\omega_{p_1}\, d\omega_{p_2})^{1/2}(4\pi Z\alpha)^2 u^*_{p_1\sigma_1} i\gamma_5\boldsymbol{\sigma} \cdot \mathbf{V} u_{p_2\sigma_2}, \tag{3-14.28}$$

where

$$\mathbf{V} = \int \frac{(d\mathbf{p})}{(2\pi)^3} \frac{1}{(\mathbf{p}_1 - \mathbf{p})^2} \frac{\mathbf{p} - \tfrac{1}{2}(\mathbf{p}_1 + \mathbf{p}_2)}{\mathbf{p}^2 + m^2 - (p^0)^2 - i\epsilon} \frac{1}{(\mathbf{p} - \mathbf{p}_2)^2}. \tag{3-14.29}$$

The symmetry of this integral in $\mathbf{p}_1$ and $\mathbf{p}_2$ indicates that $\mathbf{V}$ is directed along $\mathbf{p}_1 + \mathbf{p}_2$ and we therefore write

$$\mathbf{V} = \frac{\mathbf{p}_1 + \mathbf{p}_2}{(\mathbf{p}_1 + \mathbf{p}_2)^2} S, \tag{3-14.30}$$

with

$$S = \int \frac{(d\mathbf{p})}{(2\pi)^3} \frac{1}{(\mathbf{p}_1 - \mathbf{p})^2}$$

$$\times \frac{(\mathbf{p}^2 + m^2 - (p^0)^2) - \tfrac{1}{2}(\mathbf{p}_1 - \mathbf{p})^2 - \tfrac{1}{2}(\mathbf{p}_2 - \mathbf{p})^2 + \tfrac{1}{2}(\mathbf{p}_1 - \mathbf{p}_2)^2}{\mathbf{p}^2 + m^2 - (p^0)^2 - i\epsilon}$$

$$\times \frac{1}{(\mathbf{p} - \mathbf{p}_2)^2}. \tag{3-14.31}$$

Before discussing this integral, let us observe that, when helicity states are used,

$$2mu^*_{p_1\sigma_1} i\gamma_5 \frac{\boldsymbol{\sigma} \cdot (\mathbf{p}_1 + \mathbf{p}_2)}{(\mathbf{p}_1 + \mathbf{p}_2)^2} u_{p_2\sigma_2} = \frac{\sigma_1 + \sigma_2}{4|\mathbf{p}|\cos^2 \tfrac{1}{2}\theta} 2mu^*_{p_1\sigma_1} i\gamma_5 u_{p_2\sigma_2}$$

$$= \begin{cases} \sigma_1 = \sigma_2: & 1/\cos\tfrac{1}{2}\theta, \\ \sigma_1 = -\sigma_2: & 0, \end{cases} \tag{3-14.32}$$

which means that the correction is confined to transitions in which the helicity does not change. Now the last term of (3-14.31), with the constant factor $\tfrac{1}{2}(\mathbf{p}_1 - \mathbf{p}_2)^2$, can be identified as altering the phase associated with helicity-preserving transitions. Put another way, this term is imaginary and, to the accuracy with which we are working, it does not interfere with the principal contribution to the scattering matrix and can be neglected, along with the imaginary parts of the other terms. The remaining real terms of S, the only significant ones for the cross section modification are

$$\int \frac{(d\mathbf{p})}{(2\pi)^3} \frac{1}{(\mathbf{p}_1 - \mathbf{p})^2} \frac{1}{(\mathbf{p} - \mathbf{p}_2)^2} = \int (d\mathbf{x}) e^{-i\mathbf{p}_1 \cdot \mathbf{x}} \frac{1}{4\pi|\mathbf{x}|} \frac{1}{4\pi|\mathbf{x}|} e^{i\mathbf{p}_2 \cdot \mathbf{x}}$$

$$= \frac{1}{8} \frac{1}{|\mathbf{p}_1 - \mathbf{p}_2|} = \frac{1}{16|\mathbf{p}|} \frac{1}{\sin\tfrac{1}{2}\theta}, \tag{3-14.33}$$

and the two equal integrals illustrated by

$$\text{Re} \int \frac{(dp)}{(2\pi)^3} \frac{1}{\mathbf{p}^2 + m^2 - (p^0)^2 - i\epsilon} \frac{1}{|\mathbf{p} - \mathbf{p}_2|^2} = \text{Re} \int (dx) \frac{e^{i|\mathbf{p}_2||\mathbf{x}|}}{4\pi|\mathbf{x}|} \frac{1}{4\pi|\mathbf{x}|} e^{-i\mathbf{p}_2 \cdot \mathbf{x}}$$

$$= \frac{1}{4\pi} \int_0^\infty d|\mathbf{x}| \cos |\mathbf{p}| \, |\mathbf{x}| \, \frac{\sin |\mathbf{p}| \, |\mathbf{x}|}{|\mathbf{p}| \, |\mathbf{x}|}$$

$$= \frac{1}{16|\mathbf{p}|}, \qquad (3\text{-}14.34)$$

where the return to coordinate space has been advantageous. Also utilized is the three-dimensional momentum integral

$$\int \frac{(dp)}{(2\pi)^3} \frac{e^{i\mathbf{p} \cdot \mathbf{x}}}{\mathbf{p}^2 + m^2 - (p^0)^2 - i\epsilon} = \frac{e^{i((p^0)^2 - m^2)^{1/2}|\mathbf{x}|}}{4\pi|\mathbf{x}|}. \qquad (3\text{-}14.35)$$

The modification in the transition amplitude for scattering without helicity change is conveyed by

$$\frac{\cos \tfrac{1}{2}\theta}{\sin^2 \tfrac{1}{2}\theta} \to \frac{1}{\sin^2 \tfrac{1}{2}\theta} \left[\cos \tfrac{1}{2}\theta - \frac{\pi}{2} Z\alpha q \left(1 - \frac{m^2}{(p^0)^2} \right)^{1/2} \frac{1}{\cos \tfrac{1}{2}\theta} \sin \tfrac{1}{2}\theta (1 - \sin \tfrac{1}{2}\theta) \right], \qquad (3\text{-}14.36)$$

and the corresponding differential cross section for unpolarized particles is

$$\frac{d\sigma}{d\Omega} = \frac{1}{4} Z^2 \alpha^2 \left[\frac{p^0}{(p^0)^2 - m^2} \right]^2 \frac{1}{\sin^4 \tfrac{1}{2}\theta} \left[\cos^2 \tfrac{1}{2}\theta + \left(\frac{m}{p^0} \right)^2 \sin^2 \tfrac{1}{2}\theta \right.$$

$$\left. - \pi Z\alpha q \left(1 - \frac{m^2}{(p^0)^2} \right)^{1/2} \sin \tfrac{1}{2}\theta (1 - \sin \tfrac{1}{2}\theta) \right]. \qquad (3\text{-}14.37)$$

In this result, and in the structure of the coordinate integral of (3–14.33), we recognize a mechanism that is common to spin 0 and spin $\tfrac{1}{2}$. The $\sin^2 \tfrac{1}{2}\theta$ correction term is specific to spin $\tfrac{1}{2}$. Another way of writing the last, square bracket, factor emphasizes the association of the correction term with helicity-preserving transitions:

$$\left[1 - \pi Z\alpha q \left(1 - \frac{m^2}{(p^0)^2} \right)^{1/2} \frac{\sin \tfrac{1}{2}\theta}{1 + \sin \tfrac{1}{2}\theta} \right] \cos^2 \tfrac{1}{2}\theta + \left(\frac{m}{p^0} \right)^2 \sin^2 \tfrac{1}{2}\theta. \quad (3\text{-}14.38)$$

Before continuing the discussion, we shall evaluate the imaginary part of S, which played no role in the cross section calculation. On recognizing that

$$\tfrac{1}{2}[(\mathbf{p}_1 - \mathbf{p}_2)^2 - (\mathbf{p}_1 - \mathbf{p})^2 - (\mathbf{p}_2 - \mathbf{p})^2] = -(\mathbf{p}_1 - \mathbf{p}) \cdot (\mathbf{p}_2 - \mathbf{p}), \quad (3\text{-}14.39)$$

we have

$$\text{Im} \, S = - \int \frac{(dp)}{(2\pi)^3} \pi \delta(\mathbf{p}^2 - ((p^0)^2 - m^2)) \frac{(\mathbf{p}_1 - \mathbf{p}) \cdot (\mathbf{p}_2 - \mathbf{p})}{(\mathbf{p}_1 - \mathbf{p})^2 (\mathbf{p}_2 - \mathbf{p})^2}$$

$$= - \frac{1}{4\pi} ((p^0)^2 - m^2)^{-1/2} I, \qquad (3\text{-}14.40)$$

where

$$I = \int \frac{d\Omega}{4\pi} \frac{(\mathbf{n}_1 - \mathbf{n}) \cdot (\mathbf{n}_2 - \mathbf{n})}{(\mathbf{n}_1 - \mathbf{n})^2 (\mathbf{n}_2 - \mathbf{n})^2} \tag{3-14.41}$$

integrates the vector $\mathbf{n}$ over the unit sphere. The unit vectors $\mathbf{n}_1$, $\mathbf{n}_2$ specify the directions of $\mathbf{p}_1$, $\mathbf{p}_2$. The integral can be written as

$$I = \frac{1}{2} \int \frac{d\Omega}{4\pi} \frac{\cos^2 \tfrac{1}{2}\theta - \mathbf{n} \cdot \dfrac{\mathbf{n}_1 + \mathbf{n}_2}{2}}{\left(1 - \mathbf{n} \cdot \dfrac{\mathbf{n}_1 + \mathbf{n}_2}{2}\right)^2 - \left(\mathbf{n} \cdot \dfrac{\mathbf{n}_1 - \mathbf{n}_2}{2}\right)^2}, \tag{3-14.42}$$

where $\tfrac{1}{2}(\mathbf{n}_1 + \mathbf{n}_2)$ and $\tfrac{1}{2}(\mathbf{n}_1 - \mathbf{n}_2)$ are perpendicular vectors of length $\cos \tfrac{1}{2}\theta$ and $\sin \tfrac{1}{2}\theta$, respectively. Basing a choice of spherical coordinates on them gives

$$I = \frac{1}{2} \int_{-1}^{1} \frac{1}{2} \, d\mu \int_{0}^{2\pi} \frac{1}{2\pi} \, d\varphi \, \frac{\cos \tfrac{1}{2}\theta \, (\cos \tfrac{1}{2}\theta - \mu)}{(1 - \mu \cos \tfrac{1}{2}\theta)^2 - (1 - \mu^2) \sin^2 \tfrac{1}{2}\theta \cos^2 \varphi}$$

$$= \frac{1}{4} \int_{-1}^{1} d\mu \, \frac{\cos \tfrac{1}{2}\theta}{1 - \mu \cos \tfrac{1}{2}\theta} \frac{\cos \tfrac{1}{2}\theta - \mu}{|\cos \tfrac{1}{2}\theta - \mu|} = \frac{1}{2} \log \frac{1}{\sin \tfrac{1}{2}\theta}. \tag{3-14.43}$$

Thus, the complete evaluation of S is

$$S = \tfrac{1}{16}((p^0)^2 - m^2)^{-1/2} \left[\frac{1}{\sin \tfrac{1}{2}\theta} - 1 - i\frac{2}{\pi} \log \frac{1}{\sin \tfrac{1}{2}\theta} \right]. \tag{3-14.44}$$

The complex structure of S, expressing a relative phase shift between helicity-preserving and helicity-changing transitions, has a physical implication that can best be appreciated by relinquishing the helicity description. With an unspecified choice of the v_σ spinors, we present the transition matrix as

$$\langle 1_{p_1\sigma_1 q} | T | 1_{p_2\sigma_2 q} \rangle = 2m(d\omega_{p_1} \, d\omega_{p_2})^{1/2} \left[-\frac{Ze^2 q}{(\mathbf{p}_1 - \mathbf{p}_2)^2} \right] v_{\sigma_1}^* M v_{\sigma_2}, \tag{3-14.45}$$

where

$$M = f + ig\boldsymbol{\sigma} \cdot \mathbf{n}_1 \times \mathbf{n}_2 \tag{3-14.46}$$

and

$$f = (p^0/m) \cos^2 \tfrac{1}{2}\theta + \sin^2 \tfrac{1}{2}\theta - 8\pi Z\alpha q \sin^2 \tfrac{1}{2}\theta \frac{(p^0)^2 - m^2}{m} S,$$

$$g = \frac{p^0 - m}{2m} - 4\pi Z\alpha q \tan^2 \tfrac{1}{2}\theta \frac{(p^0)^2 - m^2}{m} S. \tag{3-14.47}$$

The computation of the total differential cross section for an arbitrary initial spin involves

$$\sum_{\sigma_1} |v_{\sigma_1}^* M v_{\sigma_2}|^2 = v_{\sigma_2}^*(f^* - ig^*\boldsymbol{\sigma} \cdot \mathbf{n}_1 \times \mathbf{n}_2)(f + ig\boldsymbol{\sigma} \cdot \mathbf{n}_1 \times \mathbf{n}_2)v_{\sigma_2}$$

$$= |f|^2 + |g|^2 \sin^2 \theta + i(f^*g - g^*f)\langle \boldsymbol{\sigma} \cdot \mathbf{n}_1 \times \mathbf{n}_2 \rangle_{\sigma_2}. \tag{3-14.48}$$

Accordingly, if f/g is a complex number, there is an explicit dependence on the initial spin, provided it has a nonvanishing expectation value in the direction perpendicular to the plane of scattering. A state of definite helicity does not have that property; it requires a linear combination of the helicity states.

Reciprocal to the dependence of the differential cross section on the initial spin is the appearance of polarization in the scattered particles for an initially unpolarized beam. At a given scattering angle, the average final spin is

$$\langle \sigma \rangle = \frac{v_{\sigma_2}^*(f^* - ig^*\sigma \cdot \mathbf{n}_1 \times \mathbf{n}_2)\sigma(f + ig\sigma \cdot \mathbf{n}_1 \times \mathbf{n}_2)v_{\sigma_2}}{|f|^2 + |g|^2 \sin^2 \theta}$$

$$= p\nu, \tag{3-14.49}$$

where ν is the unit vector normal to the scattering plane,

$$\mathbf{n}_1 \times \mathbf{n}_2 = \sin\theta\nu, \tag{3-14.50}$$

and

$$p = -2 \frac{|f|^2 \sin\theta}{|f|^2 + |g|^2 \sin^2\theta} \operatorname{Im}\left(\frac{g}{f}\right). \tag{3-14.51}$$

In the special situations expressed by $f = \pm ig \sin\theta$, the polarization is complete: $p = \pm 1$. Notice that the same polarization parameter expresses the relative dependence of the scattering differential cross section on the initial spin:

$$1 + p\langle \sigma \cdot \nu \rangle. \tag{3-14.52}$$

This effect can be demonstrated experimentally by a double scattering arrangement, with the first scattering act polarizing the particles and the second one detecting that polarization. If we designate them as a and b, the insertion of the polarization produced by the first deflection into the cross section for the second scattering gives the relative factor

$$1 + p_a p_b \nu_a \cdot \nu_b. \tag{3-14.53}$$

The observational sign that neither p_a nor p_b is zero thus comes from a dependence of the final intensity upon the relative orientation of the two scattering planes. In particular, if the two planes are the same geometrically but deflections in opposite senses are compared, $\nu_b = \pm \nu_a$, the ratio of the intensity for deflections in the same sense to that for the opposite sense is

$$(1 + p_a p_b)/(1 - p_a p_b). \tag{3-14.54}$$

This is greater than one when the individual scattering acts are identical, $p_a = p_b$. The preference for successive deflection in the same sense will be maintained with any choice of individual scattering angles if, as in the present discussion, the polarization parameter has a definite sign at all angles:

$$p = -Z\alpha q \frac{m}{p^0}\left(1 - \frac{m^2}{(p^0)^2}\right)^{1/2} \frac{\tan^3 \tfrac{1}{2}\theta}{1 + (m/p^0)^2 \tan^2 \tfrac{1}{2}\theta} \log \frac{1}{\sin^2 \tfrac{1}{2}\theta}. \tag{3-14.55}$$

We turn now to examples of the class of phenomena in which both simple and extended photon sources are involved. These are charged particle interactions with fixed charges, in which photons are emitted or absorbed. The simplest illustrations are contained in W_{22}. They are single photon emission during scattering in a Coulomb field and the creation of a pair of charged particles by a photon passing through a Coulomb field. The relevant part of W_{22} for a spin 0 particle that emits a photon during a collision with a massive particle of charge Ze is

$$W_{22} \rightarrow \int (dx)(dx')\phi_1(x)[2eqpA_1(x)\Delta_+(x - x')2eqpA_Z(x')$$

$$+ 2eqpA_Z(x)\Delta_+(x - x')2eqpA_1(x')]\phi_2(x')$$

$$- \int (dx)\phi_1(x)2e^2A_1(x)A_Z(x)\phi_2(x), \qquad (3\text{-}14.56)$$

where $A_Z^\mu(x)$ indicates the vector potential associated with the charge Ze. Using the form of the latter that is stated in (3-14.2), the transition matrix element appears as

$$\langle 1_{p_1 q} 1_{k_1 \lambda_1} | T | 1_{p_2 q} \rangle = (d\omega_{p_1}\, d\omega_{k_1}\, d\omega_{p_2})^{1/2} \frac{Ze^2}{(p_1 + k_1 - p_2)^2}\, 2e\, e_{k_1\lambda_1}^{\mu *} V_{\mu\nu} n^\nu,$$

$$(3\text{-}14.57)$$

where

$$V_{\mu\nu} = \frac{p_{1\mu}p_{2\nu}}{k_1 p_1} - \frac{p_{2\mu}p_{1\nu}}{k_1 p_2} - g_{\mu\nu} \qquad (3\text{-}14.58)$$

and n^μ is the unit time-like vector that has the single component $n^0 = 1$ in the rest frame of the charge Ze. Energy conservation takes the form

$$p_1^0 + k_1^0 - p_2^0 = -(p_1 + k_1 - p_2)n = 0; \qquad (3\text{-}14.59)$$

it is used in verifying that the effective photon emission source is conserved, for this is the algebraic property

$$k_1^\mu V_{\mu\nu} n^\nu = (p_2 - p_1 - k_1)n = 0. \qquad (3\text{-}14.60)$$

First let us observe the simplifications that appear for soft photons, where the photon momentum $\mathbf{k}$ is negligible compared to $\mathbf{p}_1 - \mathbf{p}_2$ and $p_1^0 \simeq p_2^0$. This gives

$$\langle 1_{p_1 q} 1_{k_1 \lambda_1} | T | 1_{p_2 q} \rangle$$

$$= (d\omega_{p_1}\, d\omega_{p_2})^{1/2} \left[-\frac{2Ze^2 qp_2^0}{(\mathbf{p}_1 - \mathbf{p}_2)^2} \right] (d\omega_{k_1})^{1/2} eqe_{k_1\lambda_1}^* \left[\frac{p_1}{k_1 p_1} - \frac{p_2}{k_1 p_2} \right], \quad (3\text{-}14.61)$$

which is the transition matrix element for scattering in the Coulomb field, multiplied by the probability amplitude for photon emission by the source that represents the instantaneous transition of the charge eq from velocity p_2/m to velocity p_1/m. This conforms with expectation. We should remark, however, for future reference, that the complete neglect of the photon mechanical properties at sufficiently low frequencies is justified for finite particle deflection angles, but requires more careful consideration when the deflection angle is very small.

Closely related to the soft photon situation, but distinct from it, is the low energy or nonrelativistic limit. Here, the photon momentum is negligible but any fraction of the initial kinetic energy can be radiated as the photon energy

$$k^0 = \frac{1}{2m}\mathbf{p}_2^2 - \frac{1}{2m}\mathbf{p}_1^2 = \frac{1}{2m}(\mathbf{p}_2 + \mathbf{p}_1)\cdot(\mathbf{p}_2 - \mathbf{p}_1) \ll |\mathbf{p}_1 - \mathbf{p}_2|. \quad (3\text{-}14.62)$$

Using the gauge $e^0_{k_1\lambda_1} = 0$, the transition matrix element simplifies to

$$\langle 1_{p_1q}1_{k_1\lambda_1}|T|1_{p_2q}\rangle = (d\omega_{p_1}\,d\omega_{p_2})^{1/2}\left[\frac{2Ze^2}{(\mathbf{p}_1 - \mathbf{p}_2)^2}\right](d\omega_{k_1})^{1/2}\frac{e}{k^0}\mathbf{e}^*_{k_1\lambda_1}\cdot(\mathbf{p}_1 - \mathbf{p}_2),$$
$$(3\text{-}14.63)$$

and the differential cross section for specified polarization, emission directions, and photon energy is (unnecessary labels are omitted)

$$d\sigma = d\Omega\,d\Omega_k\frac{dk^0}{k^0}\frac{|\mathbf{p}_1|}{|\mathbf{p}_2|}\frac{1}{\pi^2}\frac{Z^2\alpha^3}{[(\mathbf{p}_1 - \mathbf{p}_2)^2]^2}|\mathbf{e}\cdot(\mathbf{p}_1 - \mathbf{p}_2)|^2. \quad (3\text{-}14.64)$$

The successive operations of summing over polarizations and then integrating over photon emission directions reduce it to

$$d\sigma \rightarrow d\Omega\,d\Omega_k\frac{dk^0}{k^0}\frac{|\mathbf{p}_1|}{|\mathbf{p}_2|}\frac{1}{\pi^2}\frac{Z^2\alpha^3}{[(\mathbf{p}_1 - \mathbf{p}_2)^2]^2}\left[(\mathbf{p}_1 - \mathbf{p}_2)^2 - \left(\frac{\mathbf{k}}{k^0}\cdot(\mathbf{p}_1 - \mathbf{p}_2)\right)^2\right]$$
$$\rightarrow d\Omega\frac{dk^0}{k^0}\frac{|\mathbf{p}_1|}{|\mathbf{p}_2|}\frac{8}{3\pi}\frac{Z^2\alpha^3}{(\mathbf{p}_1 - \mathbf{p}_2)^2}, \quad (3\text{-}14.65)$$

and the further integration over all particle scattering angles gives a cross section for the photon energy distribution:

$$d\sigma = \frac{dk^0}{k^0}\frac{16}{3}\frac{Z^2\alpha^3}{|\mathbf{p}_2|^2}\log\frac{|\mathbf{p}_2| + |\mathbf{p}_1|}{|\mathbf{p}_2| - |\mathbf{p}_1|}. \quad (3\text{-}14.66)$$

It is also interesting to consider the differential cross per unit solid angle $d\Omega$ that is integrated over all photon energies, from the minimum detectable energy k^0_{min} to the maximum energy set by the initial kinetic energy $T = \mathbf{p}_2^2/2m$. This is

$$\frac{d\sigma}{d\Omega} = \frac{8}{3\pi}\frac{Z^2\alpha^3}{|\mathbf{p}_2|^2}\int_0^{(1-k^0_{min}/T)^{1/2}}\frac{2x\,dx}{1 - x^2}\frac{x}{1 - 2x\cos\theta + x^2}, \quad (3\text{-}14.67)$$

which uses the integration variable

$$x = \frac{|\mathbf{p}_1|}{|\mathbf{p}_2|} = \left(1 - \frac{k^0}{T}\right)^{1/2}. \quad (3\text{-}14.68)$$

The integral can be evaluated in general, most simply by factoring the denominator into $1 - x$, $1 + x$, $1 - xe^{i\theta}$, $1 - xe^{-i\theta}$, but we shall only present the result for the circumstance

$$k^0_{\min} \ll T, \tag{3-14.69}$$

where

$$\frac{d\sigma}{d\Omega} = \frac{2}{3\pi} \frac{Z^2 \alpha^3}{|\mathbf{p}_2|^2} \frac{1}{\sin^2 \tfrac{1}{2}\theta} \left[\log \frac{4T}{k^0_{\min}} - (\pi - \theta) \tan \tfrac{1}{2}\theta - \frac{\cos \theta}{\cos^2 \tfrac{1}{2}\theta} \log \frac{1}{\sin \tfrac{1}{2}\theta} \right]. \tag{3-14.70}$$

This is not valid for arbitrarily small angles, however. In contrast with the apparent singularity at $\theta = 0$, (3–14.67) there yields

$$\frac{d\sigma}{d\Omega} (\theta = 0) \simeq \frac{4}{3\pi} \frac{Z^2 \alpha^3}{|\mathbf{p}_2|^2} \frac{1}{(k^0_{\min}/2T)^2}. \tag{3-14.71}$$

The applicability of (3–14.70) requires that $\theta \gg k^0_{\min}/T$. Still another caveat must be mentioned. As in the discussion of nonradiative scattering, W_{22} is only the first of an infinite series of processes that contribute to the emission of a photon during deflection by the Coulomb field. But unlike elastic scattering in a Coulomb field, these additional processes do alter the cross section, particularly at low energies and large Z. We do not intend to go into this matter here, however.

Let us return to the transition matrix element (3–14.57) and note the following expression for a differential cross section that still refers to the detailed energy distribution:

$$d\sigma = d\omega_{p_1} d\omega_{k_1} 2\pi \, \delta(p_1^0 + k_1^0 - p_2^0) \frac{1}{2|\mathbf{p}_2|} \frac{Z^2 e^4}{[(\mathbf{p}_1 + \mathbf{k}_1 - \mathbf{p}_2)^2]^2} 4e^2 |e^{\mu *}_{k_1 \lambda_1} V_{\mu\nu} n^\nu|^2, \tag{3-14.72}$$

where the summation over the polarization of the emitted photon is given by

$$\sum_{\lambda_1} |e^\mu_{k_1\lambda_1} V_{\mu\nu} n^\nu|^2 = \frac{np_1}{k_1 p_1} \frac{np_2}{k_1 p_2} (p_1 + k_1 - p_2)^2 - m^2 \left(\frac{np_2}{k_1 p_1} - \frac{np_1}{k_1 p_2} \right)^2 - 1. \tag{3-14.73}$$

Here is another, invariant way to write the differential cross section:

$$|\mathbf{p}_2| \, d\sigma = \int d\omega_{p_1} d\omega_{k_1} \frac{(dk_2)}{(2\pi)^3} (2\pi)^4 \, \delta(p_1 + k_1 - p_2 - k_2) \frac{\delta(nk_2)}{(k_2^2)^2} 2Z^2 (4\pi\alpha)^3$$
$$\times |e^{\mu *}_{k_1\lambda_1} V_{\mu\nu} n^\nu|^2, \tag{3-14.74}$$

although we have not troubled to introduce the invariant equivalent of the initial particle momentum. The four-dimensional delta function states that

$$k_2 = p_1 + k_1 - p_2, \tag{3-14.75}$$

and, in the rest frame of n^μ, which is the coordinate system of physical interest,

$$\delta(nk_2) = \delta(p_1^0 + k_1^0 - p_2^0), \qquad k_2^2 = (\mathbf{p}_1 + \mathbf{k}_1 - \mathbf{p}_2)^2, \qquad (3\text{-}14.76)$$

which regains (3-14.72). But the expression (3-14.74) has a suggestive character, for processes resembling particle-photon scattering are being considered. Of course, the incident photons are virtual, since $k_2^2 > 0$. Nevertheless, this point of view has practical advantages at high energies. Viewed in a suitable coordinate system, a major fraction of the differential cross section can be evaluated in terms of the properties of real photons.

In the physical coordinate system, the incident particle is considered to move along the z-axis with velocity $v \simeq +1$, so that

$$p_2^0 = \gamma m, \qquad p_{2z} = \gamma m v, \qquad \gamma = (1 - v^2)^{-1/2} \gg 1. \qquad (3\text{-}14.77)$$

Now think of the coordinate system in which the particle is at rest initially and the charge Ze moves along the z-axis with velocity $-v$. In this frame the vector n^μ has the components $[(0, x, y, z)]$

$$n^\mu = \gamma(1, 0, 0, -v). \qquad (3\text{-}14.78)$$

The requirement $nk_2 = 0$, which asserts the static nature of the field in the physical, or Z-attached, coordinate system, becomes in the particle rest frame

$$k_2^0 + vk_z = 0, \qquad (3\text{-}14.79)$$

and therefore

$$k_2^2 = k_T^2 + k_2^{0^2}[(1 - v^2)/v^2], \qquad (3\text{-}14.80)$$

where

$$k_T^2 = k_{2x}^2 + k_{2y}^2. \qquad (3\text{-}14.81)$$

Thus, in circumstances for which

$$1 - v^2 = 1/\gamma^2 \ll 1, \qquad (3\text{-}14.82)$$

and k_T^2 is sufficiently small, it would seem that the virtual photons could be approximated by real ones.

There is one apparent difficulty, however. Playing the role of polarization vector for the incident photon is the vector n^μ, which is indeed such that $nk_2 = 0$. But we should expect that the polarization vector of a real photon is, or can be chosen, without time component or component along the propagation direction, which is here the negative z-axis. This suggests performing what should be a gauge transformation:

$$n^\mu \to n^\mu - (\gamma/k_2^0)k_2^\mu, \qquad (3\text{-}14.83)$$

which is constructed to have vanishing time component in the particle rest frame. The z or longitudinal component of the new vector is then

$$-\gamma v + (\gamma/v) = 1/(\gamma v) \ll 1 \qquad (3\text{-}14.84)$$

and, provided

$$\gamma k_T / k_2^0 \gg 1/\gamma, \tag{3-14.85}$$

the transformed vector will be predominantly the multiple $-\gamma k_T / k_2^0$ of the transverse unit vector $\mathbf{k}_T / k_T$, which acts as the incident photon polarization vector. But all this is contingent on the magnitude of the additional term introduced by the transformation (3-14.83), which is proportional to

$$e_{k_1\lambda_1}^{\mu *} V_{\mu\nu} k_2^\nu = e_{k_1\lambda_1}^* p_1 \frac{k_2 p_2}{k_1 p_1} - e_{k_1\lambda_1}^* p_2 \frac{k_2 p_1}{k_1 p_2} - e_{k_1\lambda_1}^* k_2. \tag{3-14.86}$$

Now,

$$k_2 p_2 = k_1 p_1 - \tfrac{1}{2} k_2^2, \qquad k_2 p_1 = k_1 p_2 + \tfrac{1}{2} k_2^2 \tag{3-14.87}$$

and therefore

$$e_{k_1\lambda_1}^{\mu *} V_{\mu\nu} k_2^\nu = -\left(\frac{e_{k_1\lambda_1}^* p_1}{k_1 p_1} + \frac{e_{k_1\lambda_1}^* p_2}{k_1 p_2} \right) \frac{1}{2} k_2^2, \tag{3-14.88}$$

which indicates that the substitution of real photons for the virtual photons will be justified if suitable upper limits are placed on $k_2^2 \simeq k_T^2$. A suggestion of the magnitude of this upper limit is obtained by comparing, in the gauge $e_{k_1\lambda_1}^* p_2 = 0$, the particle rest frame values

$$e_{k_1\lambda_1}^{\mu *} V_{\mu\nu} \left(n^\nu - \frac{\gamma}{k_2^0} k_2^\nu \right) = \left(e_{k_1\lambda_1}^* \cdot \frac{\mathbf{k}_T}{k_T} \right) \frac{\gamma k_T}{k_2^0} \tag{3-14.89}$$

and

$$e_{k_1\lambda_1}^{\mu *} V_{\mu\nu} k_2^\nu \frac{\gamma}{k_2^0} = -\frac{e_{k_1\lambda_1}^* p_1}{k_1 p_1} \frac{1}{2} k_T^2 \frac{\gamma}{k_2^0} \simeq -\frac{e_{k_1\lambda_1}^* \cdot \mathbf{k}_2}{k_2 p_2} \frac{1}{2} k_T^2 \frac{\gamma}{k_2^0}$$

$$= -e_{k_1\lambda_1 z}^* \frac{k_T}{2m} \frac{\gamma k_T}{k_2^0}, \tag{3-14.90}$$

namely

$$k_T \lesssim m. \tag{3-14.91}$$

We shall confine the discussion to the differential cross section that gives the energy spectrum of the emitted photons in the Z coordinate system. Since we are now firmly established in two different coordinate systems moving relative to each other at practically the speed of light, a few notational distinctions are needed. The Z frame photon energy will be denoted by

$$K = -nk_1 = \gamma(k_1^0 + vk_{1z}) \simeq \gamma k_1^0 (1 - \cos \theta), \tag{3-14.92}$$

where θ is the photon scattering angle in the particle frame. The kinematics of the photon scattering process in that reference frame, as derived from

$$0 = (p_2 + k_2 - k_1)^2 + m^2 = -2m(k_2^0 - k_1^0) + 2k_1^0 k_2^0 (1 - \cos \theta), \tag{3-14.93}$$

is expressed by

$$k_1^0 = \frac{k_2^0}{1 + (k_2^0/m)(1 - \cos\theta)} \cdot \qquad (3\text{-}14.94)$$

Some derived relations involving K and the Z frame particle energies

$$E_2 = \gamma m, \qquad E_1 = E_2 - K \qquad (3\text{-}14.95)$$

are

$$\frac{K}{E_2} = \frac{(k_2^0/m)(1 - \cos\theta)}{1 + (k_2^0/m)(1 - \cos\theta)}, \qquad \frac{E_1}{E_2} = \frac{k_1^0}{k_2^0}, \qquad (3\text{-}14.96)$$

and

$$\frac{K}{E_1} = \frac{k_2^0}{m}(1 - \cos\theta). \qquad (3\text{-}14.97)$$

The latter shows that the incident photon energies k_2^0 that can produce a scattered photon of energy K (two different coordinate systems are used here) will be restricted by

$$k_2^0 > \tfrac{1}{2}m\frac{K}{E_1} \cdot \qquad (3\text{-}14.98)$$

Also useful is the differential relation

$$\frac{E_2}{E_1}\frac{dK}{E_1} = \frac{k_2^0}{m}\sin\theta\, d\theta. \qquad (3\text{-}14.99)$$

Considered in the rest frame of the incident particle, the differential cross section for photon-particle scattering is

$$d\sigma = \int d\omega_{p_1}\, d\omega_{k_1}(2\pi)^4\, \delta(p_1 + k_1 - p_2 - k_2)\frac{(8\pi\alpha)^2}{4mk_2^0}\frac{1}{2}\sum_{\lambda_1\lambda_2}|\mathbf{e}_{k_1\lambda_1}^*\cdot\mathbf{e}_{k_2\lambda_2}|^2,$$

$$(3\text{-}14.100)$$

where the factor in the denominator arises from division by the photon flux multiplied by the particle density, $2k_2^0\, d\omega_{k_2}2m\, d\omega_{p_2}$. The polarization summation and average is

$$\tfrac{1}{2}\sum_{\lambda_1\lambda_2}|\mathbf{e}_{k_1\lambda_1}^*\cdot\mathbf{e}_{k_2\lambda_2}|^2 = \tfrac{1}{2}\sum_{\lambda_2}\left[1 - \left(\frac{\mathbf{k}_1}{k_1^0}\cdot\mathbf{e}_{k_2\lambda_2}\right)^2\right] = \tfrac{1}{2}(1 + \cos^2\theta), \quad (3\text{-}14.101)$$

which also follows from (3–12.124), with $n = p_2/m$. The final momentum integration can be performed with the aid of the kinematical relation

$$\frac{dk_1^0}{p_1^0 k_1^0} = \frac{d(p_1^0 + k_1^0)}{mk_2^0}, \qquad (3\text{-}14.102)$$

and

$$d\sigma = 2\pi \sin \theta \, d\theta \, \frac{\alpha^2}{m^2} \left(\frac{k_1^0}{k_2^0}\right)^2 \frac{1}{2} (1 + \cos^2 \theta), \qquad (3\text{-}14.103)$$

which could also be produced by transformation from (3–12.117), the center of mass expression for the differential cross section. The form in which we shall use this differential cross section is obtained from the relations of (3–14.96, 97, 99) as

$$[m k_2^0 \, d\sigma] = 2\pi \alpha^2 \, \frac{dK}{E_2} \frac{1}{2} \left[1 + \left(1 - \frac{mK}{E_1 k_2^0}\right)^2\right]. \qquad (3\text{-}14.104)$$

The real photon contribution to the differential cross section (3–14.74) is

$$|\mathbf{p}_2| \, d\sigma\Big]_{\text{real}} = 8\pi \alpha Z^2 \int \frac{(dk_2)}{(2\pi)^3} \frac{\delta(nk_2)}{(k_2^2)^2} \gamma^2 \frac{k_T^2}{(k_2^0)^2} [m k_2^0 \, d\sigma], \qquad (3\text{-}14.105)$$

or, with

$$(dk_2) = \pi \, dk_T^2 \, dk_{2z} \, dk_2^0, \qquad (3\text{-}14.106)$$

$$|\mathbf{p}_2| \, d\sigma\Big]_{\text{real}} = \frac{\alpha Z^2}{\pi} \int \frac{dk_2^0}{(k_2^0)^2} \frac{k_T^2 \, dk_T}{(k_T^2 + (k_2^0)^2/\gamma^2)^2} \frac{E_2}{m} [m k_2^0 \, d\sigma]. \qquad (3\text{-}14.107)$$

If we change the scale of the incident photon energy in the following manner,

$$k_2^0 = (mK/2E_1)x, \qquad (3\text{-}14.108)$$

so that x ranges from 1 to ∞, this reads

$$d\sigma\Big]_{\text{real}} = 4\frac{Z^2\alpha^3}{m^2} \frac{E_1}{E_2} \frac{dK}{K} \int_1^\infty \frac{dx}{x^2} \left(1 - \frac{2}{x} + \frac{2}{x^2}\right) \int_0^{k_{\max}^2} dk_T^2 \frac{k_T^2}{\left[k_T^2 + \left(\frac{m^2 K}{2E_1 E_2} x\right)^2\right]^2}.$$
$$(3\text{-}14.109)$$

According to the restriction (3–14.85), the k_T^2 integral should be stopped at a lower limit that is a fraction of $(k_2^0/\gamma)^2 = (m^2 Kx/2E_1 E_2)^2$, one that is large compared with $1/\gamma^2$, say $1/\gamma$, but a negligible error is introduced by extending the integral down to zero. The value of the integral, under the conditions

$$\frac{k_{\max}}{m} \gg \frac{mK}{2E_1 E_2}, \qquad (3\text{-}14.110)$$

is

$$\int_0^{k_{\max}^2} dk_T^2 \frac{k_T^2}{\left[k_T^2 + \left(\frac{m^2 K}{2E_1 E_2} x\right)^2\right]^2} = \left(2 \log\left(\frac{2E_1 E_2}{mK}\right) - 1\right) + 2 \log\left(\frac{k_{\max}}{mx}\right)$$
$$(3\text{-}14.111)$$

and then

$$d\sigma\Big]_{\text{real}} = \frac{16}{3}\frac{Z^2\alpha^3}{m^2}\frac{E_1}{E_2}\frac{dK}{K}\left[\log\left(\frac{2E_1E_2}{mK}\right) - \frac{1}{2} + \left(\log\frac{k_{\text{max}}}{m} - \frac{13}{12}\right)\right].$$

$$(3\text{–}14.112)$$

The suggestion implicit in (3–14.91), $k_{\text{max}} \lesssim m$, is that no significant interactions occur for larger values of the transverse momentum. We propose to examine this question.

A quick indication of the quantitatively correct result is obtained if one merely accepts that the effective replacement for k_{max} is independent of K. Then it suffices to compare (3–14.112) with the differential cross section appropriate to soft photons. This discussion takes place entirely in the Z-attached physical coordinate system. The polarization summation in the differential cross section derived from (3–14.61) is

$$\sum_{\lambda_1}\left|e^*_{k_1\lambda_1}\left(\frac{p_1}{k_1p_1} - \frac{p_2}{k_1p_2}\right)\right|^2 = -\frac{m^2}{(k_1p_1)^2} - \frac{m^2}{(k_1p_2)^2} - \frac{2p_1p_2}{k_1p_1\,k_1p_2}.$$

$$(3\text{–}14.113)$$

This expression contains the only reference to the direction of the emitted photon, and we shall integrate over all solid angles. Removing the factor $1/K^2$, that integral is

$$\int d\Omega_k\left[-\frac{m^2}{(E_1 - \mathbf{n}\cdot\mathbf{p}_1)^2} - \frac{m^2}{(E_2 - \mathbf{n}\cdot\mathbf{p}_2)^2} + 2\,\frac{E_1E_2 - |\mathbf{p}_1|\,|\mathbf{p}_2|\cos\theta}{(E_1 - \mathbf{n}\cdot\mathbf{p}_1)(E_2 - \mathbf{n}\cdot\mathbf{p}_2)}\right],$$

$$(3\text{–}14.114)$$

where $\mathbf{n}$ is now the unit propagation vector of the photon, and high energy, soft photon simplifications have not yet been introduced. We first observe that

$$\int\frac{d\Omega_k}{(E - \mathbf{n}\cdot\mathbf{p})^2} = \int_{-1}^{1}\frac{2\pi\,d\mu}{(E - |\mathbf{p}|\mu)^2} = \frac{4\pi}{m^2}. \qquad (3\text{–}14.115)$$

To integrate the term containing two denominators, it is useful to write $(E_1 \simeq E_2 = E)$

$$\frac{1}{E - \mathbf{p}_1\cdot\mathbf{n}}\frac{1}{E - \mathbf{p}_2\cdot\mathbf{n}} = \frac{1}{\mathbf{p}_1\cdot\mathbf{n} - \mathbf{p}_2\cdot\mathbf{n}}\left(\frac{1}{E - \mathbf{p}_1\cdot\mathbf{n}} - \frac{1}{E - \mathbf{p}_2\cdot\mathbf{n}}\right)$$

$$= \int_{-1}^{1}\tfrac{1}{2}\,dv\,\frac{1}{\left(E - \mathbf{p}_1\cdot\mathbf{n}\dfrac{1+v}{2} - \mathbf{p}_2\cdot\mathbf{n}\dfrac{1-v}{2}\right)^2},$$

$$(3\text{–}14.116)$$

as one can verify immediately. Then we have

$$\int \frac{d\Omega_k}{\left[E - \left(\mathbf{p}_1 \frac{1+v}{2} + \mathbf{p}_2 \frac{1-v}{2}\right)\cdot\mathbf{n}\right]^2} = \frac{4\pi}{E^2 - \left(\mathbf{p}_1 \frac{1+v}{2} + \mathbf{p}_2 \frac{1-v}{2}\right)^2}$$

$$= \frac{4\pi}{m^2 + |\mathbf{p}|^2 \sin^2 \tfrac{1}{2}\theta (1 - v^2)}.$$

$$(3\text{-}14.117)$$

The v integration can also be performed, of course, but it is preferable to leave it as it stands.

The soft photon differential cross section is

$$d\sigma = \frac{Z^2\alpha^3}{E^2} \frac{\sin\theta\, d\theta}{\sin^4 \tfrac{1}{2}\theta} \frac{dK}{K}$$

$$\times \left[\left(1 + 2\frac{E^2}{m^2}\sin^2 \tfrac{1}{2}\theta\right)\int_0^1 dv \frac{1}{1 + \frac{E^2}{m^2}\sin^2 \tfrac{1}{2}\theta(1 - v^2)} - 1\right],$$

$$(3\text{-}14.118)$$

which still needs to be integrated over the deflection angle θ. But now we must recall the warning that the soft photon simplifications need to be qualified for very small angles. In contrast to the singularity of (3-14.118) at $\theta = 0$, the minimum value attained by $(\mathbf{p}_1 + \mathbf{k}_1 - \mathbf{p}_2)^2$, which occurs for scattering and emission in the forward direction, is

$$\left[E_1 - \frac{m^2}{2E_1} + K - \left(E_2 - \frac{m^2}{2E_2}\right)\right]^2 = \left(\frac{m^2 K}{2E_1 E_2}\right)^2.$$

$$(3\text{-}14.119)$$

It would be more accurate to replace $(\mathbf{p}_1 - \mathbf{p}_2)^2$ by

$$(\mathbf{p}_1 - \mathbf{p}_2)^2 \to 4E^2 \sin^2 \tfrac{1}{2}\theta + \left(\frac{m^2 K}{2E_1 E_2}\right)^2,$$

$$(3\text{-}14.120)$$

and we recognize a characteristic aspect of (3-14.109). It is unnecessary to incorporate these refinements, however. They are the content of the real photon computation. The soft photon evaluation need only be applied at angles such that

$$2E \sin \tfrac{1}{2}\theta \gg \frac{m^2 K}{2E_1 E_2},$$

$$(3\text{-}14.121)$$

where (3-14.118) can be used without correction. Employing the variable

$$y = (E/m)\sin \tfrac{1}{2}\theta,$$

$$(3\text{-}14.122)$$

we begin the integration at a conservative upper limit to the real photon discussion, $k_{max} \ll m$, and thus

$$y_{min} = (k_{max}/2m) \ll 1. \tag{3-14.123}$$

This gives

$$d\sigma \Big]_{virtual} = 4 \frac{Z^2\alpha^3}{m^2} \frac{dK}{K} \int_{y_{min}}^{\infty} \frac{dy}{y^3} \left[(1 + 2y^2) \int_0^1 dv \frac{1}{1 + y^2(1 - v^2)} - 1 \right],$$

$$\tag{3-14.124}$$

where the y integral simplifies to

$$\int_{y_{min}}^{\infty} \frac{dy}{y} \int_0^1 dv \frac{1 + v^2}{1 + y^2(1 - v^2)} = \frac{4}{3} \log \frac{2m}{k_{max}} + \frac{1}{2} \int_0^1 dv(1 + v^2) \log \frac{1}{1 - v^2}$$

$$= \frac{4}{3} \left(\log \frac{m}{k_{max}} + \frac{13}{12} \right) \tag{3-14.125}$$

and

$$d\sigma \Big]_{virtual} = \frac{16}{3} \frac{Z^2\alpha^3}{m^2} \frac{dK}{K} \left(\log \frac{m}{k_{max}} + \frac{13}{12} \right). \tag{3-14.126}$$

When the real photon contribution (3-14.112) is considered under soft photon conditions, $E_1 \simeq E_2$, the addition of the two parts precisely cancels $\log (m/k_{max}) + \frac{13}{12}$, and the inference is that, generally,

$$d\sigma = \frac{16}{3} \frac{Z^2\alpha^3}{m^2} \frac{E_1}{E_2} \frac{dK}{K} \left[\log \left(\frac{2E_1 E_2}{mK} \right) - \frac{1}{2} \right]. \tag{3-14.127}$$

The inference is correct, as we shall verify by repeating the virtual photon calculation without using soft photon simplifications.

But Harold interrupts.

H. : I have not forgotten that you decided to omit all reference to historical matters, but your use of a parametric device for combining denominators prompts me to ask about a small historical point. The technique of introducing parameters to unite denominator products into a single denominator is invariably ascribed to Feynman in the literature. Is it not true, however, that the usual intent of that device, to replace space-time integrations by invariant parametric integrals, was earlier exploited by you in a related exponential version, and that the elementary identity combining two denominators, (3-14.116) in fact, appears quite explicitly in a paper of yours, published in the same issue that contains Feynman's contribution?

S. : Yes.

In the physical coordinate system, at high energy, radiation processes occur predominantly near the forward or longitudinal direction. We express this through a decomposition into longitudinal and transverse components, as illustrated by the invariant combinations

$$-k_1p_2 = KE_2 - \left(K - \frac{k_{1T}^2}{2K}\right)\left(E_2 - \frac{m^2}{2E_2}\right),$$

$$-k_1p_1 = KE_1 - \left(K - \frac{k_{1T}^2}{2K}\right)\left(E_1 - \frac{m^2 + p_{1T}^2}{2E_1}\right) - \mathbf{k}_{1T} \cdot \mathbf{p}_{1T},$$

(3–14.128)

which are written in terms of the transverse momentum supplied by the stationary charge

$$\mathbf{k}_{2T} = \mathbf{k}_{1T} + \mathbf{p}_{1T},$$ (3–14.129)

as

$$-k_1p_2 = (E_2/2K)D_2, \qquad -k_1p_1 = (E_2/E_1)(E_2/2K)D_1, \quad (3\text{–}14.130)$$

where

$$D_2 = (\mathbf{p}_{1T} - \mathbf{k}_{2T})^2 + (mK/E_2)^2, \qquad D_1 = \left(\mathbf{p}_{1T} - \frac{E_1}{E_2}\mathbf{k}_{2T}\right)^2 + (mK/E_2)^2.$$

(3–14.131)

The virtual photon contribution to the differential cross section that is produced by (3–14.72, 73) is

$$d\sigma\bigg]_{\text{virtual}} = \frac{Z^2\alpha^3}{\pi^2}\frac{E_1}{E_2}\frac{dK}{K}\left(\frac{K}{E_2}\right)^2\int\frac{(d\mathbf{k}_{2T})}{(k_{2T}^2)^2}(d\mathbf{p}_{1T})\left[-\frac{m^2}{D_1^2}-\frac{m^2}{D_2^2}+\frac{k_T^2+2m^2}{D_1D_2}\right].$$

(3–14.132)

Using appropriate variable translations, we have

$$\int\frac{(d\mathbf{p}_{1T})}{D_1^2}=\int\frac{(d\mathbf{p}_{1T})}{D_2^2}=\int_0^\infty\frac{\pi\,dp_T^2}{[p_T^2+(mK/E_2)^2]^2}=\pi\left(\frac{E_2}{mK}\right)^2 \quad (3\text{–}14.133)$$

and, combining denominators with the aid of the v parameter,

$$\int\frac{(d\mathbf{p}_{1T})}{D_1D_2}=\int_{-1}^1\tfrac12\,dv\int_0^\infty\frac{\pi\,dp_T^2}{[p_T^2+(K/E_2)^2(m^2+\frac14 k_T^2(1-v^2))]^2}$$

$$=\pi\left(\frac{E_2}{mK}\right)^2\int_0^1 dv\,\frac{1}{1+(k_T^2/4m^2)(1-v^2)}. \quad (3\text{–}14.134)$$

The introduction of the variable

$$y = k_T/2m$$ (3–14.135)

then gives

$$d\sigma\bigg]_{\text{virtual}} = 4\frac{Z^2\alpha^3}{m^2}\frac{E_1}{E_2}\frac{dK}{K}\int_{v_{\min}}^\infty\frac{dy}{y^3}\left[(1+2y^2)\int_0^1 dv\,\frac{1}{1+y^2(1-v^2)}-1\right],$$

(3–14.136)

which does indeed differ from (3–14.124) only by the factor E_1/E_2 that is needed to combine properly with the general real photon contribution (3–14.112) and produce (3–14.127).

To give an analogous discussion for spin $\frac{1}{2}$ particles requires, first, the explicit form of the electron-photon differential cross section in the rest frame of the incident electron. That is available to us through transformation of (3–13.117), the center of mass cross section, but there is some interest in a direct derivation. The transition matrix element is

$$\langle 1_{k_1\lambda_1} 1_{p_1\sigma_1 q} | T | 1_{k_2\lambda_2} 1_{p_2\sigma_2 q} \rangle = 2m (d\omega_{p_1}\, d\omega_{k_1}\, d\omega_{p_2}\, d\omega_{k_2})^{1/2} e^2$$

$$\times u_1^* \gamma^0 \left[\gamma \cdot e_1^* \frac{1}{\gamma(p_2 + k_2) + m} \gamma \cdot e_2 + \gamma \cdot e_2 \frac{1}{\gamma(p_2 - k_1) + m} \gamma \cdot e_1^* \right] u_2,$$

$$(3\text{–}14.137)$$

which uses purely spatial polarization vectors and a simplified notation. The matrix factor in square brackets reduces to

$$\gamma \cdot e_1^* \frac{m + \gamma^0(m + k_2^0) - \gamma \cdot k_2}{-2mk_2^0} \gamma \cdot e_2 + \gamma \cdot e_2 \frac{m + \gamma^0(m - k_1^0) + \gamma \cdot k_1}{2mk_1^0} \gamma \cdot e_1^*$$

$$\rightarrow \frac{1}{2m} \{\gamma \cdot e_1^*, \gamma \cdot e_2\} - \frac{i}{2m} [\gamma \cdot e_1^* \sigma \cdot n_2 \times e_2 + \gamma \cdot e_2 \sigma \cdot n_1 \times e_1^*], \quad (3\text{–}14.138)$$

where the latter exploits the fact that u_2 is an eigenvector of γ^0 with eigenvalue $+1$, and introduces the notation $n_{1,2}$ for the unit propagation vectors of the photons. Considering real polarization vectors for simplicity, the transition matrix element becomes

$$(d\omega_{p_1}\, d\omega_{k_1}\, d\omega_{p_2}\, d\omega_{k_2})^{1/2} 2e^2 u_1^* [-e_1 \cdot e_2 + \tfrac{1}{2}\gamma_5 (\sigma \cdot e_1 \sigma \cdot n_2 \times e_2$$

$$+ \sigma \cdot e_2 \sigma \cdot n_1 \times e_1)] u_2. \quad (3\text{–}14.139)$$

If the γ_5 term and the spinors are omitted we get the corresponding spin 0 expression. The summation of the transition probability over final spins can be performed with the aid of (3–13.110), giving the spinor factor

$$u_2^* [-e_1 \cdot e_2 - \tfrac{1}{2}\gamma_5 (\sigma \cdot n_2 \times e_2 \sigma \cdot e_1 + \sigma \cdot n_1 \times e_1 \sigma \cdot e_2)] \frac{m\gamma^0 + p_1^0 + i\gamma_5\sigma \cdot p_1}{2m}$$

$$\times [-e_1 \cdot e_2 + \tfrac{1}{2}\gamma_5 (\sigma \cdot e_1 \sigma \cdot n_2 \times e_2 + \sigma \cdot e_2 \sigma \cdot n_1 \times e_1)] u_2, \quad (3\text{–}14.140)$$

where the kinematics of the collision determine

$$p_1 = k_2 - k_1, \qquad p_1^0 = m + k_2^0 - k_1^0. \quad (3\text{–}14.141)$$

The matrix product is reduced by omitting all terms that contain a γ_5 factor, since u_2 is a γ^0 eigenvector, or a σ factor, the latter expressing the averaging over all initial spins. A quite short calculation then gives the following for

(3–14.140):

$$(e_1 \cdot e_2)^2 + \frac{k_2^0 - k_1^0}{4m}$$
$$\times [1 - \cos\theta(e_1 \cdot e_2)^2 + e_1 \cdot e_2 n_1 \cdot e_2 n_2 \cdot e_1 - e_1 \times e_2 \cdot n_1 e_1 \times e_2 \cdot n_2],$$

$$(3–14.142)$$

where

$$n_1 \cdot n_2 = \cos\theta. \qquad (3–14.143)$$

The apparent dependence of the second term on the polarization vectors disappears on invoking the identity

$$[(e_1 \times e_2) \times n_1] \cdot [(e_1 \times e_2) \times n_2]$$
$$= (e_1 \times e_2)^2 \cos\theta - e_1 \times e_2 \cdot n_1 e_1 \times e_2 \cdot n_2$$
$$= -e_1 \cdot e_2 n_1 \cdot e_2 n_2 \cdot e_1, \qquad (3–14.144)$$

and we get

$$(e_1 \cdot e_2)^2 + \frac{k_2^0 - k_1^0}{4m}(1 - \cos\theta) = (e_1 \cdot e_2)^2 + \frac{1}{4}\left(\frac{k_1^0}{k_2^0} + \frac{k_2^0}{k_1^0} - 2\right),$$

$$(3–14.145)$$

replacing $(e_1 \cdot e_2)^2$ in the spin 0 cross section. When summed and averaged over photon polarizations the differential cross section that appears in place of (3–14.103) is

$$d\sigma = 2\pi \sin\theta \, d\theta \, \frac{\alpha^2}{m^2}\left(\frac{k_1^0}{k_2^0}\right)^2 \frac{1}{2}\left[\frac{k_1^0}{k_2^0} + \frac{k_2^0}{k_1^0} - 1 + \cos^2\theta\right]. \quad (3–14.146)$$

For the purpose of evaluating the photon emission cross section, this is used as

$$[mk_2^0 \, d\sigma] = 2\pi\alpha^2 \frac{dK}{E_2} \frac{1}{2}\left[\frac{E_1}{E_2} + \frac{E_2}{E_1} - 1 + \left(1 - \frac{mK}{E_1 k_2^0}\right)^2\right], \quad (3–14.147)$$

according to the second relation of (3–14.96) and the spin 0 result (3–14.104). The corresponding modification of (3–14.109, 111) is

$$d\sigma\bigg]_{\text{real}} = 4\frac{Z^2\alpha^3}{m^2}\frac{E_1}{E_2}\frac{dK}{K}\int_1^\infty \frac{dx}{x^2}\left[\frac{1}{2}\left(\frac{E_1}{E_2} + \frac{E_2}{E_1}\right) - \frac{2}{x} + \frac{2}{x^2}\right]$$
$$\times \left[2\log\left(\frac{2E_1E_2}{mK}\right) - 1 + 2\log\left(\frac{k_{\max}}{mx}\right)\right]$$
$$= 4\frac{Z^2\alpha^3}{m^2}\frac{E_1}{E_2}\frac{dK}{K}\left[\left(\frac{E_1}{E_2} + \frac{E_2}{E_1} - \frac{2}{3}\right)\left(\log\frac{2E_1E_2}{mK} - \frac{1}{2}\right)\right.$$
$$\left. - \left(\frac{E_1}{E_2} + \frac{E_2}{E_1} - \frac{2}{3}\right)\left(\log\frac{m}{k_{\max}} + 1\right) - \frac{1}{9}\right]. \quad (3–14.148)$$

The virtual photon contribution inferred from W_{22} has the following high energy appearance:

$$d\sigma\bigg]_{\text{virtual}} = \frac{Z^2\alpha^3}{\pi^2} \frac{dK}{K} \frac{m^2}{E_1 E_2} \int \frac{(d\mathbf{k}_{2T})(d\mathbf{p}_{1T})}{(k_{2T}^2)^2} |u_1^* \gamma^0 M u_2|^2, \quad (3\text{-}14.149)$$

with

$$M = \boldsymbol{\gamma} \cdot \mathbf{e}_1^* \frac{m - \gamma(p_1 + k_1)}{2k_1 p_1} \gamma^0 - \gamma^0 \frac{m - \gamma(p_2 - k_1)}{2k_1 p_2} \boldsymbol{\gamma} \cdot \mathbf{e}_1^*, \quad (3\text{-}14.150)$$

which is to be summed over the photon polarization and the final electron spin, and averaged over the initial spin. As in the discussion of photon-electron scattering at small angles, transitions with electron helicity changes are significant. The calculation can be performed advantageously by methods that have already been illustrated, using photon helicity states and the photon emission direction for reference, and expressing the electron helicity states with the aid of suitable rotation matrices. We shall only give the results here, which are classified with respect to helicity change:

$$\frac{1}{2}\sum_{\lambda_1\sigma_1\sigma_2} |u_1^* \gamma^0 M u_2|^2_{\text{no}} = 2\frac{E_1}{E_2} \frac{E_1^2 + E_2^2}{m^2} \left(\frac{K}{E_2}\right)^2 \left[-\frac{m^2}{D_1^2} - \frac{m^2}{D_2^2} + \frac{2m^2 + k_T^2}{D_1 D_2}\right],$$

$$\frac{1}{2}\sum_{\lambda_1\sigma_1\sigma_2} |u_1^* \gamma^0 M u_2|^2_{\text{yes}} = 2\frac{E_1}{E_2} \frac{(E_1 - E_2)^2}{m^2} \left(\frac{K}{E_2}\right)^2 \left[\frac{m}{D_1} - \frac{m}{D_2}\right]^2. \tag{3-14.151}$$

For soft photons, helicity changes are relatively negligible and the spin 0 structure is reproduced. The slightly different integral associated with helicity changes is

$$\int_{\nu_{\min}}^{\infty} \frac{dy}{y^3}\left[1 - \int_0^1 dv \frac{1}{1 + y^2(1 - v^2)}\right]$$

$$= \int_0^1 dv \int_{\nu_{\min}}^{\infty} \frac{dy}{y} \frac{1 - v^2}{1 + y^2(1 - v^2)}$$

$$= \frac{2}{3}\log\frac{1}{y_{\min}} - \frac{1}{2}\int_0^1 dv(1 - v^2)\log(1 - v^2)$$

$$= \frac{2}{3}\left(\log\frac{m}{k_{\max}} + 1\right) - \frac{1}{9}. \tag{3-14.152}$$

The individual cross sections are

$$d\sigma\bigg]_{\text{virtual, no}} = 2\frac{Z^2\alpha^3}{m^2} \frac{E_1}{E_2} \frac{dK}{K} \left(\frac{E_1}{E_2} + \frac{E_2}{E_1}\right) \frac{4}{3}\left(\log\frac{m}{k_{\max}} + \frac{13}{12}\right),$$

$$\tag{3-14.153}$$

$$d\sigma\bigg]_{\text{virtual, yes}} = 2\frac{Z^2\alpha^3}{m^2} \frac{E_1}{E_2} \frac{dK}{K} \left(\frac{E_1}{E_2} + \frac{E_2}{E_1} - 2\right)\left[\frac{2}{3}\left(\log\frac{m}{k_{\max}} + 1\right) - \frac{1}{9}\right],$$

which add to

$$d\sigma\Big]_{virtual} = 4\,\frac{Z^2\alpha^3}{m^2}\,\frac{E_1}{E_2}\,\frac{dK}{K}\left[\left(\frac{E_1}{E_2}+\frac{E_2}{E_1}-\frac{2}{3}\right)\left(\log\frac{m}{k_{max}}+1\right)+\frac{1}{9}\right].$$
$$(3\text{-}14.154)$$

This virtual photon part and the real photon contribution of (3-14.148) combine to give the final high energy form of the differential cross section for photon emission by an electron deflected in a Coulomb field,

$$d\sigma = 4\,\frac{Z^2\alpha^3}{m^2}\,\frac{E_1}{E_2}\,\frac{dK}{K}\left(\frac{E_1}{E_2}+\frac{E_2}{E_1}-\frac{2}{3}\right)\left(\log\frac{2E_1E_2}{mK}-\frac{1}{2}\right).\quad(3\text{-}14.155)$$

The process that converts a photon into a pair of oppositely charged particles, in the neighborhood of a stationary charge, is related by crossing transformations to the reaction just considered. These transformations are

$$k_1 \to -k_2, \qquad p_2 \to -p_1'. \qquad (3\text{-}14.156)$$

In order to reconstruct the absolute squared transition matrix element, we take the known differential cross section for photon emission, $d\sigma_{emiss.}$, referring to definite spins and polarizations, and form

$$d\sigma_{emiss.}2|\mathbf{p}_2|\,d\omega_{p_2} = d\sigma_{emiss.}E_2^2\,dE_2 4\pi/(2\pi)^3, \qquad (3\text{-}14.157)$$

where the second high energy version also emphasizes that we are interested only in the energy specification of the particles. Under the crossing transformation, the kinematical factor $d\omega_{p_1}\,d\omega_{k_1}\,d\omega_{p_2}$ becomes $d\omega_{p_1}\,d\omega_{p_1'}\,d\omega_{k_2}$. The differential cross section referring to an incident photon beam would require division by $2K\,d\omega_{k_2}$, $K = k_2^0$, and thus

$$d\sigma_{emiss.}E_2^2\,dE_2 \to d\sigma_{absorp.}K^2\,dK, \qquad (3\text{-}14.158)$$

apart from the spurious minus sign that accompanies this formal substitution, with spin $\frac{1}{2}$ particles. When cross sections involving summations over final helicities and averages over initial ones are used, appropriate corrections must be made for the different weight factors. With spin $\frac{1}{2}$ particles this is not required, in these reactions involving one initial particle and two final particles, since both electron and photon have two helicity states. For spin 0 particles, however, the photon emission cross section, summed over the photon polarization, contains an additional factor of 2 relative to the photon absorption cross section, where photon polarizations are averaged. The implied pair production cross sections are

$$\text{spin 0:}\quad d\sigma = \frac{8}{3}\,\frac{Z^2\alpha^3}{m^2}\,\frac{EE'}{K^2}\,\frac{dE'}{K}\left(\log\frac{2EE'}{mK}-\frac{1}{2}\right)$$

$$(3\text{-}14.159)$$

$$\text{spin }\tfrac{1}{2}\text{:}\quad d\sigma = 4\,\frac{Z^2\alpha^3}{m^2}\,\frac{E^2+E'^2+2/3EE'}{K^2}\,\frac{dE'}{K}\left(\log\frac{2EE'}{mK}-\frac{1}{2}\right),$$

where unnecessary causal labels are omitted, together with charge indices, since the partitioning of the photon energy,

$$K = E + E',\qquad (3\text{-}14.160)$$

does not depend upon the specific charge assignments.

3–15 H-PARTICLES

In using extended photon sources to represent heavy charged particles, the limitations of the interaction skeleton description have been transcended, to the point that new kinds of particles are encountered. They are idealized versions of composite systems. Since hydrogenic atoms are the most familiar example, they will be termed H-particles.

We first consider a static source distribution $J^\mu(\mathbf{x})$, and the associated vector potential $A^\mu(\mathbf{x})$ in some convenient gauge. The time translational invariance of Green's functions, Δ_+ for example, is conveyed by

$$\Delta_+(x, x') = \int_{-\infty}^{\infty} \frac{dp^0}{2\pi}\, e^{-ip^0(x^0 - x^{0\prime})}\Delta_+(\mathbf{x}, \mathbf{x}', p^0),\qquad (3\text{-}15.1)$$

where

$$[-(p^0 - eqA^0(\mathbf{x}))^2 + (-i\nabla - eq\mathbf{A}(\mathbf{x}))^2 + m^2 - i\epsilon]\Delta_+(\mathbf{x}, \mathbf{x}', p^0) = \delta(\mathbf{x} - \mathbf{x}').$$
$$(3\text{-}15.2)$$

The B. E. symmetry of this zero spin Green's function appears as

$$\Delta_+(\mathbf{x}', \mathbf{x}, -p^0)^T = \Delta_+(\mathbf{x}, \mathbf{x}', p^0),\qquad (3\text{-}15.3)$$

in which matrix transposition is applied to the charge indices. Eigenfunctions, solutions of the homogeneous Green's function equation, assist in constructing the Green's function. They are labeled by energy and charge values, supplemented by other quantum numbers which are usually related to angular momentum. The homogeneous equation and its complex conjugate are

$$[-(p^{0\prime} - eq'A^0(\mathbf{x}))^2 + (-i\nabla - eq'\mathbf{A}(\mathbf{x}))^2 + m^2]\phi_{p^{0\prime}q'a'}(\mathbf{x}) = 0,$$
$$[-(p^{0\prime} - eq'A^0(\mathbf{x}))^2 + (i\nabla - eq'\mathbf{A}(\mathbf{x}))^2 + m^2]\phi_{p^{0\prime}q'a'}(\mathbf{x})^* = 0.\qquad (3\text{-}15.4)$$

Notice that the joint sign reversal of $p^{0\prime}$ and q' interchanges the forms of the two differential operators. Hence the eigenfunctions can be so chosen that

$$\phi_{p^{0\prime}q'a'}(\mathbf{x})^* = \phi_{-p^{0\prime}\,-q'\,a'^*}(\mathbf{x}),\qquad (3\text{-}15.5)$$

where a'^* indicates a related set of quantum numbers. If a' includes a magnetic quantum number, for example, a'^* refers to the negative of that quantum number. It is possible to choose the eigenfunctions in a way that identifies a'^* with a'. While not usually convenient for individual problems that choice simplifies general discussions. To avoid confusion, we shall also understand that $p^{0\prime}$ is a positive quantity unless there is a specific indication otherwise.

Another consequence of the equation pair (3–15.4) is the formal integral relation

$$(p^{0\prime} - p^{0\prime\prime}) \int (d\mathbf{x}) \phi_{p^{0\prime}q^\prime a^\prime}(\mathbf{x})^* (p^{0\prime} + p^{0\prime\prime} - 2eq^\prime A^0(\mathbf{x})) \phi_{p^{0\prime\prime}q^\prime a^{\prime\prime}}(\mathbf{x}) = 0,$$

$$(3\text{–}15.6)$$

which is incorporated in the statement of orthonormality:

$$\int (d\mathbf{x}) \phi_{p^{0\prime}q^\prime a^\prime}(\mathbf{x})^* (p^{0\prime} + p^{0\prime\prime} - 2eq^\prime A^0(\mathbf{x})) \phi_{p^{0\prime\prime}q^{\prime\prime}a^{\prime\prime}}(\mathbf{x}) = \delta_{p^{0\prime}q^\prime a^\prime, p^{0\prime\prime}q^{\prime\prime}a^{\prime\prime}}.$$

$$(3\text{–}15.7)$$

In the absence of the static source, this property is obeyed by the known eigenfunctions associated with small momentum cells,

$$\phi_{pq}(\mathbf{x}) = \phi_{p^0 qp}(\mathbf{x}) e^{-ip^0 x^0}, \qquad \phi_{p^0 qp}(\mathbf{x}) = (d\omega_p)^{1/2} \varphi_q e^{i\mathbf{p}\cdot\mathbf{x}}, \qquad (3\text{–}15.8)$$

according to the orthonormality of the φ_q and the clarification of the spatial normalization given in Eqs. (3–6.24, 25). Here the momentum vector **p** plays the role of the quantum numbers a, which also specify the energy value. There is an analogue of (3–15.6) in which the homogeneous equation obeyed by $\phi_{p^{0\prime\prime}q^\prime a^{\prime\prime}}(\mathbf{x})$ is replaced by the inhomogeneous Green's function equation,

$$(p^{0\prime} - p^0) \int (d\mathbf{x}) \phi_{p^{0\prime}q^\prime a^\prime}(\mathbf{x})^* (p^{0\prime} + p^0 - 2eq^\prime A^0(\mathbf{x})) \Delta_+(\mathbf{x}, \mathbf{x}^\prime, p^0)$$

$$= \phi_{p^{0\prime}q^\prime a^\prime}(\mathbf{x}^\prime)^*. \quad (3\text{–}15.9)$$

Sufficiently near a particular energy eigenvalue $p^{0\prime}$, supposed to be isolated, however slightly, from all the others, the Green's function is dominated by the corresponding eigenfunctions, and (3–15.9) implies that

$$p^0 \sim p^{0\prime}:$$

$$\Delta_+(\mathbf{x}, \mathbf{x}^\prime, p^0) \sim \sum_{q^\prime a^\prime} \frac{\phi_{p^{0\prime}q^\prime a^\prime}(\mathbf{x}) \phi_{p^{0\prime}q^\prime a^\prime}(\mathbf{x}^\prime)^*}{p^{0\prime} - p^0}. \qquad (3\text{–}15.10)$$

The analogous behavior in the neighborhood of $-p^{0\prime}$, demanded by (3–15.3), is

$$p^0 \sim -p^{0\prime}:$$

$$\Delta_+(\mathbf{x}, \mathbf{x}^\prime, p^0) \sim \sum_{q^\prime a^\prime} \frac{\phi_{p^{0\prime}q^\prime a^\prime}(\mathbf{x})^* \phi_{p^{0\prime}q^\prime a^\prime}(\mathbf{x}^\prime)}{p^{0\prime} + p^0}. \qquad (3\text{–}15.11)$$

A representation of the Green's function that is valid near any part of the physical energy spectrum, or its negative, is given by

$$\Delta_+(\mathbf{x}, \mathbf{x}^\prime, p^0) = \sum_{p^{0\prime}q^\prime a^\prime} \left[\frac{\phi_{p^{0\prime}q^\prime a^\prime}(\mathbf{x}) \phi_{p^{0\prime}q^\prime a^\prime}(\mathbf{x}^\prime)^*}{p^{0\prime} - p^0 - i\epsilon} + \frac{\phi_{p^{0\prime}q^\prime a^\prime}(\mathbf{x})^* \phi_{p^{0\prime}q^\prime a^\prime}(\mathbf{x}^\prime)}{p^{0\prime} + p^0 - i\epsilon} \right],$$

$$(3\text{–}15.12)$$

in which we have also exhibited the appropriate use of the parameter $\epsilon \to +0$ in order to satisfy the time boundary condition—positive (negative) frequencies for positive (negative) time differences. This is verified directly:

$$\int_{-\infty}^{\infty} \frac{dp^0}{2\pi} \frac{e^{-ip^0(x^0-x^{0\prime})}}{p^{0\prime} - p^0 - i\epsilon} = \begin{cases} x^0 > x^{0\prime}: & i\Delta_{p0\prime}^{(+)}(x^0 - x^{0\prime}), \\ x^0 < x^{0\prime}: & 0; \end{cases}$$

$$\int_{-\infty}^{\infty} \frac{dp^0}{2\pi} \frac{e^{-ip^0(x^0-x^{0\prime})}}{p^{0\prime} + p^0 - i\epsilon} = \begin{cases} x^0 > x^{0\prime}: & 0, \\ x^0 < x^{0\prime}: & i\Delta_{p0\prime}^{(-)}(x^0 - x^{0\prime}), \end{cases}$$

(3-15.13)

where

$$\Delta_{p0\prime}^{(+)}(x^0 - x^{0\prime}) = \Delta_{p0\prime}^{(-)}(x^{0\prime} - x^0) = e^{-ip^{0\prime}(x^0-x^{0\prime})}. \qquad (3\text{-}15.14)$$

When the eigenfunctions (3-15.8) are inserted in (3-15.12), the known form of the free particle propagation function is recovered. If we now allow $p^{0\prime}$ to assume both positive and negative values, the Green's function can be presented more compactly as

$$\Delta_+(\mathbf{x}, \mathbf{x}', p^0) = \sum_{(\pm)p^{0\prime}q'a'} \phi_{p^{0\prime}q'a'}(\mathbf{x})\Delta_{p^{0\prime}}(p^0)\phi_{-p^{0\prime}-q'a'}(\mathbf{x}') \quad (3\text{-}15.15)$$

where $(\pm)$ signals the extended meaning of $p^{0\prime}$, and

$$\Delta_{p^{0\prime}}(p^0) = \frac{1}{|p^{0\prime}| - i\epsilon - \epsilon(p^{0\prime})p^0} = \Delta_{-p^{0\prime}}(-p^0). \qquad (3\text{-}15.16)$$

Not to be confused with the infinitesimal parameter ϵ is $\epsilon(p^{0\prime})$, stating the algebraic sign of $p^{0\prime}$. Another way of writing this function, in which the scale of $\epsilon \to +0$ has been changed, is

$$\Delta_{p^{0\prime}}(p^0) = \frac{\epsilon(p^{0\prime})}{p^{0\prime}(1 - i\epsilon) - p^0}. \qquad (3\text{-}15.17)$$

The time-dependent version of (3-15.15) is

$$\Delta_+(x, x') = \sum_{(\pm)p^{0\prime}q'a'} \phi_{p^{0\prime}q'a'}(\mathbf{x})\Delta_{p^{0\prime}}(x^0 - x^{0\prime})\phi_{-p^{0\prime}-q'a'}(\mathbf{x}'), \quad (3\text{-}15.18)$$

where

$$\Delta_{p^{0\prime}}(x^0 - x^{0\prime}) = \Delta_{-p^{0\prime}}(x^{0\prime} - x^0)$$
$$= \eta(x^0 - x^{0\prime})\eta(p^{0\prime})i\Delta_{p0\prime}^{(+)}(x^0 - x^{0\prime})$$
$$+ \eta(x^{0\prime} - x^0)\eta(-p^{0\prime})i\Delta_{-p0\prime}^{(-)}(x^0 - x^{0\prime}). \qquad (3\text{-}15.19)$$

We are interested, in this section, only in that portion of the energy spectrum which is inaccessible to a free particle: $|p^{0\prime}| < m$. Such states can exist, localized in the neighborhood of the source, if there is a force of attraction between the particle and source, of sufficient strength and range. In the familiar situation of the long-ranged Coulomb interaction between oppositely signed charges, no minimum strength is required, and an unlimited number of such bound states exists. These are the H-particles. What are the emission and absorption sources for H-particles?

The insertion of the Green's function (3–15.18) into the source coupling term

$$W = \tfrac{1}{2}\int (dx)(dx')K(x)\Delta_+(x, x')K(x') \qquad (3\text{–}15.20)$$

gives

$$W = \sum_{(\pm)p^{0\prime}q'a'} \tfrac{1}{2}\int dx^0\, dx^{0\prime} K_{-p^{0\prime} -q'a'}(x^0)\Delta_{p^{0\prime}}(x^0 - x^{0\prime})K_{p^{0\prime}q'a'}(x^{0\prime}). \qquad (3\text{–}15.21)$$

The time dependent quantities

$$K_{p^{0\prime}q'a'}(x^0) = \int (d\mathbf{x})\phi_{-p^{0\prime} -q'a'}(\mathbf{x})K(\mathbf{x}x^0)$$

$$= K_{-p^{0\prime} -q'a'}(x^0)^* \qquad (3\text{–}15.22)$$

are sources associated with the particular H-particle labeled by $p^{0\prime}$, q', with a' appearing as an additional index analogous to spin. That these sources refer only to time conveys the immobility of the very massive H-particles. The repeated operation of these sources will inject any number of particles into bound states. Since no account is being given of the interactions among the particles, we shall be concerned only with the properties of a single particle, bound to the source and forming an H-particle. Nevertheless, it is desirable to verify that probability requirements are satisfied in the dynamically simplified many-particle situation.

The usual consideration of a causal arrangement of emission and absorption sources leads to

$$\langle 0_+|0_-\rangle^K = \langle 0_+|0_-\rangle^{K_1} \exp\Big[\sum_{p^{0\prime}q'a'} iK^*_{1p^{0\prime}q'a'}iK_{2p^{0\prime}q'a'}\Big]\langle 0_+|0_-\rangle^{K_2}, \qquad (3\text{–}15.23)$$

where

$$K_{p^{0\prime}q'a'} = \int dx^0 e^{ip^{0\prime}x^0}K_{p^{0\prime}q'a'}(x^0) = K^*_{-p^{0\prime} -q'a'}, \qquad (3\text{–}15.24)$$

and the causal arrangement restricts the energy summation in (3–15.23) to the physical, positive values. The multiparticle states produced by the causal analysis of the vacuum amplitude have the usual construction in terms of source products, and probability normalization implies that

$$|\langle 0_+|0_-\rangle^K|^2 = \exp\Big[-\sum_{p^{0\prime}q'a'}|K_{p^{0\prime}q'a'}|^2\Big]. \qquad (3\text{–}15.25)$$

The direct verification of this property employs the relation

$$-i\Delta_{p^{0\prime}}(x^0 - x^{0\prime}) + (-i\Delta_{-p^{0\prime}}(x^0 - x^{0\prime}))^* = \Delta_{p^{0\prime}}^{(+)}(x^0 - x^{0\prime}). \qquad (3\text{–}15.26)$$

For spin $\tfrac{1}{2}$ particles, the transform Green's function introduced by writing

$$G_+(x, x') = \int_{-\infty}^{\infty} \frac{dp^0}{2\pi} e^{-ip^0(x^0 - x^{0\prime})}G_+(\mathbf{x}, \mathbf{x}', p^0) \qquad (3\text{–}15.27)$$

has the F. D. symmetry

$$[\gamma^0 G_+(\mathbf{x}', \mathbf{x}, -p^0)]^T = -\gamma^0 G_+(\mathbf{x}, \mathbf{x}', p^0) \tag{3-15.28}$$

and obeys

$$[-\gamma^0(p^0 - eqA^0(\mathbf{x})) + \boldsymbol{\gamma} \cdot (-i\boldsymbol{\nabla} - eq\mathbf{A}(\mathbf{x})) + m - i\epsilon]$$
$$\times G_+(\mathbf{x}, \mathbf{x}', p^0) = \delta(\mathbf{x} - \mathbf{x}'). \tag{3-15.29}$$

Eigenfunctions are defined by the homogeneous equations

$$[-\gamma^0(p^{0\prime} - eq'A^0(\mathbf{x})) + \boldsymbol{\gamma} \cdot (-i\boldsymbol{\nabla} - eq'\mathbf{A}(\mathbf{x})) + m]\psi_{p^{0\prime}q'a'}(\mathbf{x}) = 0,$$
$$\psi_{p^{0\prime}q'a'}(\mathbf{x})\,{}^*\gamma^0[-\gamma^0(p^{0\prime} - eq'A^0(\mathbf{x})) + \boldsymbol{\gamma} \cdot (-i\boldsymbol{\nabla}^T - eq'\mathbf{A}(\mathbf{x})) + m] = 0, \tag{3-15.30}$$

where

$$\chi(\mathbf{x})\boldsymbol{\nabla}^T = -\boldsymbol{\nabla}\chi(\mathbf{x}). \tag{3-15.31}$$

The two equations are related by the Hermitian character of the matrices $\gamma^0\gamma^\mu$. If, instead, attention is paid to the imaginary nature of the matrices γ^μ, we infer the eigenfunction connection (for suitable choices of a')

$$\psi_{p^{0\prime}q'a'}(\mathbf{x})^* = \psi_{-p^{0\prime}-q'a'}(\mathbf{x}). \tag{3-15.32}$$

The eigenfunctions are normalized in accordance with the integral property

$$\int (d\mathbf{x})\psi_{p^{0\prime}q'a'}(\mathbf{x})\,{}^*\psi_{p^{0\prime\prime}q''a''}(\mathbf{x}) = \delta_{p^{0\prime}q'a',\,p^{0\prime\prime}q''a''}. \tag{3-15.33}$$

Another integral relation,

$$(p^{0\prime} - p^0)\int (d\mathbf{x})\psi_{p^{0\prime}q'a'}(\mathbf{x})^* G_+(\mathbf{x}, \mathbf{x}', p^0) = \psi_{p^{0\prime}q'a'}(\mathbf{x}')^*\gamma^0, \tag{3-15.34}$$

combined with the symmetry requirement of (3–15.28), leads to the Green's function construction ($p^{0\prime} > 0$):

$$G_+(\mathbf{x}, \mathbf{x}', p^0) = \sum_{p^{0\prime}q'a'} \left[\frac{\psi_{p^{0\prime}q'a'}(\mathbf{x})\psi_{p^{0\prime}q'a'}(\mathbf{x}')^*\gamma^0}{p^{0\prime} - p^0 - i\epsilon} - \frac{\psi_{p^{0\prime}q'a'}(\mathbf{x})^*\psi_{p^{0\prime}q'a'}(\mathbf{x}')\gamma^0}{p^{0\prime} + p^0 - i\epsilon} \right]$$
$$\tag{3-15.35}$$

or, more compactly,

$$G_+(\mathbf{x}, \mathbf{x}', p^0) = \sum_{(\pm)p^{0\prime}q'a'} \psi_{p^{0\prime}q'a'}(\mathbf{x})G_{p^{0\prime}}(p^0)\psi_{-p^{0\prime}-q'a'}(\mathbf{x}')\gamma^0, \tag{3-15.36}$$

where

$$G_{p^{0\prime}}(p^0) = -G_{-p^{0\prime}}(-p^0) = \frac{1}{p^{0\prime}(1 - i\epsilon) - p^0}. \tag{3-15.37}$$

The explicit time dependence of the Green's function is given by

$$G_+(x, x') = \sum_{(\pm)p^{0\prime}q'a'} \psi_{p^{0\prime}q'a'}(\mathbf{x})G_{p^{0\prime}}(x^0 - x^{0\prime})\psi_{-p^{0\prime}-q'a'}(\mathbf{x}')\gamma^0, \tag{3-15.38}$$

in which

$$G_{p^0{}'}(x^0 - x^{0'}) = -G_{-p^0{}'}(x^{0'} - x^0)$$
$$= \eta(x^0 - x^{0'})\eta(p^{0'})i\Delta_{p^0{}'}^{(+)}(x^0 - x^{0'})$$
$$- \eta(x^{0'} - x^0)\eta(-p^{0'})i\Delta_{-p^0{}'}^{(-)}(x^0 - x^{0'}). \quad (3\text{-}15.39)$$

Time dependent H-particle sources are defined by

$$\eta_{p^0{}'q'a'}(x^0) = \int (d\mathbf{x})\psi_{p^0{}'q'a'}(\mathbf{x})^*\gamma^0\eta(\mathbf{x}x^0)$$
$$= -\eta_{-p^0{}' -q'a'}(x^0)^*, \quad (3\text{-}15.40)$$

and

$$W = \tfrac{1}{2}\int (dx)(dx')\eta(x)\gamma^0 G_+(x, x')\eta(x')$$
$$= \sum_{(\pm)p^0{}'q'a'} \tfrac{1}{2}\int dx^0\, dx^{0'}\eta_{p^0{}'q'a'}(x^0)^* G_{p^0{}'}(x^0 - x^{0'})\eta_{p^0{}'q'a'}(x^{0'}). \quad (3\text{-}15.41)$$

With a causal arrangement of sources, the vacuum amplitude becomes

$$\langle 0_+|0_-\rangle^\eta = \langle 0_+|0_-\rangle^{\eta_1} \exp\left[\sum_{p^0{}'q'a'} i\eta_1{}^*_{p^0{}'q'a'}i\eta_2{}_{p^0{}'q'a'}\right]\langle 0_+|0_-\rangle^{\eta_2}, \quad (3\text{-}15.42)$$

where we have introduced

$$\eta_{p^0{}'q'a'} = \int dx^0 e^{ip^{0'}x^0}\eta_{p^0{}'q'a'}(x^0)$$
$$= -\eta^*_{-p^0{}' -q'a'}. \quad (3\text{-}15.43)$$

The completeness of the multiparticle states, which have the usual source product representation, implies that $(p^{0'} > 0)$

$$|\langle 0_+|0_-\rangle^\eta|^2 = \exp\left[-\sum_{p^0{}'q'a'} \eta^*_{p^0{}'q'a'}\eta_{p^0{}'q'a'}\right]. \quad (3\text{-}15.44)$$

Direct calculation from (3–15.41), with the aid of the relation

$$-iG_{p^0{}'}(x^0 - x^{0'}) - (-iG_{-p^0{}'}(x^0 - x^{0'}))^* = \epsilon(p^{0'})\Delta_{p^0{}'}^{(+)}(x^0 - x^{0'}), \quad (3\text{-}15.45)$$

gives

$$|\langle 0_+|0_-\rangle^\eta|^2 = \exp\left[-\tfrac{1}{2}\sum_{(\pm)p^0{}'q'a'} \epsilon(p^{0'})\eta^*_{p^0{}'q'a'}\eta_{p^0{}'q'a'}\right]$$
$$= \exp\left[-\sum_{p^0{}'q'a'} \eta^*_{p^0{}'q'a'}\eta_{p^0{}'q'a'}\right], \quad (3\text{-}15.46)$$

since, in accordance with F. D. statistics,

$$-\eta^*_{-p^0{}' -q'a'}\eta_{-p^0{}' -q'a'} = -\eta_{p^0{}'q'a'}\eta^*_{p^0{}'q'a'}$$
$$= \eta^*_{p^0{}'q'a'}\eta_{p^0{}'q'a'}. \quad (3\text{-}15.47)$$

Now, let the static source that represents a heavy charged particle be supplemented by a simple photon source. The terms in W that contain one such photon source describe processes in which, through transitions between different H-particles, a single photon is emitted or absorbed. Using two photon sources, we describe transitions that result in two photons being emitted or absorbed, and also transitions in which a photon is scattered with, or without, an accompanying H-particle transition. And so on. It is convenient to use the characterization of electromagnetic processes in which all interactions refer directly to the charged particles, with the electromagnetic model of the particle source transformed into a gauge restriction on the vector potential. We recall the space-like choice [Eq. (3-10.49)]

$$if^{\mu}(k) = \frac{k^{\mu} + n^{\mu}nk}{k^2 + (nk)^2}.$$

(3-15.48)

The static source defines a coordinate system in which n^{μ} can be chosen to have only a time component. Then $f^{\mu}(k)$ has only spatial components, which are proportional to the vector $\mathbf{k}$, and the gauge condition reads:

$$\mathbf{k} \cdot \mathbf{A}(k) = 0; \qquad \mathbf{\nabla} \cdot \mathbf{A}(x) = 0.$$

(3-15.49)

This gauge is called the radiation gauge, since the property of transversality to the momentum vector is characteristic of the polarization vectors associated with photons. It has also, but less appropriately, been termed the Coulomb gauge. As is most evident from the three-dimensional form of the second-order Maxwell differential equations,

$$-\nabla^2 A^0(x) = J^0(x) + \partial_0 \mathbf{\nabla} \cdot \mathbf{A}(x),$$
$$-\partial^2 \mathbf{A}(x) + \mathbf{\nabla}\mathbf{\nabla} \cdot \mathbf{A}(x) = \mathbf{J}(x) - \mathbf{\nabla}\partial_0 A^0(x),$$

(3-15.50)

the scalar potential $A^0(x)$ in the radiation gauge is necessarily given by the instantaneous Coulomb potential of the charge distribution,

$\mathbf{\nabla} \cdot \mathbf{A}(x) = 0$:

$$A^0(x) = \int (dx') \mathfrak{D}(\mathbf{x} - \mathbf{x}') J^0(\mathbf{x}'x^0).$$

(3-15.51)

But the converse is not true. If it is required that $A^0(x)$ shall be the instantaneous Coulomb potential, presumably the intent of a Coulomb gauge, the inference is that the time derivative of $\mathbf{\nabla} \cdot \mathbf{A}(x)$ must vanish. No restriction is thereby placed on any static component of the vector potential, $\mathbf{A}(x)$.

It is the radiation gauge to which the static potential $A^{\mu}(\mathbf{x})$ refers. The vector potential $\mathbf{A}(\mathbf{x})$ can be used to represent the field of nuclear magnetic dipole moments, leading to the hyperfine structure of H-particles. In the following, however, attention will be confined to the static charge density and its scalar potential. This avoids notational conflicts with the potentials that are associated with the simple photon sources. The latter are only needed far from the emission or detection sources. There, they reduce to the vector potential $\mathbf{A}(x)$, which is related to its sources by

$$-\partial^2 \mathbf{A}(x) = \mathbf{J}_T(x), \qquad \mathbf{\nabla} \cdot \mathbf{A}(x) = \mathbf{\nabla} \cdot \mathbf{J}_T(x) = 0, \qquad (3\text{-}15.52)$$

where

$$\mathbf{J}_T(x) = \mathbf{J}(x) - \mathbf{\nabla}\partial_0 \int (d\mathbf{x}')\mathfrak{D}(\mathbf{x} - \mathbf{x}')J^0(\mathbf{x}'x^0) \qquad (3\text{-}15.53)$$

is the transverse or divergenceless part of $\mathbf{J}(x)$ and, indeed,

$$\mathbf{\nabla} \cdot \mathbf{J}_T(x) = \mathbf{\nabla} \cdot \mathbf{J}(x) + \partial_0 J^0(x) = 0. \qquad (3\text{-}15.54)$$

The solution of (3-15.52) in regions that are causally intermediate between emission and detection sources is, of course,

$$\mathbf{A}(x) = \sum_{k\lambda} [\mathbf{A}_{k\lambda}(x)iJ_{2k\lambda} + iJ^*_{1k\lambda}\mathbf{A}_{k\lambda}(x)^*], \qquad (3\text{-}15.55)$$

with

$$\mathbf{A}_{k\lambda}(x) = (d\omega_k)^{1/2}\mathbf{e}_{k\lambda}e^{ikx}, \qquad \mathbf{\nabla} \cdot \mathbf{A}_{k\lambda}(x) = 0. \qquad (3\text{-}15.56)$$

We now distinguish $\Delta_+(x, x')$, which contains the static scalar potential $A^0(\mathbf{x})$, from $\Delta^A_+(x, x')$. The latter also describes the effect of the vector potential $\mathbf{A}(x)$ that represents photons. The differential equation for the Green's function $\Delta^A_+(x, x')$ can be presented as $(\mathbf{p} = -i\mathbf{\nabla})$

$$[-(i\partial_0 - eqA^0(\mathbf{x}))^2 - \nabla^2 + m^2 - i\epsilon]\Delta^A_+(x, x')$$
$$= \delta(x - x') + (2eq\mathbf{p} \cdot \mathbf{A}(x) - e^2\mathbf{A}(x)^2)\Delta^A_+(x, x'). \qquad (3\text{-}15.57)$$

The use of the Green's function $\Delta_+(x, x')$ converts this into an integral equation,

$$\Delta^A_+(x, x') = \Delta_+(x, x') + \int (dx_1)\Delta_+(x, x_1)[2eq\mathbf{p}_1 \cdot \mathbf{A}(x_1) - e^2\mathbf{A}(x_1)^2]\Delta^A_+(x_1, x'),$$
$$(3\text{-}15.58)$$

which can be solved by an iteration procedure. All this is entirely analogous to the discussion of (3-12.27) and as there, the successive interaction terms, $W_{21}, W_{22}, \ldots$ are most compactly expressed with the aid of the particle field,

$$\phi(x) = \int (dx')\Delta_+(x, x')K(x'). \qquad (3\text{-}15.59)$$

Thus,

$$W_{21} = \tfrac{1}{2}\int (dx)\phi(x)2eq\mathbf{p} \cdot \mathbf{A}(x)\phi(x), \qquad (3\text{-}15.60)$$

$$W_{22} = \tfrac{1}{2}\int (dx)(dx')\phi(x)2eq\mathbf{p} \cdot \mathbf{A}(x)\Delta_+(x, x')2eq\mathbf{p}' \cdot \mathbf{A}(x')\phi(x')$$
$$- \tfrac{1}{2}\int (dx)\phi(x)e^2\mathbf{A}(x)^2\phi(x), \qquad (3\text{-}15.61)$$

and so forth.

The particle field $\phi(x)$ is related to H-particle sources by

$$\phi(x) = \sum_{(\pm)p^{0\prime}q'a'} \phi_{p^{0\prime}q'a'}(\mathbf{x})\int dx^{0\prime}\Delta_{p^{0\prime}}(x^0 - x^{0\prime})K_{p^{0\prime}q'a'}(x^{0\prime}). \qquad (3\text{-}15.62)$$

At a time that is causally intermediate between the actions of emission and absorption sources, this becomes ($p^{0'} > 0$)

$$\phi(\mathbf{x}, x^0) = \sum_{p^{0'}q'a'} [\phi_{p^{0'}q'a'}(\mathbf{x})e^{-ip^{0'}x^0}iK_{2p^{0'}q'a'} + iK^*_{1p^{0'}q'a'}\phi_{p^{0'}q'a'}(\mathbf{x})^* e^{ip^{0'}x^0}].$$

(3–15.63)

Accordingly, the transition matrix element for a process in which the H-particle labeled $p^{0''}a''$ transforms into the H-particle denoted by $p^{0'}a'$ (charge specifications are omitted since only one sign of charge, say q', will be bound to the static source), with the emission of a photon, $k\lambda$, is

$$\langle 1_{p^{0'}a'}1_{k\lambda}|T|1_{p^{0''}a''}\rangle = (d\omega_k)^{1/2}2eq'\int(dx)\phi_{p^{0'}a'}(\mathbf{x})^*\mathbf{p}\cdot\mathbf{e}^*_{k\lambda}e^{-i\mathbf{k}\cdot\mathbf{x}}\phi_{p^{0''}a''}(\mathbf{x}).$$

(3–15.64)

The transition probability per unit time is given by

$$2\pi\,\delta(p^{0'}+k^0-p^{0''})\frac{d\Omega(k^0)^2\,dk^0}{(2\pi)^3 2k^0}\,4\pi\alpha\left|2\int(dx)\phi^*_{p^{0'}a'}(\mathbf{x})\mathbf{p}\cdot\mathbf{e}^*_{k\lambda}e^{-i\mathbf{k}\cdot\mathbf{x}}\phi_{p^{0''}a''}(\mathbf{x})\right|^2.$$

(3–15.65)

Since energy conservation precisely specifies the photon energy,

$$p^{0''} > p^{0'}: \qquad\qquad k^0 = p^{0''} - p^{0'}, \qquad\qquad (3\text{–}15.66)$$

there is no need to refer to the energy distribution. We shall also carry out the polarization summation and the integration over all photon emission directions, using, for simplicity of illustration, the nonrelativistic situation. Here the photon momentum, but not its energy, is negligible, and the particle eigenfunctions are related to the conventionally normalized nonrelativistic wavefunctions $\psi_{E'a'}(\mathbf{x})$ by

$$p^{0'} = m + E': \qquad \phi_{p^{0'}a'}(\mathbf{x}) \simeq \frac{1}{(2m)^{1/2}}\,\psi_{E'a'}(\mathbf{x}), \qquad (3\text{–}15.67)$$

in virtue of the normalization condition (3–15.7). Using standard matrix element notation, this gives

$$2\int(dx)\phi_{p^{0'}a'}(\mathbf{x})^*\mathbf{p}\cdot\mathbf{e}^*_{k\lambda}e^{-i\mathbf{k}\cdot\mathbf{x}}\phi_{p^{0''}a''}(\mathbf{x}) \simeq \mathbf{e}^*_{k\lambda}\cdot\langle E'a'|\mathbf{p}/m|E''a''\rangle. \quad (3\text{–}15.68)$$

Since polarization vectors realize only two of the three orthonormal unit vectors, the polarization summation and integration over emission directions produce the following expression for the transition probability per unit time:

$$A_{p^{0'}a'\leftarrow p^{0''}a''} = \tfrac{2}{3}\alpha k^0|\langle E'a'|\mathbf{p}/m|E''a''\rangle|^2$$
$$= \tfrac{2}{3}\alpha(k^0)^3|\langle E'a'|\mathbf{x}|E''a''\rangle|^2, \qquad (3\text{–}15.69)$$

which also uses the matrix connection between the particle velocity and position vectors. This is the probability per unit time for spontaneous photon emission in a transition between the specified H-particle states and, as such, is Einstein's A-coefficient. A less specific A-coefficient that refers only to energy is summed over a' and averaged over a''.

When a photon is incident on an H-particle in the state $p^{0''}a''$, a transition to the state $p^{0'}a'$ can occur if

$$k^0 = p^{0'} - p^{0''}. \tag{3-15.70}$$

The transition matrix element is

$$\langle 1_{p^{0'}a'} | T | 1_{p^{0''}a''} 1_{k\lambda} \rangle = (d\omega_k)^{1/2} 2eq' \int (d\mathbf{x}) \phi_{p^{0'}a'}(\mathbf{x})^* \mathbf{p} \cdot \mathbf{e}_{k\lambda} e^{i\mathbf{k}\cdot\mathbf{x}} \phi_{p^{0''}a''}(\mathbf{x})$$

$$\simeq (d\omega_k)^{1/2} eq' \mathbf{e}_{k\lambda} \cdot \langle E'a' | \mathbf{p}/m | E''a'' \rangle, \tag{3-15.71}$$

which implies the transition probability per unit time:

$$2\pi\, \delta(E' - E'' - k^0)\, d\omega_k 4\pi\alpha(k^0)^2 |\mathbf{e} \cdot \langle E'a' | \mathbf{x} | E''a'' \rangle|^2. \tag{3-15.72}$$

After integration over the sharply selected photon energy, this can be expressed as the Einstein B-coefficient, which relates the transition probability per unit time to the photon energy density per unit angular frequency range: $k^0(2k^0\, d\omega_k/dk^0)$. Averaging over the incident photon polarization and direction of motion gives the nonrelativistic expression

$$B_{p^{0'}a' \leftarrow p^{0''}a''} = \frac{4\pi^2}{3}\, \alpha |\langle E'a' | \mathbf{x} | E''a'' \rangle|^2. \tag{3-15.73}$$

The simplicity of the ratio, for a given pair of H-particles,

$$A/B = (k^0)^3/\pi^2, \tag{3-15.74}$$

is a reminder that, apart from the kinematical factors involved in the definitions, the transition probabilities for single photon emission and absorption are interchanged by the photon crossing transformation, $k^\mu \to -k^\mu$. The emission and absorption rates are equal when the definitions refer to single photons of definite polarization. And, as we learned long ago in the simpler context of a probe source, if n photons of the appropriate frequency are present initially, the absorption rate is multiplied by n and the emission rate by $n + 1$. The latter represents the combination of stimulated and spontaneous emission processes.

The analogous discussion for spin $\frac{1}{2}$ particles begins with the Green's function differential equation

$$[-\gamma^0(i\partial_0 - eqA^0(\mathbf{x})) + \boldsymbol{\gamma} \cdot (1/i)\boldsymbol{\nabla} + m - i\epsilon] G_+^A(x, x')$$
$$= \delta(x - x') + eq\boldsymbol{\gamma} \cdot \mathbf{A}(x) G_+^A(x, x') \tag{3-15.75}$$

and the equivalent integral equation

$$G_+^A(x, x') = G_+(x, x') + \int (dx_1) G_+(x, x_1) eq\boldsymbol{\gamma} \cdot \mathbf{A}(x_1) G_+^A(x_1, x'). \tag{3-15.76}$$

The successive photon interaction terms are exhibited, with the aid of the particle field

$$\psi(x) = \int (dx')G_+(x, x')\eta(x'), \qquad (3\text{-}15.77)$$

as [compare Eq. (3-12.24)]

$$W_{21} = \tfrac{1}{2}\int (dx)\psi(x)\gamma^0 eq\gamma \cdot \mathbf{A}(x)\psi(x)$$

$$W_{22} = \tfrac{1}{2}\int (dx)(dx')\psi(x)\gamma^0 eq\gamma \cdot \mathbf{A}(x)G_+(x, x')\gamma \cdot \mathbf{A}(x')\psi(x'), \qquad (3\text{-}15.78)$$

and so on. The particle field can be expressed in terms of H-particle sources,

$$\psi(x) = \sum_{(\pm)p^{0'}q'a'} \psi_{p^{0'}q'a'}(\mathbf{x}) \int dx^{0'} G_{p^{0'}}(x^0 - x^{0'})\eta_{p^{0'}q'a'}(x^{0'}), \qquad (3\text{-}15.79)$$

and, at a time intermediate between the operations of emission and absorption sources, is given by

$$\psi(\mathbf{x}, x^0) = \sum_{p^{0'}q'a'} [\psi_{p^{0'}q'a'}(\mathbf{x})e^{-ip^{0'}x^0}i\eta_{2p^{0'}q'a'} + i\eta^*_{1p^{0'}q'a'}\psi_{p^{0'}q'a'}(\mathbf{x})^*e^{ip^{0'}x^0}]. \qquad (3\text{-}15.80)$$

The transition matrix element for single photon emission is

$$\langle 1_{p^{0'}a'}1_{k\lambda}|T|1_{p^{0''}a''}\rangle = (d\omega_k)^{1/2}eq'\int (dx)\psi_{p^{0'}a'}(\mathbf{x})^*\gamma^0\gamma \cdot \mathbf{e}^*_{k\lambda}e^{-i\mathbf{k}\cdot\mathbf{x}}\psi_{p^{0''}a''}(\mathbf{x}). \qquad (3\text{-}15.81)$$

One can combine the eigenfunction differential equations in the manner of (3-6.67) to produce the integral identity

$$0 = \int (dx)\psi_{p^{0'}a'}(\mathbf{x})^*\gamma^0 \left[\gamma \cdot \mathbf{e}^* e^{-i\mathbf{k}\cdot\mathbf{x}} - \left(\frac{\mathbf{p}}{m}\right) \cdot \mathbf{e}^* e^{-i\mathbf{k}\cdot\mathbf{x}} + \left(\frac{1}{2m}\right) i\boldsymbol{\sigma} \cdot \mathbf{k} \times \mathbf{e}^* e^{-i\mathbf{k}\cdot\mathbf{x}} \right.$$

$$\left. + \left(\frac{k^0}{2m}\right)\gamma^0\gamma \cdot \mathbf{e}^* e^{-i\mathbf{k}\cdot\mathbf{x}} \right]\psi_{p^{0''}a''}(\mathbf{x}). \qquad (3\text{-}15.82)$$

In a nonrelativistic limit, where the $\psi_{p^{0'}a'}(\mathbf{x})$ are approximate eigenvectors of γ^0, the last term, containing the matrix $\gamma^0\gamma = i\gamma_5\boldsymbol{\sigma}$, is relatively negligible. It would be inconsistent to retain the $\boldsymbol{\sigma} \cdot \mathbf{k} \times \mathbf{e}^*$ contribution while replacing $e^{-i\mathbf{k}\cdot\mathbf{x}}$ by unity, where it multiplies $\mathbf{p} \cdot \mathbf{e}^*$. The next term of an expansion is

$$-(\mathbf{p}/m) \cdot \mathbf{e}^*(-i\mathbf{k} \cdot \mathbf{x}) = \tfrac{1}{2}i[(\mathbf{p}/m) \cdot \mathbf{e}^*\mathbf{k} \cdot \mathbf{x} + \mathbf{x} \cdot \mathbf{e}^*\mathbf{k} \cdot (\mathbf{p}/m)]$$

$$+ (1/2m)i\mathbf{x} \times \mathbf{p} \cdot \mathbf{k} \times \mathbf{e}^*, \qquad (3\text{-}15.83)$$

and one recognizes the orbital contribution to the magnetic moment, which adds to the spin magnetic moment in the manner represented by $g = 2$. If we neglect this magnetic dipole radiation, and the related electric quadrupole radiation, which is the other term on the right-hand side of (3-15.83), there remains only the radiation of the electric dipole moment, $eq'\mathbf{x}$. This is radiation associated with accelerated charges, and is independent of spin. Indeed, with the simplifications implied by retaining only the first two terms of (3-15.82), and replacing spin $\frac{1}{2}$ eigenfunctions by nonrelativistic wave functions, in accordance with the normalization (3–15.33), we have

$$\int (d\mathbf{x})\psi_{p^{0\prime}a'}(\mathbf{x})^* \gamma^0 \boldsymbol{\gamma} \cdot \mathbf{e}^* e^{-i\mathbf{k}\cdot\mathbf{x}} \psi_{p^{0\prime\prime}a''}(\mathbf{x}) \simeq \mathbf{e}^* \cdot \langle E'a'|\mathbf{p}/m|E''a''\rangle. \quad (3\text{–}15.84)$$

This coincides with the corresponding spin 0 limit, (3–15.68). A similar consideration, related by the photon crossing transformation, applies to the absorption process.

We shall discuss W_{22} only in the context of photon scattering. Using the spin 0 structure (3–15.61), we insert the causal field decomposition

$$\phi = \phi_1 + \phi_2, \qquad \mathbf{A} = \mathbf{A}_1 + \mathbf{A}_2, \quad (3\text{–}15.85)$$

and isolate the terms of interest:

$$\begin{aligned}
W_{22} \to \int (dx)(dx')\phi_1(x)[2eq\mathbf{p} \cdot \mathbf{A}_1(x)\Delta_+(x, x')2eq\mathbf{p}' \cdot \mathbf{A}_2(x') \\
+ 2eq\mathbf{p} \cdot \mathbf{A}_2(x)\Delta_+(x, x')2eq\mathbf{p}' \cdot \mathbf{A}_1(x')]\phi_2(x') \\
- \int (dx)\phi_1(x)2e^2\mathbf{A}_1(x) \cdot \mathbf{A}_2(x)\phi_2(x).
\end{aligned} \quad (3\text{–}15.86)$$

The general transition matrix element is

$$\langle 1_{p^{0\prime}a'}1_{k_1\lambda_1}|T|1_{p^{0\prime\prime}a''}1_{k_2\lambda_2}\rangle = (d\omega_{k_1}\,d\omega_{k_2})^{1/2}2e^2\mathbf{e}^*_{k_1\lambda_1} \cdot \mathbf{V} \cdot \mathbf{e}_{k_2\lambda_2}, \quad (3\text{–}15.87)$$

where the dyadic $\mathbf{V}$ has the components

$$\begin{aligned}
V_{kl} = \int (d\mathbf{x})(d\mathbf{x}')\phi_{p^{0\prime}a'}(\mathbf{x})^*[p_k e^{-i\mathbf{k}_1\cdot\mathbf{x}}2\Delta_+(\mathbf{x}, \mathbf{x}', p^{0\prime}+k_1^0)p_l'e^{i\mathbf{k}_2\cdot\mathbf{x}'} \\
+ p_l e^{i\mathbf{k}_2\cdot\mathbf{x}}2\Delta_+(\mathbf{x}, \mathbf{x}', p^{0\prime}-k_2^0)p_k'e^{-i\mathbf{k}_1\cdot\mathbf{x}'}]\phi_{p^{0\prime\prime}a''}(\mathbf{x}') \\
- \int (d\mathbf{x})\phi_{p^{0\prime}a'}(\mathbf{x})^*\delta_{kl}e^{i(\mathbf{k}_2-\mathbf{k}_1)\cdot\mathbf{x}}\phi_{p^{0\prime\prime}a''}(\mathbf{x}).
\end{aligned} \quad (3\text{–}15.88)$$

The Δ_+ symbols are the transform Green's function $\Delta_+(\mathbf{x}, \mathbf{x}', p^0)$, with p^0 assigned one of the values

$$p^{0\prime} + k_1^0 = p^{0\prime\prime} + k_2^0, \qquad p^{0\prime} - k_2^0 = p^{0\prime\prime} - k_1^0. \quad (3\text{–}15.89)$$

For simplicity we consider only the nonrelativistic limit, where the photon momenta are neglected, as are the terms in $\Delta_+(\mathbf{x}, \mathbf{x}', p^0)$ having denominators $p^{0\prime} + p^0 \simeq 2m$, in contrast with the denominators $p^{0\prime} - p^0 = E' - E$. This gives (using a slightly simplified notation)

$$\begin{aligned}
2\mathbf{e}^*_1 \cdot \mathbf{V} \cdot \mathbf{e}_2 = \sum_{Ea} &\left[\frac{\langle E'a'|\mathbf{e}^*_1 \cdot \mathbf{p}/m|Ea\rangle\langle Ea|\mathbf{e}_2 \cdot \mathbf{p}/m|E''a''\rangle}{E - E'' - k_2^0} \right. \\
&\left. + \frac{\langle E'a'|\mathbf{e}_2 \cdot \mathbf{p}/m|Ea\rangle\langle Ea|\mathbf{e}^*_1 \cdot \mathbf{p}/m|E''a''\rangle}{E - E' + k_2^0} \right] \\
&- \frac{1}{m}\mathbf{e}^*_1 \cdot \mathbf{e}_2\delta_{E'a',E''a''}.
\end{aligned} \quad (3\text{–}15.90)$$

If we restrict our attention to elastic scattering ($E' = E''$, $k_1^0 = k_2^0 = k^0$), the result is a differential cross section for deflection of the photon into the solid angle $d\Omega$:

$$\frac{d\sigma}{d\Omega} = \alpha^2 \left| \sum_{Ea} \left[\frac{\langle E'a'|\mathbf{e}_1^* \cdot \mathbf{p}/m|Ea\rangle\langle Ea|\mathbf{e}_2 \cdot \mathbf{p}/m|E'a''\rangle}{E - E' - k^0} \right. \right.$$
$$\left. + \frac{\langle E'a'|\mathbf{e}_2 \cdot \mathbf{p}/m|Ea\rangle\langle Ea|\mathbf{e}_1^* \cdot \mathbf{p}/m|E'a''\rangle}{E - E' + k^0} \right]$$
$$\left. - \frac{1}{m} \mathbf{e}_1^* \cdot \mathbf{e}_2 \delta_{a'a''} \right|^2 , \tag{3-15.91}$$

from which less specific cross sections are obtained by summing over a' and averaging over a'', by summing over λ_1 and averaging over λ_2, and by integrating over all solid angles.

At photon energies that are large in comparison with H-particle binding energies, only the last term of (3–15.91) survives and, with $a' = a''$, one recognizes the Thomson cross section, which describes the scattering of (on a relativistic scale) low energy photons by a free particle of charge $\pm e$ and mass m. Another limiting connection with Thomson scattering should appear at very low frequencies, small in comparison with the energy intervals between different H-particles. When a photon of essentially zero frequency is scattered elastically by a particular H-particle, the dynamical connections with other H-particles are not in evidence, and the scattering should be described by the Thomson formula appropriate to the H-particle charge and mass. Since the latter has been idealized as infinite, the elastic scattering cross section should vanish as $k^0 \to 0$. This implies a set of relations, known as sum rules, which can be variously presented. The immediate form implied by (3–15.91) is

$$\frac{1}{m} \sum_{Ea} \frac{1}{E - E'} [\langle E'a'|p_k|Ea\rangle\langle Ea|p_l|E'a''\rangle + \langle E'a'|p_l|Ea\rangle\langle Ea|p_k|E'a''\rangle] = \delta_{kl}\delta_{a'a''}. \tag{3-15.92}$$

Here is another:

$$m \sum_{Ea} (E - E')[\langle E'a'|x_k|Ea\rangle\langle Ea|x_l|E'a''\rangle + \langle E'a'|x_l|Ea\rangle\langle Ea|x_k|E'a''\rangle] = \delta_{kl}\delta_{a'a''}, \tag{3-15.93}$$

and yet a third, intermediate form, obtained by replacing only one of the momentum matrix element factors by the corresponding coordinate matrix element,

$$\langle E'a'|p_k|Ea\rangle = -im(E - E')\langle E'a'|x_k|Ea\rangle,$$
$$\langle Ea|p_k|E'a''\rangle = im(E - E')\langle Ea|x_k|E'a''\rangle, \tag{3-15.94}$$

is

$$-i \sum_{Ea} [\langle E'a'|x_k|Ea\rangle\langle Ea|p_l|E'a''\rangle - \langle E'a'|p_l|Ea\rangle\langle Ea|x_k|E'a''\rangle] = \delta_{kl}\delta_{a'a''}. \tag{3-15.95}$$

The last version shows the mathematical origin of the sum rules; they are matrix elements of the commutation relation

$$(1/i)[x_k, p_l] = \delta_{kl}. \tag{3–15.96}$$

The elementary origin of the sum rules does not detract from their significance as conditions of consistency for the phenomenological particle description of composite systems. That is emphasized by removing the idealization of infinite mass to obtain the necessary result involving the charge $(Z - 1)e$ and mass M of the H-particle, viewed as a composite of the two particles with charge and mass assignments given by $-e$, m (electron) and Ze, $M - m$ (nucleus). The scattering amplitude that appears in (3–15.91) describes the processes in which the electron absorbs the incident photon and emits the scattered photon. To this will now be added the representation of the processes in which the nucleus alone performs these acts, and of those in which both particles are involved. Although we have not developed the relevant general formalism, the necessary modifications here are quite clear. The matrix product terms of (3–15.91) describe two successive interactions with the electric current, to which both particles now make contributions:

$$-e\left(\frac{\mathbf{p}}{m}\right) \to -e\left(\frac{\mathbf{p}_{\text{el.}}}{m}\right) + \frac{Ze\mathbf{p}_{\text{nucl.}}}{(M - m)}$$

$$= \left(-\frac{e}{m} - \frac{Ze}{M - m}\right)\mathbf{p}, \tag{3–15.97}$$

where

$$\mathbf{p} = \mathbf{p}_{\text{el.}} = -\mathbf{p}_{\text{nucl.}} \tag{3–15.98}$$

is the relative momentum in the center of mass frame. In addition, there is a contribution in which the scattering takes place in one act that is associated with an individual particle; it is extended by

$$-\frac{e^2}{m} \to -\frac{e^2}{m} - \frac{(Ze)^2}{M - m}. \tag{3–15.99}$$

In carrying out the reduction of the matrix product, the relation between relative momentum and relative velocity is now given by the reduced mass, $m(M - m)/M$. Removing the factor of e^2, we find that what replaces the amplitude of (3–15.91) for a realistic H-particle as $k^0 \to 0$ is, apart from the polarization vector product,

$$m\left(1 - \frac{m}{M}\right)\left(\frac{1}{m} + \frac{Z}{M - m}\right)^2 - \frac{1}{m} - \frac{Z^2}{M - m} = -\frac{(Z - 1)^2}{M}. \tag{3–15.100}$$

This is just what is demanded by the phenomenological H-particle description.

Let us introduce these realistic modifications in (3–15.91), while retaining an arbitrary value for k^0. The sum rule (3–15.92), together with another sum rule that expresses the null value of the commutator $[x_k, x_l]$, can be used to rewrite the cross section as

$$\frac{d\sigma}{d\Omega} = \alpha^2 \left| (k^0)^2 \sum_{Ea} \left[\frac{\langle E'a'|e_1^* \cdot d|Ea\rangle\langle Ea|e_2 \cdot d|E'a''\rangle}{E - E' - k^0} \right. \right.$$
$$\left. + \frac{\langle E'a'|e_2 \cdot d|Ea\rangle\langle Ea|e_1^* \cdot d|E'a''\rangle}{E - E' + k^0} \right]$$
$$\left. - \frac{(Z-1)^2}{M} e_1^* \cdot e_2 \delta_{a'a''} \right|^2. \qquad (3\text{-}15.101)$$

Here, $-ed$ is the internal electric dipole moment of the system in which all position vectors refer to the center of mass vector

$$\mathbf{R} = \frac{1}{M} [m\mathbf{x}_{el.} + (M - m)\mathbf{x}_{nucl.}]. \qquad (3\text{-}15.102)$$

Thus,

$$-ed = -e(\mathbf{x}_{el.} - \mathbf{R}) + Ze(\mathbf{x}_{nucl.} - \mathbf{R})$$
$$= -e\left[1 - \frac{m}{M} + Z\frac{m}{M}\right]\mathbf{x}, \qquad (3\text{-}15.103)$$

which relates d to the relative position vector

$$\mathbf{x} = \mathbf{x}_{el.} - \mathbf{x}_{nucl.}. \qquad (3\text{-}15.104)$$

In this version, the low frequency behavior characteristic of the H-particle is explicit, while the disclosure of the constituents at high frequencies is assured by the sum rules, which here produce the combination

$$-\frac{(Z-1)^2}{M} - \frac{M}{m(M-m)}\left[1 + (Z-1)\frac{m}{M}\right]^2 = -\frac{1}{m} - \frac{Z^2}{M-m}. \qquad (3\text{-}15.105)$$

It is the amplitudes for individual scattering by the two particles that are added, and not their cross sections, since this simplified treatment neglects the photon momentum and thereby assumes that the photon wavelength is large compared to the particle separation. That restriction is easily removed by inserting the relative phase factor and, with increasing frequency, the coherence between the two scattering amplitudes disappears.

It is also possible to derive (3-15.101) directly, by using a different gauge which is specifically adapted to the long wavelength regime. If the electric field of the photons is homogeneous over the interior of the H-particle and processes involving the magnetic field are negligible, a suitable choice of potentials is

$$A^0(\mathbf{x}, x^0) = -\mathbf{x} \cdot \mathbf{E}(\mathbf{R}, x^0), \quad A(\mathbf{x}, x^0) = 0, \qquad (3\text{-}15.106)$$

where

$$\mathbf{E}(\mathbf{R}, x^0) = \sum_{k\lambda} (d\omega_k)^{1/2} ik^0 [e_{k\lambda} \exp(i\mathbf{k} \cdot \mathbf{R} - ik^0 x^0) iJ_{2k\lambda}$$
$$- iJ_{1k\lambda}^* e_{k\lambda}^* \exp(-i\mathbf{k} \cdot \mathbf{R} + ik^0 x^0)]. \qquad (3\text{-}15.107)$$

The scalar potential couples to the charge density,

$$-\int(dx)j^0(x)A^0(x) = \mathbf{E}\cdot[-e\mathbf{d}+(Z-1)e\mathbf{R}]. \qquad (3\text{-}15.108)$$

Transitions between different H-particles are excited by the internal dipole moment term, and this contribution to photon scattering reproduces the summation term of (3-15.101). The external dipole moment $(Z-1)e\mathbf{R}$ affects only the motion of the given H-particle. The scattering amplitude that it provides is given by the diagonal matrix element of the operator

$$(Z-1)^2(k^0)^2\left[\mathbf{e}_1^*\cdot\mathbf{R}\,\frac{1}{(P^2/2M)-k^0}\,\mathbf{e}_2\cdot\mathbf{R}+\mathbf{e}_2\cdot\mathbf{R}\,\frac{1}{(P^2/2M)+k^0}\,\mathbf{e}_1^*\cdot\mathbf{R}\right]$$

$$=(Z-1)^2(k^0)^2\left[\mathbf{e}_1^*\cdot\mathbf{R}\mathbf{e}_2\cdot\mathbf{R}\,\frac{1}{(P^2/2M)-k^0}+\frac{1}{(P^2/2M)+k^0}\,\mathbf{e}_2\cdot\mathbf{R}\mathbf{e}_1^*\cdot\mathbf{R}\right]$$

$$+(Z-1)^2\frac{(k^0)^2}{M}\,i\left[\mathbf{e}_1^*\cdot\mathbf{R}\mathbf{e}_2\cdot\mathbf{P}\,\frac{1}{((P^2/2M)-k^0)^2}\right.$$

$$\left.-\frac{1}{((P^2/2M)+k^0)^2}\,\mathbf{e}_2\cdot\mathbf{P}\mathbf{e}_1^*\cdot\mathbf{R}\right]. \qquad (3\text{-}15.109)$$

The reference to the rest frame, the state of zero momentum, reduces this to

$$\frac{(Z-1)^2}{M}\,i[\mathbf{e}_1^*\cdot\mathbf{R},\mathbf{e}_2\cdot\mathbf{P}]=-\frac{(Z-1)^2}{M}\,\mathbf{e}_1^*\cdot\mathbf{e}_2, \qquad (3\text{-}15.110)$$

which completes the derivation of (3-15.101) since there is no joint effect of the different kinds of dipole moments.

The analogue of (3-15.86) for a spin $\frac{1}{2}$ particle is

$$W_{22}\to\int(dx)(dx')\psi_1(x)\gamma^0[eq\boldsymbol{\gamma}\cdot\mathbf{A}_1(x)G_+(x,x')eq\boldsymbol{\gamma}\cdot\mathbf{A}_2(x')$$

$$+eq\boldsymbol{\gamma}\cdot\mathbf{A}_2(x)G_+(x,x')eq\boldsymbol{\gamma}\cdot\mathbf{A}_1(x')]\psi_2(x'), \qquad (3\text{-}15.111)$$

and the dyadic that replaces (3-15.88) has the components

$$V_{kl}=\tfrac{1}{2}\int(dx)(dx')\psi_{p^{0\prime}a'}(x)^*\gamma^0[\gamma_k e^{-i\mathbf{k}_1\cdot\mathbf{x}}G_+(x,x',p^{0\prime}+k_1^0)\gamma_l e^{i\mathbf{k}_2\cdot\mathbf{x}'}$$

$$+\gamma_l e^{i\mathbf{k}_2\cdot\mathbf{x}}G_+(x,x',p^{0\prime}-k_2^0)\gamma_k e^{-i\mathbf{k}_1\cdot\mathbf{x}'}]\psi_{p^{0\prime\prime}a''}(x'). \qquad (3\text{-}15.112)$$

Again, we only consider the nonrelativistic limit. But this time the terms in the Green's function with denominators $p^{0\prime}+p^0\simeq2m$ cannot be neglected. We shall need the explicit statement of completeness for the eigenfunctions. It can be inferred by comparing the high energy limit of the Green's function $G_+(\mathbf{x},\mathbf{x}',p^0)$,

$$\underset{p^0\to\infty}{\text{Lim}}\,[-\gamma^0p^0G_+(\mathbf{x},\mathbf{x}',p^0)]=\delta(\mathbf{x}-\mathbf{x}'), \qquad (3\text{-}15.113)$$

with the construction (3-15.35), which gives

$$\sum_{p^{0\prime}q^{\prime}a^{\prime}} [\psi_{p^{0\prime}q^{\prime}a^{\prime}}(\mathbf{x})\psi_{p^{0\prime}q^{\prime}a^{\prime}}(\mathbf{x}^{\prime})^* + \psi_{p^{0\prime}q^{\prime}a^{\prime}}(\mathbf{x})^*\psi_{p^{0\prime}q^{\prime}a^{\prime}}(\mathbf{x}^{\prime})] = \delta(\mathbf{x} - \mathbf{x}^{\prime}). \quad (3\text{-}15.114)$$

We exploit this relation by writing the Green's function as

$$G_+(\mathbf{x}, \mathbf{x}^{\prime}, p^0) = \sum_{p^{0\prime}q^{\prime}a^{\prime}} \psi_{p^{0\prime}q^{\prime}a^{\prime}}(\mathbf{x})\psi_{p^{0\prime}q^{\prime}a^{\prime}}(\mathbf{x}^{\prime})^*\gamma^0$$

$$\times \left(\frac{1}{p^{0\prime} - p^0 - i\epsilon} + \frac{1}{p^{0\prime} + p^0 - i\epsilon}\right)$$

$$- \sum_{p^{0\prime}q^{\prime}a^{\prime}} (\psi_{p^{0\prime}q^{\prime}a^{\prime}}(\mathbf{x})\psi_{p^{0\prime}q^{\prime}a^{\prime}}(\mathbf{x}^{\prime})^* + \psi_{p^{0\prime}q^{\prime}a^{\prime}}(\mathbf{x})^*\psi_{p^{0\prime}q^{\prime}a^{\prime}}(\mathbf{x}^{\prime}))\gamma^0$$

$$\times \frac{1}{p^{0\prime} + p^0 - i\epsilon} \qquad\qquad (3\text{-}15.115)$$

and then introducing the nonrelativistic simplification $p^{0\prime} + p^0 \simeq 2m$. The correction to $(p^{0\prime} - p^0)^{-1}$ is negligible. But the last term of (3-15.115) becomes $-(2m)^{-1}\gamma^0 \delta(\mathbf{x} - \mathbf{x}^{\prime})$, and this supplies the following addition to $2V_{kl}$:

$$-\frac{1}{2m} \int (d\mathbf{x})\psi_{p^{0\prime}a^{\prime}}(\mathbf{x})^*\gamma^0(\gamma_k\gamma^0\gamma_l + \gamma_l\gamma^0\gamma_k)\psi_{p^{0\prime\prime}a^{\prime\prime}}(\mathbf{x}) \simeq -\frac{1}{m}\delta_{kl}\delta_{E^{\prime}a^{\prime},E^{\prime\prime}a^{\prime\prime}};$$

$$(3\text{-}15.116)$$

it is the Thomson term. Introducing the nonrelativistic equivalence of $\gamma^0\boldsymbol{\gamma}$ to $\mathbf{p}/m$, the complete structure of the spin 0 result [Eq. (3-15.90)] is realized, as one would expect. The situation is simpler when the gauge (3-15.106) is used. No matrices appear and the nonrelativistic reduction of the eigenfunctions to wave functions can be performed directly, with the justifiable neglect of the $1/2m$ term in the Green's function. The immediate result is (3-15.101) (without the $1/M$ term, of course, since we have been using the source description of the charge Ze).

We now have before us some simple physical situations in which the incompleteness of the skeletal description of photon interactions becomes evident. Spontaneous emission is described as proceeding at a constant rate, even though the initial H-particle supply would be exhausted after a sufficient lapse of time. Under conditions of exact 'resonance,' $k^0 + E^{\prime} = E$, the photon scattering cross section is predicted to be infinite, which is always unacceptable as the answer to a physical question. In the next section we shall identify the significant phenomena that are omitted in the skeletal description, and remedy these difficulties.

3-16 INSTABILITY AND MULTIPARTICLE EXCHANGE

Although the need to describe unstable particles as naturally as stable ones is one of the motivations in devising the theory of sources, the H-particle provides our first encounter with unstable particles. The distinction between stable and unstable particles is a matter of time scale. Within suitably restricted time intervals, the mechanism producing particle instability is ineffective and the stable particle description is applicable, provided, of course, that enough time is still available for the accurate determination of the characteristic particle properties. Otherwise, no single-particle description is meaningful. The H-particles supply examples of stable and unstable particles. The particle of minimum energy is stable. Those of greater energy are capable of emitting one or more photons, thereby transforming themselves eventually into the absolutely stable variety. The initial description of H-particles assumed their stability, and is applicable over a restricted time scale. The description is false for very long time intervals because it asserts that weak H-particle sources emit and absorb single H-particles that propagate unaltered between these acts. But, given enough time, an unstable H-particle will transform itself into another H-particle and a photon. These two particles are also capable of recombining to form a single H-particle. Thus, a description of the coupling between weak, causally arranged H-particle sources that does not refer to the real existence of two or more particles propagating between them is physically incomplete. It is the inclusion of such multiparticle exchanges between sources and the consideration of some of the physical consequences that will occupy us in this section.

The first task is the identification of effective sources for the emission and the absorption of an H-particle and a photon. This is analogous to the discussion of Section 3–11. The description of a noninteracting photon and H-particle is given by (using the spin 0 example)

$$\langle 0_+|0_-\rangle^{JK} = 1 + \cdots + i\int (d\xi)(d\xi')J_1^\mu(\xi)D_+(\xi - \xi')J_{2\mu}(\xi')$$

$$\times\; i\int (dx)(dx')K_1(x)\Delta_+(x, x')K_2(x') + \cdots$$

$$= 1 + \cdots + i\int (d\xi)A_1^\mu(\xi)J_{2\mu}(\xi)\int (dx)\phi_1(x)K_2(x') + \cdots .$$

$$(3\text{–}16.1)$$

Comparison with the vacuum amplitude term describing single photon emission, as contained in (3–15.60), gives

$$i\mathbf{J}_2(\xi)K_2(x)\Big|_{\text{eff.}} = \delta(x - \xi)2eq\mathbf{p}\phi_2(x), \qquad (3\text{–}16.2)$$

and the same form applies, with appropriate causal labels, to the absorption of a photon and an H-particle. Since this effective photon source is meant to be multiplied by a vector potential in the radiation gauge, its appearance is simplified in comparison with the structure of (3–11.15). On replacing $J_1^\mu(\xi)K_1(x)$ and $J_{2\mu}(\xi')K_2(x')$ in (3–16.1) by these effective combinations, we obtain a description of the causal coupling between H-particle sources that is mediated by the exchange of an H-particle and a photon, under physical conditions of noninteraction. But, to be consistent with the use of the radiation gauge, we must first change the tensor that couples the vector photon sources, in relation to the exchange of a particular photon,

$$g^{\mu\nu} \rightarrow \sum_\lambda e^\mu_{k\lambda} e^{\nu*}_{k\lambda}. \tag{3-16.3}$$

The polarization vector summation is a spatial dyadic,

$$\sum_\lambda \mathbf{e}_{k\lambda}\mathbf{e}^*_{k\lambda} = 1 - \frac{\mathbf{kk}}{(k^0)^2}, \tag{3-16.4}$$

which extracts the transverse parts of the multiplying currents. Thus, the coupling term in the vacuum amplitude is

$$\int (dx)(dx')\phi_1(x)2eq\mathbf{p}_T D_+(x-x')\Delta_+(x, x') \cdot 2eq\mathbf{p}'_T\phi_2(x'). \tag{3-16.5}$$

An understanding of the causal situation is required before the necessary control can be exercised. As we see in Eqs. (3-16.2) and (3-16.5), the distribution of the two-particle emission and absorption sources is characterized by the fields $\phi_2(x')$ and $\phi_1(x)$. Suppose the H-particle emission source supplies the correct energy to create a particle that is capable of spontaneous transition to other H-particles of lesser energy, with accompanying photon emission. This process occurs at a steady rate throughout the subsequent history of the particle; it has no effective localization in time. Therefore, in order to exert a causal, temporal control over the act of two-particle emission and the subsequent absorption, we must use the H-particle sources in the extended sense. Then the sources emit and absorb virtual H-particles, which cannot exist far from their sources and, consequently, are transmuted into or are produced by a real photon and a real H-particle near these sources. This is what gives us the ability to influence where (when) the acts take place. The quantitative equivalent of these remarks is contained in the expression for the field,

$$\phi(x) = \sum_{(\pm)p^{0'}q'a'} \phi_{p^{0'}q'a'}(\mathbf{x}) \int_{-\infty}^{\infty} \frac{dP^0}{2\pi} e^{-iP^0x^0} \frac{\epsilon(p^{0'})}{p^{0'}(1-i\epsilon) - P^0} K_{p^{0'}q'a'}(P^0), \tag{3-16.6}$$

which has no propagation characteristics if the sources $K_{p^{0'}q'a'}(P^0)$ vanish for $P^0 = p^{0'}$. As indicated in this formula, we shall use the symbol P^0 to denote the energy that is injected by an extended H-particle source, converted into a real H-particle and a real photon, and finally absorbed by an extended H-particle detection source.

Through the use of extended sources, then, we ensure that the fields ϕ_1 and ϕ_2 have supports that are causally related. This permits us to use the causal forms of the propagation functions in (3–16.5),

$$x^0 > x^{0'}:$$

$$D_+(x - x')\Delta_+(x, x')$$
$$= i^2 \int d\omega_k \sum_{+p^{0'}q'a'} e^{i\mathbf{k} \cdot \mathbf{x}} \phi_{p^{0'}q'a'}(\mathbf{x}) e^{-i(p^{0'} + k^{0'})(x^0 - x^{0'})} \phi_{p^{0'}q'a'}(\mathbf{x}')^* e^{-i\mathbf{k} \cdot \mathbf{x}'}, \quad (3\text{–}16.7)$$

where, as indicated, only positive values of $p^{0'}$ appear. The resulting form of the vacuum amplitude coupling term is

$$- \sum_{\substack{(\pm)p^{0'}a' \\ (\pm)p^{0''}a''}} \int_{-\infty}^{\infty} \frac{dP^0}{2\pi} K_{1p^{0'}a'}(P^0)^*$$

$$\times \frac{\epsilon(p^{0'})}{p^{0'} - P^0} \Gamma_{p^{0'}a', p^{0''}a''}(P^0) \frac{\epsilon(p^{0''})}{p^{0''} - P^0} K_{2p^{0''}a''}(P^0) \quad (3\text{–}16.8)$$

with

$$\Gamma_{p^{0'}a', p^{0''}a''}(P^0) = \Gamma_{p^{0''}a'', p^{0'}a'}(P^0)^*$$
$$= 2\pi \sum_{+p^0a} \int d\omega_k \, \delta(p^0 + k^0 - P^0)$$
$$\times \left[\int (d\mathbf{x}) \phi_{p^{0'}a'}(\mathbf{x})^* 2eq' \mathbf{p}_T e^{i\mathbf{k} \cdot \mathbf{x}} \phi_{p^0a}(\mathbf{x}) \right]$$
$$\cdot \left[\int (d\mathbf{x}') \phi_{p^0a}(\mathbf{x}')^* e^{-i\mathbf{k} \cdot \mathbf{x}'} 2eq' \mathbf{p}_T \phi_{p^{0''}a''}(\mathbf{x}') \right], \quad (3\text{–}16.9)$$

which are the elements of a positive Hermitian matrix.

The additional coupling between H-particle sources can be expressed as a modification of the propagation function $\Delta_{p^{0'}}(x^0 - x^{0'})$, which now becomes a matrix, defined generally by

$$W_2 = \sum_{\substack{(\pm)p^{0'}a' \\ (\pm)p^{0''}a''}} \tfrac{1}{2} \int dx^0 \, dx^{0'} K_{-p^{0'}a'}(x^0) \bar{\Delta}_{p^{0'}a', p^{0''}a''}(x^0 - x^{0'}) K_{p^{0''}a''}(x^{0'}),$$

$$(3\text{–}16.10)$$

where

$$\bar{\Delta}_{p^{0'}a', \, p^{0''}a''}(x^0 - x^{0'}) = \bar{\Delta}_{-p^{0''}a'', \, -p^{0'}a'}(x^{0'} - x^0). \quad (3\text{–}16.11)$$

The emission and absorption sources of the vacuum amplitude term (3–16.8) occur in the combination

$$-K_{1p^{0'}a'}(P^0)^* K_{2p^{0''}a''}(P^0)$$
$$= -\left[\int dx^0 K_{1p^{0'}a'}(x^0)^* e^{-iP^0x^0} \right]\left[\int dx^{0'} e^{iP^0x^{0'}} K_{2p^{0''}a''}(x^{0'}) \right]$$
$$= i \int dx^0 \, dx^{0'} K_{1\,-p^{0'}a'}(x^0)[i e^{-iP^0(x^0 - x^{0'})}] K_{2p^{0''}a''}(x^{0'}). \quad (3\text{–}16.12)$$

We recognize in the central factor the propagation function $\Delta_{P^0}(x^0 - x^{0'})$, evaluated under the causal restriction $x^0 > x^{0'}$. The space-time extrapolation of this structure is performed under the guidance of the symmetry property (3–16.11). It suffices to define $\Gamma_{p^{0\prime}a^\prime,\,p^{0\prime\prime}a^{\prime\prime}}(P^0)$ for negative values of P^0:

$$\Gamma_{-p^{0\prime\prime}a^{\prime\prime},-p^{0\prime}a^\prime}(-P^0) = \Gamma_{p^{0\prime}a^\prime,p^{0\prime\prime}a^{\prime\prime}}(P^0), \qquad (3\text{–}16.13)$$

and the symmetry property is then satisfied by

$$\bar{\Delta}_{p^{0\prime}a^\prime,p^{0\prime\prime}a^{\prime\prime}}(x^0 - x^{0\prime}) = \delta_{p^{0\prime}a^\prime,p^{0\prime\prime}a^{\prime\prime}}\Delta_{p^{0\prime}}(x^0 - x^{0\prime})$$

$$+ \int \frac{dP^0}{2\pi} \frac{\epsilon(p^{0\prime})}{p^{0\prime} - P^0}\,\Gamma_{p^{0\prime}a^\prime,p^{0\prime\prime}a^{\prime\prime}}(P^0)$$

$$\times \frac{\epsilon(p^{0\prime\prime})}{p^{0\prime\prime} - P^0}\,\Delta_{P^0}(x^0 - x^{0\prime}). \qquad (3\text{–}16.14)$$

The effective limitations on P^0 must also be removed if this propagation function is to be meaningful for arbitrary sources. The P^0 integral can be defined to simulate the initial consideration of extended sources, by excluding neighborhoods of the values $p^{0\prime}$ and $p^{0\prime\prime}$. If this is done symmetrically about these values and then the limit of arbitrarily small excluded intervals considered, with a special provision for $p^{0\prime} = p^{0\prime\prime}$,

$$\frac{1}{(p^{0\prime} - P^0)^2} = -\frac{d}{dp^{0\prime}} \frac{1}{p^{0\prime} - P^0}, \qquad (3\text{–}16.15)$$

the result is to use the principal value of the singular P^0 integral. In contrast to other recipes that assign complex values to singular integrals, this procedure has the satisfactory feature of preserving the essential association of complex numbers with the propagation function $\Delta_{P^0}(x^0 - x^{0\prime})$.

For a more explicit test of these extrapolations, we examine how the simple propagation function is modified, by choosing $p^{0\prime}a^\prime = p^{0\prime\prime}a^{\prime\prime}$, and considering $x^0 > x^{0\prime}$, $p^{0\prime} > 0$:

$$-i\bar{\Delta}_{p^{0\prime}a^\prime,p^{0\prime}a^\prime}(x^0 - x^{0\prime}) = e^{-ip^{0\prime}(x^0-x^{0\prime})} - \int \frac{dP^0}{2\pi}\,\Gamma_{p^{0\prime}a^\prime,p^{0\prime}a^\prime}(P^0)$$

$$\times \left(\frac{d}{dp^{0\prime}} \frac{1}{p^{0\prime} - P^0}\right) e^{-iP^0(x^0-x^{0\prime})}. \qquad (3\text{–}16.16)$$

The physically interesting regime begins after a time lapse of many periods, $p^{0\prime}(x^0 - x^{0\prime}) \gg 1$. Then the integral is dominated by the immediate neighborhood of the singularity at $P^0 = p^{0\prime}$, and one can introduce a simplification by replacing $\Gamma(P^0)$ with

$$\gamma_{p^{0\prime}a^\prime} = \Gamma_{p^{0\prime}a^\prime,p^{0\prime}a^\prime}(p^{0\prime}) \geq 0. \qquad (3\text{–}16.17)$$

The principal value integral is effectively computed as

$$P\int_{-\infty}^{\infty} \frac{dP^0}{2\pi} \frac{e^{-iP^0(x^0-x^{0\prime})}}{p^{0\prime} - P^0} = \tfrac{1}{2}ie^{-ip^{0\prime}(x^0-x^{0\prime})}, \qquad (3\text{–}16.18)$$

$$\frac{d}{dp^{0\prime}} P\int_{-\infty}^{\infty} \frac{dP^0}{2\pi} \frac{e^{-iP^0(x^0-x^{0\prime})}}{p^{0\prime} - P^0} = \tfrac{1}{2}(x^0 - x^{0\prime})e^{-ip^{0\prime}(x^0-x^{0\prime})}. \qquad (3\text{–}16.19)$$

The result is

$$-i\bar{\Delta}_{p^{0\prime}a^\prime,\,p^{0\prime}a^\prime}(x^0 - x^{0\prime}) = [1 - \tfrac{1}{2}\gamma_{p^{0\prime}a^\prime}(x^0 - x^{0\prime})]e^{-ip^{0\prime}(x^0 - x^{0\prime})}, \quad (3\text{-}16.20)$$

which introduces an amplitude that diminishes in time, without altering the time varying phase. This is in accord with the phenomenological viewpoint of source theory. The H-particle energies that have been accurately identified over the finite time intervals $\gamma_{p^{0\prime}a^\prime}(x^0 - x^{0\prime}) \ll 1$ do not change their values when the time scale is enlarged. We are not concerned here with examining how the theoretical understanding of the energy spectrum changes as we move to another level of dynamical description. For our present purposes the numbers $p^{0\prime}$ are given, whether by theory or by experiment is immaterial.

The unit value of the absolute square of $\exp[-ip^{0\prime}(x^0 - x^{0\prime})]$ represents the certainty with which a stable particle will be found in the same energy state after any lapse of time. The square of the amplitude factor in (3-16.20) describes the changing probability that an unstable H-particle ($\gamma_{p^{0\prime}a^\prime} > 0$) shall still exist after the time interval $x^0 - x^{0\prime} = t$,

$$(1 - \tfrac{1}{2}\gamma_{p^{0\prime}a^\prime}t)^2 = 1 - \gamma_{p^{0\prime}a^\prime}t + (\tfrac{1}{2}\gamma_{p^{0\prime}a^\prime}t)^2. \quad (3\text{-}16.21)$$

There is an initial decrease, at a rate given by $\gamma_{p^{0\prime}a^\prime}$. But this result becomes unsatisfactory at larger time values. The persistence probability of the H-particle, according to (3-16.21), reaches zero at a finite time; it then increases and eventually becomes larger than unity. The probability formula is evidently limited in physical applicability to small values of $\gamma_{p^{0\prime}a^\prime}t$.

What is still missing in the physical account is this: We began with an extended H-particle source emitting a virtual H-particle that quickly transformed into a real H-particle and a photon. This situation endured until both particles reached the neighborhood of the extended detection source where they recombined to form a virtual H-particle that was absorbed. But, given enough time, the recombination to form a virtual H-particle can occur far from detection sources with this excitation rapidly decomposing back into real particles. The cycle can be repeated many times before the virtual H-particle is finally absorbed by the detection source. Otherwise expressed, the fields appearing in the coupling term (3-16.5) originate, not only directly in the sources, but also indirectly through other, effective sources which are associated with the virtual H-particles that form far from the sources through the propagation of real particles.

The qualitative description in the last sentence is given a quantitative meaning by the following integral equation for the field $\phi_{p^\prime a^\prime}(x^0)$:

$$\phi_{p^{0\prime}a^\prime}(x^0) = \int dx^{0\prime}\Delta_{p^{0\prime}}(x^0 - x^{0\prime})K_{p^{0\prime}a^\prime}(x^{0\prime})$$
$$+ \int dx_1^0 \, dx^{0\prime}\Delta_{p^{0\prime}}(x^0 - x_1^0) \sum_{p^{0\prime\prime}a^{\prime\prime}} \Pi_{p^{0\prime}a^\prime,\,p^{0\prime\prime}a^{\prime\prime}}(x_1^0 - x^{0\prime})\phi_{p^{0\prime\prime}a^{\prime\prime}}(x^{0\prime}),$$

$$(3\text{-}16.22)$$

where the matrix function Π describes the mechanism whereby, for the last time, a virtual H-particle goes through the cycle of transforming into a real

H-particle and photon, then back into a (not necessarily the same) virtual H-particle that is detected by the probe source used to define the field. The exciting field that appears in the integral expression summarizes the effect of the initial source excitation and of the unlimited repetitions of these reversible conversions and is, therefore, considering all $p^{0'}a'$ together, the very field that is being constructed. This point of view is similar to a multiple scattering analysis in terms of the last collision. If this integral equation were to be solved by iteration, we would indeed be considering successively more elaborate repetitions of the same basis process. The comparison with the known description of one such action then identifies the matrix Π. This comparison is facilitated by writing.

$$\phi_{p^{0'}a'}(x^0) = \sum_{p^{0''}a''} \int dx^{0'} \bar{\Delta}_{p^{0'}a',p^{0''}a''}(x^0 - x^{0'}) K_{p^{0''}a''}(x^{0'}), \quad (3\text{-}16.23)$$

where the modified propagation function obeys the integral equation

$$\bar{\Delta}_{p^{0'}a',p^{0''}a''}(x^0 - x^{0'}) = \delta_{p^{0'}a',p^{0''}a''} \Delta_{p^{0'}}(x^0 - x^{0'})$$
$$+ \sum_{p_1^0 a_1} \int dx_1^0 \, dx_2^0 \Delta_{p^{0'}}(x^0 - x_1^0) \Pi_{p^{0'}a',p_1^0 a_1}(x_1^0 - x_2^0)$$
$$\times \bar{\Delta}_{p_1^0 a_1,p^{0''}a''}(x_2^0 - x^{0'}). \quad (3\text{-}16.24)$$

The identification of (3–16.14) with the first two terms of the iterative solution of (3–16.24), gives, using transform propagation functions,

$$\Delta_{p^{0'}}(p^0) \Pi_{p^{0'}a',p^{0''}a''}(p^0) \Delta_{p^{0''}}(p^0)$$
$$= \int \frac{dP^0}{2\pi} \frac{\epsilon(p^{0'})}{p^{0'} - P^0} \Gamma_{p^{0'}a',p^{0''}a''}(P^0) \frac{\epsilon(p^{0''})}{p^{0''} - P^0} \Delta_{P^0}(p^0). \quad (3\text{-}16.25)$$

The corresponding form of the integral equation (3–16.24) can be presented as

$$(p^{0'}(1 - i\epsilon) - p^0)\bar{\Delta}_{p^{0'}a',p^{0''}a''}(p^0) - (p^{0'} - p^0)\epsilon(p^{0'})$$
$$\times \int \frac{dP^0}{2\pi} \sum_{p_1^0 a_1} \frac{\Gamma_{p^{0'}a',p_1^0 a_1}(P^0)}{(p^{0'} - P^0)(p_1^0 - P^0)} \Delta_{P^0}(p^0)(p_1^0 - p^0)\bar{\Delta}_{p_1^0 a_1,p^{0''}a''}(p^0)$$
$$= \epsilon(p^{0'})\delta_{p^{0'}a',p^{0''}a''}. \quad (3\text{-}16.26)$$

Although these are rather general equations we shall produce only an approximate solution that is applicable to ordinary circumstances, as indicated by the specialization

$$\Gamma_{p^{0'}a',p^{0'}a''}(P^0) = \delta_{a'a''} \Gamma_{p^{0'}}(P^0). \quad (3\text{-}16.27)$$

Such statements express the rotational invariance of isolated systems, when the a' are identified as angular momentum quantum numbers. Only equal energies are considered in (3–16.27) since attention is also restricted to the dominant elements of the propagation matrix:

$$\bar{\Delta}_{p^{0'}a',p^{0''}a''}(p^0) \simeq \delta_{p^{0'}p^{0''}} \delta_{a'a''} \bar{\Delta}_{p^{0'}}(p^0). \quad (3\text{-}16.28)$$

The resulting simplified equation is

$$\left[p^{0\prime}(1 - i\epsilon) - p^0 + (p^{0\prime} - p^0)^2 \int \frac{dP^0}{2\pi} \frac{\Gamma_{p^{0\prime}}(P^0)}{P^0(1 - i\epsilon) - p^0} \, \epsilon(p^{0\prime})\epsilon(P^0) \right.$$

$$\left. \times \frac{d}{dp^{0\prime}} \frac{1}{p^{0\prime} - P^0} \right] \bar{\Delta}_{p^{0\prime}}(p^0) = \epsilon(p^{0\prime}), \quad (3\text{–}16.29)$$

which is consistent with the symmetry

$$\bar{\Delta}_{-p^{0\prime}}(-p^0) = \bar{\Delta}_{p^{0\prime}}(p^0). \tag{3–16.30}$$

This symmetry is maintained when $\epsilon(p^{0\prime})\epsilon(P^0)$ is replaced by unity, as is justified by the predominance of the contributions for $P^0 \sim p^{0\prime}$. The inference that the integral is only of interest for $p^0 \sim p^{0\prime}$ would seem to be contradicted by the factor $(p^{0\prime} - p^0)^2$, which vanishes strongly under just these circumstances.
  To see which tendency prevails, we approximate $\Gamma_{p^{0\prime}}(P^0)$ by

$$\Gamma_{p^{0\prime}}(p^{0\prime}) = \Gamma_{-p^{0\prime}}(-p^{0\prime}) = \gamma_{p^{0\prime}} \geq 0 \tag{3–16.31}$$

and consider the integral

$$\int_{-\infty}^{\infty} \frac{dP^0}{2\pi} \frac{1}{P^0(1 - i\epsilon) - p^0} \frac{1}{2}\left(\frac{1}{p^{0\prime} - P^0 - i\eta} + \frac{1}{p^{0\prime} - P^0 + i\eta} \right), \tag{3–16.32}$$

which uses a complex equivalent of the principal value of integrals, according to [Eq. (2–1.62)]

$$\frac{1}{x - i\eta}\bigg|_{\eta \to +0} = P\frac{1}{x} + \pi i \, \delta(x). \tag{3–16.33}$$

The integral is evaluated by closing the contour at infinity in either half-plane, as is convenient, with the result

$$\frac{1}{2} i \frac{\epsilon(p^0)}{p^{0\prime} - p^0 - i\eta\epsilon(p^0)} = \frac{1}{2} i \frac{\epsilon(p^0)}{p^{0\prime} - p^0} - \frac{\pi}{2} \delta(p^{0\prime} - p^0). \tag{3–16.34}$$

The imaginary term in (3–16.34) is more directly inferred by writing (3–16.32) as

$$\int_{-\infty}^{\infty} \frac{dP^0}{2\pi} \left[\frac{1}{P^0 - p^0} + \pi i \epsilon(p^0) \, \delta(P^0 - p^0) \right] \frac{1}{p^{0\prime} - P^0}$$

$$= P \int_{-\infty}^{\infty} \frac{dP^0}{2\pi} \frac{1}{P^0 - p^0} \frac{1}{p^{0\prime} - P^0} + \frac{1}{2} i \frac{\epsilon(p^0)}{p^{0\prime} - p^0}. \tag{3–16.35}$$

Thus, the structure that appears in (3–16.29) is

$$\gamma_{p^{0\prime}}(p^{0\prime} - p^0)^2 \left[-\frac{\pi}{2} \frac{d}{dp^{0\prime}} \delta(p^{0\prime} - p^0) - \frac{1}{2} i \frac{\epsilon(p^0)}{(p^{0\prime} - p^0)^2} \right] = -\tfrac{1}{2} i \epsilon(p^{0\prime})\gamma_{p^{0\prime}};$$

$$\tag{3–16.36}$$

the factor $(p^{0'} - p^0)^2$ does indeed suppress the real part of the integral, but not its imaginary part.

If $\gamma_{p^0} \neq 0$, the finite imaginary term maintains the sign of the infinitesimal imaginary quantity, $-ip^0\epsilon$, and the latter is superfluous in the resulting approximate equation:

$$[p^{0'} - \tfrac{1}{2}i\epsilon(p^{0'})\gamma_{p^0} - p^0]\bar{\Delta}_{p^0}(p^0) = \epsilon(p^{0'}). \qquad (3\text{-}16.37)$$

The implied time behavior is

$$\bar{\Delta}_{p^0}(x^0 - x^{0'})$$

$$= \int_{-\infty}^{\infty} \frac{dp^0}{2\pi} \exp[-ip^0(x^0 - x^{0'})] \frac{\epsilon(p^{0'})}{p^{0'} - \tfrac{1}{2}i\epsilon(p^{0'})\gamma_{p^0} - p^0}$$

$$= \begin{cases} x^0 > x^{0'}: & \eta(p^{0'})i\exp[-ip^{0'}(x^0 - x^{0'})]\exp[-\tfrac{1}{2}\gamma_{p^0}(x^0 - x^{0'})] \\ x^0 < x^{0'}: & \eta(-p^{0'})i\exp[-ip^{0'}(x^0 - x^{0'})]\exp[-\tfrac{1}{2}\gamma_{p^0}(x^{0'} - x^0)], \end{cases}$$

$$(3\text{-}16.38)$$

consistent with the symmetry

$$\bar{\Delta}_{-p^0}(x^{0'} - x^0) = \bar{\Delta}_{p^0}(x^0 - x^{0'}). \qquad (3\text{-}16.39)$$

The time-dependent complex phase factor continues to identify the energy $p^{0'} > 0$, but the variable amplitude $1 - \tfrac{1}{2}\gamma_{p^0}t$, $t = x^0 - x^{0'} > 0$, has been replaced by $\exp(-\tfrac{1}{2}\gamma_{p^0}t)$. This is quite satisfactory since the implied probability, $\exp(-\gamma_{p^0}t)$, never exceeds unity and decreases monotonically to zero with increasing time. The exponential function generalizes the linear decrease of probability over short time intervals, extending it from the initial instant to arbitrary later times, according to

$$e^{-\gamma(t+\Delta t)} \simeq (1 - \gamma\,\Delta t)e^{-\gamma t}, \qquad (3\text{-}16.40)$$

where $\gamma\,\Delta t \ll 1$. Nonrelativistic approximations can be introduced in (3-16.9), and we arrive at a simple explicit formula for γ_{p^0} [compare (3-15.65)],

$$\gamma_{p^0} = \sum_{E<E',a} A_{Ea\leftarrow E'a'}. \qquad (3\text{-}16.41)$$

This statement is not restricted to the nonrelativistic limit, of course, for it equates the initial rate at which the probability of persistence decreases to the corresponding rate at which transitions are made to H-particles of lower energy.

It is interesting to verify that this balance of probability persists at arbitrary times. We first consider the most elementary situation, an unstable H-particle, designated II, which can only radiate down to the stable H-particle of lowest energy, labeled I, as indicated by

$$\gamma_{II} = A_{I\leftarrow II}. \qquad (3\text{-}16.42)$$

The probability that H-particle II still exists at a time t after its creation is $\exp(-\gamma_{II}t) < 1$. Is this loss of probability compensated by the probability that H-particle I exists, accompanied by a photon? To evaluate the latter probability we must extend the formula for W_{21}, Eq. (3–15.60), to the new situation of unstable particles. From the viewpoint of H-particles this formula describes two stages of emissions and absorptions. After the creation of the initial H-particle it propagates until the moment that the photon is emitted when it ceases to exist. At that instant the final H-particle is created and is eventually detected. Evidently, we must now use the modified propagation function to represent these voyages between emission and absorption acts. Sources are introduced to describe what is common to all emission and absorption mechanisms of a particular type, and the corresponding propagation function is of general applicability. Individual realistic mechanisms will also have specific features that need additional characterization. This we shall discuss later. In the present essentially nonrelativistic situation, however, the description of the mechanism for H-particle transmutation and photon emission requires no significant correction and the introduction of modified propagation functions is quite sufficient for our purposes—as we shall see.

What has just been said is that (3–15.60) continues to apply with the changed meaning of field given by (3–16.23) and the simplification (3–16.28). In the situation being considered, the causally labeled H-particle field $\phi_2(x)$ that appears in

$$W_{21} \to \int (dx)\phi_1(x)2eq\mathbf{p} \cdot \mathbf{A}_1(x)\phi_2(x) \qquad (3\text{–}16.43)$$

is given by [we use the nonrelativistic approximation (3–15.67) but retain the relativistic origin of energy]

$$\phi_2(x) = (2m)^{-1/2}\psi_{II}(\mathbf{x}) \int dx^{0\prime} \bar{\Delta}_{II}(x^0 - x^{0\prime})K_{2\,II}(x^{0\prime})$$

$$= (2m)^{-1/2}\psi_{II}(\mathbf{x}) \int dx^{0\prime} \exp[-i(m + E_{II})(x^0 - x^{0\prime})$$

$$- \tfrac{1}{2}\gamma_{II}(x^0 - x^{0\prime})]iK_{2\,II}(x^{0\prime})$$

$$= (2m)^{-1/2}\psi_{II}(\mathbf{x}) \exp[-i(m + E_{II})(x^0 - t_2) - \tfrac{1}{2}\gamma_{II}(x^0 - t_2)]iK_{2\,II}.$$

$$(3\text{–}16.44)$$

The time t_2 is a fiducial point within the source $K_{2\,II}(x^{0\prime})$, and correspondingly the definition of H-particle emission source appears as

$$K_{2\,II} = \int dx^{0\prime} \exp[-i(m + E_{II})(t_2 - x^{0\prime}) - \tfrac{1}{2}\gamma_{II}(t_2 - x^{0\prime})]K_{2\,II}(x^{0\prime}).$$

$$(3\text{–}16.45)$$

The use of a reference point that is interior to the source rather than arbitrarily chosen is always possible and can be useful in identifying the mechanical properties of states. It becomes mandatory in describing unstable particles.

The coupling of emission and absorption sources for a particular unstable H-particle is given by

$$i\int dx^0\,dx^{0\prime}K_1(x^0)^*\bar{\Delta}(x^0 - x^{0\prime})K_2(x^{0\prime})$$
$$= \int dx^0\,dx^{0\prime}iK_1(x^0)^*\exp[-i(m+E)(x^0 - x^{0\prime}) - \tfrac{1}{2}\gamma(x^0 - x^{0\prime})]iK_2(x^{0\prime})$$
$$= iK_1^*\exp[-i(m+E)(t_1 - t_2) - \tfrac{1}{2}\gamma(t_1 - t_2)]iK_2, \qquad (3\text{-}16.46)$$

where, illustrated by particle II,

$$K_{1\,\mathrm{II}}^* = \int dx^0 K_{1\,\mathrm{II}}(x^0)^*\exp[-i(m+E_\mathrm{II})(x^0 - t_1) - \tfrac{1}{2}\gamma_\mathrm{II}(x^0 - t_1)], \qquad (3\text{-}16.47)$$

and t_1 is a reference time that locates the source $K_1(x^0)$. The factorization of (3-16.46) clearly separates the three stages of emission, propagation for the interval $t = t_1 - t_2$, and absorption. The possibility of describing unstable particles as stable over short time intervals must certainly apply to the largest values assumed by $x^0 - t_1$ and $t_2 - x^{0\prime}$, which are limited to displacements within the sources. Thus the decay factors in the H-particle source definitions (3-16.45) and (3-16.47) can be omitted, and these sources play the same role as with stable particles. In this way, then, the weakening of the coupling (3-16.46) with increasing time interval between the sources, owing to the instability of the particle, is specifically associated with the process of propagation only.

The same kind of description is used for particle I, even though $\gamma_\mathrm{I} = 0$, and the probability amplitude deduced from (3-16.43) is

$$\langle 1_\mathrm{I} 1_{k^0\lambda} t_1 | 1_\mathrm{II} t_2 \rangle = i(d\omega_k)^{1/2}eq'\langle \mathrm{I}|\mathbf{e}^* \cdot \mathbf{p}/m|\mathrm{II}\rangle$$
$$\times \int_{t_2}^{t_1} dx^0 \exp[-i(m+E_\mathrm{I})(t_1 - x^0)]\exp[-ik^0(t_1 - x^0)]$$
$$\times \exp[-i(m+E_\mathrm{II})(x^0 - t_2) - \tfrac{1}{2}\gamma_\mathrm{II}(x^0 - t_2)]. \qquad (3\text{-}16.48)$$

The initial and final times are now explicit in the specification of states, although only $t = t_1 - t_2$ is significant, as we have emphasized by using t_1 as the reference time for the photon field. The time integration is evaluated as

$$\frac{\exp[-i(m+E_\mathrm{II})t - \tfrac{1}{2}\gamma_\mathrm{II}t] - \exp[-i(m+E_\mathrm{I}+k^0)t]}{i(E_\mathrm{I}+k^0 - E_\mathrm{II}) - \tfrac{1}{2}\gamma_\mathrm{II}}, \qquad (3\text{-}16.49)$$

and the transition probability, summed over photon polarizations and emission directions, but still differential in the photon energy, becomes

$$\gamma_\mathrm{II}\frac{k^0}{k_{\mathrm{I\,II}}^0}\frac{dk^0}{2\pi}\frac{1 - 2e^{-(1/2)\gamma_\mathrm{II}t}\cos(k^0 - k_{\mathrm{I\,II}}^0)t + e^{-\gamma_\mathrm{II}t}}{(k^0 - k_{\mathrm{I\,II}}^0)^2 + (\tfrac{1}{2}\gamma_\mathrm{II})^2}, \qquad (3\text{-}16.50)$$

where
$$k_{I\ II}^0 = E_{II} - E_I \tag{3-16.51}$$
and, of course,
$$\gamma_{II} = \tfrac{4}{3}\alpha k_{I\ II}^0 |\langle I|\mathbf{p}/m|II\rangle|^2. \tag{3-16.52}$$

The total probability is produced by carrying out the k^0 integration. That is approximated under the assumption of weak instability, $\gamma_{II} \ll k_{I\ II}^0$, by replacing $k^0/k_{I\ II}^0$ with unity and evaluating the integrals as

$$\int_{-\infty}^{\infty} \frac{dk^0}{2\pi} \frac{\cos\,(k^0 - k_{I\ II}^0)t}{(k^0 - k_{I\ II}^0)^2 + (\tfrac{1}{2}\gamma_{II})^2} = \frac{1}{\gamma_{II}} e^{-(1/2)\gamma_{II}t}, \tag{3-16.53}$$

together with the specialization to $t = 0$. In this way we obtain the probability of finding H-particle I after time t as

$$1 - e^{-\gamma_{II}t}, \tag{3-16.54}$$

which is the required value. The spectral distribution of the emitted photon is also exhibited on evaluating (3-16.50) at a time $\gamma_{II}t \gg 1$, such that the radiative transition has certainly occurred. The result,

$$dk^0 \frac{\gamma_{II}}{2\pi} \frac{1}{(k^0 - k_{I\ II}^0)^2 + (\tfrac{1}{2}\gamma_{II})^2}, \tag{3-16.55}$$

is the familiar Lorentzian shape that identifies the decay constant γ_{II}, the reciprocal of the mean lifetime, with the width of the spectral line at half-maximum.

This is the shape of a spectral line emitted in a transition to the stable H-particle. But what if the final H-particle is also unstable? Now consider a third H-particle III, which can only decay into II, with the subsequent transmutation of the latter into the stable variety I. In this situation two photons are emitted and we must use W_{22} to describe the process. There are two analogous terms in the relevant probability amplitude which are related by the B. E. symmetry of the photons. But, apart from the special circumstance $k_{I\ II}^0 \simeq k_{II\ III}^0$, only one of these terms is appreciable depending upon which of the photons has its frequency near $k_{I\ II}^0$ while the other frequency is close to $k_{II\ III}^0$. Thus, it suffices to regard the photons as distinguishable through their frequencies and use only one of these terms. The probability amplitude for the whole process is

$$\langle 1_1 1_{k^0\lambda} 1_{k^{0'}\lambda'} t_1 | 1_{III} t_2 \rangle$$
$$= i(d\omega_k\, d\omega_{k'})^{1/2} e^2 \langle I|\mathbf{e}^* \cdot \mathbf{p}/m|II\rangle\langle II|\mathbf{e}'^* \cdot \mathbf{p}/m|III\rangle$$
$$\times \int_{t_2}^{t_1} dx^0\, dx^{0'} \exp[-i(m + E_I)(t_1 - x^0)]\exp[-ik^0(t_1 - x^0)]$$
$$\times \exp[-i(m + E_{II})(x^0 - x^{0'}) - \tfrac{1}{2}\gamma_{II}(x^0 - x^{0'})]\eta(x^0 - x^{0'})$$
$$\times \exp[-ik^{0'}(t_1 - x^{0'})]\exp[-i(m + E_{III})(x^{0'} - t_2) - \tfrac{1}{2}\gamma_{III}(x^{0'} - t_2)], \tag{3-16.56}$$

where the time propagation functions detail the successive causal acts of the drama. In writing this expression, we have proceeded as though the H-particles of types II and III were unique, although additional indices a_{II}, a_{III} are necessary. These details can be inserted and do not affect the results, under the physical circumstances indicated in (3–16.27). The x^0 time integration is the one already performed, with $x^{0'}$ supplying the lower limit instead of t_2. The time integral factor of (3–16.26) is, therefore,

$$\int_{t_2}^{t_1} dx^{0'} \frac{e^{-i(m+E_{II})(t_1-x^{0'})-\frac{1}{2}\gamma_{II}(t_1-x^{0'})} - e^{-i(m+E_I+k^0)(t_1-x^{0'})}}{i(k^0 - k^0_{I\,II}) - \frac{1}{2}\gamma_{II}} e^{-ik^{0'}(t_1-x^{0'})}$$
$$\times\, e^{-i(m+E_{III})(x^{0'}-t_2)-\frac{1}{2}\gamma_{III}(x^{0'}-t_2)}. \quad (3\text{–}16.57)$$

The next integration poses no difficulties, but we shall be content to evaluate it only for such large t, $\gamma_{II}t \gg 1$, $\gamma_{III}t \gg 1$, that it represents the completed process of cascading decay. Only the exponential containing E_I contributes, in contrast with the one containing E_{II}, to give the value

$$\frac{1}{i(k^0 - k^0_{I\,II}) - \frac{1}{2}\gamma_{II}} \frac{1}{i(k^0 + k^{0'} - k^0_{I\,III}) - \frac{1}{2}\gamma_{III}} e^{-i(m+E_I+k^0+k^{0'})t}.$$
$$(3\text{–}16.58)$$

The implied transition probability that refers only to the spectral distribution of the photons is

$$dk^0\, \frac{\gamma_{II}}{2\pi} \frac{1}{(k^0 - k^0_{I\,II})^2 + (\frac{1}{2}\gamma_{II})^2}\, dk^{0'}\, \frac{\gamma_{III}}{2\pi} \frac{1}{(k^0 + k^{0'} - k^0_{I\,III})^2 + (\frac{1}{2}\gamma_{III})^2}.$$
$$(3\text{–}16.59)$$

The successive emissions are not independent. It is the energy $k^0 + E_I$ of the photon and particle into which II decays, rather than the energy E_{II}, that determines the spectral distribution of $k^{0'}$. When one integrates over $k^{0'}$, the result is just (3–16.55), which means that H-particle II is certainly produced at some time by decay from III, after which the previous discussion applies. The answer to the question concerning the spectral distribution of the photon radiated in a transition between unstable H-particles is obtained by integrating over k^0. It is instructive to write this integral as ($E = k^0 + E_I$)

$$dk^{0'} \int dE\, dE'\, \frac{(1/2\pi)\gamma_{II}}{(E - E_{II})^2 + (\frac{1}{2}\gamma_{II})^2}\, \delta(k^{0'} + E - E')\, \frac{(1/2\pi)\gamma_{III}}{(E' - E_{III})^2 + (\frac{1}{2}\gamma_{III})^2},$$
$$(3\text{–}16.60)$$

which describes an energy-conserving radiative transition between two energy distributions of Lorentzian shape that have the widths γ_{II} and γ_{III}, respectively. According to an elementary contour integral evaluation, the resulting spectral distribution is also of Lorentzian shape,

$$dk^{0\prime} \frac{\gamma_{\mathrm{II}} + \gamma_{\mathrm{III}}}{2\pi} \frac{1}{(k^{0\prime} - k_{\mathrm{II\ III}}^{0})^{2} + \left(\dfrac{\gamma_{\mathrm{II}} + \gamma_{\mathrm{III}}}{2}\right)^{2}}, \qquad (3\text{–}16.61)$$

with a width given by the sum of the individual H-particle widths. This conclusion is particularly transparent if one recognizes that the double energy integral of (3–16.60) is equivalent to a single time integral:

$$dk^{0\prime} \int_{-\infty}^{\infty} \frac{dt}{2\pi} \exp[iE_{\mathrm{II}}t - \tfrac{1}{2}\gamma_{\mathrm{II}}|t|] \exp[ik^{0\prime}t] \exp[-iE_{\mathrm{III}}t - \tfrac{1}{2}\gamma_{\mathrm{III}}|t|]. \quad (3\text{–}16.62)$$

It would be hard not to suspect the existence of another approach that is capable of producing this formula directly. We shall find it, not surprisingly, in the time cycle description.

But, first, let us give an analogous discussion of photon scattering, in order to verify that the unphysical infinite cross section at exact resonance has been removed by the explicit recognition of H-particle instability. Elastic scattering by the stable H-particle I will be considered. Then it suffices to introduce modified H-particle propagation functions in (3–15.88), which will be used only in the nonrelativistic limit and in the gauge of (3–15.106). The significant change is the introduction in (3–15.101) (we ignore the $1/M$ term) of the substitution

$$E - (E_{\mathrm{I}} + k^{0}) \rightarrow E - \tfrac{1}{2}i\Gamma_{E}(E_{\mathrm{I}} + k^{0}) - (E_{\mathrm{I}} + k^{0}), \qquad (3\text{–}16.63)$$

while $E - (E_{\mathrm{I}} - k^{0})$ remains unaltered. To understand this it is necessary to be somewhat more general than (3–16.37), where $\Gamma_{p^{0\prime}}(P^{0})$ is considered only for $P^{0} = p^{0\prime}$. We return to (3–16.29) and proceed as in (3–16.35), but with $\Gamma_{p^{0\prime}}(P^{0})$ retained, and get

$$(p^{0\prime} - p^{0})^{2} \int \frac{dP^{0}}{2\pi} \frac{\Gamma_{p^{0\prime}}(P^{0})}{P^{0}(1 - i\epsilon) - p^{0}} \frac{d}{dp^{0\prime}} \frac{1}{p^{0\prime} - P^{0}}$$

$$= (p^{0\prime} - p^{0})^{2} P \int \frac{dP^{0}}{2\pi} \frac{\Gamma_{p^{0\prime}}(P^{0})}{P^{0} - p^{0}} \frac{d}{dp^{0\prime}} \frac{1}{p^{0\prime} - P^{0}} - \tfrac{1}{2}i\epsilon(p^{0})\Gamma_{p^{0\prime}}(p^{0}), \quad (3\text{–}16.64)$$

showing the general form of the imaginary term. This distinction is unnecessary near resonance, $p^{0} \sim p^{0\prime}$, or $E_{\mathrm{I}} + k^{0} \sim E$, but it is needed far from resonance conditions. Otherwise we should have, incorrectly, added an imaginary term to $E - (E_{\mathrm{I}} - k^{0})$, whereas

$$\Gamma_{E}(E_{\mathrm{I}} - k^{0}) = 0, \qquad (3\text{–}16.65)$$

since no photon emission can occur if the total energy is less than E_{I}.

H-particle II becomes strongly excited when

$$k^{0} + E_{\mathrm{I}} \simeq E_{\mathrm{II}}: \qquad\qquad k^{0} \simeq k_{\mathrm{I\ II}}^{0}. \qquad (3\text{–}16.66)$$

Under these circumstances the dominant contribution to the differential cross section of (3–15.101) is

$$\frac{d\sigma}{d\Omega} = \alpha^2 (k^0_{I\ II})^4 \left| \sum_a \frac{\langle I|e_1^* \cdot d|IIa\rangle\langle IIa|e_2 \cdot d|I\rangle}{k^0_{I\ II} - k^0 - \frac{1}{2}i\gamma_{II}} \right|^2 . \tag{3–16.67}$$

This differential cross section for specified polarizations is replaced by the total cross section on summing over final polarizations and directions and averaging over the initial polarization (and direction). Recalling that

$$\gamma_{II} = \frac{\alpha}{2\pi} (k^0_{I\ II})^3 \int d\Omega \sum_\lambda |\langle I|e_\lambda^* \cdot d|IIa\rangle|^2 , \tag{3–16.68}$$

together with the orthogonality stated in (3–16.27), we find that

$$\sigma = \frac{4\pi}{(k^0_{I\ II})^2} \frac{1}{2} g_{II} \frac{(\frac{1}{2}\gamma_{II})^2}{(k^0 - k^0_{I\ II})^2 + (\frac{1}{2}\gamma_{II})^2} , \tag{3–16.69}$$

where g_{II} is the multiplicity of particle II, the number of different values assumed by a_{II}. The form of the cross section at exact resonance,

$$\sigma_{\text{res.}} = \frac{4\pi}{(k^0_{I\ II})^2} \frac{g_{II}}{2} , \tag{3–16.70}$$

is typical of any resonant scattering process. The basic resonant cross section is $4\pi\lambda^2$, here $4\pi/(k^0_{I\ II})^2$, which is multiplied by the number of resonant states, g_{II}, and divided by the multiplicity of the initial particles. That is just the factor of 2, referring to the two photon polarizations, since H-particle I has been assumed to be unique.

The promise to exhibit another and more direct derivation of (3–16.62) will be fulfilled, even to the point of generalizing this formula so that it refers to any pair of unstable H-particles, which are capable of decaying in other sequences than III → II → I. Here is the statement of the more general problem. The arbitrary unstable H-particle III is created near time zero. It can decay to a particular unstable H-particle II as well as in other ways, and these secondary unstable particles continue the cascade until the stable particle I is reached. What is the differential probability for finding a photon of frequency $k^0 \simeq k^0_{II\ III}$, without reference to the other photons of different frequency that are also emitted? For a specified polarization, that probability is expressed by

$$\sum_{\{n\}} |\langle 1_I \{n + 1_{k\lambda}\} + |1_{III} - \rangle|^2 , \tag{3–16.71}$$

which assumes a time interval long enough to have the probability attain its final value. Let us supply two additional factors, $iJ_{k\lambda}^*$, the probability amplitude for detecting the photon $k\lambda$, and $-iJ_{k\lambda}$, its complex conjugate. This produces a quantity that can be presented as

$$\sum_{\{n\}} \langle \text{III}_- | 1_{\text{I}} \{n + 1_{k\lambda}\}_+ \rangle \langle 1_{\text{I}} \{n + 1_{k\lambda}\}_- | 1_{\text{I}} \{n\}_+ \rangle^J$$

$$\times \langle 1_{\text{I}} \{n\}_+ | 1_{\text{I}} \{n + 1_{k\lambda}\}_- \rangle^J \langle 1_{\text{I}} \{n + 1_{k\lambda}\}_+ | 1_{\text{III}-} \rangle. \quad (3\text{–}16.72)$$

Appearing here are successive stages of a time cycle, in which two analogous photon sources act, one on the forward time path and the other on the return path. Thus, we are now interested in the time cycle generalization of W_{22}.

To use a consistent nonrelativistic description, one should subtract m from the frequencies in the propagation function (3–16.38). This reduces the positive frequencies to nonrelativistic energies, for $x^0 > x^{0'}$, but converts the negative frequencies to values $\simeq -2m$, for $x^0 < x^{0'}$. The latter produce negligible contributions to time integrals, and the nonrelativistic version of (3–16.38) is, accordingly,

$$\bar{\Delta}_{E'}(t - t') = i\eta(t - t') \exp[-iE'(t - t') - \tfrac{1}{2}\gamma_{E'}(t - t')]. \quad (3\text{–}16.73)$$

When the gauge (3–15.106, 107) is used, the structure of iW_{22}, written in a simplified matrix notation, becomes

$$iW_{22} = ie^2 \int dt\, dt_1\, dt_2\, dt' K^*(t) i\eta(t - t_1) \exp[-iE(t - t_1) - \tfrac{1}{2}\gamma(t - t_1)] \mathbf{x} \cdot \mathbf{E}(t_1)$$

$$\times i\eta(t_1 - t_2) \exp[-iE(t_1 - t_2) - \tfrac{1}{2}\gamma(t_1 - t_2)] \mathbf{x} \cdot \mathbf{E}(t_2) i\eta(t_2 - t')$$

$$\times \exp[-iE(t_2 - t') - \tfrac{1}{2}\gamma(t_2 - t')] K(t'). \quad (3\text{–}16.74)$$

The transition to the time cycle is made after time t_2. Time t_1 is now encountered on the return path, which is certainly 'later' than t_2, and $\eta(t_1 - t_2)$ is replaced by unity. Also, time t is reached 'after' time t_1 and $\eta(t - t_1)$ must be replaced by $\eta(t_1 - t)$. Since both t and t_1 refer to the return time path, there is no sign change in the integral. The proper treatment of the γ terms is fixed by the physical necessity of maintaining the damping, the weakening of the coupling with increasing time interval. All this gives the substitution:

$$iW_{22} \to e^2 \int dt\, dt_1\, dt_2\, dt' K^*(t) \eta(t_1 - t) \exp[-iE(t - t_1) - \tfrac{1}{2}\gamma(t_1 - t)] \mathbf{x} \cdot \mathbf{E}(t_1)$$

$$\times \exp[-iE(t_1 - t_2) - \tfrac{1}{2}\gamma|t_1 - t_2|] \mathbf{x} \cdot \mathbf{E}(t_2) \eta(t_2 - t')$$

$$\times \exp[-iE(t_2 - t') - \tfrac{1}{2}\gamma(t_2 - t')] K(t'). \quad (3\text{–}16.75)$$

With the H-particle sources operating in the vicinity of $t = 0$,

$$\int dt' e^{iEt' + (1/2)\gamma t'} K(t') \simeq \int dt' e^{iEt'} K(t') = K,$$

$$\int dt K^*(t) e^{-iEt + (1/2)\gamma t} \simeq \int dt K^*(t) e^{-iEt} = K^*, \quad (3\text{–}16.76)$$

and the coefficient of $(-iK_{\mathrm{III}}^*)(iK_{\mathrm{III}})$ supplies the required time cycle quantity, in the form

$$\langle\mathrm{III}_-|\mathrm{III}_-\rangle^{J_-(-),J_-(+)} = e^2 \int_0^\infty dt_1\, dt_2\, \exp[iE_{\mathrm{III}}t_1 - \tfrac{1}{2}\gamma_{\mathrm{III}}t_1]$$

$$\times\, \langle\mathrm{III}|\mathbf{x}\cdot\mathbf{E}(t_1)\, \exp[-iE(t_1 - t_2) - \tfrac{1}{2}\gamma|t_1 - t_2|]$$

$$\times\, \mathbf{x}\cdot\mathbf{E}(t_2)|\mathrm{III}\rangle\, \exp[-iE_{\mathrm{III}}t_2 - \tfrac{1}{2}\gamma_{\mathrm{III}}t_2]. \quad (3\text{-}16.77)$$

We must still extract the coefficient of $iJ_{k\lambda}^*$ and of $-iJ_{k\lambda}$ from $\mathbf{E}(t_2)$ and $\mathbf{E}(t_1)$, respectively. According to (3-15.107), the first of these is

$$(d\omega_k)^{1/2}(-ik^0)\mathbf{e}^*e^{ik^0t_2}, \quad (3\text{-}16.78)$$

and its complex conjugate, evaluated at t_1, applies on the reverse time path. Since the restriction $k^0 \simeq k_{\mathrm{II}\,\mathrm{III}}^0$ picks out the contribution from the specific H-particle II, the desired probability, as it is deduced from (3-16.77), is

$$e^2\, d\omega_k(k_{\mathrm{II}\,\mathrm{III}}^0)^2 \sum_a |\langle\mathrm{II}a|\mathbf{x}\cdot\mathbf{e}^*|\mathrm{III}\rangle|^2 \int_0^\infty dt_1\, dt_2\, \exp[iE_{\mathrm{II}}(t_2 - t_1) - \tfrac{1}{2}\gamma_{\mathrm{II}}|t_2 - t_1|]$$

$$\times\, \exp[ik^0(t_2 - t_1)]\, \exp[-iE_{\mathrm{III}}(t_2 - t_1) - \tfrac{1}{2}\gamma_{\mathrm{III}}(t_1 + t_2)]. \quad (3\text{-}16.79)$$

With $dk^0/2\pi$ removed, the factor in front of the double time integral, summed over polarizations, is the A-coefficient for the decay III → II. In the interests of a more uniform notation we shall now denote it by $\gamma_{\mathrm{II}\,\mathrm{III}}$. To simplify the time integrals we introduce new variables:

$$t = t_2 - t_1, \quad (3\text{-}16.80)$$

which ranges from $-\infty$ to ∞, and $t_<$, the smaller of the two times, which varies from 0 to ∞. Then, the transformations

$$t_1 + t_2 = 2t_< + |t| \quad (3\text{-}16.81)$$

and

$$\int_0^\infty dt_1\, dt_2 f(t) e^{-(1/2)\gamma_{\mathrm{III}}(t_1 + t_2)} = \int_0^\infty dt_< e^{-\gamma_{\mathrm{III}}t_<} \int_{-\infty}^\infty dt f(t) e^{-(1/2)\gamma_{\mathrm{III}}|t|}$$

$$(3\text{-}16.82)$$

give the desired probability as

$$dk^0(\gamma_{\mathrm{II}\,\mathrm{III}}/\gamma_{\mathrm{III}}) \int_{-\infty}^\infty \frac{dt}{2\pi}\, \exp[iE_{\mathrm{II}}t - \tfrac{1}{2}\gamma_{\mathrm{II}}|t|]\, \exp[ik^0t]\, \exp[-iE_{\mathrm{III}}t - \tfrac{1}{2}\gamma_{\mathrm{III}}|t|].$$

$$(3\text{-}16.83)$$

When III can only radiate to II, $\gamma_{\mathrm{II}\,\mathrm{III}} = \gamma_{\mathrm{III}}$, and we have reproduced (3-16.62). More generally, the probability of emitting any frequency in the neighborhood of $k_{\mathrm{II}\,\mathrm{III}}^0$ is $(\gamma_{\mathrm{II}\,\mathrm{III}}/\gamma_{\mathrm{III}}) < 1$, according to

$$\int_{-\infty}^\infty \frac{dk^0}{2\pi}\, e^{i(k^0 - k_{\mathrm{II}\,\mathrm{III}}^0)t} = \delta(t), \quad (3\text{-}16.84)$$

and this expresses the competition between the specified transition and all others that III can undergo. The sum of these fractions over all decay modes of III is equal to unity.

The time cycle extension of W_{22} also gives a direct derivation of the resonance scattering cross section (3-16.69), or, rather, its generalization in which II becomes an H-particle that can decay in ways other than down to the stable particle I, and σ becomes the corresponding total cross section. A photon is incident on I, and eventually one again finds I, accompanied by one or more photons. The total probability for these phenomena, with a given interaction time, is

$$\sum_{\{n\}} |\langle 1_I\{n\}_+|1_I1_{k\lambda-}\rangle|^2 = \sum_{\{n\}} \langle 1_I1_{k\lambda-}|1_I\{n\}_+\rangle\langle 1_I\{n\}_+|1_I1_{k\lambda-}\rangle. \quad (3\text{-}16.85)$$

When the initial particles are introduced by appropriate sources, two of each kind, this becomes a time cycle vacuum amplitude, described by iW_{22}. The result is obtained from (3-16.77) by replacing III with the stable I, and using the field of an incoming photon instead of (3-16.78):

$$d\omega_k e^2 (k^0)^2 \int_0^\infty dt_1\, dt_2\, \exp[iE_I t_1]\langle I|\mathbf{x}\cdot\mathbf{e}^* \exp[ik^0 t_1]\exp[-iE(t_1-t_2)-\tfrac{1}{2}\gamma|t_1-t_2|]$$
$$\times\, \mathbf{x}\cdot\mathbf{e}\, \exp[-ik^0 t_2]|I\rangle \exp[-iE_I t_2], \quad (3\text{-}16.86)$$

or

$$d\omega_k e^2 (k^0)^2 \sum_{\mathrm{IIa}} \int_0^\infty dt_1\, dt_2\, \exp[-i(E_I + k^0 - E_{II})(t_2 - t_1)]$$
$$\times\, \exp[-\tfrac{1}{2}\gamma_{II}|t_2 - t_1|]|\langle \mathrm{IIa}|\mathbf{x}\cdot\mathbf{e}|I\rangle|^2. \quad (3\text{-}16.87)$$

The integrand depends only upon the time variable $t = t_2 - t_1$, and the integration over $t_<$ is identified with the duration of the interaction. The total cross section is found by dividing the photon flux $2k^0\, d\omega_k$ into the transition probability per unit time. On recognizing that

$$e^2 (k^0_{I\,II})^3 \sum_a |\langle \mathrm{IIa}|\mathbf{x}\cdot\mathbf{e}|I\rangle|^2 = \pi g_{II}\gamma_{I\,II}, \quad (3\text{-}16.88)$$

we get the dominant contribution to the cross section, for $k^0 \simeq k^0_{I\,II}$, as

$$\sigma_{\text{tot.}} = \frac{\pi}{(k^0_{I\,II})^2}\, \frac{1}{2}\, g_{II}\gamma_{I\,II} \int_{-\infty}^\infty dt\, \exp[-i(k^0 - k^0_{I\,II})t - \tfrac{1}{2}\gamma_{II}|t|]$$

$$= \frac{4\pi}{(k^0_{I\,II})^2}\, \frac{1}{2}\, g_{II}\, \frac{\gamma_{I\,II}}{\gamma_{II}}\, \frac{(\tfrac{1}{2}\gamma_{II})^2}{(k^0 - k^0_{I\,II})^2 + (\tfrac{1}{2}\gamma_{II})^2}. \quad (3\text{-}16.89)$$

When $\gamma_{I\,II} = \gamma_{II}$ we regain (3-16.69). The additional factor, $\gamma_{I\,II}/\gamma_{II} < 1$, evidently represents the diminished ability to excite II directly from I, which is the reciprocal aspect of the fractional probability for reaching I directly from II in decay. From this point of view, the elastic scattering cross section should be obtained from the total cross section by multiplying the latter with an additional $\gamma_{I\ II}/\gamma_{II}$ factor. That is indeed the result deduced from (3–16.67) when the quantity contained in (3–16.68) is given its general interpretation as the partial width $\gamma_{I\ II}$:

$$\sigma_{\text{elast.}} = \frac{4\pi}{(k_{I\ II}^0)^2} \frac{1}{2} g_{II} \frac{(\tfrac{1}{2}\gamma_{I\ II})^2}{(k^0 - k_{I\ II}^0)^2 + (\tfrac{1}{2}\gamma_{II})^2}. \qquad (3\text{–}16.90)$$

The above discussion is incomplete since no mention has been made of the additional term in W_{22} that is demanded by the crossing symmetry of the photons. It is produced by reversing the sign of k^0. This term is certainly nonresonant. But, more important is the appearance of the initial energy as $E_I - k^0$; the value that should be assigned to the damping constant of H-particle II is not γ_{II} but zero, as in (3–16.65). Then the resulting time integral gives $\delta(k_{I\ II}^0 + k^0) = 0$.

All the developments of this section have used the example of spinless particles that are bound to form H-particles. A similar treatment for spin $\tfrac{1}{2}$ particles would run in exact parallel, with occasional insertions or deletions of $\epsilon(p^0)$ factors, for example, to represent the changed statistics. The nonrelativistic results are identical.

The natural instability of H-particles has directed attention to the necessity of considering multiparticle exchanges, in addition to single-particle propagation. It is a complementary aspect of the principle of space-time uniformity that couplings identified through the examination of real processes continue to be meaningful when applied to virtual processes. This says that multiparticle exchanges are significant, although the energy to produce several real particles may not be available. Thus, the next stage of dynamical evolution is the systematic generalization of all single-particle exchanges between sources to those involving two particles, including their unlimited repetition. Before embarking on this massive program, however, we shall give a relatively brief discussion of the gravitational version of such concepts as primitive interactions and gauge invariance.

3–17 THE GRAVITATIONAL FIELD

The field associated with massless particles of helicity ± 2 has not yet been given an independent discussion. We refer back to (2–4.24), but use the mechanical measure of $T_{\mu\nu}$, according to (2–4.33):

$$W(T) = \frac{\kappa}{2} \int (dx)(dx')[T^{\mu\nu}(x)D_+(x - x')T_{\mu\nu}(x') - \tfrac{1}{2}T(x)D_+(x - x')T(x')],$$
$$\partial_\mu T^{\mu\nu}(x) = 0. \qquad (3\text{–}17.1)$$

The symmetrical tensor field $h_{\mu\nu}(x)$ is defined by

$$\delta W(T) = \int (dx)\, \delta T^{\mu\nu}(x)h_{\mu\nu}(x), \qquad (3\text{–}17.2)$$

subject to the source restriction

$$\partial_\mu\,\delta T^{\mu\nu}(x) = 0. \tag{3–17.3}$$

The corresponding arbitrariness in the identification of $h_{\mu\nu}(x)$ is exhibited in

$$h_{\mu\nu}(x) = \kappa\int(dx')D_+(x-x')[T_{\mu\nu}(x')-\tfrac{1}{2}g_{\mu\nu}T(x')] + \partial_\mu\xi_\nu(x) + \partial_\nu\xi_\mu(x). \tag{3–17.4}$$

Contraction of indices in the tensor gives

$$\begin{aligned}h(x) &= g_{\mu\nu}h^{\mu\nu}(x)\\ &= -\kappa\int(dx')D_+(x-x')T(x') + 2\partial_\mu\xi^\mu(x),\end{aligned} \tag{3–17.5}$$

and therefore

$$\begin{aligned}h_{\mu\nu}(x) - \tfrac{1}{2}g_{\mu\nu}h(x) &= \kappa\int(dx')D_+(x-x')T_{\mu\nu}(x')\\ &+ \partial_\mu\xi_\nu(x) + \partial_\nu\xi_\mu(x) - g_{\mu\nu}\partial_\lambda\xi^\lambda(x).\end{aligned} \tag{3–17.6}$$

The introduction of the source restriction through the divergence of this equation isolates the aspect of the field $h_{\mu\nu}(x)$ that is governed by the arbitrary $\xi_\mu(x)$ vector,

$$\partial_\mu(h^{\mu\nu}(x) - \tfrac{1}{2}g^{\mu\nu}h(x)) = \partial^2\xi^\nu(x). \tag{3–17.7}$$

Returning to (3–17.4), we deduce

$$-\partial^2 h_{\mu\nu}(x) + \partial_\mu\partial^2\xi_\nu(x) + \partial_\nu\partial^2\xi_\mu(x) = \kappa(T_{\mu\nu}(x) - \tfrac{1}{2}g_{\mu\nu}T(x)), \tag{3–17.8}$$

which is the second-order differential field equation,

$$-\partial^2 h_{\mu\nu}(x) + \partial_\mu\partial^\lambda h_{\lambda\nu}(x) + \partial_\nu\partial^\lambda h_{\mu\lambda}(x) - \partial_\mu\partial_\nu h(x) = \kappa(T_{\mu\nu}(x) - \tfrac{1}{2}g_{\mu\nu}T(x)). \tag{3–17.9}$$

It is also the form of the equation obtained by placing $m = 0$ in Eq. (3–3.19). Contracting the indices in (3–17.9), or converting (3–17.5) to a differential equation, implies

$$-\partial^2 h(x) + \partial_\mu\partial_\nu h^{\mu\nu}(x) = -\tfrac{1}{2}\kappa T(x), \tag{3–17.9a}$$

from which we derive another version of the differential field equations,

$$\begin{aligned}-\partial^2 h_{\mu\nu}(x) + \partial_\mu\partial^\lambda h_{\lambda\nu}(x) &+ \partial_\nu\partial^\lambda h_{\mu\lambda}(x) - \partial_\mu\partial_\nu h(x)\\ &- g_{\mu\nu}(-\partial^2 h(x) + \partial_\kappa\partial_\lambda h^{\kappa\lambda}(x)) = \kappa T_{\mu\nu}(x);\end{aligned} \tag{3–17.10}$$

it is (3–3.17), with $m = 0$. The structure of the left-hand side of this equation is such that its divergence is identically zero. The vanishing divergence of the source tensor now appears as an algebraic consequence of the field equations.

Since the arbitrariness of the vector $\xi_\mu(x)$ is still maintained in these field equations, they are unaffected by a redefinition of the field $h_{\mu\nu}(x)$ having the form

$$h_{\mu\nu}(x) \rightarrow h_{\mu\nu}(x) + \partial_\mu \xi_\nu(x) + \partial_\nu \xi_\mu(x), \tag{3-17.11}$$

which is a gravitational gauge transformation.

The following definition, analogous to (3–3.23),

$$\Gamma_{\mu\nu\lambda}(x) = \Gamma_{\nu\mu\lambda}(x) = \partial_\mu h_{\nu\lambda}(x) + \partial_\nu h_{\mu\lambda}(x) - \partial_\lambda h_{\mu\nu}(x), \tag{3-17.12}$$

with its consequence

$$\Gamma_\mu(x) = \Gamma_{\mu\lambda}{}^\lambda(x) = \partial_\mu h(x), \tag{3-17.13}$$

provides a first-order form of the field equations [(3–3.22), with $m = 0$]

$$\partial^\lambda \Gamma_{\mu\nu\lambda}(x) - \partial_\nu \Gamma_\mu(x) = \kappa\left(T_{\mu\nu}(x) - \tfrac{1}{2}g_{\mu\nu}T(x)\right). \tag{3-17.14}$$

The gauge invariance of the left-hand side of (3–17.14) is not realized through the invariance of $\Gamma_{\mu\nu\lambda}$, but rather

$$\Gamma_{\mu\nu\lambda}(x) \rightarrow \Gamma_{\mu\nu\lambda}(x) + 2\partial_\mu\partial_\nu \xi_\lambda(x), \tag{3-17.15}$$

and

$$\Gamma_\mu(x) \rightarrow \Gamma_\mu(x) + 2\partial_\mu\partial_\lambda \xi^\lambda(x). \tag{3-17.16}$$

Note, however, that these gauge transformation responses do not involve first derivatives of the $\xi_\lambda(x)$.

Another system of first-order differential equations [it is (3–3.20, 21), with $m = 0$] is produced by the definitions

$$\omega_{\mu\lambda\nu}(x) = -\omega_{\nu\lambda\mu}(x) = \partial_\mu h_{\lambda\nu}(x) - \partial_\nu h_{\lambda\mu}(x) \tag{3-17.17}$$

and

$$\omega_\mu(x) = \omega_{\mu\lambda}{}^\lambda(x) = \partial_\mu h(x) - \partial^\lambda h_{\mu\lambda}(x), \tag{3-17.18}$$

namely,

$$\partial^\lambda \omega_{\mu\nu\lambda}(x) - \partial_\nu \omega_\mu(x) = \kappa\left(T_{\mu\nu}(x) - \tfrac{1}{2}g_{\mu\nu}T(x)\right). \tag{3-17.19}$$

The response of these fields to gauge transformations is given by

$$\omega_{\mu\lambda\nu}(x) \rightarrow \omega_{\mu\lambda\nu}(x) + \partial_\lambda\left(\partial_\mu \xi_\nu(x) - \partial_\nu \xi_\mu(x)\right) \tag{3-17.20}$$

and

$$\omega_\mu(x) \rightarrow \omega_\mu(x) + \partial_\mu\partial_\lambda \xi^\lambda(x) - \partial^2 \xi_\mu(x). \tag{3-17.21}$$

It is observed that the divergence of the vector field $\omega_\mu(x)$ is gauge invariant. A comparison of the form of the divergence inferred from (3–17.18) with (3–17.9a) shows that

$$\partial_\mu \omega^\mu(x) = \tfrac{1}{2}\kappa T(x), \tag{3-17.22}$$

which is also the contraction of (3–17.19), since

$$\omega_\nu{}^\nu{}_\lambda(x) = -\omega_\lambda{}^\nu{}_\nu(x) = -\omega_\lambda(x). \tag{3-17.23}$$

An analogous use of (3–17.14), however, introduces a new vector field,

$$\lambda\Gamma(x) = \Gamma^\nu{}_{\nu\lambda}(x) = 2\partial^\nu h_{\nu\lambda}(x) - \partial_\lambda h(x),\qquad(3\text{–}17.24)$$

such that

$$\Gamma_\lambda(x) - \lambda\Gamma(x) = 2\omega_\lambda(x).\qquad(3\text{–}17.25)$$

This is a contraction of the tensor relations

$$\Gamma_{\mu\nu\lambda}(x) - \Gamma_{\lambda\nu\mu}(x) = 2\omega_{\mu\nu\lambda}(x).\qquad(3\text{–}17.26)$$

Another connection between the two third-rank tensors, which implies this one, is

$$\Gamma_{\mu\nu\lambda}(x) = \omega_{\mu\nu\lambda}(x) + \partial_\nu h_{\mu\lambda}(x).\qquad(3\text{–}17.27)$$

The successive stages involved in producing an action expression to represent the first-order field equations (3–17.12, 14) are

$$W = \tfrac{1}{2}\int(dx)T^{\mu\nu}h_{\mu\nu} = \tfrac{1}{2}\int(dx)(T^{\mu\nu} - \tfrac{1}{2}g^{\mu\nu}T)(h_{\mu\nu} - \tfrac{1}{2}g_{\mu\nu}h)$$

$$= \frac{1}{2\kappa}\int(dx)(\partial^\lambda\Gamma_{\mu\nu\lambda} - \partial_\nu\Gamma_\mu)(h^{\mu\nu} - \tfrac{1}{2}g^{\mu\nu}h)$$

$$= -\frac{1}{2\kappa}\int(dx)[\Gamma^{\lambda\mu\nu}\Gamma_{\mu\nu\lambda} - {}^\lambda\Gamma\Gamma_\lambda],\qquad(3\text{–}17.28)$$

where the last version introduces the relation

$$\partial^\lambda h^{\mu\nu}(x) = \tfrac{1}{2}(\Gamma^{\lambda\mu\nu}(x) + \Gamma^{\lambda\nu\mu}(x)),\qquad(3\text{–}17.29)$$

its contraction

$$\partial_\nu h^{\mu\nu}(x) = \tfrac{1}{2}(\Gamma^\mu(x) + {}^\mu\Gamma(x)),\qquad(3\text{–}17.30)$$

and (3–17.13). The action is

$$W = \int(dx)[T^{\mu\nu}h_{\mu\nu} + \mathcal{L}(h,\Gamma)]\qquad(3\text{–}17.31)$$

with

$$\kappa\mathcal{L}(h,\Gamma) = -(h^{\mu\nu} - \tfrac{1}{2}g^{\mu\nu}h)(\partial^\lambda\Gamma_{\mu\nu\lambda} - \partial_\nu\Gamma_\mu) - \tfrac{1}{2}(\Gamma^{\lambda\mu\nu}\Gamma_{\mu\nu\lambda} - {}^\lambda\Gamma\Gamma_\lambda).$$
$$(3\text{–}17.32)$$

Apart from a divergence term, this Lagrange function is the analogue of (3–5.41), with $m = 0$. For simplicity, we do not include a source for the third-rank tensor field, in contrast with (3–5.40). The stationary requirement for variations of $h^{\mu\nu}$, or $h^{\mu\nu} - \tfrac{1}{2}g^{\mu\nu}h$, recovers (3–17.14), and variations of $\Gamma_{\mu\nu\lambda}$, after rearrangements indicated by the structure of (3–5.45), reproduce (3–17.12). The Lagrange function is not gauge invariant,

$$\kappa\mathcal{L} \to \kappa\mathcal{L} - \partial^\lambda[(\partial^\mu\xi^\nu + \partial^\nu\xi^\mu - g^{\mu\nu}\partial_\kappa\xi^\kappa)\Gamma_{\mu\nu\lambda}] + \partial_\nu[(\partial^\mu\xi^\nu + \partial^\nu\xi^\mu - g^{\mu\nu}\partial_\kappa\xi^\kappa)\Gamma_\mu],$$
$$(3\text{–}17.33)$$

but the action is invariant. If we relinquish the use of $\Gamma_{\mu\nu\lambda}$ as an independent variable and let it be defined by (3–17.12), an appropriate Lagrange function is

$$\kappa\mathcal{L}(h) = \tfrac{1}{2}(\Gamma^{\lambda\mu\nu}\Gamma_{\mu\nu\lambda} - {}^{\lambda}\Gamma\Gamma_{\lambda}). \tag{3–17.34}$$

When written out as a quadratic function of the first derivatives of $h_{\mu\nu}(x)$, this Lagrange function differs from that of Eq. (3–5.99) only in the last term:

$$\partial_{\mu}h^{\mu\nu}\partial^{\lambda}h_{\lambda\nu} \rightarrow \partial^{\lambda}h^{\mu\nu}\partial_{\mu}h_{\lambda\nu}, \tag{3–17.35}$$

which illustrates the freedom to add divergence terms. This alternative has already been noted in Eqs. (3–5.31, 32).

The similar development that is based on the first-order differential equations (3–17.17, 19) starts with

$$W = \frac{1}{2\kappa}\int(dx)(\partial^{\lambda}\omega_{\mu\nu\lambda} - \partial_{\nu}\omega_{\mu})(h^{\mu\nu} - \tfrac{1}{2}g^{\mu\nu}h)$$

$$= \frac{1}{2\kappa}\int(dx)(\omega^{\lambda\mu\nu}\omega_{\lambda\nu\mu} - \omega^{\lambda}\omega_{\lambda}), \tag{3–17.36}$$

where the second version uses the substitutions

$$-\omega_{\mu\nu\lambda}\partial^{\lambda}h^{\mu\nu} = \omega_{\mu\nu\lambda}\omega^{\nu\mu\lambda} = \omega^{\lambda\mu\nu}\omega_{\lambda\nu\mu}. \tag{3–17.37}$$

The Lagrange function is obtained as

$$\kappa\mathcal{L}(h, \omega) = -(h^{\mu\nu} - \tfrac{1}{2}g^{\mu\nu}h)(\partial^{\lambda}\omega_{\mu\nu\lambda} - \partial_{\nu}\omega_{\mu}) + \tfrac{1}{2}(\omega^{\lambda\mu\nu}\omega_{\lambda\nu\mu} - \omega^{\lambda}\omega_{\lambda}). \tag{3–17.38}$$

Apart from divergence terms, it is (3–5.34) with $m = 0$. Again, the source coupled to the third-rank tensor will not be used. The variation of $h^{\mu\nu} - \tfrac{1}{2}g^{\mu\nu}h$ reproduces (3–17.19) and that of $\omega_{\lambda\mu\nu}$ yields (3–17.17) after rearrangements that are indicated by the structure of (3–5.38). The response of this Lagrange function to gauge transformations is

$$\kappa\mathcal{L} \rightarrow \kappa\mathcal{L} - \partial^{\lambda}[(\partial^{\mu}\xi^{\nu} + \partial^{\nu}\xi^{\mu} - g^{\mu\nu}\partial_{\kappa}\xi^{\kappa})\omega_{\mu\nu\lambda}] + \partial_{\nu}[(\partial^{\mu}\xi^{\nu} + \partial^{\nu}\xi^{\mu} - g^{\mu\nu}\partial_{\kappa}\xi^{\kappa})\omega_{\mu}], \tag{3–17.39}$$

which assures the invariance of the action. When $\omega_{\lambda\mu\nu}$ loses its independent status and is defined by (3–17.17), the Lagrange function can be chosen as

$$\kappa\mathcal{L}(h) = -\tfrac{1}{2}(\omega^{\lambda\mu\nu}\omega_{\lambda\nu\mu} - \omega^{\lambda}\omega_{\lambda}), \tag{3–17.40}$$

which is the quadratic function of the first derivatives of $h_{\mu\nu}(x)$ that is produced by averaging the two alternatives of (3–17.35):

$$\kappa\mathcal{L}(h) = -\tfrac{1}{2}(\partial^{\lambda}h^{\mu\nu}\partial_{\lambda}h_{\mu\nu} - \partial^{\lambda}h\partial_{\lambda}h) - \partial_{\mu}h^{\mu\nu}\partial_{\nu}h$$

$$+ \tfrac{1}{2}(\partial_{\mu}h^{\mu\nu}\partial^{\lambda}h_{\lambda\nu} + \partial^{\lambda}h^{\mu\nu}\partial_{\mu}h_{\lambda\nu}). \tag{3–17.41}$$

The stress tensor $t^{\mu\nu}(x)$ has been given a kinematical definition, which is not unique, through the response to infinitesimal coordinate deformations,

$$\delta_{\text{coord.}} W = -\int (dx) t^{\mu\nu}(x) \partial_\mu \, \delta x_\nu(x). \qquad (3\text{–}17.42)$$

It acquires a dynamical definition by imitating the role of the graviton source $T^{\mu\nu}(x)$. This is indicated, in the response of an action expression to infinitesimal gauge transformations, by adding $t^{\mu\nu}$ to $T^{\mu\nu}$,

$$\delta_{\text{gauge}} W = \int (dx) \left(T^{\mu\nu}(x) + t^{\mu\nu}(x) \right) 2 \partial_\mu \, \delta\xi_\nu(x). \qquad (3\text{–}17.43)$$

The two concepts are identified by requiring invariance of the action under the unified gauge-coordinate transformation

$$2 \, \delta\xi_\nu(x) = \delta x_\nu(x). \qquad (3\text{–}17.44)$$

Thus, infinitesimal coordinate transformations induce the infinitesimal gauge transformations

$$\begin{aligned} \delta h_{\mu\nu}(x) &= \tfrac{1}{2} \left(\partial_\mu \, \delta x_\nu(x) + \partial_\nu \, \delta x_\mu(x) \right), \\ \delta\Gamma_{\mu\nu\lambda}(x) &= \partial_\mu \partial_\nu \, \delta x_\lambda(x) \end{aligned} \qquad (3\text{–}17.45)$$

and

$$\delta\omega_{\mu\lambda\nu}(x) = \tfrac{1}{2} \partial_\lambda \left(\partial_\mu \, \delta x_\nu(x) - \partial_\nu \, \delta x_\mu(x) \right). \qquad (3\text{–}17.46)$$

The use of the total stress tensor $T^{\mu\nu} + t^{\mu\nu}$, as the factor of $h_{\mu\nu}$ in the action, is the introduction of a primitive interaction. Some modification of $t^{\mu\nu}$ is needed since it is not conserved inside particle sources, and a gravitational model of particle sources must be introduced. But let us defer the discussion of that question and proceed with the development, which is modeled so closely on the electromagnetic one, in order to reach the point of divergence between the two very different physical systems. Consider the example of spinless particles, using the simplest stress tensor form, Eq. (3–7.8),

$$t_{\mu\nu} = \partial_\mu \phi \partial_\nu \phi + g_{\mu\nu} \mathcal{L}(\phi), \qquad \mathcal{L}(\phi) = -\tfrac{1}{2} [\partial^\lambda \phi \partial_\lambda \phi + m^2 \phi^2]. \qquad (3\text{–}17.47)$$

As in the electromagnetic analogue, the coupling term $t_{\mu\nu} h^{\mu\nu}$ will be combined with the particle Lagrange function, to form

$$\begin{aligned} \mathcal{L}(\phi, h) &= \mathcal{L}(\phi) + t_{\mu\nu} h^{\mu\nu} \\ &= -\tfrac{1}{2} \partial_\mu \phi [g^{\mu\nu}(1 + h) - 2h^{\mu\nu}] \partial_\nu \phi - \tfrac{1}{2}(1 + h) m^2 \phi^2. \end{aligned} \qquad (3\text{–}17.48)$$

At this stage in the electromagnetic discussion, facilitated by the use of a slightly different Lagrange function, the gauge covariant derivative $\partial_\mu - ieqA_\mu$ appeared, and one verified invariance under the whole Abelian group of gauge transformations. The related gravitational situation is only partly produced by the substitution $ieqA_\mu \to h_{\mu\nu} \partial^\nu$; there is also a gravitational coupling that does not refer to derivatives. But, much more significant is what underlies the replacement of the single matrix eq by the four differential operators $(1/i)\partial_\nu$.

The general coordinate transformation group is non-Abelian, as indicated by

$$[\delta_1 x^\mu(x)\partial_\mu, \ \delta_2 x^\nu(x)\partial_\nu] = (\delta_1 x^\mu(x)\partial_\mu \ \delta_2 x^\lambda(x) - \delta_2 x^\mu(x)\partial_\mu \ \delta_1 x^\lambda(x))\partial_\lambda, \quad (3\text{--}17.49)$$

and the extension of invariance under infinitesimal transformations to cover the whole group is not trivial.

It is instructive to examine just how infinitesimal coordinate transformation invariance comes about. The transformations associated with

$$\bar{x}^\mu = x^\mu - \delta x^\mu(x) \tag{3--17.50}$$

describe the scalar nature of the particle field,

$$\bar\phi(\bar x) = \phi(x), \tag{3--17.51}$$

and give the induced gauge transformation of the graviton field,

$$\bar h_{\mu\nu}(\bar x) = h_{\mu\nu}(x) + \tfrac{1}{2}(\partial_\mu \ \delta x_\nu(x) + \partial_\nu \ \delta x_\mu(x)), \tag{3--17.52}$$

including

$$\bar h(\bar x) = h(x) + \partial_\mu \ \delta x^\mu(x). \tag{3--17.53}$$

The invariance of the mass term in the action is stated by

$$\int (dx)(1 + h(x))\phi(x)^2 = \int (dx)(1 + \bar h(x))\bar\phi(x)^2$$
$$= \int (d\bar x)(1 + \bar h(\bar x))\bar\phi(\bar x)^2, \tag{3--17.54}$$

which is satisfied if

$$(dx)(1 + h(x)) = (d\bar x)(1 + \bar h(\bar x)). \tag{3--17.55}$$

For the infinitesimal transformation (3–17.50), the transformation law of volume elements becomes

$$(d\bar x) = (dx) \det (\partial \bar x^\mu/\partial x^\nu)$$
$$= (dx)(1 - \partial_\mu \ \delta x^\mu(x)), \tag{3--17.56}$$

and it is required that

$$1 + h(x) = (1 - \partial_\mu \ \delta x^\mu(x))(1 + h(x) + \partial_\mu \ \delta x^\mu(x)) \tag{3--17.57}$$

or

$$h(x) = h(x) - h(x)\partial_\mu \ \delta x^\mu(x). \tag{3--17.58}$$

This can only mean that $h(x)$ is restricted to be a very small quantity, permitting $h\partial_\mu \ \delta x^\mu$ to be neglected as a second-order object.

The situation is similar for the quadratic derivative term of the action. We first notice that

$$\partial_\mu\phi[g^{\mu\nu}(1 + h) - 2h^{\mu\nu})\partial_\nu\phi \simeq (1 + h)\partial_\mu\phi(g^{\mu\nu} - 2h^{\mu\nu})\partial_\nu\phi \tag{3--17.59}$$

isolates the factor $1 + h$ that compensates the transformation behavior of (dx). Then,

$$g^{\mu\nu}(x) = g^{\mu\nu} - 2h^{\mu\nu}(x) \tag{3--17.60}$$

must transform appropriately to produce a scalar combination:

$$\bar\partial_\mu\bar\phi(\bar x)\bar g^{\mu\nu}(\bar x)\bar\partial_\nu(\bar\phi\bar x) = \partial_\mu\phi(x)g^{\mu\nu}(x)\partial_\nu\phi(x), \quad (3\text{–}17.61)$$

or

$$\bar g^{\mu\nu}(\bar x) = g^{\kappa\lambda}(x)\partial_\kappa\bar x^\mu\partial_\lambda\bar x^\nu. \quad (3\text{–}17.62)$$

This property characterizes $g^{\mu\nu}(x)$ as a contravariant tensor of the second rank under general coordinate transformations. For infinitesimal transformations, that transformation law becomes

$$\delta g^{\mu\nu}(x) = \delta x^\lambda(x)\partial_\lambda g^{\mu\nu}(x) - g^{\kappa\nu}(x)\partial_\kappa \delta x^\mu(x) - g^{\mu\lambda}(x)\partial_\lambda \delta x^\nu(x), \quad (3\text{–}17.63)$$

which does reduce to the first statement of (3–17.45), if one neglects second-order quantities by replacing $g^{\mu\nu}(x)$ with $g^{\mu\nu}$, on the right-hand side.

The tensor $g_{\mu\nu}(x)$, inverse to $g^{\mu\nu}(x)$,

$$g_{\mu\lambda}(x)g^{\lambda\nu}(x) = \delta_\mu^\nu, \quad (3\text{–}17.64)$$

has the transformation law of a covariant tensor of the second rank,

$$\bar g_{\mu\nu}(\bar x) = g_{\kappa\lambda}(x)\bar\partial_\mu x^\kappa\bar\partial_\nu x^\lambda. \quad (3\text{–}17.65)$$

The implied behavior of the determinant

$$g(x) = \det g_{\mu\nu}(x) \quad (3\text{–}17.66)$$

is

$$\bar g(\bar x) = g(x)[\det(\partial x^\mu/\partial\bar x^\nu)]^2, \quad (3\text{–}17.67)$$

and therefore

$$(-\bar g(\bar x))^{1/2}(d\bar x) = (-g(x))^{1/2}(dx). \quad (3\text{–}17.68)$$

It is consistent to regard this as the generalization of (3–17.55) for, under the weak field conditions to which the latter refers,

$$g_{\mu\nu}(x) \simeq g_{\mu\nu} + 2h_{\mu\nu}(x) \quad (3\text{–}17.69)$$

and

$$(-g(x))^{1/2} \simeq 1 + h(x). \quad (3\text{–}17.70)$$

Accordingly, to ensure invariance of the action under arbitrary coordinate transformations, the Lagrange function of (3–17.48), which is appropriate to weak gravitational fields, should be replaced by

$$\mathcal{L}(\phi(x), g(x)) = -(-g(x))^{1/2}\tfrac{1}{2}[\partial_\mu\phi(x)g^{\mu\nu}(x)\partial_\nu\phi(x) + m^2\phi(x)^2]. \quad (3\text{–}17.71)$$

We must find a similar generalization of the weak field form of the gravitational Lagrange function,

$$\kappa\mathcal{L}(h, \Gamma) = -(h^{\mu\nu} - \tfrac{1}{2}g^{\mu\nu}h)(\partial_\lambda\Gamma_{\mu\nu}^\lambda - \partial_\nu\Gamma_{\mu\lambda}^\lambda) + \tfrac{1}{2}g^{\mu\nu}(\Gamma_{\mu\nu}^\lambda\Gamma_{\lambda\kappa}^\kappa - \Gamma_{\mu\kappa}^\lambda\Gamma_{\nu\lambda}^\kappa), \quad (3\text{–}17.72)$$

which is (3–17.32) with all reference to third-rank tensors stated in terms of

$$\Gamma_{\mu\nu}{}^\lambda \equiv \Gamma_{\mu\nu}^\lambda = \Gamma_{\nu\mu}^\lambda. \quad (3\text{–}17.73)$$

One recognizes in $h^{\mu\nu} - \tfrac{1}{2}g^{\mu\nu}h$ part of the weak field evaluation

$$\tfrac{1}{2}(-g(x))^{1/2}g^{\mu\nu}(x) \simeq \tfrac{1}{2}g^{\mu\nu} - (h^{\mu\nu}(x) - \tfrac{1}{2}g^{\mu\nu}h(x)). \qquad (3\text{-}17.74)$$

The missing constant term can be added in (3-17.72) since it changes the Lagrange function by a divergence. Then the strong field generalization is clearly indicated:

$$2\kappa\mathcal{L}(g(x),\,\Gamma(x)) = (-g(x))^{1/2}g^{\mu\nu}(x)R_{\mu\nu}(x), \qquad (3\text{-}17.75)$$

with

$$R_{\mu\nu} = \partial_\lambda\Gamma^\lambda_{\mu\nu} - \partial_\nu\Gamma^\lambda_{\mu\lambda} + \Gamma^\lambda_{\mu\nu}\Gamma^\kappa_{\lambda\kappa} - \Gamma^\lambda_{\mu\kappa}\Gamma^\kappa_{\nu\lambda}. \qquad (3\text{-}17.76)$$

This will indeed contribute an invariant action if $g^{\mu\nu}R_{\mu\nu}$ is a scalar with respect to arbitrary coordinate transformations. The required covariant tensor behavior of $R_{\mu\nu}(x)$ must emerge from the transformation law of the three-index symbol $\Gamma^\lambda_{\mu\nu}(x)$. The latter should resemble a third-rank tensor but cannot be entirely of this nature, according to the weak field transformation of (3-17.45) which contains second derivatives with respect to coordinates. A suitable generalization is stated by

$$\Gamma^\kappa_{\mu\nu}(\bar{x})\bar{\partial}_\kappa x^\lambda = \Gamma^\lambda_{\rho\sigma}(x)\partial_\mu x^\rho \partial_\nu x^\sigma + \bar{\partial}_\mu\bar{\partial}_\nu x^\lambda. \qquad (3\text{-}17.77)$$

This transformation law is such that a coordinate covariant derivative of first-rank contravariant vectors can be defined:

$$\nabla_\nu V^\mu(x) = (\partial_\nu + \Gamma_\nu(x))^\mu_\lambda V^\lambda(x) = \partial_\nu V^\mu(x) + \Gamma^\mu_{\nu\lambda}(x)V^\lambda(x). \qquad (3\text{-}17.78)$$

The matrix notation facilitates the consideration of

$$[\nabla_\mu,\,\nabla_\nu]^\kappa_\lambda V^\lambda = R_{\mu\nu}{}^\kappa{}_\lambda\, V^\lambda, \qquad (3\text{-}17.79)$$

where

$$R_{\mu\nu}{}^\kappa{}_\lambda = \partial_\mu\Gamma^\kappa_{\nu\lambda} - \partial_\nu\Gamma^\kappa_{\mu\lambda} + \Gamma^\kappa_{\mu\rho}\Gamma^\rho_{\nu\lambda} - \Gamma^\kappa_{\nu\rho}\Gamma^\rho_{\mu\lambda} \qquad (3\text{-}17.80)$$

is indeed a fourth-rank tensor, which is antisymmetrical in μ and ν. We can now recognize the tensor character of

$$R_{\mu\nu} = R_{\lambda\nu}{}^\lambda{}_\mu. \qquad (3\text{-}17.81)$$

The notion of covariant derivative, identified with ordinary differentiation for scalars, is extended to first-rank covariant vectors by the requirement

$$\partial_\nu(V_{1\mu}V^\mu_2) = (\nabla_\nu V_{1\mu})V^\mu_2 + V_{1\mu}(\nabla_\nu V^\mu_2), \qquad (3\text{-}17.82)$$

whence

$$\nabla_\nu V_\mu = \partial_\nu V_\mu - V_\lambda\Gamma^\lambda_{\mu\nu}, \qquad (3\text{-}17.83)$$

and to arbitrary tensors by generalizing the differentiation rule for products. As an application, we note that $\delta\Gamma^\lambda_{\mu\nu}(x)$, which is any infinitesimal change of the $\Gamma^\lambda_{\mu\nu}(x)$, does transform as a tensor, and

$$\begin{aligned}
\delta R_{\mu\nu} &= [\partial_\lambda\,\delta\Gamma^\lambda_{\mu\nu} + \Gamma^\kappa_{\lambda\kappa}\,\delta\Gamma^\lambda_{\mu\nu} - \delta\Gamma^\kappa_{\lambda\nu}\Gamma^\lambda_{\mu\kappa} - \delta\Gamma^\lambda_{\mu\kappa}\Gamma^\kappa_{\lambda\nu}] - [\partial_\nu\,\delta\Gamma^\lambda_{\mu\lambda} - \delta\Gamma^\kappa_{\lambda\kappa}\Gamma^\lambda_{\mu\nu}] \\
&= \nabla_\lambda\,\delta\Gamma^\lambda_{\mu\nu} - \nabla_\nu\,\delta\Gamma^\lambda_{\mu\lambda}.
\end{aligned} \qquad (3\text{-}17.84)$$

The covariant derivative of $g(x)$ is defined by the determinantal differentiation formula,

$$\nabla_\lambda g(x) = g(x)g^{\mu\nu}(x)\nabla_\lambda g_{\mu\nu}(x)$$
$$= g(x)g^{\mu\nu}(x)[\partial_\lambda g_{\mu\nu}(x) - 2g_{\mu\kappa}(x)\Gamma^\kappa_{\lambda\nu}(x)] \qquad (3\text{-}17.85)$$
$$= \partial_\lambda g(x) - 2g(x)\Gamma^\nu_{\lambda\nu}(x),$$

or

$$\nabla_\lambda(-g(x))^{1/2} = \partial_\lambda(-g(x))^{1/2} - (-g(x))^{1/2}\Gamma^\nu_{\lambda\nu}(x). \qquad (3\text{-}17.86)$$

A simple consequence is the divergence formula

$$\nabla_\lambda[(-g(x))^{1/2}V^\lambda(x)] = \partial_\lambda[(-g(x))^{1/2}V^\lambda(x)]. \qquad (3\text{-}17.87)$$

These results are used in applying the stationary action principle to variations of $\Gamma^\lambda_{\mu\nu}$, as it appears in $\mathcal{L}(g,\Gamma)$. The vanishing coefficient of $\delta\Gamma^\lambda_{\mu\nu}$ in δW states that $[g^{\mu\nu}$ is $g^{\mu\nu}(x)]$

$$\nabla_\lambda[(-g)^{1/2}g^{\mu\nu}] - \delta^\nu_\lambda\nabla_\kappa[(-g)^{1/2}g^{\mu\nu}] = 0, \qquad (3\text{-}17.88)$$

which implies

$$\nabla_\lambda[(-g)^{1/2}g^{\mu\nu}] = 0. \qquad (3\text{-}17.89)$$

From the latter property one derives, successively, the vanishing of the covariant derivatives for $g(x)$, $g^{\mu\nu}(x)$, and $g_{\mu\nu}(x)$. The last statement,

$$0 = \nabla_\lambda g_{\mu\nu}(x) = \partial_\lambda g_{\mu\nu}(x) - g_{\kappa\nu}(x)\Gamma^\kappa_{\lambda\mu}(x) - g_{\mu\kappa}(x)\Gamma^\kappa_{\lambda\nu}(x), \qquad (3\text{-}17.90)$$

leads to the explicit construction

$$\Gamma^\lambda_{\mu\nu}(x) = g^{\lambda\kappa}(x)\tfrac{1}{2}[\partial_\mu g_{\nu\kappa}(x) + \partial_\nu g_{\mu\kappa}(x) - \partial_\kappa g_{\mu\nu}(x)], \qquad (3\text{-}17.91)$$

which is the strong field generalization of (3–17.12). The weak field version of (3–17.90) appears in (3–17.29). As one can verify directly, the vanishing covariant derivative of $g(x)$ implies, according to (3–17.86), that

$$\Gamma^\nu_{\lambda\nu}(x) = \partial_\lambda \log (-g(x))^{1/2}, \qquad (3\text{-}17.92)$$

which generalizes (3–17.13). This form ensures that $R_{\mu\nu}$, as defined in (3–17.76), is a symmetrical tensor.

The variation of $g^{\mu\nu}(x)$ in the purely gravitational contribution to the action induces

$$\delta_g \int (dx)\mathcal{L}(g(x),\Gamma(x)) = \frac{1}{2\kappa}\int(dx)(-g(x))^{1/2}\,\delta g^{\mu\nu}(x)G_{\mu\nu}(x), \qquad (3\text{-}17.93)$$

where

$$G_{\mu\nu} = R_{\mu\nu} - \tfrac{1}{2}g_{\mu\nu}R, \qquad R = g^{\mu\nu}R_{\mu\nu}, \qquad (3\text{-}17.94)$$

and we have used the determinantal property

$$\delta(-g)^{1/2} = \tfrac{1}{2}(-g)^{1/2}g^{\mu\nu}\,\delta g_{\mu\nu} = -\tfrac{1}{2}(-g)^{1/2}g_{\mu\nu}\,\delta g^{\mu\nu}. \qquad (3\text{-}17.95)$$

The tensor $G_{\mu\nu}$ obeys a differential identity, which is a consequence of the coordinate invariance of the gravitational action term. We first note the infinitesimal response of $g_{\mu\nu}(x)$, analogous to (3–17.63),

$$
\begin{aligned}
\delta g_{\mu\nu} &= \delta x^\lambda \partial_\lambda g_{\mu\nu} + g_{\kappa\nu} \partial_\mu \, \delta x^\kappa + g_{\mu\kappa} \partial_\nu \, \delta x^\kappa \\
&= \partial_\mu (g_{\kappa\nu} \, \delta x^\kappa) + \partial_\nu (g_{\mu\kappa} \, \delta x^\kappa) - 2\Gamma^\kappa_{\mu\nu} g_{\kappa\lambda} \, \delta x^\lambda \\
&= \nabla_\mu (g_{\kappa\nu} \, \delta x^\kappa) + \nabla_\nu (g_{\mu\kappa} \, \delta x^\kappa),
\end{aligned}
\tag{3–17.96}
$$

which generalizes the weak field gauge transformation of (3–17.45). On writing

$$
\delta g^{\mu\nu} G_{\mu\nu} = -\delta g_{\mu\nu} G^{\mu\nu},
\tag{3–17.97}
$$

where

$$
G^{\mu\nu} = g^{\mu\kappa} G_{\kappa\lambda} g^{\lambda\nu},
\tag{3–17.98}
$$

we conclude from the invariance of the action that

$$
\nabla_\mu G^{\mu\nu}(x) = 0.
\tag{3–17.99}
$$

The variation of $g^{\mu\nu}(x)$ in the matter part of the action defines a tensor $t_{\mu\nu}(x)$ that generalizes the stress tensor,

$$
\delta_g \int (dx) \mathcal{L}(\phi, g) = -\tfrac{1}{2} \int (dx)(-g)^{1/2} \, \delta g^{\mu\nu} t_{\mu\nu}.
\tag{3–17.100}
$$

Inspection of the Lagrange function (3–17.71) shows that

$$
t_{\mu\nu}(x) = \partial_\mu \phi(x) \partial_\nu \phi(x) + g_{\mu\nu}(x)(-g(x))^{-1/2} \mathcal{L}(\phi(x), g(x)).
\tag{3–17.101}
$$

We also note the generalization of (3–7.9), in source-free regions.

$$
t = g^{\mu\nu} t_{\mu\nu} = -m^2 \phi^2 - (-g)^{-1/2} \partial_\mu [(-g)^{1/2} g^{\mu\nu} \partial_\nu \tfrac{1}{2} \phi^2].
\tag{3–17.102}
$$

When the stationary property with respect to ϕ variations is invoked, the coordinate invariance of this action term leads, as before, to a differential statement,

$$
\nabla_\mu t^{\mu\nu}(x) = 0.
\tag{3–17.103}
$$

Another form of this generalized local conservation law is

$$
\partial_\mu [(-g)^{1/2} t^{\mu\nu}] = -(-g)^{1/2} \Gamma^\nu_{\kappa\lambda} t^{\kappa\lambda}.
\tag{3–17.104}
$$

The field equation deduced by varying $g^{\mu\nu}$ in the complete action

$$
W = \int (dx)[\mathcal{L}(\phi, g) + \mathcal{L}(g, \Gamma)]
\tag{3–17.105}
$$

is

$$
G_{\mu\nu}(x) = \kappa t_{\mu\nu}(x),
\tag{3–17.106}
$$

or

$$
R_{\mu\nu}(x) = \kappa (t_{\mu\nu}(x) - \tfrac{1}{2} g_{\mu\nu}(x) t(x));
\tag{3–17.107}
$$

it is Einstein's gravitational field equation The stress tensor divergence condition (3–17.103) appears again, now as an identity demanded by the structure of the gravitational field equation.

The replacement of spin 0 particles as the model of gravitating matter by other integer spin particles is relatively straightforward. A rather special but interesting example is provided by photons. The Lagrange function

$$\mathcal{L} = -\tfrac{1}{2}F^{\mu\nu}(\partial_\mu A_\nu - \partial_\nu A_\mu) + \tfrac{1}{4}F^{\mu\nu}F_{\mu\nu} \qquad (3\text{–}17.108)$$

is immediately generalized to realize invariance under arbitrary coordinate transformations, while maintaining electromagnetic gauge invariance, by writing

$$\mathcal{L}(A_\mu, F^{\mu\nu}, g) = (-g)^{1/2}[-\tfrac{1}{2}F^{\mu\nu}(\partial_\mu A_\nu - \partial_\nu A_\mu) + \tfrac{1}{4}F^{\mu\nu}g_{\mu\kappa}g_{\nu\lambda}F^{\kappa\lambda}]. \qquad (3\text{–}17.109)$$

At points not occupied by electromagnetic sources, the implied field equations are

$$\begin{aligned} F_{\mu\nu}(x) &= g_{\mu\kappa}(x)g_{\nu\lambda}(x)F^{\kappa\lambda}(x) \\ &= \partial_\mu A_\nu(x) - \partial_\nu A_\mu(x) \\ &= \nabla_\mu A_\nu(x) - \nabla_\nu A_\mu(x) \end{aligned} \qquad (3\text{–}17.110)$$

and

$$\partial_\nu[(-g(x))^{1/2}F^{\mu\nu}(x)] = (-g(x))^{1/2}\nabla_\nu F^{\mu\nu}(x) = 0. \qquad (3\text{–}17.111)$$

The stress tensor derived by varying $g_{\mu\nu}(x)$ in (3–17.109) is

$$t^{\mu\nu}(x) = F^{\mu\lambda}(x)F^\nu{}_\lambda(x) - g^{\mu\nu}(x)\tfrac{1}{4}F^{\kappa\lambda}(x)F_{\kappa\lambda}(x), \qquad (3\text{–}17.112)$$

where all contravariant and covariant indices are related by means of the tensor $g_{\mu\nu}(x)$. Another instructive derivation can be given, by redefining $F^{\mu\nu}$ to absorb $(-g)^{1/2}$, which produces the Lagrange function

$$\mathcal{L} = -\tfrac{1}{2}F^{\mu\nu}(\partial_\mu A_\nu - \partial_\nu A_\mu) + \tfrac{1}{4}F^{\mu\nu}(-g)^{-1/2}g_{\mu\kappa}g_{\nu\lambda}F^{\kappa\lambda}; \qquad (3\text{–}17.113)$$

all reference to $g_{\mu\nu}$ is concentrated in the last term. The stress tensor (3–17.112) is regained, divided by $(-g)$ to conform with the altered meaning of $F^{\mu\nu}$. But now we can see something very clearly: $(-g)^{-1/2}g_{\mu\kappa}g_{\nu\lambda}$ is homogeneous of degree zero in the components of $g_{\mu\nu}(x)$, which is to say that the Lagrange function (3–17.113) is invariant under the transformation

$$g_{\mu\nu}(x) \rightarrow \lambda(x)g_{\mu\nu}(x), \qquad (3\text{–}17.114)$$

for arbitrary $\lambda(x)$. The implication, for an infinitesimal deviation of $\lambda(x)$ from unity, is

$$\delta_\lambda \int (dx)\mathcal{L} = \tfrac{1}{2}\int (dx)(-g)^{1/2}t^{\mu\nu}g_{\mu\nu}\,\delta\lambda = 0 \qquad (3\text{–}17.115)$$

or

$$t(x) = g_{\mu\nu}(x)t^{\mu\nu}(x) = 0, \qquad (3\text{–}17.116)$$

which is true. It is evident that we are now considering a generalization of the conformal transformations that were originally introduced through the consideration of isotropic dilations [(3–7.153)]. Incidentally, while the alternative Lagrange function of (3–17.113) was helpful in recognizing conformal invariance, one can also use (3–17.109), combining the conformal transformation (3–17.114) with the field transformations

$$F^{\mu\nu}(x) \to \lambda(x)^{-2}F^{\mu\nu}(x), \qquad A_\mu(x) \to A_\mu(x) \qquad (3\text{–}17.117)$$

to attain the invariance of $\mathcal{L}$.

As we have noted earlier in connection with (3–7.168), the kinematical arbitrariness in stress tensors must be considered in testing for conformal invariance. The arbitrariness can be placed in a dynamical context, akin to the electromagnetic procedure illustrated in (3–10.63). Returning to the weak gravitational field situation, we examine the possibility of replacing any given stress tensor $t^{\mu\nu}$ by [Eqs. (3–7.83, 84, 85)]

$$t^{\mu\nu} + \partial_\kappa\partial_\lambda m^{\mu\nu,\,\kappa\lambda} \qquad (3\text{–}17.118)$$

where $m^{\mu\nu,\,\kappa\lambda}$ is symmetrical in μ and ν, in κ and λ, and obeys

$$m^{\mu\nu,\,\kappa\lambda} + m^{\kappa\nu,\,\lambda\mu} + m^{\lambda\nu,\,\mu\kappa} = 0. \qquad (3\text{–}17.119)$$

The addition to the Lagrange function term $t^{\mu\nu}$ is, effectively,

$$m^{\mu\nu,\,\kappa\lambda}\partial_\kappa\partial_\lambda h_{\mu\nu} = -\tfrac{1}{3}m^{\mu\nu,\,\kappa\lambda}R_{\mu\kappa\nu\lambda} \qquad (3\text{–}17.120)$$

where

$$R_{\mu\kappa\nu\lambda} = \partial_\mu\partial_\lambda h_{\kappa\nu} + \partial_\nu\partial_\kappa h_{\lambda\mu} - \partial_\kappa\partial_\lambda h_{\mu\nu} - \partial_\mu\partial_\nu h_{\kappa\lambda} \qquad (3\text{–}17.121)$$

is produced by using the cyclic property (3–17.119),

$$m^{\mu\nu,\,\kappa\lambda}\partial_\kappa\partial_\lambda h_{\mu\nu} = -m^{\mu\nu,\,\kappa\lambda}(\partial_\mu\partial_\lambda h_{\kappa\nu} + \partial_\nu\partial_\kappa h_{\lambda\mu})$$
$$= m^{\mu\nu,\,\kappa\lambda}\partial_\mu\partial_\nu h_{\kappa\lambda}. \qquad (3\text{–}17.122)$$

Also contained here, in

$$(m^{\mu\nu,\,\kappa\lambda} - m^{\kappa\lambda,\,\mu\nu})\partial_\kappa\partial_\lambda h_{\mu\nu} = 0, \qquad (3\text{–}17.123)$$

is the requirement that $m^{\mu\nu,\,\kappa\lambda}$ be symmetrical in the two pairs of indices, as already found in particular situations [(3–7.88), (3–7.111), (3–7.137)].

The four index object $R_{\mu\kappa\nu\lambda}$ has many symmetries. It is antisymmetrical in μ and κ, in ν and λ, and symmetrical in the two pairs, $\mu\kappa$ and $\nu\lambda$. The sum of the three terms obtained by cyclic permutation, with one index held fixed, equals zero. As the notation betrays, $R_{\mu\kappa\nu\lambda}$ is the weak field version of the tensor derived from (3–17.80):

$$R_{\mu\kappa\nu\lambda} = g_{\nu\sigma}[\partial_\mu\Gamma^\sigma_{\kappa\lambda} - \partial_\kappa\Gamma^\sigma_{\mu\lambda} + \Gamma^\sigma_{\mu\rho}\Gamma^\rho_{\kappa\lambda} - \Gamma^\sigma_{\kappa\rho}\Gamma^\rho_{\mu\lambda}]$$
$$= \tfrac{1}{2}[\partial_\mu\partial_\lambda g_{\kappa\nu} + \partial_\nu\partial_\kappa g_{\lambda\mu} - \partial_\kappa\partial_\lambda g_{\mu\nu} - \partial_\mu\partial_\nu g_{\kappa\lambda}]$$
$$- g_{\rho\sigma}(\Gamma^\rho_{\kappa\lambda}\Gamma^\sigma_{\mu\nu} - \Gamma^\rho_{\mu\lambda}\Gamma^\sigma_{\kappa\nu}), \qquad (3\text{–}17.124)$$

which also possesses all these symmetry properties.

We conclude from these results that possible additional terms in the Lagrange function of matter have the form

$$-\tfrac{1}{3}(-g(x))^{1/2}m^{\mu\nu,\kappa\lambda}(x)R_{\mu\kappa\nu\lambda}(x), \qquad (3\text{-}17.125)$$

where $m^{\mu\nu,\kappa\lambda}$ is a tensor, referring to the matter field and the gravitational field, that has the symmetries previously noted. An illustration for spin 0, generalized from (3-7.88), is

$$m^{\mu\nu,\kappa\lambda}(x) = [g^{\mu\nu}(x)g^{\kappa\lambda}(x) - \tfrac{1}{2}g^{\kappa\nu}(x)g^{\mu\lambda}(x) - \tfrac{1}{2}g^{\kappa\mu}(x)g^{\lambda\nu}(x)]\tfrac{1}{6}\phi(x)^2. \quad (3\text{-}17.126)$$

For definiteness, the coefficient $\tfrac{1}{6}$ is chosen so that the new stress tensor, in the absence of the gravitational field,

$$t_{\mu\nu} = \partial_\mu\phi\partial_\nu\phi - \tfrac{1}{2}g_{\mu\nu}(\partial^\lambda\phi\,\partial_\lambda\phi + m^2\phi^2) - \tfrac{1}{6}(\partial_\mu\partial_\nu\phi^2 - g_{\mu\nu}\,\partial^2\phi^2) \quad (3\text{-}17.127)$$

has the property

$$t = -m^2\phi^2, \qquad (3\text{-}17.128)$$

and vanishes for $m = 0$. When (3-17.126) is used in the preceding equation we encounter

$$R_{\mu\kappa\nu\lambda}g^{\mu\nu}g^{\kappa\lambda} = R_{\mu\kappa}{}^\mu{}_\lambda g^{\kappa\lambda} = R \qquad (3\text{-}17.129)$$

and the modified spin 0 Lagrange function is

$$\mathcal{L}(\phi, g) = -(-g)^{1/2}\tfrac{1}{2}[\partial_\mu\phi\,g^{\mu\nu}\partial_\nu\phi + (m^2 + \tfrac{1}{6}R)\phi^2]. \qquad (3\text{-}17.130)$$

It would be interesting to verify that this system, with $m = 0$, is conformally invariant, in the sense of (3-17.114) supplemented by an appropriate response for $\phi(x)$. One sees that

$$\phi(x) \rightarrow (\lambda(x))^{-1/2}\phi(x) \qquad (3\text{-}17.131)$$

is suitable, if $\lambda(x)$ is constant. To complete the test it is sufficient to consider an infinitesimal variation of $\lambda(x)$ from unity, $\delta\lambda(x)$. Then,

$$\delta_\lambda\mathcal{L}(\phi, g) = \tfrac{1}{4}(-g)^{1/2}g^{\mu\nu}\partial_\nu\phi^2\partial_\mu\,\delta\lambda - \tfrac{1}{12}\phi^2(-g)^{1/2}g^{\mu\nu}\delta_\lambda R_{\mu\nu}, \quad (3\text{-}17.132)$$

where

$$(-g)^{1/2}g^{\mu\nu}\,\delta R_{\mu\nu} = \partial_\kappa[(-g)^{1/2}g^{\mu\nu}\,\delta\Gamma^\kappa_{\mu\nu}] - \partial_\nu[(-g)^{1/2}g^{\mu\nu}\,\delta\Gamma^\kappa_{\mu\kappa}] \quad (3\text{-}17.133)$$

is to be computed from

$$\delta g_{\mu\nu} = \delta\lambda g_{\mu\nu}. \qquad (3\text{-}17.134)$$

Involved here are

$$\delta\Gamma^\kappa_{\mu\kappa} = \delta[\partial_\mu \log(-g)^{1/2}] = 2\partial_\mu\,\delta\lambda \qquad (3\text{-}17.135)$$

and

$$g^{\mu\nu}\,\delta\Gamma^\kappa_{\mu\nu} = -g^{\kappa\mu}\partial_\mu\,\delta\lambda, \qquad (3\text{-}17.136)$$

which gives

$$\delta_\lambda\mathcal{L} = \tfrac{1}{4}(-g)^{1/2}g^{\mu\nu}\partial_\nu\phi^2\partial_\mu\,\delta\lambda + \tfrac{1}{4}\phi^2\partial_\nu[(-g)^{1/2}g^{\mu\nu}\partial_\mu\,\delta\lambda]$$
$$= \partial_\nu[\tfrac{1}{4}\phi^2(-g)^{1/2}g^{\mu\nu}\partial_\mu\,\delta\lambda], \tag{3-17.137}$$

thus assuring the invariance of the action, for $m=0$, under the group of conformal transformations.

Harold interjects a question.

H. Your preoccupation with conformal transformations in the context of what is customarily called general relativity makes me suspect that you intend to give a source theory setting for some of the more recent attempts to enlarge the framework of general relativity. I am thinking particularly of the ideas of P. Jordan and Brans-Dicke (B-D), and of Dicke's related efforts to establish a discrepancy between the residual perihelion precession of Mercury and the Einstein prediction. The B-D proposal is based on Mach's principle which, while a very intriguing notion, is devoid of immediate observational content. Can one suggest possibilities of modifying the Einstein theory on somewhat more physical grounds?

S. That is indeed my intention.

Let us begin by asking whether, through some extension of the theory, conformal invariance could be made an exact symmetry property. Certainly the mass term of (3–17.130) can be multiplied by $\sigma(x)^2$, where $\sigma(x)$ is a new scalar field that responds to conformal transformations as

$$\sigma(x) \to (\lambda(x))^{-1/2}\sigma(x). \tag{3-17.138}$$

Furthermore, the conformal response of the gravitational Lagrange function (3–17.75) could also be compensated by multiplication with $\sigma(x)^2$, at least for constant $\lambda(x)$ which leaves $R_{\mu\nu}$ unchanged. And, when one recognizes that $(-g)^{1/2}R\sigma^2$ is part of the conformally invariant Lagrange function (3–17.130) [with $m=0$ and $\phi\to\sigma$] the generalization to arbitrary $\lambda(x)$ is clear, leading to the complete conformally invariant

$$\mathcal{L}(g,\sigma,\phi) = \frac{1}{2\kappa}(-g)^{1/2}[R\sigma^2 + 6\partial_\mu\sigma g^{\mu\nu}\partial_\nu\sigma]$$
$$- (-g)^{1/2}\tfrac{1}{2}[\partial_\mu\phi g^{\mu\nu}\partial_\nu\phi + (m^2\sigma^2 + \tfrac{1}{6}R)\phi^2], \tag{3-17.139}$$

where $R = g^{\mu\nu}R_{\mu\nu}$ retains its meaning in terms of the $g_{\mu\nu}$ and their derivatives. It would seem that we have acquired a new massless particle of spin 0, represented by the scalar field $\sigma(x)$. But something is amiss. In a weak field situation, with

$$\sigma(x) \simeq 1 + \varphi(x), \tag{3-17.140}$$

the dominant φ terms in this Lagrange function are

$$\frac{3}{\kappa}\partial^\mu\varphi\partial_\mu\varphi + \varphi\left(\frac{R}{\kappa} - m^2\phi^2\right). \tag{3-17.141}$$

The φ derivative term has the wrong sign. And the source of the φ field, proportional to $R + \kappa t$, vanishes according to (3-17.107). All this indicates that the φ field does not describe a physical excitation. It can be transformed away, by introducing the conformal transformation with $(\lambda(x))^{1/2}$ equal to $\sigma(x)$, which reduces the latter to unity.

Nevertheless, the conformal invariant version is valuable in pointing out a new direction. As we have been learning in high energy particle physics, nature does not always select what we, in our ignorance, would judge to be the most symmetrical and harmonious possibility. Perhaps the formal invariance under conformal transformations is broken in such a way that a massless, zero spin particle does exist. Despite the principle of nonlocality for massless particles, one cannot object to such a particle on experimental grounds if it interacts with matter sufficiently more weakly than a graviton. In order to realize this suggestion, we must add an additional contribution to the Lagrange function that effectively reverses the sign of the σ derivative term and assigns it an arbitrary coefficient. That is not enough, however, for the σ field would still have no source; it is necessary to destroy the combination $R + \kappa t$. This can be done arbitrarily, but the possibilities are illustrated by two elementary alternative procedures: remove the σ^2 factor that multiplies m^2; remove the σ^2 factor that multiplies R. The first procedure gives a version of the B-D theory. The second one has the same practical consequences, and seems somewhat simpler. It is described below.

The modified Lagrange function is

$$\mathcal{L}(g, \sigma, \phi) = \frac{1+\alpha}{2\kappa} (-g)^{1/2} \left[R - \frac{2}{\alpha} \partial_\mu \sigma g^{\mu\nu} \partial_\nu \sigma \right]$$
$$- (-g)^{1/2} \tfrac{1}{2} [\partial_\mu \phi g^{\mu\nu} \partial_\nu \phi + (m^2 \sigma^2 + \tfrac{1}{6} R)\phi^2] \quad (3\text{-}17.142)$$

where $\alpha > 0$ is a new empirical constant. The factor $1 + \alpha$ is introduced in order to retain the original physical significance of κ. This Lagrange function leads to the following field equations:

$$G^{\mu\nu} = \frac{\kappa}{1+\alpha} t^{\mu\nu}, \quad (3\text{-}17.143)$$

$$-(-g)^{-1/2} \partial_\mu [(-g)^{1/2} g^{\mu\nu} \partial_\nu \sigma^2] = \frac{\alpha\kappa}{1+\alpha} t, \quad (3\text{-}17.144)$$

where $t_{\mu\nu}$ is the total stress tensor, adding to the matter contribution that of the σ field,

$$t_{\mu\nu} = t_{m\mu\nu} + \frac{2}{\alpha} \frac{1+\alpha}{\kappa} (\partial_\mu \sigma \partial_\nu \sigma - \tfrac{1}{2} g_{\mu\nu} \partial_\kappa \sigma g^{\kappa\lambda} \partial_\lambda \sigma),$$
$$t = t_m - \frac{2}{\alpha} \frac{1+\alpha}{\kappa} \partial_\kappa \sigma g^{\kappa\lambda} \partial_\lambda \sigma. \quad (3\text{-}17.145)$$

The equations (3–17.143–145) are independent of the model used for matter, provided the matter part of the Lagrange function has been made conformally invariant by the local introduction of the σ field, implying

$$\delta_\lambda \int (dx)\mathcal{L}_m = \int (dx)\left[\tfrac{1}{2}(-g)^{1/2}t_m\,\delta\lambda - \frac{\partial \mathcal{L}_m}{\partial \sigma}\frac{1}{2}\sigma\,\delta\lambda\right] = 0 \quad (3\text{–}17.146)$$

and

$$\sigma\frac{\partial \mathcal{L}_m}{\partial \sigma} = (-g)^{1/2}t_m. \qquad (3\text{–}17.147)$$

The field equation (3–17.144) first appears as

$$-(-g)^{-1/2}\partial_\mu[(-g)^{1/2}g^{\mu\nu}\partial_\nu\sigma] = \frac{1}{2}\frac{\alpha\kappa}{1+\alpha}t_m\frac{1}{\sigma}, \qquad (3\text{–}17.148)$$

which is rewritten in the stated form by eliminating t_m. It is also obtained directly by applying the action principle to the conformal response of the noninvariant Lagrange function,

$$\delta_\lambda \int (dx)\mathcal{L} = \frac{1+\alpha}{2\kappa}\int (dx)(-g)^{1/2}\left[R\,\delta\lambda + \frac{1}{\alpha}g^{\mu\nu}\partial_\nu\sigma^2\partial_\mu\,\delta\lambda\right] = 0, \qquad (3\text{–}17.149)$$

giving

$$-(-g)^{-1/2}\partial_\mu[(-g)^{1/2}g^{\mu\nu}\partial_\nu\sigma^2] = -\alpha R = \frac{\alpha\kappa}{1+\alpha}t. \qquad (3\text{–}17.150)$$

The quickest way to draw the practical consequences of the modified theory is by returning to the source procedures of Section 2–4, now supplemented by a term referring to spin 0 particles. The respective spin (helicity) 2 and spin 0 sources are appropriately normalized as $[\kappa/(1+\alpha)]^{1/2}T^{\mu\nu}$ and $[\tfrac{1}{2}\alpha\kappa/(1+\alpha)]^{1/2}T$, where the latter is based on the field $\tfrac{1}{2}(\sigma^2 - 1)$. This produces

$$W = \frac{1}{2}\frac{\kappa}{1+\alpha}\int (dx)(dx')[T^{\mu\nu}(x)D_+(x - x')T_{\mu\nu}(x') - \tfrac{1}{2}T(x)D_+(x - x')T(x')$$
$$+ \tfrac{1}{2}\alpha T(x)D_+(x - x')T(x')], \quad (3\text{–}17.151)$$

and the interaction energy with a fixed body of mass M, replacing (2–4.36, 37), is

$$E_{\text{int.}}(x^0) = -\frac{\kappa}{8\pi}M\int (dx)\frac{1}{|\mathbf{x}|}\left[t^{00}(\mathbf{x}, x^0) + \frac{1-\alpha}{1+\alpha}t_{kk}(\mathbf{x}, x^0)\right]. \qquad (3\text{–}17.152)$$

Under the circumstances $t_{kk} \ll t^{00}$, the Newtonian potential energy is retained, along with the gravitational red shift. For light, with $t_{kk} = t^{00}$, the deflection and the slowing of the speed of light are reduced by the factor $1/(1+\alpha)$. In discussing perihelion precession, the kinetic energy correction factor $1 + (2T/m)$ is changed into $1 + [(1-\alpha)/(1+\alpha)](2T/m)$, which gives

$$V_{\text{eff.}} = V - \left(1 + 2\frac{1-\alpha}{1+\alpha}\right)\left(\frac{V^2}{m}\right), \qquad (3\text{–}17.153)$$

and the perihelion precession is reduced by the factor $(1 - \frac{1}{3}\alpha)/(1 + \alpha)$. A convenient presentation of the correction factors is

light phenomena: $1 - [\alpha/(1 + \alpha)]$;

perihelion precession: $1 - \frac{4}{3}[\alpha/(1 + \alpha)]$.

$$(3\text{–}17.154)$$

At the time of writing, measurements on the time delays in radar echoes from Venus have produced a result that is 0.9 ± 0.2 of that expected from the tensor, or Einstein, theory. This limits the parameter α to 0.1 ± 0.2. It has also been claimed that 8 percent of the Mercury perihelion precession can be assigned to a solar mass quadrupole moment, leaving 92 percent to be accounted for by the scalar-tensor modification of the Einstein theory. This gives

$$\alpha = 0.06. \qquad (3\text{–}17.155)$$

Until independent evidence for the solar quadrupole moment is forthcoming, perhaps from continued observation of the asteroid Icarus, the question whether a weakly coupled, massless, spin 0 particle exists must still be considered *sub judice*.

Now that a scalar-tensor theory of gravitation has been devised without reference to Mach's principle, perhaps a word about this cosmic speculation is in order. It is a natural hypothesis from the source theory viewpoint, for it asserts that the weak field decompositions,

$$g_{\mu\nu}(x) \simeq g_{\mu\nu} + 2h_{\mu\nu}(x), \qquad \sigma(x) \simeq 1 + \varphi(x), \qquad (3\text{–}17.156)$$

identify the fields of nearby sources—$2h_{\mu\nu}(x)$, $\varphi(x)$—and the fields of very distant sources—$g_{\mu\nu}$, 1. To draw a qualitative inference from this idea without having to use weak field approximations throughout the cosmos, we consider an averaged situation in which the 'mass of the universe,' M, and the 'radius of the universe,' R, provide the only scales for fields and sources. This is expressed by

$$g_{\mu\nu}(x) = \gamma_{\mu\nu}(x/R), \qquad \sigma(x) = s(x/R)$$
$$t_{\mu\nu}(x) = (M/R^3)\tau_{\mu\nu}(x/R), \qquad t = (M/R^3)\tau(x/R),$$

$$(3\text{–}17.157)$$

where the functions $\gamma_{\mu\nu}$, s, $\tau_{\mu\nu}$, τ are all of order unity which is to be judged here on a logarithmic scale. Then, if we exhibit only the scale factors on opposite sides of the two field equations (3–17.143, 144), they read

$$\frac{1}{R^2} \sim \frac{\kappa}{1 + \alpha} \frac{M}{R^3}, \qquad \frac{1}{R^2} \sim \frac{\alpha\kappa}{1 + \alpha} \frac{M}{R^3}. \qquad (3\text{–}17.158)$$

The implications are

$$\alpha \sim 1, \qquad (3\text{–}17.159)$$

logarithmically consistent with $\alpha = 0.6 \times 10^{-1}$, and

$$\kappa M/R \sim 1. \qquad (3\text{–}17.160)$$

The latter is a well-known empirical connection between the conventional orders of magnitude, $R \sim 10^{28}$ cm, $M \sim 10^{54}$ g $\sim 10^{91}$ cm^{-1}, and the gravitational constant $\kappa \sim 10^{-64}$ cm^2 [Eq. (2–4.40)]. If this relation is viewed as characteristic of Mach's principle, it cannot be said of source theory that the situation is qualitatively altered by the introduction of the scalar field, for (3–17.160) also applies to the pure tensor theory.

The suggestion that the value of α would be significantly restricted by a common solution of the two field equations is not quite borne out on examining a most elementary model. To describe it we use geometrical language (for the first time) and characterize the space as a homogeneous, isotropic, three-dimensionally flat space. This is one of the Friedmann models:

$$-g_{\mu\nu} \, dx^\mu \, dx^\nu = dt^2 - S(t)^2 (dx_k)^2. \tag{3–17.161}$$

To be consistent with the purely time-dependent tensor field, the scalar field is also of that character, $\sigma(t)$. If it is assumed that the matter stress tensor has only the energy component, $t_{00} = \rho_m$, the field equations imply

$$3 \frac{\dot{S}^2}{S^2} = \frac{1}{\alpha} \dot{\sigma}^2 + \frac{\kappa}{1+\alpha} \rho_m,$$

$$-2 \frac{\ddot{S}}{S} - \frac{\dot{S}^2}{S^2} = \frac{1}{\alpha} \dot{\sigma}^2, \tag{3–17.162}$$

$$\frac{\sigma}{S^3} \frac{d}{dt} (S^3 \dot{\sigma}) = -\frac{1}{2} \frac{\alpha \kappa}{1+\alpha} \rho_m,$$

where the dot designates time derivative. We shall be content to pick out a particular solution:

$$S(t) = (t/T)^{1/3}, \qquad \sigma(t) = (\alpha/3)^{1/2} \log (t/t_0), \qquad \rho_m = 0, \tag{3–17.163}$$

where T indicates the present era, and

$$t_0 = T e^{-(3/\alpha)^{1/2}}. \tag{3–17.164}$$

This solution describes the matter density as negligible compared to the energy density contributed by the σ field, which, evaluated at the present era, is

$$\rho = \frac{1+\alpha}{3\kappa T^2} = \frac{3(1+\alpha)}{\kappa} H^2, \tag{3–17.165}$$

where H is the Hubble expansion parameter

$$H = \dot{S}/S = 1/(3T). \tag{3–17.166}$$

Here is an illustration of (3–17.160), with $R \sim T$, $M \sim \rho T^3$. The currently accepted value of $H \sim 2.4 \times 10^{-18}$ sec^{-1} implies $\rho \sim 1.0 \times 10^{-29}$ g/cm^3, assuming that α is fairly small compared to unity. Presumably the simplifying feature of the model, $\rho_m = 0$, means that the matter density is at least an order of magnitude less than this value of ρ, which is not inconsistent with the observational data. The only sensitive dependence on α occurs in t_0, the time at which the laws of physics become qualitatively similar to those now prevailing, in that $\sigma(t) > 0$, $t > t_0$. To the extent that there is evidence for the maintenance of these laws over a significant fraction of the age of the universe, α is correspondingly bounded from above. The nominal value $\alpha = 0.06$ gives $t_0 \sim 10^{-3}T$.

The ease with which integer spin Lagrange functions acquire general coordinate invariance by suitably introducing $g_{\mu\nu}(x)$ does not extend to particles of integer $+\frac{1}{2}$ spin. To appreciate the difference let us follow the earlier weak field procedure, now using the spin $\frac{1}{2}$ Lagrange function

$$\mathcal{L}(\psi) = -\tfrac{1}{2}\bar{\psi}\gamma^0[\gamma^\mu(1/i)\partial_\mu + m]\psi \qquad (3\text{-}17.167)$$

and the stress tensor [(3–7.120)]

$$t_{\mu\nu} = \tfrac{1}{2}\bar{\psi}\gamma^0\tfrac{1}{2}[\gamma_\mu(1/i)\partial_\nu + \gamma_\nu(1/i)\partial_\mu]\psi + g_{\mu\nu}\mathcal{L} \qquad (3\text{-}17.168)$$

to form

$$\mathcal{L}(\psi, h) = \mathcal{L}(\psi) + t_{\mu\nu}h^{\mu\nu} \simeq -(1+h)\tfrac{1}{2}\bar{\psi}\gamma^0[\gamma_\mu(g^{\mu\nu} - h^{\mu\nu})(1/i)\partial_\nu + m]\psi. \qquad (3\text{-}17.169)$$

Appearing here is, not

$$g^{\mu\nu}(x) \simeq g^{\mu\nu} - 2h^{\mu\nu}(x), \qquad (3\text{-}17.170)$$

but something resembling the square root of this combination,

$$g^{\mu\nu}(x) \simeq (g^{\mu\kappa} - h^{\mu\kappa}(x))g_{\kappa\lambda}(g^{\nu\lambda} - h^{\nu\lambda}(x)). \qquad (3\text{-}17.171)$$

The necessary generalizations can be carried out, however, if we distinguish the vector index in $g^{\mu\nu} - h^{\mu\nu}(x)$ that is associated with the coordinate derivative ∂_ν from the vector index that is tied to the matrices γ_μ. To emphasize this we henceforth use latin letters to indicate a local Minkowski coordinate system, which will be retained for the description of spin. That notational distinction is used in writing the generalization of (3–71.171) as

$$g^{\mu\nu}(x) = e^{\mu a}(x)g_{ab}e^{\nu b}(x) = e^{\mu a}(x)e^\nu_a(x), \qquad (3\text{-}17.172)$$

which is maintained under general coordinate transformations if

$$\bar{e}^\mu_a(\bar{x}) = e^\nu_a(x)\partial_\nu\bar{x}^\mu. \qquad (3\text{-}17.173)$$

There is an independent invariance under local Lorentz transformations:

$$\bar{e}^{\mu a}(x) = l^a{}_b(x)e^{\mu b}(x), \qquad (3\text{-}17.174)$$

where

$$l^a{}_c(x)g_{ab}l^b{}_d(x) = g_{cd}. \qquad (3\text{-}17.175)$$

With the definition

$$e_{\mu a}(x) = g_{\mu\nu}(x)e^\nu_a(x) \qquad (3\text{-}17.176)$$

we deduced from (3–17.172) that

$$e^{\mu a}(x)e_{\nu a}(x) = \delta^\mu_\nu \qquad (3\text{-}17.177)$$

and

$$g_{\mu\nu}(x) = e_\mu^a(x)g_{ab}e_\nu^b(x) = e_\mu^a(x)e_{\nu a}(x). \tag{3-17.178}$$

Regarding the first form of the last equation as a matrix product, we infer the determinantal relation

$$-g(x) = (\det e_{\mu a}(x))^2 \tag{3-17.179}$$

or

$$(-g(x))^{1/2} = \det e_{\mu a}(x) \equiv e(x). \tag{3-17.180}$$

The indicated provisional generalization of (3-17.169) is

$$\mathcal{L}(\psi, e) = -e(x)\tfrac{1}{2}\bar\psi(x)\gamma^0[\gamma^a e_a^\nu(x)(1/i)\partial_\nu + m]\psi(x). \tag{3-17.181}$$

The corresponding action is certainly invariant under arbitrary coordinate transformations if the field $\psi(x)$ is transformed as a scalar,

$$\bar\psi(\bar x) = \psi(x). \tag{3-17.182}$$

But we shall also require invariance under arbitrary local Lorentz transformations, for which (3-17.181) is inadequate.

To appreciate the physical significance of the last requirement, let us consider the response of the matter action to variations of $e_a^\mu(x)$:

$$\delta_e \int (dx)\mathcal{L}(\psi, e) = -\int (dx)e(x)\,\delta e_a^\mu(x)t_\mu^a(x), \tag{3-17.183}$$

which defines $t_\mu^a(x)$. An infinitesimal local Lorentz transformation is

$$\delta e_a^\mu(x) = \delta\omega_{ab}(x)e^{\mu b}(x), \tag{3-17.184}$$

with

$$\delta\omega_{ab}(x) = -\delta\omega_{ba}(x). \tag{3-17.185}$$

Invariance with respect to arbitrary transformations of this type thus demands that

$$t^{ba}(x) = e^{\mu b}(x)t_\mu^a(x) = t^{ab}(x), \tag{3-17.186}$$

which is also the symmetry property

$$t_{\mu\nu}(x) = e_{\mu a}(x)e_{\nu b}(x)t^{ab}(x) = t_{\nu\mu}(x). \tag{3-17.187}$$

The tensor that is required to be symmetrical is indeed the stress tensor of matter. This follows on writing

$$\delta e_a^\mu t_\mu^a = \delta e_a^\mu e^{\nu a}t_{\mu\nu} = \tfrac{1}{2}\delta(e_a^\mu e^{\nu a})t_{\mu\nu}, \tag{3-17.188}$$

which reproduces the stress tensor definition (3-7.100),

$$\delta_e \int (dx)\mathcal{L} = \delta_g \int (dx)\mathcal{L} = -\tfrac{1}{2}\int (dx)(-g)^{1/2}\,\delta g^{\mu\nu}t_{\mu\nu}. \tag{3-17.189}$$

The response of $\psi(x)$ to the local Lorentz transformation (3-17.174) is

$$\bar\psi(x) = L(l(x))\psi(x), \tag{3-17.190}$$

where

$$L^T \gamma^0 L = \gamma^0, \qquad L^{-1} \gamma^a L = l^a_b \gamma^b. \tag{3-17.191}$$

It is the action of the coordinate derivatives on $L(l(x))$ that disturbs the local invariance of the Lagrange function (3–17.181). A coordinate displacement induces an infinitesimal Lorentz transformation and an associated field transformation:

$$l^a_b(x + dx) = l^a_c(x)[\delta^c_b + d\omega^c_b(x)],$$
$$L(l(x + dx)) = L(l(x))[1 + \tfrac{1}{2} i \, d\omega_{ab}(x) \tfrac{1}{2} \sigma^{ab}], \tag{3-17.192}$$

in which

$$d\omega_{ab} = -d\omega_{ba} = l^c_a \, dl_{cb}, \tag{3-17.193}$$

and therefore

$$L^{-1} \partial_\mu L = \tfrac{1}{4} i l^c_a \partial_\mu l_{cb} \sigma^{ab}. \tag{3-17.194}$$

In order to compensate this effect the coordinate derivative in the Lagrange function is replaced by

$$\partial_\mu - \tfrac{1}{4} i \omega_{a\mu b} \sigma^{ab}, \tag{3-17.195}$$

where $\omega_{a\mu b}(x)$ behaves like a covariant vector with respect to general coordinate transformations, and responds to local Lorentz transformations in such a manner that

$$L^{-1}(\partial_\mu - \tfrac{1}{4} i \bar\omega_{a\mu b} \sigma^{ab}) L = \partial_\mu - \tfrac{1}{4} i \omega_{a\mu b} \sigma^{ab}. \tag{3-17.196}$$

The required transformation law is

$$\bar\omega_{a'\mu b'} l^{a'}_a l^{b'}_b = \omega_{a\mu b} + l^c_a \partial_\mu l_{cb}. \tag{3-17.197}$$

A fundamental mixed tensor is defined by the commutator

$$[\partial_\mu - \tfrac{1}{4} i \omega_{a\mu b}(x) \sigma^{ab}, \partial_\nu - \tfrac{1}{4} i \omega_{c\nu d}(x) \sigma^{cd}] = \tfrac{1}{4} i R_{\mu\nu ab}(x) \sigma^{ab}. \tag{3-17.198}$$

It is

$$-R_{\mu\nu ab} = \partial_\mu \omega_{a\nu b} - \partial_\nu \omega_{a\mu b} - \omega_{a\mu}{}^c \omega_{c\nu b} + \omega_{a\nu}{}^c \omega_{c\mu b}, \tag{3-17.199}$$

which has the character of an antisymmetrical tensor of the second rank for general coordinate transformations, and of an antisymmetrical tensor of the second rank with respect to local Lorentz transformations. A scalar in both senses is constructed by

$$e^{\mu a}(x) e^{\nu b}(x) R_{\mu\nu ab}(x) = R(x), \tag{3-17.200}$$

and provides the basis for a gravitational Lagrange function:

$$2\kappa \mathcal{L}(e, \omega) = e e^{\mu a} e^{\nu b} R_{\mu\nu ab}. \tag{3-17.201}$$

In the weak field limit, where the linguistic distinction of the indices is removed and

$$e^{\mu a}(x) \simeq g^{\mu a} - h^{\mu a}(x), \tag{3-17.202}$$

400   Fields   **Chap. 3**

this Lagrange function reduces to (3–17.38), apart from a divergence term.

The application of the stationary action principle to variations of $\omega_{a\mu b}$ in $\mathcal{L}(e, \omega)$ gives

$$\partial_\nu[e(e^{\mu a}e^{\nu b} - e^{\mu b}e^{\nu a})] - \omega^a{}_{\nu c}[e(e^{\mu c}e^{\nu b} - e^{\mu b}e^{\nu c})] - \omega^b{}_{\nu c}[e(e^{\mu a}e^{\nu c} - e^{\mu c}e^{\nu a})] = 0. \tag{3–17.203}$$

The following definition of a quantity that is not a tensor,

$$\Omega_a{}^\mu{}_b = e_a^\nu \partial_\nu e_b^\mu - e_b^\nu \partial_\nu e_a^\mu = -\Omega_b{}^\mu{}_a, \tag{3–17.204}$$

together with

$$\Omega_{abc} = e_{\mu b}\Omega_a{}^\mu{}_c, \qquad \omega_{abc} = e_b^\mu \omega_{a\mu c}, \tag{3–17.205}$$

enables one to present this equation as

$$\Omega_{abc} + \omega_{acb} - \omega_{cab} + g_{bc}\lambda_a - g_{ab}\lambda_c = 0, \tag{3–17.206}$$

where

$$\lambda_a = e^{-1}\partial_\nu(e e_a^\nu) - \omega_{ab}{}^b = -(\Omega_{ab}{}^b + \omega_{ab}{}^b), \tag{3–17.207}$$

according to the determinantal formula

$$\partial_\nu e = e e^{\mu a}\partial_\nu e_{\mu a}. \tag{3–17.208}$$

On contracting b and c in (3–17.206) we get various multiples of λ_a, implying that

$$\lambda_a = 0 \tag{3–17.209}$$

and reducing (3–17.206) to

$$\Omega_{abc} = \omega_{cab} - \omega_{acb}, \tag{3–17.210}$$

from which $\lambda_a = 0$ is still recovered. The last relation is solved by

$$\omega_{abc} = \tfrac{1}{2}[\Omega_{cab} + \Omega_{bca} - \Omega_{abc}]; \tag{3–17.211}$$

it is the strong field generalization of (3–17.17).

Another weak field property, Eq. (3–17.27), is generalized by defining

$$\Gamma_{abc}(x) = \omega_{abc}(x) - e_{\nu c}(x)e_b^\mu(x)\partial_\mu e_a^\nu(x)$$
$$= \Gamma_{bac}(x), \tag{3–17.212}$$

which symmetry property expresses the relation of (3–17.210). The additional definition

$$\Gamma^\lambda_{\mu\nu} = e_\mu^a e_\nu^b \Gamma_{abc}e^{\lambda c} = \Gamma^\lambda_{\nu\mu} \tag{3–17.213}$$

enables one to write (3–17.212) as

$$\partial_\lambda e_a^\nu + \Gamma^\nu_{\lambda\kappa}e_a^\kappa - \omega_{a\lambda b}e^{\nu b} = 0. \tag{3–17.214}$$

Multiplication of this equation by $e^{\mu a}$, followed by symmetrization in μ and ν, removes the $\omega_{a\lambda b}$ term and gives

$$\partial_\lambda g^{\mu\nu} + \Gamma^\mu_{\lambda\kappa}g^{\kappa\nu} + \Gamma^\nu_{\lambda\kappa}g^{\mu\kappa} = 0. \tag{3–17.215}$$

This is recognized as the statement that the covariant derivative of $g^{\mu\nu}$ vanishes, and identifies $\Gamma^\lambda_{\mu\nu}$ with the quantities of (3-17.91), known usually as Christoffel symbols. After this, it is abundantly clear that the two objects defined in (3-17.124) and (3-17.199) are connected by

$$R_{\mu\nu\kappa\lambda} = R_{\mu\nu ab}e^a_\kappa e^b_\lambda,\qquad(3\text{-}17.216)$$

where the correctness of the algebraic sign can be verified in the weak field limit. Thus, the two gravitational Lagrange functions, $\mathcal{L}(g, \Gamma)$ and $\mathcal{L}(e, \omega)$ are identical.

Returning to the spin $\frac{1}{2}$ Lagrange function (3-17.181), we insert the co-ordinate derivative generalization stated in (3-17.195) and obtain

$$\mathcal{L}(\psi, e) = -e\tfrac{1}{2}\psi\gamma^0[\gamma^a e^\nu_a(-i\partial_\nu - \tfrac{1}{4}\omega_{b\nu c}\sigma^{bc}) + m]\psi\qquad(3\text{-}17.217)$$

where, it should be noted, the total anticommutativity of the field extracts the antisymmetrical part of the matrices $\gamma^0\gamma^a\sigma^{bc}$. This removes the terms with $a = b$ or $a = c$, which are proportional to the symmetrical matrices $\gamma^0\gamma^a$. Then, since

$$a \neq b, c: \qquad\qquad \gamma^a\sigma^{bc} = \epsilon^{abcd}i\gamma_d\gamma_5,\qquad(3\text{-}17.218)$$

we can write the Lagrange function as

$$\mathcal{L}(\psi, e) = -e\tfrac{1}{2}\psi\gamma^0[\gamma^a e^\nu_a(1/i)\partial_\nu + {}^*\omega^a i\gamma_a\gamma_5 + m]\psi,\qquad(3\text{-}17.219)$$

where

$$ {}^*\omega^d = \tfrac{1}{4}\epsilon^{abcd}\omega_{abc}.\qquad(3\text{-}17.220)$$

The notation $\mathcal{L}(\psi, e)$ might have been elaborated as $\mathcal{L}(\psi, e, \omega)$. When this structure is added to the gravitational Lagrange function (3-17.201), in which e^μ_a and $\omega_{a\nu b}$ are used as independent variables, the $\omega_{a\nu b}$ dependence of $\mathcal{L}(\psi, e, \omega)$ produces an additional term in (3-17.203), which removes the symmetry property noted in (3-17.212). This is a natural form of the theory. But we shall prefer, for simplicity only, to regard ω_{abc} as defined by (3-17.211) and therefore not subject to redefinition through the appearance of ω_{abc} in the matter Lagrange function. In order to identify the stress tensor $t_{\mu\nu}$ directly, we consider the special variation

$$\delta e^\mu_a = \tfrac{1}{2}\delta g^{\mu\nu}e_{\nu a},\qquad(3\text{-}17.221)$$

which is consistent with the construction of $g^{\mu\nu}$ from the e^μ_a, and gives

$$\delta_e \int (dx)\mathcal{L}(\psi, e) = -\int (dx)e\,\delta e^\mu_a t^a_\mu$$
$$= -\tfrac{1}{2}\int (dx)(-g)^{1/2}\,\delta g^{\mu\nu}t_{\mu\nu}.\qquad(3\text{-}17.222)$$

A little calculation shows that

$$\delta\,{}^*\omega^d = -\delta g^{\mu\nu}\tfrac{1}{8}\epsilon^{abcd}e_{\mu a}\omega_{b\nu c},\qquad(3\text{-}17.223)$$

and thus,

$$t_{\mu\nu} = \tfrac{1}{2}\bar{\psi}\gamma^0\gamma^a(1/2i)(e_{\mu a}\partial_\nu + e_{\nu a}\partial_\mu)\psi$$
$$- \tfrac{1}{4}\epsilon^{abcd}\tfrac{1}{2}\bar{\psi}\gamma^0 i\gamma_d\gamma_5\psi\tfrac{1}{2}(e_{\mu a}\omega_{b\nu c} + e_{\nu a}\omega_{b\mu c}) + g_{\mu\nu}e^{-1}\mathcal{L}, \quad (3\text{-}17.224)$$

where the first two terms can be united through the reintroduction of the spinor covariant derivative (3–17.195).

The scalar derived from this tensor is

$$t = -m\tfrac{1}{2}\bar{\psi}\gamma^0\psi + 3e^{-1}\mathcal{L}. \quad (3\text{-}17.225)$$

A simplification can be made through the use of the field equation implied by the Lagrange function. But what is required here can be obtained more directly by applying the action principle to the particular field variation

$$\delta\psi(x) = -\tfrac{3}{4}\,\delta\lambda(x)\psi(x). \quad (3\text{-}17.226)$$

The response of the matter action term is

$$\delta_\lambda \int (dx)\mathcal{L}(\psi, e) = -\tfrac{3}{2}\int (dx)\mathcal{L}\,\delta\lambda(x), \quad (3\text{-}17.227)$$

since no contribution involving $\partial_\mu\,\delta\lambda$ appears, owing to the symmetry of the matrices $\gamma^0\gamma^a$. The conclusion is that the Lagrange function vanishes (at points not occupied by particle sources) and

$$t = -m\tfrac{1}{2}\bar{\psi}\gamma^0\psi. \quad (3\text{-}17.228)$$

The last consideration is intimately related to the possibility of exhibiting a conformally invariant matter Lagrange function through the introduction of the scalar field $\sigma(x)$.

Conformal transformations on the tetrad of vector fields $e_\mu^a(x)$ appear as

$$e_\mu^a(x) \rightarrow (\lambda(x))^{1/2}e_\mu^a(x) \quad (3\text{-}17.229)$$

or, using infinitesimal transformations,

$$\delta e_\mu^a = \tfrac{1}{2}\,\delta\lambda e_\mu^a, \qquad \delta e_a^\mu = -\tfrac{1}{2}\,\delta\lambda e_a^\mu \quad (3\text{-}17.230)$$

and

$$\delta e = 2\,\delta\lambda e. \quad (3\text{-}17.231)$$

Some derived conformal responses are

$$\delta\omega_{abc} = -\tfrac{1}{2}\,\delta\lambda\omega_{abc} + \tfrac{1}{2}(e_a^\nu g_{bc} - e_c^\nu g_{ab})\partial_\nu\,\delta\lambda \quad (3\text{-}17.232)$$

and

$$\delta\,{}^*\omega_a = -\tfrac{1}{2}\delta\lambda\,{}^*\omega_a. \quad (3\text{-}17.233)$$

The appropriate conformal behavior of $\psi(x)$ is already stated in (3–17.226). It is such that the replacement of m by $m\sigma$ in (3–17.219) suffices to produce a conformally invariant Lagrange function. The derivation of the scalar t through the σ-dependence of $\mathcal{L}(\psi, e, \sigma)$ according to (3-17.147) then gives (3-17.228) directly, with $m\sigma$ substituted for m.

The temptation to extend this discussion to arbitrary multispinor fields will be resisted. Instead, we turn to the long deferred topic of the gravitational model of particle and graviton sources. It is worth appreciating why it is that we have managed thus far without examining this question. The arena of gravitational phenomena is confined essentially to astronomical bodies, which are beyond our experimental control. Accordingly, we have no overt use for sources, which give idealized expression to the experimenter's ability to manipulate the physical situation being studied. The graviton source concept has already fulfilled its primary mission by serving as the model upon which the coordinate invariant dynamical theory has been erected. Nevertheless, some remarks are called for, although, as the preceding comment indicates, they can be limited to the use of particle and graviton sources under the weak field gravitational conditions prevailing in terrestrial experiments. The following brief analysis need not be applicable to experiments conducted on a spaceship in close orbit about a rapidly spinning neutron star.

The first point at issue has previously been raised for charged particle sources. The generalized stress tensor conservation law, (3-17.103, 104), will fail inside particle sources unless one recognizes the pre-existence of the energy and momentum that is transferred to the emitted particle. There is, however, no electromagnetic analogue to the graviton source problem. Photons are electrically neutral, whereas gravitons carry energy-momentum which must also be transferred rather than created within the source. Just as an explicit A_μ-dependence was introduced to provide the correct gauge transformation behavior of charged particle sources, these source problems can be viewed as the search for the explicit $g_{\mu\nu}$-dependence that will give the various sources the correct response to general coordinate transformations.

The simplest example is a scalar source $K(x)$, appearing in the action through the term

$$\int (dx)(-g(x))^{1/2}K(x)\phi(x). \tag{3-17.234}$$

We must replace x^μ by a functional of the $g_{\mu\nu}$, $x^\mu(x, g)$, such that under a general coordinate transformation

$$x^\mu(\bar{x}, \bar{g}) = x^\mu(x, g), \tag{3-17.235}$$

for then

$$\int (dx)(-\bar{g}(x))^{1/2}K(x(x,\bar{g}))\bar{\phi}(x) = \int (d\bar{x})(-\bar{g}(\bar{x}))^{1/2}K(x(\bar{x},\bar{g}))\bar{\phi}(\bar{x})$$

$$= \int (dx)(-g(x))^{1/2}K(x(x,g))\phi(x) \tag{3-17.236}$$

shows the required dynamical equivalence of the fields $g_{\mu\nu}(x)$, $\phi(x)$ and $\bar{g}_{\mu\nu}(x)$, $\bar{\phi}(x)$. Under weak field conditions we write

$$x^\mu(x, g) = x^\mu + X^\mu(h), \qquad (3\text{-}17.237)$$

where the invariance property (3–17.235), stated for infinitesimal transformations $(\bar{x}^\mu = x^\mu - \delta x^\mu)$, requires that

$$\delta X^\mu(h) = \delta x^\mu. \qquad (3\text{-}17.238)$$

This must hold as a consequence of the gauge transformation (3–17.45). A solution is

$$X^\lambda(h) = \int (dx')(dx'')f^\mu(x - x')f^\nu(x' - x'')\Gamma^\lambda_{\mu\nu}(x'')$$

$$= 2\int (dx')f_\nu(x - x')h^{\nu\lambda}(x') - \partial^\lambda \int (dx')(dx'')f^\mu(x - x')f^\nu(x' - x'')h_{\mu\nu}(x''), \qquad (3\text{-}17.239)$$

where $f^\mu(x - x')$ is one of the familiar class of functions obeying

$$\partial_\mu f^\mu(x - x') = \delta(x - x'). \qquad (3\text{-}17.240)$$

The matter stress tensor that is now derived, stated in the absence of the gravitation field for simplicity, is the conserved object

$$t^{\mu\nu}_{\text{cons.}}(x) = t^{\mu\nu}(x) + g^{\mu\nu}K(x)\phi(x)$$

$$- \int (dx')[f^\mu(x - x')\partial^{\nu'}K(x') + f^\nu(x - x')\partial^{\mu'}K(x')]\phi(x')$$

$$+ \int (dx')(dx'')f^\mu(x - x')f^\nu(x' - x'')\partial^{\lambda''}(\phi(x'')\partial''_\lambda K(x'')), \qquad (3\text{-}17.241)$$

where $t^{\mu\nu}$ is the tensor given in (3–7.8), which is such that

$$\partial_\mu(t^{\mu\nu} + g^{\mu\nu}K\phi) = \phi\partial^\nu K. \qquad (3\text{-}17.242)$$

The possibility of graviton emission occurring directly from the matter source is exhibited, for single graviton radiation, by writing the coordinate invariant form of the source term (3–17.234) as

$$\int (dx)(1 + h)\phi(K + X^\mu(h)\partial_\mu K). \qquad (3\text{-}17.243)$$

Assuming that the graviton detection sources do not overlap the K support region, one can use the source-free, weak gravitational field equations (3–17.13, 14) to derive

$$\partial_\lambda X^\lambda = \int (dx')(dx'')f^\mu(x - x')f^\nu(x' - x'')\partial''_\mu\partial''_\nu h(x'')$$

$$= h(x). \qquad (3\text{-}17.244)$$

This enables one to present (3–17.243) as

$$\int (dx)\phi[K + \partial_\mu(X^\mu(h)K)] = \int (dx)K[\phi - X^\mu(h)\partial_\mu\phi]. \quad (3\text{--}17.245)$$

Alternatively, one might have begun with the last form, where the additional term serves to remove the response of $\phi(x)$ to infinitesimal coordinate transformations. Similar discussions can be given for any other type of matter source and field, with appropriate attention to their transformation properties.

The weak field form of the graviton source term in the action is

$$\int (dx)T_0^{\mu\nu}(x)h_{\mu\nu}(x) \rightarrow \tfrac{1}{2}\int (dx)T_0^{\mu\nu}(x)g_{\mu\nu}(x), \quad (3\text{--}17.246)$$

where

$$\partial_\mu T_0^{\mu\nu}(x) = 0, \quad (3\text{--}17.247)$$

and a constant is added to arrive at the second version of (3–17.246). The physical property to be represented is that the radiation of an additional graviton can accompany the working of a graviton source as well as a matter source. The mathematical problem is the removal of the response of $g_{\mu\nu}(x)$ to infinitesimal coordinate transformations, apart from gradient terms—gauge transformations—which do not contribute in (3–17.246). If we use the symbol $\sim$ to indicate identity apart from gradient terms, the response of $g_{\mu\nu}$ to infinitesimal coordinate transformations, Eq. (3–17.96), is expressed by

$$\delta g_{\mu\nu} \sim -2\Gamma_{\mu\nu}^\lambda g_{\kappa\lambda}\,\delta x^\kappa. \quad (3\text{--}17.248)$$

Then, applying the weak field statement (3–17.45),

$$\delta(g_{\mu\nu} + 2\Gamma_{\mu\nu}^\lambda X_\lambda) \sim 2(\partial_\mu\partial_\nu\,\delta x^\lambda)X_\lambda \sim -[\partial_\mu\,\delta x^\lambda\partial_\nu X_\lambda + \partial_\mu X^\lambda\partial_\nu\,\delta x_\lambda]$$
$$= -\delta[\partial_\mu X^\lambda\partial_\nu X_\lambda], \quad (3\text{--}17.249)$$

which gives the required generalization of (3–17.246):

$$\tfrac{1}{2}\int (dx)T_0^{\mu\nu}[g_{\mu\nu} + 2\Gamma_{\mu\nu}^\lambda X_\lambda + \partial_\mu X^\lambda\partial_\nu X_\lambda]. \quad (3\text{--}17.250)$$

The source $T^{\mu\nu}(x)$ that is now derived through variation of $g_{\mu\nu}(x)$ is

$$T^{\mu\nu} = T_0^{\mu\nu} + X^\lambda\partial_\lambda T_0^{\mu\nu} - T_0^{\lambda\nu}\partial_\lambda X^\mu - T_0^{\mu\lambda}\partial_\lambda X^\nu$$
$$- \int (dx')\big(f^\mu(x - x')\tau^\nu(x') + f^\nu(x - x')\tau^\mu(x')\big)$$
$$+ \int (dx')(dx'')f^\mu(x - x')f^\nu(x' - x'')\partial_\lambda''\tau^\lambda(x''), \quad (3\text{--}17.251)$$

where

$$\tau^\lambda = (\Gamma_{\alpha\beta}^\lambda - \partial_\alpha\partial_\beta X^\lambda)T_0^{\alpha\beta}. \quad (3\text{--}17.252)$$

Through its dependence on the gravitational field to the required accuracy, $T^{\mu\nu}$ does respond appropriately to infinitesimal coordinate transformations, and obeys the divergence equation (3–17.104). Finally, it should be said that, as in the photon situation, the consideration of additional radiation from the sources can be avoided by adopting an equivalent gauge. The gravitational gauge condition is

$$\int (dx') f^{\mu}(x - x') h^{f}_{\mu\nu}(x') = 0, \qquad (3\text{-}17.253)$$

leading to the vanishing of the X_{λ} terms.

This volume closes with a short exchange between Harold and the author.

H. How can it be the end of the book? You have hardly begun. There are any number of additional topics I should like to see developed from the viewpoint of source theory. And think of the field day you will give the reviewers, who usually prefer to list all the subjects not included in a volume rather than discuss what it does contain.

S. Quite true. But we have now reached the point of transition to the next dynamical level. And, since this volume is already of a reasonable size, and many of the ideas of source theory are in it, if hardly fully developed and applied, it seems better to put it before the public as the first volume of a series. Hopefully, the next volume will be prepared in time to meet the growing demand for more Source Theory.

APPENDIX

HOW TO READ VOLUME I

The first volume was described as a research document, and a textbook. Unfortunately, the beginning student was given no guidelines to tell him into which category a particular section fell. Accordingly, here are some suggestions for a first encounter with source theory, and relativistic quantum mechanics.

a) In Chapter 1, omit Section 1–4.
b) In Section 2–1, the derivation of the Lorentz transformation behavior of the source function from that of states can be omitted. It is sufficiently evident from the form of Eq. (2–1.35), for example, that $K(x)$ is a scalar function.
c) Omit the multi-particle generalizations of the vacuum amplitude in Section 2–2. They are of interest primarily in many-particle applications, which are not yet at the center of attention.
d) Section 2–5 need be read only to appreciate the general linear transformation of sources and its relation to spin, together with the ·possibility of composing arbitrary spins from more elementary ones.
e) The discussion in Section 2–6 that begins with Eq. (2–6.24) can be omitted by recognizing directly that (2–6.26) is the covariant generalization of the projection matrix $\frac{1}{2}(1 + \rho_3)$, which selects a definite parity in the rest frame and, thereby, the two components appropriate to spin $\frac{1}{2}$.
f) Omit the multi-particle generalizations in Section 2–7.
g) It is sufficient, in Section 2–8, to read the discussion of spin $\frac{3}{2}$.
h) In Sections 3–1 and 3–2, omit multi-particle generalizations.
i) Omit the discussion of spins 3 and $\frac{5}{2}$ in Section 3–3.
j) In Section 3–4, restrict attention to multispinors of ranks 2 and 3.
k) The spin limitations already noted should be continued in Section 3–5.
l) The rambling discussion about the arbitrariness of stress tensors that appears in Section 3–7 should only be skimmed.
m) The lengthy account of magnetic charge and its conceivable relevance to hadronic behavior [Sections 3–8, 3–9] is optional. However, don't miss the debut of Harold on p. 240, nor the remarks on mass normalization [p. 247].
n) Most of Section 3–17 is optional reading, particularly the discussions of broken conformal invariance, cosmology, and spin gravitational coupling.

Finally, we add two minor comments about specific topics in Volume I.

1. The discussion of Eq. (1–1.44) does not make clear that commutativity of the two displacement operators remains an alternative possibility (the numerical coefficient zero cannot be changed to unity by redefining the operators).
2. The forms of Lagrange functions that yield first order differential equations were merely stated in the text. The genesis of these expressions might be clarified by this illustration for spin 0. Beginning with the second order form [Eq. (3–5.12)]

$$\mathscr{L} = -\tfrac{1}{2}\partial^\mu \phi \, \partial_\mu \phi - \tfrac{1}{2}m^2\phi^2, \tag{A-1}$$

we introduce the independent vector field ϕ_μ by adding to $\mathscr{L}$ the term

$$\tfrac{1}{2}(\phi^\mu - \partial^\mu\phi)(\phi_\mu - \partial_\mu\phi). \tag{A-2}$$

The nature of the system is not changed thereby since, on extending the action principle to ϕ_μ, we learn that $\phi_\mu - \partial_\mu\phi$ vanishes, apart from a possible source term. But, on adding (A–1) and (A–2), the squares of the first derivatives cancel, producing the Lagrange function (3–15.16), from which the first-order field equations follow. This procedure is the analogue of one for ordinary mechanics that begins with the quadratic Lagrangian

$$L = \tfrac{1}{2}\dot{q}m\dot{q} - v(q) \tag{A-3}$$

where m is a non-singular symmetrical matrix, and introduces the independent variables p by adding

$$-\tfrac{1}{2}(p - m\dot{q})m^{-1}(p - m\dot{q}). \tag{A-4}$$

The sum of (A–3) and (A–4),

$$L = p\dot{q} - H, \qquad H = \tfrac{1}{2}pm^{-1}p + v(q), \tag{A-5}$$

yields the equivalent first-order Hamiltonian description.

Index